THE ROLE OF NITRIC OXIDE IN HEART FAILURE

THE ROLE OF NITRIC OXIDE IN HEART FAILURE

Edited by

Bodh I. Jugdutt
Cardiology Division
Department of Medicine
University of Alberta, Edmonton
Alberta, Canada

Reprinted from *Heart Failure Reviews*, Volume 7:4 (2002) and Volume 8:1 (2003)

KLUWER ACADEMIC PUBLISHERS
Boston / Dordrecht / London

Distributors for North, Central and South America:
Kluwer Academic Publishers
101 Philip Drive
Assinippi Park
Norwell, Massachusetts 02061 USA
Telephone (781) 871-6600
Fax (781) 681-9045
E-Mail <kluwer@wkap.com>

Distributors for all other countries:
Kluwer Academic Publishers Group
Post Office Box 322
3300 AH Dordrecht, THE NETHERLANDS
Telephone 31 786 576 000
Fax 31 786 576 254
E-Mail <services@wkap.nl>

Electronic Services <http://www.wkap.nl>

Library of Congress Cataloging-in-Publication Data

The role of nitric oxide in heart failure / edited by Bodh I. Jugdutt.
p. cm.
Articles reprinted from Heart failure reviews, 2002-2003.
ISBN 1-4020-7736-X (alk. paper)
1. Heart failure—Pathophysiology. 2. Nitric oxide—Pathophysiology. 3. Nitric Oxide—Physiological effect. I. Jugdutt, Bodh I. II. Heart failure reviews.
[DNLM: 1. Cardiac Output, Low—physiopathology—Collected Works. 2. Nitric Oxide—therapeutic use—Collected Works. WG 210 R7453 2004]
RC685.C53R655 2004
616.1'29061—dc22 2003064143

Printed on acid-free paper.
Printed in the United States of America

Dedication

To Catherine and Bernadine

HEART FAILURE REVIEWS

Role of NO in Heart Failure:

Table of Contents

FOREWORD

Nitric oxide (NO) is one of the most remarkable biological molecules characterized at the turn of XX century. Small in size, but ubiquitous and complex in its network of molecular, cellular and organ interactions, NO plays an important role in mammalian and non-mammalian biology. The scope and versatility of NO actions on the biological systems range from neuromodulation and regulation of cardiovascular homeostasis to cell signaling and immunity. An impressive array of physiologic and pathologic reactions are controlled by or associated with changes in the biological specter of NO.

The "career" of NO in science started from a handful of publications written before 1986 to more than 52,000 contributions at the end of 2002 (Pubmed, National Library of Medicine). More than 15,000 publications deal with the effects of NO in the cardiovascular system. The NO field has been recently knighted with the 1998 Nobel Prize in Medicine awarded to R.F. Furchgott, F. Murad and L.J. Ignarro.

This book attempts to present the state-of-the-art knowledge on the role of NO in cardiovascular homeostasis and in the pathogenesis of heart failure and diseases that underlie the development of heart failure. In the United States, heart failure, a multifactorial pathophysiologic state, which complicates the clinical course of many cardiovascular diseases, is responsible for close to 1,000,000 hospital admissions and 40,000 deaths annually. The book has 20 chapters highlighting the role of NO in the pathophysiology and therapeutics of heart failure. Book chapters have been written both by clinicians and basic scientists to emphasize the translational character of the contemporary NO research. A broad cross-section of presented topics includes discussion of basic cardiovascular physiology, as well as of pathophysiology and "cutting edge" pharmacology of heart failure. Considerable attention has been paid to interactions between NO and other players on the cardiovascular arena including peptides and proteins such as angiotensins, natriuretic peptides, matrix metalloproteinases, and free radicals including superoxide and peroxynitrite.

In the age of proteomics and combinatorial chemistry, the elucidation of interactions between NO synthase and other bioactive enzymes and proteins may contribute to better understanding and treatment of heart failure.

Marek W. Radomski

NO Preface

Research in the field of nitric oxide (NO) has grown exponentially since 1998, when three prominent researchers were recognized by Nobel prize awards. With the expansion in knowledge and the growing number of publications on the role of NO in several disciplines, it has become necessary to separately review the current knowledge of the role of NO in the pathophysiology and therapeutics of the major cardiovascular diseases, such as coronary artery disease and atherosclerosis, acute coronary syndromes and myocardial infarction, and heart failure. The emerging literature on the therapeutic implications of NO in pulmonary hypertension is also reviewed. This first volume is a compilation of a series of comprehensive review articles on selected topics that were published in the Journal of Heart Failure Reviews. The 20 excellent articles were contributed by some of the leading basic and clinician scientists in the field. The articles are well balanced and address clinically pertinent issues. The book should serve as an excellent reference source for basic science and medical students, medical residents and postdoctoral fellows, pharmacologists, physicians, cardiologists, cardiovascular surgeons, endocrinologists, nephrologists, and neurologists. It is our hope that the book will become a companion for cardiovascular researchers and scientists working in the field of NO, the role of NO in cardiovascular pathophysiology, and the development of NO-related targets for cardiovascular therapeutics. We would welcome the comments and suggestions of the readers as we plan future volumes.

*Acknowledgements*

I thank Catherine Jugdutt for her assistance during all stages of preparation of this project and Melissa Ramondetta for her continuous guidance.

Part I
NO and Pathophysiology of Heart Failure

Nitric Oxide in Heart Failure: Friend or Foe

Bodh I. Jugdutt
Cardiology Division of the Department of Medicine, University of Alberta, Edmonton, Alberta, Canada

Abstract. **Nitric oxide (NO) is a controversial molecule. It is either beneficial or deleterious. As with NO donors, one reason for this duality is related to the dose. Small doses are highly beneficial, maintaining blood flow in vessels and blood pressure, and protecting against foreign invaders. In high doses, it results in hypotension, forms peroxynitrite which is cytotoxic, and contributes to heart failure.**

Key Words. **nitric oxide, nitroglycerin, ischemia, infarction, heart failure**

Introduction

Nearly 135 years have elapsed since the Scottish physician, Sir Lauder Brunton, first proposed amyl nitrate for the relief of angina pectoris in 1867 [1]. It was not until 12 decades later that the underlying mechanism, that nitrates act as exogenous donors of nitric oxide ($NO^\bullet$ or NO in short) and it is NO that is the biological messenger which causes direct dilation of coronary arteries, was elucidated [2–7]. In 1980, Furchgott and Zawadski [2] discovered the endogenous endothelium-derived relaxation factor, and this was shown to be NO by Palmer, Ferrige and Moncada in 1987 [3]. Over the last 15 years, the NO literature has experienced exponential growth (from only 7 papers on endogenous NO in 1987 to over 3000 by 1994 and over 45,000 by 2002) and has been associated with considerable growth in knowledge.

In 1992, NO was named molecule of the year [8]. In 1998, the importance of NO was underscored by the award of Nobel prizes to 3 outstanding cardiovascular investigators, namely Robert Furchgott, Louis Ignarro and Ferid Murad.

Since 1998, the field of NO research has continued to expand rapidly and has contributed a wealth of new knowledge about the pathobiochemistry of NO, the pathobiology of NO synthases, the clinical pharmacology of nitrovasodilators, the roles of NO in cardiovascular disease, and the application of NO in research and clinical settings. Also in 1998, NO gained popularity as Pfizer marketed sildenafil (Viagra) for erectile dysfunction. This drug inhibits NO breakdown by the enzyme cGMP-specific phosphodiesterase 5, thereby enhancing the action of NO (vasodilation and smooth muscle relaxation) and facilitating penile erection in patients with impotence.

Although the importance of NO was first described in the cardiovascular system, there is continuing controversy about the effects of NO in cardiovascular disease: good or bad, beneficial or deleterious, friend or foe?

In fact, NO is a ubiquitous and controversial molecule [5,6]. It is a noxious, unstable gas, and a byproduct of automobile exhaust, electrical power stations and lightning. It is generated in tissues and participates in several important physiological functions in the body (Table 1), including vasodilation, neurotransmission and elimination of pathogens [5–8]. It also plays a role in the pathophysiology of several diseases in the cardiovascular system and other organs (Table 2). Cumulative evidence suggests that the biologic actions of NO in the heart are quite complex (Fig. 1).

The NO molecule has 2 important properties that may in part explain its duality: it is uncharged and has an unpaired electron. As an uncharged molecule, NO can diffuse freely across cell membranes and the unpaired electron makes it a reactive free radical. NO interacts readily with oxygen free radicals (O_2^-), which may be regarded as a beneficial detoxification system for the potentially injurious superoxide on the one hand, and harmful on the other, as this reaction results in the formation of peroxynitrite ($ONOO^-$) which is cytotoxic. Whether NO is cytotoxic or cytoprotective in specific settings may depend on the form in which NO is delivered or transported [9].

Another reason for the duality is that NO is produced by 2 major classes of NO synthase (NOS) enzymes: on the one hand, 2 constitutive isoforms found mainly in endothelial cells (eNOS) and neurons (nNOS), release small amounts of NO for

Address for correspondence: Dr. Bodh I. Jugdutt, 2C2.43 Walter Mackenzie Health Sciences Centre, Division of Cardiology, Department of Medicine, University of Alberta, Edmonton, Alberta, Canada T6G 2R7. Tel.: (780) 407-7729; Fax: (780) 437-3546; E-mail: bjugdutt@ualberta.ca

Table 1. Physiologic functions of NO in the cardiovascular system

- Maintain vascular smooth muscle relaxation
- Regulate vascular tone
- Regulate blood flow to tissues
- Regulate blood pressure
- Regulate myocardial contractility
- Regulate endothelial integrity and permeability
- Regulate vascular cell proliferation
- Regulate endothelium-leucocyte interaction
- Inhibit platelet aggregation and adhesion
- Exert an overall antiatherogenic effect

Table 2. Pathology associated with NO

Cardiovascular
- Ischemic heart disease
- Hypertension
- Hypercholesterolemia
- Atherosclerosis
- Diabetes mellitus
- Ischemia-reperfusion
- Disease risk in female gender
- Arterial restenosis
- Heart failure
- Septic shock

Other systems
- Cerebrovascular disease and stroke
- Memory disorders
- Alzheimer's disease
- Multiple sclerosis
- Parkinson's disease
- Pulmonary hypertension
- Chronic hepatitis
- Liver failure
- Ulcerative colitis
- Eclampsia
- Schistosomiasis
- Renal disease
- Impotence

short periods of time to signal adjacent cells; on the other hand, an inducible isoform, iNOS found in macrophages releases large amounts of NO for prolonged periods of time to destroy bacteria and parasites. An excess of NO can lead to host cell damage, neurotoxicity during strokes, hypotension associated with sepsis, more ischemia-reperfusion injury, and cell death in chronic heart failure.

Several important concepts have evolved about the actions of NO. Physiologically, a continuous release of NO in small quantities by endothelial cells causes vascular smooth muscle relaxation and maintains the vasculature in a state of active vasodilation. A basal low level of NO acts as an endogenous autoregulator of blood flow to tissues (such as the heart, brain, kidney, lung and gastrointestinal tract) in response to local changes. NO mediates additional vasodilation during ischemia and reperfusion. Other locally released factors such as bradykinin and acetylcholine induce NO release in some vessels. The autonomic nervous system also controls NO release into vasculature. NO plays a role in the regulation of blood pressure. The beneficial effects of angiotensin-converting enzyme inhibitors are in part due to increased bradykinin and NO.

Defects in the regulation of NOS can lead to hypertension and vasospasm. Endothelial dysfunction and defects in NO production have been implicated in several diseases (e.g. diabetes mellitus and atherosclerosis) and been postulated in several other diseases (e.g. Prinzmetal angina, hepatorenal syndrome, Raynaud's disease, eclampsia). NO reduces blood clotting and inhibits platelet aggregation [10] and adhesion [11], and lack of NO may favor thrombosis. NO acts as a neurotransmitter and mediates penile erection via neurotransmitter and vasodilator effects, and lack of NO may cause impotence. NO plays a role in the immune system and too little NO could lead to immunodeficiency. Low doses of NO may delay the progression of heart failure.

NO overproduction may be the cause of idiopathic orthostatic hypotension. NO mediates hypotension in sepsis. Macrophages and neutrophils produce both NO and superoxide, suggesting that peroxynitrite may be the mechanism of cytotoxicity by these cells in ischemia-reperfusion. NO exerts negative chronotropic and negative inotropic effects on the heart, and excess NO may cause LV dysfunction and contribute to heart failure. Large amounts of NO from macrophage iNOS is cytotoxic and kills or inhibits pathogens (e.g. bacteria, fungi, parasites) and tumor cells by mechanisms which include inhibition of ATP production and DNA, or DNA damage. Excess NO can damage normal host cells as in autoimmune disease. It is therefore important that macrophage iNOS is transcriptionally mediated and precisely controlled.

NO from different sources can modulate regional cardiac contractility. The cellular distribution of NOS and partitioning of NOS isoforms within cellular compartments of cardiac myocytes are complex. NO generated from eNOS in vascular endothelium can diffuse into cardiac myocytes and act in paracrine fashion to alter function and metabolism. NO generated in the cardiac myocyte can also act in autocrine fashion to modulate its function. Low doses of NO increase contractility through a cAMP dependent mechanism [12] while high doses of NO reduce contractility by a cGMP dependent mechanism [12]. NO also shifts substrate catabolism from free fatty acids to lactate during heart failure [13].

NO donors such as nitrates have been used to treat angina pectoris and hypertensive crises, for afterload reduction in heart failure, and for

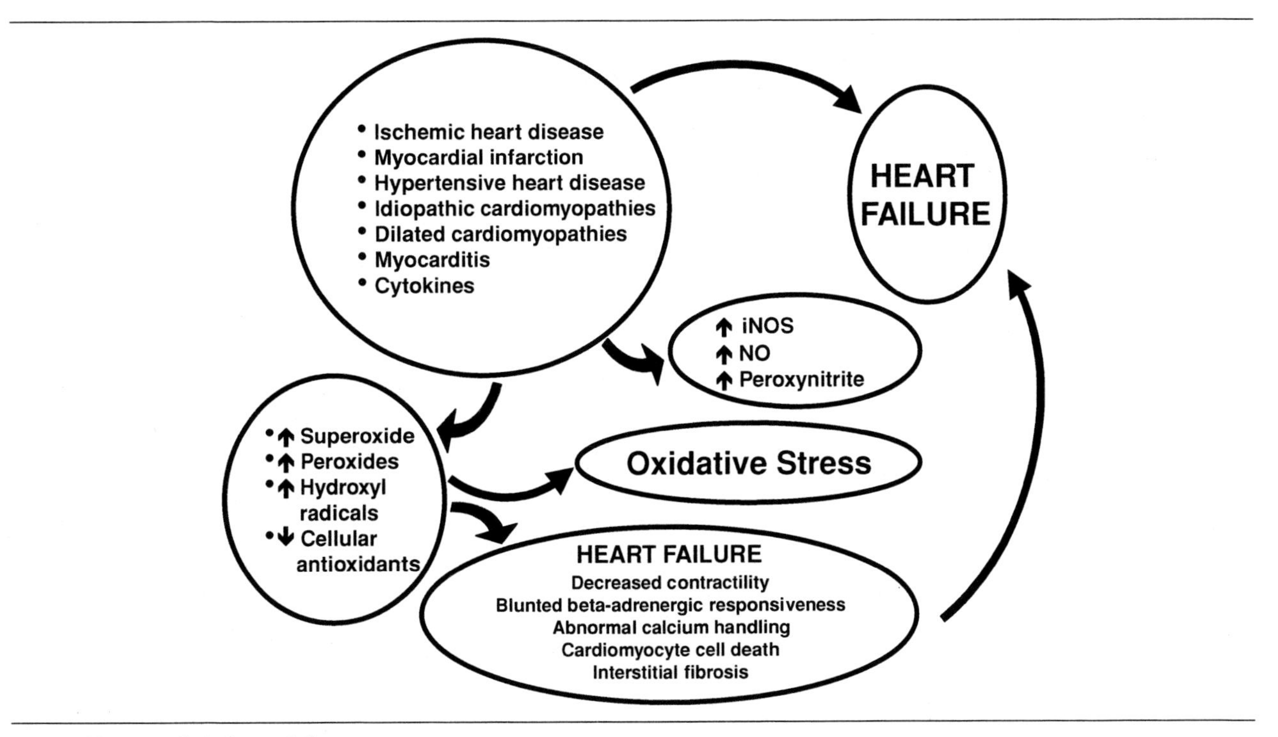

Fig. 1. *Nitric oxide in heart failure.*

the acute management of myocardial infarction [1,14–18]. It was known for decades that too much nitrate resulted in hypotension and development of nitrate tolerance [19,20]. The addages that "a little is good, but more is not better" or "too much of a good thing might not be good" have proven to be very true of NO and NO donors in clinical settings. Low dose intravenous nitroglycerin in acute myocardial infarction was shown to be highly beneficial in limiting left ventricular dysfunction, remodeling, other complications and mortality [16]. Although evidence of hemodynamic tolerance developed in 24% of the patients at the low doses used, overall beneficial effects were not abolished [17]. However, much higher doses of intravenous nitroglycerin were shown to cause hypotension and a paradoxical *J*-curve effect, with loss of some of the beneficial effects on collateral blood flow and infarct size in the canine model [15]. Nitroglycerin at very low doses (<0.2 ng/ml) was shown to produce maximal venodilation while higher doses decreased afterload without further decreases in preload [21]. Prolonged nitrate therapy allowing for a daily nitrate-free interval was also found to be beneficial after myocardial infarction [22].

With its roots in the explosives and munitions industry, nitroglycerin was considered to be a homeopathic remedy for several decades before Brunton [23]. Homeopaths like C. Hering studied 'dose and effect' and documented headache, chest tightening and palpitation following placement of nitroglycerin on the tongue. Interestingly, the homeopath W. P. Armstrong in 1882 and the famous physician W. Osler in 1897 both felt that angina was rare in the 19th century. Knowledge gleaned over the last decade supports the concept that excess exogenous NO acts as a two-edged sword. However, the tendency to use high doses of NO donors and prolonged continuous regimens for angina has continued.

Inhaled NO has been applied in pulmonary hypertension and respiratory distress syndrome in adults to relax pulmonary arteries and reduce pulmonary vascular resistance. NO has been shown to reduce infarct size [24,25], probably by inducing vasodilation, reducing inflammatory infiltration and removing other free radicals. Delivery of NO in slow, rapid or targeted release form has been proposed for pulmonary hypertension, arterial restenosis, liver failure, and schistosomiasis [26]. NO inhibitors have been used for reversing hypotension in sepsis and reducing stroke size, but they can lead to hypertension and increased clotting and are not selective.

A clear understanding of the different aspects of NO is essential in the development of novel NO donors for the treatment of myocardial ischemia and thrombosis, gene therapy for restoration of endothelial function in atherosclerosis, and the development of strategies for the prevention of nitrate tolerance, enhancing cardioprotection following coronary reperfusion and reducing reperfusion injury after myocardial infarction, and improving therapy and outcome in heart failure.

It is the purpose of this first symposium issue on NO to provide balanced, critical and authoritative presentations of pertinent basic and clinical aspects of NO research that will assist the reader in the objective management of heart failure. Andrews et al. review the chemistry of NO and the pharmacological tools used to study the role of NO in the vasculature [28]. Paulus and Bronzwaer review the myocardial contractile effects of NO pertinent to heart failure [29]. Wollert and Drexler review the role of NO in the regulation of cardiac remodeling with focus on cardiomyocyte hypertrophy and apoptosis [30]. Oyama et al. review the role of NO in cell death (necrosis and apoptosis) in heart failure [31]. Mungrue et al. review the role of NOS genes in heart failure and the important lessons learned from targeted gene ablation of the 3 types of NOS in mice [32]. Lalu et al. review the role of peroxynitrite in myocardial ischemia-reperfusion injury [33]. de Gasparo reviews the interaction between angiotensin II and NO in cardiovascular pathology and the role of AT_1 and AT_2 receptors [34]. Warnholtz et al. review the mechanisms of endothelial dysfunction induced by nitrate, an NO donor, and the role of oxidative stress [35]. Jugdutt reviews the role of NO in cardioprotection during ischemia-reperfusion [36].

Acknowledgments

We are grateful for the assistance of Catherine Jugdutt in manuscript preparation.

This study was supported in part by a grant from the Canadian Institutes for Health Research, Ottawa, Ontario.

References

1. Brunton TL. On the use of nitrate of amyl in angina pectoris. *Lancet* 1867;2:97–98.
2. Furchgott RF, Zawadski JV. The obligatory role of endothelial cells in the relaxation of arterial smooth muscle by acetylcholine. *Nature* 1980;288:373–376.
3. Palmer RMJ, Ferrige AG, Moncada S. Nitric oxide release accounts for the biological activity of endothelium-derived relaxing factor. *Nature* 1987;327:524–526.
4. Ignarro LJ, Lippton H, Edwards JC, et al. Mechanism of vascular smooth muscle relaxation by organic nitrates, nitroprusside and nitric oxide: Evidence for the involvement of S-nitrosothiols as active intermediates. *J Pharmacol Exp Ther* 1981;218:739–749.
5. Ignarro LJ. Biosynthesis and metabolism of endothelium-derived nitric oxide. *Annu Rev Pharmacol Toxicol* 1990;30:535–560.
6. Moncada S, Palmer RM, Higgs EA. Nitric oxide: Physiology, pathophysiology, and pharmacology. *Pharmacol Rev* 1991;43:109–142.
7. Luscher TF. Endogenous and exogenous nitrates and their role in myocardial ischemia. *Br J Clin Pharmacol* 1992;34(Suppl 1):29S–35S.
8. Culotta E, Korschland DE Jr. NO news is good news. *Science* 1992:258:1862–1865.
9. Stamler JS, Singel DJ, Loscalzo J. Biochemistry of nitric oxide and its redox-activated forms. *Science* 1992;258:1898–1902.
10. Radomski MW, Palmer RM, Moncada S. The anti-aggregating properties of vascular endothelium: Interaction between prostacyclin and nitric oxide. *Br J Pharmacol* 1987;92:639–646.
11. Radomski MW, Palmer RM, Moncada S. An l-arginine/nitric oxide pathway present in human platelets regulates aggregation. *Proc Natl Acad Sci USA* 1990;87:5193–5197.
12. Vila-Petroff MG, Younes A, Egan J, Lakatta EG, Sollott SJ. Activation of distinct cAMP-dependent and cGMP-dependent pathways by nitric oxide in cardiac myocytes. *Circ Res* 1999;84:1020–1031.
13. Recchia FA, McConnell PI, Bernstein RD, Vgel TR, Xu X, Hintze TH. Reduced nitric oxide production and altered myocardial metabolism during the decompensation of pacing-induced heart failure in the conscious dog. *Circ Res* 1998;83:969–979.
14. Murell W. Nitroglycerin as a remedy for angina pectoris. *Lancet* 1879;1:80–81, 113–115, 115–122, 225–227.
15. Jugdutt BI. Myocardial salvage by intravenous nitroglycerin in conscious dogs: Loss of beneficial effect with marked nitroglycerin-induced hypotension. *Circulation* 1983;68:673–684.
16. Jugdutt BI, Warnica JW. Intravenous nitroglycerin therapy to limit myocardial infarct size, expansion and complications. Effect of timing, dosage and infarct location. *Circulation* 1988;78:906–919.
17. Jugdutt BI, Warnica JW. Tolerance with low dose intravenous nitroglycerin therapy in acute myocardial infarction. *Am J Cardiol* 1989;64:581–587.
18. Jugdutt BI. Intravenous nitroglycerin unloading in acute myocardial infarction. *Am J Cardiol* 1991;68:52D–62D.
19. Stewart DD. Remarkable tolerance to nitroglycerine. *Philadelphia Polyclinic* 1888;6:43.
20. Prodger SH, Ayman D. Harmful effects of nitroglycerin: With special reference to coronary thrombosis. *Am J Med Sci* 1932;184:480–491.
21. Imhof PR, Ott B, Frankhauser P, Chu LC, Hodler J. Difference in nitroglycerin dose-response in the venous and arterial beds. *Eur J Clin Pharmacol* 1980;18:455–460.
22. Jugdutt BI. Nitrates in myocardial infarction. *Cardiovasc Drugs Ther* 1994;8:635–616.
23. Fye WB. Nitroglycerin: A homeopathic remedy. *Circulation* 1986;73:21–29.
24. Siegfried MR, Erhardt J, Rider T, Ma XL, Lefer AM. Cardioprotection and attenuation of endothelial dysfunction by organic nitric oxide donors in myocardial ischemia-reperfusion. *J Pharmacol Exp Ther* 1992;260:668–675.
25. Johnson G 3rd, Tsao PS, Lefer AM. Cardioprotective effects of authentic nitric oxide in myocardial ischemia with reperfusion. *Crit Care Med* 1991;19:244–252.
26. Keefer LK. Nitric oxide-releasing compounds: From basic research to promising drugs. *Mod Drug Discov* 1998;1:20–30.
27. Hare JM, Givertz MM, Creager MA, Colucci WS. Increased sensitivity to nitric oxide synthase inhibition in patients

with heart failure. Potentiation of β-adrenergic inotropic responsiveness. *Circulation* 1998;97:161–166.
28. Andrews KL, Triggle CR, Ellis A. NO in the vasculature: Where does it come from and what does it do? *Heart Fail Rev* 2002;7:423–445.
29. Paulus WJ, Bronzwaer JGF. Myocardial contractile effects of nitric oxide. *Heart Fail Rev* 2002;7:371–383.
30. Wollert K, Drexler H. Regulation of cardiac remodeling by nitric oxide: Focus on cardiac myocyte hypertrophy and apoptosis. *Heart Fail Rev* 2002;7:317–325.
31. Oyama J, Frantz S, Blais C, Kelly RA, Bourcier T. Nitric oxide, cell death, and heart failure. *Heart Fail Rev* 2002;7:327–334.
32. Mungrue IN, Husain M, Stewart DJ. The role of NOS in heart failure: Lessons from murine genetic models. *Heart Fail Rev* 2002;7:407–422.
33. Lalu MM, Wang W, Schultz R. Peroxynitrite in myocardial ischemia-reperfusion injury. *Heart Fail Rev* 2002;7:359–369.
34. de Gasparo M. Angiotensin II and nitric oxide interaction. *Heart Fail Rev* 2002;7:347–358.
35. Warnholtz A, Tsilimingas N, Wendt M, Munzel T. Mechanisms underlying nitrate-induced endothelial dysfunction: Insight from experimental and clinical studies. *Heart Fail Rev* 2002;7:335–345.
36. Jugdutt B. NO and cardioprotection during ischemia-reperfusion. *Heart Fail Rev* 2002;7:391–405.

NO and the Vasculature: Where Does It Come from and What Does It Do?

Karen L. Andrews, PhD, Chris R. Triggle, PhD, and Anthie Ellis, PhD

Smooth Muscle Research Group and the Canadian Institutes of Health Research Group on Regulation of Vascular Contractility, Department of Pharmacology and Therapeutics, Faculty of Medicine, University of Calgary, Hospital Drive NW, Calgary, AB, Canada

Abstract. **Nitric oxide (NO) is involved in a large number of cellular processes and dysfunctions in NO production have been implicated in many different disease states. In the vasculature NO is released by endothelial cells where it modulates the underlying smooth muscle to regulate vascular tone. Due to the unique chemistry of NO, such as its reactive and free radical nature, it can interact with many different cellular constituents such as thiols and transition metal ions, which determine its cellular actions. In this review we also discuss many of the useful pharmacological tools that have been developed and used extensively to establish the involvement of NO in endothelium-derived relaxations. In addition, the recent literature identifying a potential source of NO in endothelial cells, which is not directly derived from endothelial nitric oxide synthase is examined. Finally, the photorelaxation phenomena, which mediates the release of NO from a vascular smooth muscle NO store, is discussed.**

Key Words. **nitric oxide, vasculature, NO chemistry, photorelaxation**

Introduction

Blood vessels are lined with a monolayer of cells collectively referred to as the endothelium. The endothelium provides a large surface area for the exchange of fluids and solutes between blood and vascular tissue. Lying beneath the endothelium are multiple layers of vascular smooth muscle cells (VSMC), the contractile state of which is controlled by sympathetic nerves, circulating hormones and the endothelium. The tone of the vasculature is intricately regulated by a number of vasorelaxant factors synthesized and released from endothelial cells such as nitric oxide (NO), and prostacyclin (PGI_2) (Fig. 1). NO is synthesized by a family of oxidoreductases that bear close sequence homology to cytochrome P_{450} system enzymes [1], called NO synthases (NOS), of which three isoforms are known to exist, endothelial NOS (eNOS), inducible NOS (iNOS) and neuronal NOS (nNOS). In the endothelium, eNOS is activated by a number of agonists acting on G protein-coupled receptors (GPRCs) and by physical stimuli such as shear stress and changes in oxygen levels which leads to the activation of stretch operated non-selective cation channels (SOCC). These enzymes utilize L-arginine and molecular oxygen as substrates as well as the cofactors, nicotinamide adenine dinucleotide phosphate (NADPH), flavin adenine dinucleotide (FAD), heme, flavin adenine mononucleotide (FMN), calmodulin and tetrahydrobiopterin for the production of NO and the byproduct, L-citrulline [2]. NO elicits relaxations of adjacent VSMCs by the activation of soluble guanylate cyclase (sGC). Another endothelium-derived relaxing factor, prostacyclin (PGI_2) can also be released via the activation of GPCR and SOCC. Activation of cyclooxygenase (COX), which catalyses the conversion of arachidonic acid to produce PGI_2, which activates the prostacyclin receptor on the smooth muscle, increases cAMP levels and induces vasorelaxation.

With the discovery of the NO synthesis pathway has come the use of several NOS inhibitors to study the role of NO in endothelium-dependent vasorelaxation. They include N^G-nitro-L-arginine (L-NNA; also known as L-NA, L-NOARG and NOLA), its methyl ester, L-NAME and N^G-monomethyl-L-arginine (L-NMMA). After endothelium-dependent relaxations were observed in the presence of these inhibitors together with the COX inhibitor, indomethacin, it appeared that another relaxing factor might contribute to the modulation of vascular tone. This factor was found to induce hyperpolarization and hence, the name, endothelium-derived hyperpolarizing

Address for correspondence: Chris Triggle, Smooth Muscle Research Group and the Canadian Institutes of Health Research Group on Regulation of Vascular Contractility, Department of Pharmacology & Therapeutics, Faculty of Medicine, University of Calgary, 3330 Hospital Drive NW, Calgary, Alberta, Canada, T2N 4N1. Tel.: 1 (403) 220 5036; Fax: 1 (403) 270 9497; E-mail: triggle@ucalgary.ca

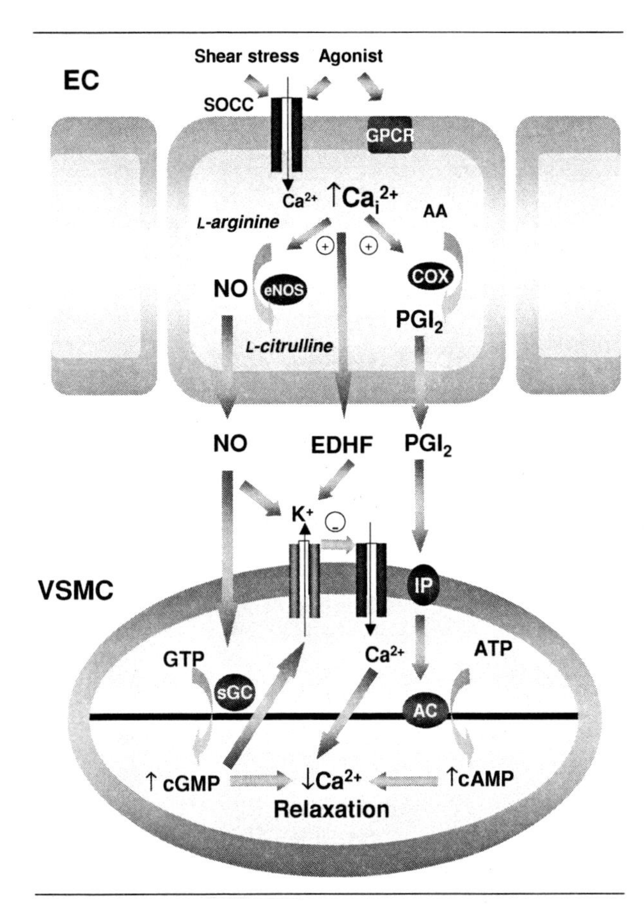

Fig. 1. Activation of a GPCR by agonists and activation of SOCC from the shear stress of blood flow in the membrane of EC increases Ca_i^{2+} levels which in turn, activates eNOS, COX as well as the synthesis / release of the putative mediator, EDHF. The activation of eNOS results in the conversion of L-arginine to L-citrulline and the liberation of NO and the activation of COX results in the conversion of AA to PGI_2. The diffusion of NO and PGI_2 to the VSMC leads to the activation of sGC and the IP and AC, which results in increases of both cGMP and cAMP, respectively and, via a number of possible mechanisms including the opening of K^+ channels and closure of Ca^{2+} channels, a decrease in Ca_i^{2+} and vasorelaxation. EDHF directly or indirectly opens K^+ channels, which results in the closure of Ca^{2+} channels on the VSMC, decreasing Ca_i^{2+} and inducing relaxation.

Abbreviations: AA, arachidonic acid; AC, adenylate cyclase; Ca^{2+}, calcium; Ca_i^{2+}, intracellular calcium; COX, cyclooxygenase; EC, endothelial cell; EDHF, endothelium derived hyperpolarizing factor; eNOS, endothelial nitric oxide synthase; GPCR, G-protein coupled receptor; IP, prostacyclin receptor; K^+ potassium; NO, nitric oxide; PGI_2, prostacyclin; SOCC, stretch operated non-selective cation channels; sGC, soluble guanylate cyclase; VSMC, vascular smooth muscle cell.

factor (EDHF) is now used (for review see [3]). Although the identity of EDHF remains elusive, it has been shown to cause vasorelaxation via activation of K^+ channels, which consequently results in VSMC hyperpolarization in a tissue- and species-dependent manner. Although there are considerable variations in the contributions of these factors in endothelium-dependent relaxations in different tissues, it appears that EDHF only makes a significant contribution in resistance arteries, while NO is the principal vasodilator in conduit arteries. NO is a nonpolar gaseous molecule that demonstrates a wide range of biological actions for functions as varied as neurotransmission and host-mediated cytotoxicity to endothelium dependent modulation. The diversity in the functions of NO can be attributed to its unique chemistry, which determines its site of action, distribution, half-life and metabolism.

Chemistry of NO

NO is a paramagnetic or free radical species, since it carries a single unpaired electron in its outer shell (in total 11 electrons). While this property of NO would explain the diversity of biological reactions that NO is involved in, it should also be stressed that NO is comparatively stable and does not possess the type of reactivity that is noted of other radical species (e.g. hydroxyl radicals). The substances that NO is known to react with include other radicals, transition metal ions and nucleophiles such as thiols (RSH) and amines.

The solubility of NO in water is quite low ($\sim 1.9 \times 10^{-3}$ M at 25°C [4]), and unlike other reactive nitrogen species, does not produce the corresponding acid in solution (i.e. does not undergo hydration). The fact that NO is not remarkably soluble in aqueous solutions is an advantage in terms of cellular access. Although, few studies have characterized the permeability of NO across phospholipid membranes, its high hydrophobicity [5] strongly suggests that it can diffuse quite easily and rapidly into cells. In an early study the production of NO was measured from a single endothelial cell using a NO-selective electrode and it was determined that the diffusion constant of NO produced to be close to 3000 $\mu m^2/s$ [5]. Considering that the thickness of the endothelial cell layer is about 0.3 μM [6], it would appear that the actions of NO would be immediate, not impeded by cellular membranes and governed only by the speed in which it reaches the target vascular smooth muscle cell. Note however, that the rapid diffusion rate of NO is not likely to be the sole determinant of its actions but rather is likely the total sum of its rate of diffusion, synthesis and reactions with cellular constituents [7].

Decomposition of NO

NO reacts with dioxygen or molecular oxygen (O_2), which like NO, is also a radical species. NO undergoes decomposition when it is autooxidized in the gas phase to form the characteristic brown gas, nitrogen dioxide (Reaction 1 – NO_2). However, since the reaction between NO and O_2 follows third order kinetics, the rate it proceeds will depend on the concentration of NO, where at high concentrations of both O_2 and NO, the reaction proceeds rapidly while at low NO concentrations the oxidation of NO will be very slow [8,9]:

$$2NO + O_2 \rightarrow 2NO_2 \quad (1)$$

$$2NO_2 \rightarrow N_2O_4 \quad (2)$$

$$NO_2 + NO \rightarrow N_2O_3 \quad (3)$$

$$N_2O_4 + N_2O \rightarrow H^+ + HNO_2 + NO_{3^-} \quad (4)$$

NO_2 either dimerises to form N_2O_4 (reaction 2) or reacts again with NO to form N_2O_3 (reaction 3), which may donate a NO moiety to various nucleophilic targets, such as thiols [10]. In the aqueous phase, this reaction proceeds slowly to eventually produce the water-soluble terminal products nitrite (NO_2^-) and nitrate (NO_3^-) (reaction 4 [11]).

The half-life of NO reported in different studies has varied depending on the experimental conditions set out in each study. For example, the half-life measurements for NO under bioassay conditions have been reported to be 3–5 sec [12,13]. Similar values were also reported by Wood and Garthwaite [14] who also commented that the highly oxygenated saline solutions used to bathe tissues may contribute to the accelerated decomposition of NO, and thus may not be an accurate reflection of the physiological state. However, since the oxidation of NO in the aqueous phase proceeds quite slowly, it is unlikely that the half-life of NO is solely determined by the rate at which it is oxidized. Kharitonov et al. [15] reported that at physiologically relevant concentrations of NO (100 nM), the theoretical first half-life of NO would be 2 hours, while Ford et al. [16] determined the half-life of NO to be approximately 100–500 s when studied at more physiological levels of oxygen. Thus, following third order kinetics, at low concentrations NO will be oxidized gradually, while at higher concentrations of NO the autooxidation reaction will proceed more rapidly. In addition, it was recently reported that while autooxidation proceeds quite slowly in the aqueous phase, the rate of this process increases dramatically when in proximity to cellular membranes [17].

Redox Forms of NO

Nitrogen oxides can exist in different redox states and the oxidation states of these nitrogen species can range between −3 to +5. The positioning of the unpaired electron in the antibonding orbital renders NO more likely to accept or lose other electrons in this orbital and in the process able to form these different oxides [9]. Ammonia (NH_3) is the most reduced of the oxides and a two electron oxidation process generates hydroxylamine (NH_2OH), which itself can be subsequently oxidized to form nitroxyl (NO^- or HNO). NO, as the free radical species, can arise when NO^- undergoes a one-electron oxidation. NO can then lose an electron to form the electrophilic nitrosonium cation (NO^+). However, this species is quite electrophilic and does not freely exist for long periods, thus it is either rapidly hydrolyzed into nitrous acid (HNO_2), which generates nitrite and nitrate ions [18], or reacts with nucleophiles to form NO adducts [8].

Reactions Between NO and Various Substances

The reactions between NO and various substances found in the extracellular milieu can determine the activity, duration of action and metabolic fate of NO. The substances known to influence the activity of NO include thiols, superoxide, and transition metal ions, notably heme iron and these all may contribute to the complex chemistry and subsequent physiological actions of NO in the vasculature (Fig. 2).

Thiols. Contrary to what was initially thought, the reaction between thiols and NO does not result in the immediate formation of S-nitrosothiols (RSNOs). Instead, NO must undergo oxidation to form N_2O_3, which can then go on to nitrosate nucleophiles such as thiols. Nitrosation involves a reaction in which a NO molecule loses an electron while attaching to a nucleophile [19]. The direct reaction between NO and thiol is more likely to form NO^- and the corresponding disulfide, a process which is dependent on the presence of transition metal ions (reaction 5 [20]). Thiols, and indeed other nucleophiles such as amines, can be nitrosated by acidified HNO_2, which can generate NO^+ (reaction 6), N_2O_3 (reaction 7), NO_2, or an alkyl nitrite (reaction 8 [21]). The nitrosation of thiols, however, is the more physiologically relevant reaction pathway, since nitrosation of amines is usually associated with pathophysiological conditions [8].

$$2RSH + NO \rightarrow NO^- + RSSR + 2H^+ \quad (5)$$

$$RSH + HNO_2 \rightarrow RSNO + H_2O \quad (6)$$

$$RSH + N_2O_3 \rightarrow RSNO + H^+ + NO_2^- \quad (7)$$

$$RS^- + RONO \rightarrow RSNO + RO^- \quad (8)$$

$$RSH + NO^+ \rightarrow RSNO + H^+ \quad (9)$$

The nitrosation of thiols actually arises from the association between a thiol and the nitrosating agent that donates NO^+ (reaction 9), however, since the existence of the NO^+ species is so fleeting

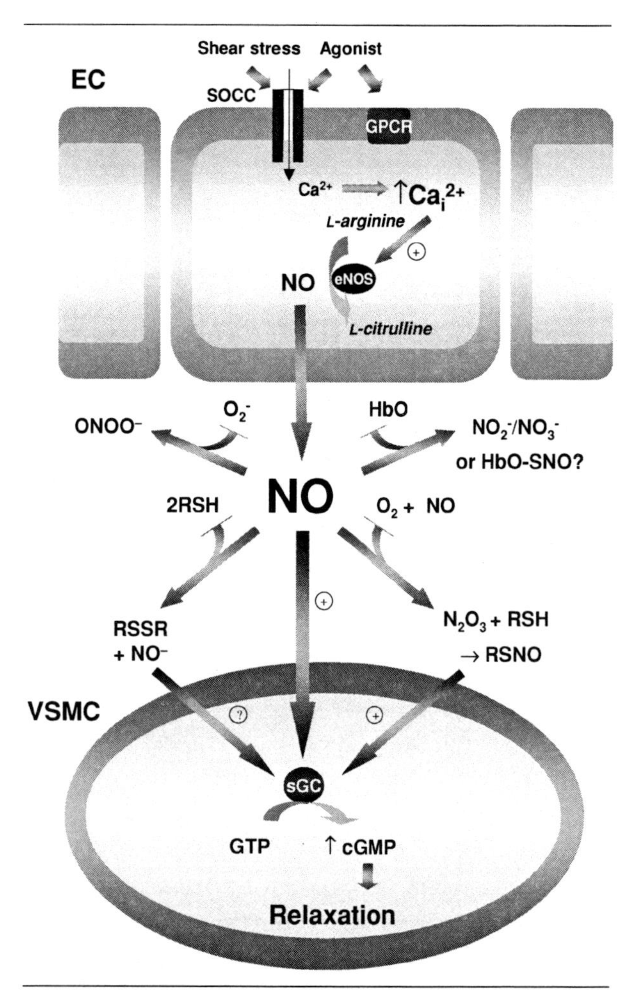

Fig. 2. Proposed metabolic fate of NO and its interactions with other molecules. Activation of a GPCR by agonists and activation of SOCC from the shear stress of blood flow in the membrane of endothelial cells increases Ca_i^{2+} levels, which in turn, activates eNOS. The activation of eNOS results in the conversion of L-arginine to L-citrulline and the liberation of NO, which diffuses across to the VSMC and activates sGC, increases in cGMP and by, a number of possible mechanisms, a decrease in Ca_i^{2+} and vasorelaxation. Alternately, NO may react with circulating HbO to oxidize NO into NO_2^- or NO_3^- or possibly form a nitrosothiol, HbO-SNO. The autooxidation of NO results in the production of NO_2, which can subsequently react with additional NO to produce N_2O_3. N_2O_3 can nitrosate RSH to form RSNOs, and induce vasorelaxation. The direct reaction between thiols and NO results in the formation of a RSSR and NO^-, which may also relax VSMCs. O_2^- produced by various enzymatic sources may react with NO to produce $ONOO^-$, which is associated with many adverse cellular processes under conditions of oxidative stress.

Abbreviations: Ca_i^{2+}, intracellular calcium; EC, endothelial cell; eNOS; endothelial nitric oxide synthase; GPCR, G-protein coupled receptor; HbO, oxyhemoglobin; NO, nitric oxide; NO^-, nitroxyl anions; NO_2^-, nitrite; NO_3^-, nitrate; N_2O_3, dinitrogen trioxide; O_2^-, superoxide; $ONOO^-$, peroxynitrite; RSNO, S-nitrosothiols; RSH, thiols; RSSR, disulfide; SOCC, stretch operated calcium channels; sGC, soluble guanylate cyclase; VSMC, vascular smooth muscle cell.

it is likely that this association is more relevant in transnitrosation reactions (see below). In the laboratory the preparation of RSNOs is a straightforward process that involves reacting solutions of thiols with the chosen nitrosating agent at a low pH [22]. The fact that an acidic environment facilitates the nitrosation of thiols also ties in with the observation by Furchgott and Jothianandan [23] who reported that millimolar concentrations of L-cysteine as the hydrochloride salt caused transient relaxations of rabbit aorta. They postulated that this effect arose because the marked acidity of the L-cysteine generated ideal conditions for it to be nitrosated by residual nitrates that contaminate the physiological bath solution. This process may potentially hold some relevance in conditions where pH balance is altered, e.g. metabolic acidosis, alkaline/acidic diuresis and may influence the rate of RSNO formation *in vivo*. Thiols also react with NO^-, however the product of this reaction does not possess significant relaxant activity, thus thiols can be regarded as NO^- scavengers [18]. The vasorelaxant activity of the NO^- donor, sodium trioxodinitrate (Angeli's salt), is substantially reduced in the presence of high concentrations of thiols such as L-cysteine [24,25]. This occurs through a two-stage reaction that results in the formation of hydroxylamine (NH_2OH), which possesses modest relaxant activity [12] and disulfide cysteine (reactions 10 and 11 [26]). However, it is important to note that while hydroxylamine may induce relaxations, its activity is dependent on catalase, whose activity can be reduced by thiols [27]:

$$H^+ + NO^- + RSH \rightarrow RSNHOH \qquad (10)$$

$$RSNHOH + RSH \rightarrow RSSR + NH_2OH \qquad (11)$$

Superoxide. NO reacts with the one-electron reduction product of oxygen, superoxide (O_2^-), at near-diffusion controlled rate to produce peroxynitrite ($ONOO^-$) (reaction 12 [28,29]). It has also been put forth recently that $ONOO^-$ could form during the reaction between molecular oxygen and NO^- (reaction 13 [30]), although this may depend on the excitation state of NO^- and has been deemed by some authors to be thermodynamically unviable [18]:

$$NO + O_2^- \rightarrow ONOO^- \qquad (12)$$

$$NO^- + O_2 \rightarrow ONOO^- \qquad (13)$$

$ONOO^-$ oxidizes proteins, DNA and lipids and consequently, may be cytotoxic to potential pathogens. Elevated oxidative stress characteristic of many cardiovascular diseases also points to the likelihood that these reactive oxygen/nitrogen species contribute to the pathogenesis of these disease states. Once formed, $ONOO^-$ can either decay into nitrate or nitrite, or in its protonated state it could potentially form the highly reactive and cytotoxic, hydroxyl radical ($OH^{\cdot}$) and NO_2 [28]. However, while these actions are related to the involvement of NO in inflammatory processes, it is possible that $ONOO^-$ could also contribute to physiological processes. For example, it has been demonstrated to induce relaxations in canine coronary arteries [31]. The formation of a NO adduct between $ONOO^-$ and glucose, or other substances with an alcohol group, was also reported to display similar activity to NO including vasorelaxation and inhibition of platelet aggregation [32,33]. There is also evidence that $ONOO^-$ may cause S-nitrosation of cellular thiols mediating a NO stimulation of soluble guanylate cyclase [34,35]. Furthermore, endothelium-dependent generation of $ONOO^-$ in response to bradykinin (BK) and the calcium ionophore, A23187 was detected from bovine aortic endothelial cells [36]. Nevertheless, the outcome of the reaction between NO and superoxide is the reduction of the relaxant activity of NO. Superoxide is formed in the mitochondria, the endoplasmic reticulum and membranes of various cells, including activated macrophages and endothelial cells [37]. Sources of superoxide in the vascular wall include NADPH oxidase [38], xanthine oxidase [39], cytochrome P_{450} [40] and cyclooxygenase [41]. Also, under certain conditions and in a cofactor-dependent manner, NOS itself may generate superoxide anions [42].

Heme Proteins. The biological actions of NO are thought to be terminated when circulating oxyhemoglobin (HbO) reacts with NO to form the complex nitrosylHbO, which is then oxidized to metHbO and NO_3^- [43]. As heme iron exists either in the ferric (Fe^{3+}) or ferrous (Fe^{2+}) state, NO may react with either reduced or oxidized heme, which consequently terminates its vasorelaxant actions [44]. However, recent evidence indicates that there may need to be a rethink of this perception, where instead the interaction between HbO and NO may in fact provide a means to preserve or sustain the actions of NO through the formation of a reversible complex with HbO [45,46]. Gow and colleagues [46] reported that allosteric modifications on oxygen-bound HbO may allow cooperative binding of NO, where NO can nitrosate thiols on HbO at the same time as it binds with oxygen without oxidizing NO into nitrite or nitrate. However, the interaction between iron and NO is, not only fundamental to its half-life, but also a vital aspect to its bioactivity, since interactions between NO and heme iron also govern its messenger functions, specifically the activation of the heme-containing enzyme guanylate cyclase. NO also participates in redox interactions with other transition metal ions such as copper and cobalt [47,48].

Endogenous Forms of NO

Due to the reasonably reactive nature of NO and its capacity to undergo redox modification by various cellular constituents, there is the distinct possibility that its cellular effects may in part be accomplished by products of these chemical interactions.

Nitroxyl Anion (NO^-). The NO^- species has been considered by some authors to be the actual mediator of the physiological functions of NO rather than the free radical form of NO ($NO^{\cdot}$) [49–55]. In the study by Schmidt and co-workers, the authors sought to directly measure NO production from purified NOS, however, only in the presence of superoxide dismutase (SOD) were they able to detect NO formation [53]. They hypothesized that the final product of NOS is therefore not NO, but NO^-, which is subsequently oxidized to NO by SOD. Alternatively, the oxidation of the intermediate of NO synthesis, *N*-hydroxy-L-arginine by catalase or hydrogen peroxide has also been shown to generate NO^- [50]. It has also been suggested that the dependency of NOS activity on the cofactor tetrahydrobiopterin enables the enzyme to synthesize $NO^{\cdot}$ rather than NO^-[56,57]. The formation of NO^- has also been reported following the reduction of NO by Cu/Zn SOD—which can be coupled to both oxidation or reduction reactions with NO [49,52], HbO [45] and cytochrome *c* [58]. NO^-generation can also arise following the decomposition of RSNOs [20], the reaction between NO and thiols [19] and may even be liberated during the reaction between thiols and RSNOs [59].

Many studies have evaluated the potential endogenous role of NO^- by comparing its actions against other forms of NO. Feelisch and coworkers ruled out the involvement of NO^- in mediating endothelium-dependent relaxations under bioassay conditions in rabbit aorta [12]. This conclusion was based on the finding that the half-life and potency of sodium nitroxyl, which was used as a source of NO^- anions, did not correspond to that of the stimulated release of NO. However, in another study, the inhibition of endothelium-dependent relaxations in rat aorta by L-cysteine was similar to its effects on relaxations elicited by a more potent donor of NO^-, Angeli's salt (sodium trioxodinitrate) but not to responses mediated by authentic NO [55]. Furthermore, other studies have also reported an inhibitory effect of thiols on endothelium-dependent relaxations [60–62]. Since thiols can be regarded as

NO^- scavengers (Reactions 10 and 11 [18]), the findings of these studies would suggest that endothelium-dependent relaxations could at least be partly mediated by NO^-.

Studies have also proposed a potential pathophysiological role for the NO^- [63,64]. This may in part relate to the possibility that it could form the reactive nitrogen species, $ONOO^-$ when it reacts with molecular oxygen [30], however this was argued by some as being a thermodynamically unfavorable pathway [18,65]. Whether or not this reaction does proceed under physiological or indeed pathophysiological reactions may be a matter of opinion; nevertheless, it appears that the cellular actions of NO^- are as complex and diverse as the non-reduced form of NO. A comparative study on the effects of $NO^{\cdot}$ and NO^- on cardiac function following induction of ischaemia-reperfusion injury in rabbits revealed that while NO, as well as the nitrosothiol, S-nitrosoglutathione (GSNO), ameliorated cardiac function, administration of Angeli's salt exacerbated the myocardial damage that resulted from inducing ischaemia [66]. Interestingly, it was shown that in the absence of the co-factor tetrahydrobiopterin NOS generates NO^- [56]. Low tetrahydrobiopterin levels have been associated with endothelial dysfunction in conditions such as diabetes, hypertension and atherosclerosis [67–69]. Therefore the elevated production of NO^- by eNOS could react with oxygen to generate $ONOO^-$ and elicit cytotoxic actions which may impact on cardiovascular function. Recent evidence also suggests that NO^- may be coupled to a redox reaction with NADPH, and thus, may potentially interfere with many NADPH-dependent reactions within the cell [70].

S-Nitrosothiols (RSNOs). Thiols may be considered as critical determinants of the activity, transport and metabolic fate of NO because NO can combine with low molecular weight thiols and thiols incorporated in proteins to form relatively stable NO adducts known as RSNOs [71]. Examples include S-nitrosocysteine (CysNO), GSNO and S-nitroso-acetyl penicillamine (SNAP). These substances under suitable conditions, can either decompose to liberate NO, and as such possess potent smooth muscle relaxant activity, or be involved in transnitrosation and S-thiolation reactions where they can affect signal transduction and enzyme activity (for review see [19]). They have also been shown to inhibit platelet aggregation, and have longer half-lives than NO [12] and more recently they have been implicated to play a critical role in stress responses such as hypoxia [72].

Some RSNOs are polar or too large to readily cross cellular membranes, therefore the passage of NO from RSNOs into cells is likely to involve a process called transnitrosation. The NO moiety is transferred between cellular thiols [17] until the intended nitrosation reaction occurs which causes conformational changes and ultimately the intended cellular effect [18]. RSNOs are usually biologically stable molecules and the S-N covalent bond is not ordinarily sensitive to homolysis except, when irradiated via strong direct light, which results in the homolytic cleavage of the sulphur nitrogen bond to yield the corresponding disulfide and the NO moiety [73,74]. The decomposition of RSNOs is also accelerated in the presence of various metal ions, with the most effective being copper ions [75] and this may be a mechanism by which NO is released from RSNOs *in vivo*. Complicating the study of these compounds is the fact that copper and iron molecules are often found in many physiological buffers. The mechanism by which copper catalyses this decomposition is not well understood, however it is thought that Cu^+, rather than Cu^{2+}, is the species responsible for this effect, since selective Cu^+ chelators were reported to prevent the release of NO from SNAP [75], attenuated relaxations induced by GSNO and SNAP in rat perfused artery [76], and reduced the anti-aggregatory effects of GSNO on platelets [77]. In a physiological setting, copper is derived from copper containing proteins, since it rarely exists in the free form under normal conditions [75]. For example, Cu/Zn SOD, but not Mn SOD, was found to promote the decomposition of GSNO and as a consequence increased the release of NO [78].

The possibility that RSNOs could account for the activity of endogenous NO was first proposed by Myers et al. [79], who reported that the agonist-induced release of NO elicited relaxations that most closely resembled those produced by S-nitrosocysteine rather than authentic NO. In another study it was shown that NO reacted with circulating albumin in plasma to produce a nitrosocompound that displayed similar properties to endothelium-derived NO, and was therefore suggested to be the actual mediator of endothelium-derived NO [80]. However, subsequent studies found distinct differences between the reactivities of RSNOs and the endogenous form of NO thus discounting this possibility [12,81,82].

More recent studies by Mayer and coworkers have suggested GSNO plays a novel role in the cardiac NO/cGMP pathway by acting as an intermediate in the formation of NO and possibly as an intracellular store for NO (Mayer, 1998). They propose that oxidation of glutathione (GSH) produces a glutathionyl radical ($GS^{\cdot}$), which reacts with NO to form GSNO, which can be decomposed by Cu^+ to liberate NO that in turn activates sGC. There also appears to be a physiological role for RSNOs in transnitrosation reactions. RSNOs have been measured in human blood, predominantly formed from circulating albumin [71,83] and in human airway smooth muscle [84]. In addition, several studies have alluded to, but never isolated, copper-containing cell-surface proteins that can catalyse the degradation of RSNOs [76,77,85]. Also the localization of protein disulfide isomerase, which catalyses thiol exchange reactions, on cellular membranes has been hypothesised to facilitate the entry of RSNOs (possibly originating from plasma bound RSNOs) inside a cell, which can then go on to transnitrosate intracellular thiols [86,87].

Dinitrosyl-Iron Complexes (DNICs). DNICs are stable molecules that display distinct electroparamagnetic spectrum signals and are capable of forming stable carriers of NO. In particular, NO binds with proteins that contain iron-sulfur cluster centres (e.g. aconitase) to form these distinct complexes, which can be quantified by electron paramagnetic resonance spectroscopy [88]. Similarly to the transnitrosation reactions that RSNOs participate in, DNICs may transfer their co-ordinated NO group to other metalloproteins, which may cause conformational changes that result in the required cellular effects [89]. DNICs display stable and potent relaxant activity of vascular smooth muscle [12,90] and have been detected in endothelial cells [91]. They can be formed *in vivo* [89] and have been suggested to account for the activity of endogenous NO [90,92]. While it appears unlikely that this would be the case [12], the thiol, *N*-acetyl-L-cysteine has been demonstrated to form a DNIC following pre-treatment of rat aorta with lipopolysaccharide to induce NO production [93]. The detection of a DNIC highlighted the possibility that a stable store of NO could arise following any number of inflammatory processes.

Other Sources of NO. Non-enzymatic formation of NO has been reported in a number of studies. The most striking demonstration of this was the detection of NO from expired air of human subjects [94], which the authors postulated arose from the highly acidic environment of the gastric lumen that converted nitrite into NO. Similarly, it was recently proposed by Modin and colleagues [95] that the acidic-metabolic vasodilatation seen under hypoxic conditions might arise following the conversion of nitrite into NO by the acidic cellular environment. The production of NO from acidified nitrite is a commonly-used approach to generate NO under experimental conditions [96] and as such it is interesting that a by-product of cellular metabolism and thus a non-enzymatic source of NO could have a physiological (or indeed pathological) role in regulating vascular tone. A non-NO synthase enzymatic source is, not surprisingly, cytochrome P_{450}, which in addition to being structurally similar to NOS, has been reported to be able to generate NO in a NADPH and oxygen-dependent manner [97,98].

Pharmacological Tools Used to Study NO-Dependent Functions

To assess vascular function, agents that interfere with the NO-dependent signaling pathway are routinely used to determine whether NO is involved in the occurrence of a particular response. For example, to elucidate whether a relaxation response in a blood vessel is mediated by NO, an inhibitor of NOS would determine if the source of this relaxant is NOS. In addition, targeting other aspects of NO-dependent signaling can provide useful information on the relaxant pathways employed by particular vessels for specific mediators. By taking advantage of what is recognized about NO chemistry, substances that are known to react with NO (e.g. superoxide anions) can be used to selectively target the NO molecule and produce particular changes in its actions. Examples of pharmacological tools that are routinely used and their actions on endogenous NO are summarized in Table 1.

Inhibition of Soluble Guanylate Cyclase
The main cellular target of NO is the enzyme sGC, located in smooth muscle cells, which catalyses the conversion of GTP into the second messenger molecule that ultimately signals relaxation, cyclic GMP. Activation of sGC arises when NO interacts with its heme moiety to form a complex that ultimately induces conformational changes to the enzyme which exposes the catalytic site to GTP [99]. The prototypical reagent used to inhibit sGC was methylene blue, which was thought to inhibit the enzyme by oxidising its critical heme group, rendering sGC with a lesser affinity for NO [100]. However, due to its long list of unspecific actions, including superoxide generation [101,102] and inhibition of NOS [103,104], its use as an inhibitor of sGC has been superceded by the more putatively selective 1H-[1,2,4]oxodiazolo[4,3,-a]quinoxalin-1-one (ODQ) [105]. Like, methylene blue, the inhibition of sGC by ODQ is also mediated through the oxidation of the heme group of the enzyme. Although thought to be selective for sGC, it has recently been found to have actions on other heme proteins, which ultimately could interfere with the activity of NOS and cytochrome P_{450} [106,107]. Nevertheless, it is the best agent available to inhibit sGC and to date a more selective compound has not emerged to replace its use. It is also important to

Table 1. *Examples of the reported effects of various modulators of the NO-dependent pathway on endothelium-dependent vasodilatation in different vascular preparations. The listed effect refers to the functional consequence of studying relaxations in the presence of the respective agent (* inhibition was observed in the absence of EDTA in hypercholestrolaemic animals but not healthy controls). The concentration of each agent refers to the concentration used in the particular study that brought about a significant effect—unless otherwise stated (i.e. IC_{50} value)*

Agents	Tissue	Concentration	Effect	Refs
NOS inhibitors				
L-NMMA	Rat aorta	0.1–100 μM	Inhibition	[150]
	Rabbit aorta	3–100 μM	Inhibition	[152]
L-NNA	Rat aorta	3–10 μM	Inhibition	[203]
	Rabbit aorta	1.5–100 μM	Inhibition	[152]
	Porcine coronary artery	10 μM	Inhibition	[204]
L-NAME	Rat aorta	0.1–100 μM	Inhibition	[150]
	Rat aorta	0.4 μM (IC_{50})	Inhibition	[205]
	Rat perfused heart	0.25 μM	Increased pressure	[206]
sGC inhibitors				
Methylene blue	Rabbit aorta	20 μM	Inhibition	[100]
	Rabbit aorta	0.4 ± 0.16 μM	Inhibition	[207]
	Rat aorta	10 μM	Inhibition	[208]
ODQ	Rat aorta	0.04 ± 0.01 μM (IC_{50})	Inhibition	[208]
	Rabbit aorta	3 μM	Inhibition	[131]
	Mouse aorta	0.3–10 μM	Inhibition	[62]
	Mouse cerebral artery	3–10 μM	Inhibition	[209]
	Bovine pulmonary artery	1 nM–30 μM	Inhibition	[210]
O_2^- generators				
Xanthine oxidase	Rat aorta	16 mU/ml	No effect	[112]
	Rabbit aorta	4.8 mU/ml	Partial inhibition	[211]
		20 mU/ml	Inhibition	[212]
Pyrogallol	Rabbit aorta	100 μM	Inhibition	[117]
	Rat aorta	100 μM	No effect	[112]
	Rabbit cerebral artery	300 μM	Inhibition	[213]
LY83583	Bovine pulmonary artery	10 μM	Inhibition	[214]
	Rat aorta	0.1 ± 0.01 μM (IC_{50})	Inhibition	[208]
Antioxidants				
SOD	Rat aorta	0.3–100 U/ml	Relaxation	[215]
Ascorbic acid	Rabbit aorta	1 mM	Partial enhancement	[121]
	Rabbit aorta	1 mM	No effect	[216]
	Rat aorta	30 μM–3 mM	No effect	[217]
Tiron	Rat aorta	0.1 μM–1 mM	Contraction	[215]
	Canine basilar artery	9.4 mM	Partial enhancement	[118]
	Rabbit aorta	1–30 mM	No effect	[211]
TEMPOL	Rabbit aorta	0.1 μM–1 mM	No effect	[215]
	Rabbit aorta	3 mM	Partial inhibition	[211]
	Rabbit renal afferent arteriole	1 mM	Inhibition	[218]
NO scavengers				
Haemoglobin	Rabbit aorta	1–10 μM	Inhibition	[100]
	Rat aorta	0.14 ± 0.05 μM (IC_{50})	Inhibition	[133]
	Porcine coronary artery	16 μM	Inhibition	[204]
	Rat mesenteric artery	1.5 μM	No effect	[161]
Hydroxocobalamin	Rat aorta	10–30 μM	Inhibition	[138]
	Mouse aorta	100 μM	Inhibition	[62]
	Porcine coronary artery	200 μM	Inhibition	[219]
Carboxy-PTIO	Rabbit aorta	100 μM	Inhibition	[128]
	Rat aorta	300 μM	Partial inhibition	[130]
	Mouse aorta	100 μM	No effect	[62]
	Rabbit renal artery	300 μM	No effect	[129]
	Rabbit pulmonary	300 μM	No effect	[129]

(*Continued on next page.*)

Table 1. (*Continued*).

Agents	Tissue	Concentration	Effect	Refs
Thiols				
L-cysteine	Rat aorta	1–3 mM	Partial inhibition	[55]
	Mouse aorta	3 mM	Inhibition	[62]
Homocysteine	Mouse pancreatic vascular bed	2 mM	Inhibition	[61]
	Rabbit aorta	1 mM	Inhibition	[60]
	Rat aorta	100 μM	Partial inhibition	[220]
N-acetyl-L-cysteine	Human forearm circulation	1 mM	No effect	[82]
Metal chelators				
EDTA	Rabbit aorta	26 μM	No effect*	[221]
	Rabbit aorta	30 μM	No effect	[216]
DETCA	Rat aorta	10–100 μM	Inhibition	[222]
	Rabbit aorta	5 mM	Inhibition	[223]
	Rabbit aorta	3–10 mM	Inhibition	[211]
	Rabbit cerebral artery	0.3–10 mM	Inhibition	[224]
	Bovine coronary artery	10 mM	Inhibition	[225]
	Canine basilar artery	7.6 mM	Inhibition	[118]
Deferoxamine	Rat aorta	1 mM	No effect	[226]
	Rat aorta	100 μM	No effect	[227]
-SH inhibitors				
N-ethylmaleimide	Rat aorta	10 μM	Inhibition	[228]
Diamide	Rabbit aorta	0.2–1 mM	Inhibition	[229]
Thimerosal	Dog coronary artery	10 μM	Inhibition	[230]
	Rabbit mesenteric artery	300 nM	Enhancement	[231]
Ebselen	Rabbit aorta	6 μM (IC_{50})	Inhibition	[144]
	Rat aorta	10 μM	Inhibition	[145]

Abbreviations: Carboxy-PTIO, 2-[4[carboxyphenyl]-4,4,5,5,-tetramethylimidazoline-1-oxyl 3-oxide; DETCA, diethyldithiocarbamate; EDTA, ethylene diamine tetraacetic acid; LY83583, 6-anilino-5,8-quinolinedione; L-NAME, N^G-nitro-L-arginine methyl ester; L-NMMA, N^G-monomethyl-L-arginine; L-NNA, N^G-nitro-L-arginine O_2^-, superoxide anion; NO, nitric oxide; NOS, nitric oxide synthase; ODQ, 1H-[1,2,4]oxodiazolo[4,3,-a]quinoxalin-1-one; sGC, soluble guanylate cyclase; -SH, sulfhydryl group; SOD, superoxide dismutase; TEMPOL, 4-hydroxy-2,2,6,6-tetramethylpiperidine-1-oxyl; Tiron, 4,5-dihydroxy-1,3-benzene disulfonic acid.

consider the finding that ODQ does not reduce relaxations to authentic NO to the same degree as it inhibits relaxations to endogenous NO or other NO donors [62]. Endothelium-dependent relaxations to acetylcholine (ACh), endothelium-independent relaxations to glyceryl trinitrate and sodium nitroprusside in the mouse aorta were almost abolished in the presence of ODQ (10 μM). On the other hand, ODQ-pretreatment could only produce parallel rightward shifts of curves to endothelium-independent relaxations to authentic NO, Angeli's salt and spermineNONOate, indicating that the contribution of sGC in mediating these responses varies with the source of NO.

Superoxide Generators

Endothelium-dependent relaxations were reported to be impaired in the presence of superoxide generators by several authors [108–110], however, these studies were reported before the endothelium-dependent factor was identified as NO. When it was established that it was in fact NO or a NO-related compound [111], it became clear that this effect occurred because superoxide anions were inactivating endogenous NO in the same way that they inactivate authentic NO. This effect also explains why contractile tone is enhanced in the presence of superoxide generators such as hypoxanthine/xanthine oxidase. The inactivation of endothelium-derived NO released under basal conditions by superoxide generators eliminates its relaxant influence over smooth muscle tone [112].

Conversely, antioxidants such as SOD cause relaxations of vascular smooth muscle by protecting basal NO from inactivation by superoxide [112–114]. In rat aorta, various antioxidant agents caused significant reductions of noradrenaline-induced contractions [115]. This effect was endothelium-dependent since following the removal of the endothelium these antioxidants did not elicit relaxations [114,115]. However, the study by Mian and Martin [112] revealed a differential susceptibility of basal NO to superoxide anions compared to that of the release induced by ACh. While the activity of basal NO was reduced by superoxide anions and enhanced by SOD, relaxations induced by acetylcholine were largely unchanged in the presence of superoxide anions. However, these observations are not consistent with earlier findings where superoxide generators did reduce ACh-induced relaxations [12,81,108,109]. This difference can be reconciled with the fact that these studies were performed using a bioassay system, where effluent from cultured endothelial cells perfused a detector vascular preparation to elicit relaxations. The time it takes the effluent to reach the detector tissue makes it subject to numerous chemical modifications (i.e. oxidation) that perhaps make the endothelial effluent more sensitive to superoxide.

It may be likely that SOD may be involved in protecting NO-dependent functions *in vivo*, since it has been detected in the vasculature, specifically binding onto the outer surface of vascular endothelial cells [116,117]. Therefore suggesting that SOD specifically serves a protective function on endothelium-derived NO, making it more resistant against inactivation by superoxide. As expected, the inhibition of this protection by SOD with the copper chelator diethyldithiocarbamate (DETCA) caused endothelium-dependent relaxations to become more sensitive to superoxide generators [112]. Similarly, pretreatment with DETCA also impaired BK- and A23187-induced relaxations in canine basilar arteries and this effect was reversed following the addition of a more selective superoxide scavenger, Tiron [118].

Ascorbate is also an antioxidant that effectively scavenges superoxide anions found in substantial levels in plasma [119]. Ascorbate has been shown to reduce blood pressure in spontaneously hypertensive rats [120] and restore endothelial-dependent function in rabbit aortic rings exposed to lipoprotein, which promotes superoxide formation [121]. However, Jackson and co-workers [122] argued that the high concentrations of ascorbate required to protect endothelium-dependent relaxations against superoxide generators in DETCA-treated rabbit aorta, implied that antioxidant protection in the vasculature may not be the main mechanism of protection.

Antioxidant substances would not be expected to greatly influence relaxations to RSNOs, since they are regarded as being less resistant to superoxide attack, or as the reaction between superoxide and RSNOs proceeds slower than that with NO [123]. As it turns out, ascorbate has a separate action on RSNOs, which does not involve oxidant protection. Ascorbate has been reported to promote the release of NO from RSNOs by reducing the RSNOs to form free sulfhydryls [124]. It has been suggested that this effect may arise following the oxidation of cupric ion (Cu^{2+}) into the cuprous ion (Cu^{+})—the copper species believed to catalyse the decomposition of RSNOs [75]—by ascorbate [125]. Alternatively, it has also been suggested that ascorbate forms a complex with copper which then targets the nitrosothiol to induce it to release NO by an as-yet-unknown mechanism [126].

NO Scavengers

As with superoxide generators, there are reported differences in the effects of NO scavengers on relaxations mediated by endothelium-derived NO when compared to authentic NO. Hydroquinone was initially regarded as a superoxide generator because it behaved similarly in that it selectively inhibited relaxations to solutions of NO gas but not to the nitrergic transmitter in the mouse anococcygeus muscle [127]. However, in the same study the authors found when measuring for chemiluminescence, hydroquinone did not generate superoxide anions and instead it quenched the chemiluminescent signal produced by NO, thus, they concluded that hydroquinone was a free radical scavenger. Therefore, the lack of selectivity of hydroquinone for free radical NO discouraged its use in subsequent studies. On the other hand, 2-[4[carboxyphenyl]-4,4,5,5,-tetramethylimidazoline-1-oxyl 3-oxide (carboxy-PTIO), as a free radical itself, was found to display free radical to radical selectivity for NO to consequently yield an imidazolineoxyl and NO_2 [128]. Endothelium-dependent relaxations in rabbit aorta were potently inhibited by carboxy-PTIO to a similar magnitude as exogenously applied solutions of NO gas [128]. Further studies in other rabbit blood vessels revealed that its effects varied depending on the particular vessel studied [129]. For example, carboxy-PTIO reduced relaxations induced by ACh in the aorta and the femoral artery. On the other hand, in renal and pulmonary arteries, carboxy-PTIO did not affect endothelium-dependent relaxations, despite relaxations being sensitive to NOS inhibition, thus ruling out an involvement of EDHF. In the rat aorta, carboxy-PTIO also discriminated between endogenous forms of NO and authentic NO (solutions of NO gas) [130]. Endothelium-dependent relaxations were significantly reduced by carboxy-PTIO only when high concentrations were used (also reported by [131]), while those to solutions of NO gas were inhibited by carboxy-PTIO at concentrations at least 10 fold less. In addition, the selectivity of carboxy-PTIO has recently come under question. It was reported that while carboxy-PTIO reduced cGMP formation in endothelial cells induced by the NO donor, diethylamine-NO and A23187 through a presumed scavenging action on these mediators, it unexpectedly enhanced cGMP formation induced by 3-morpholino-sydnonimine (SIN-1) [132]. Nonetheless, carboxy-PTIO can discriminate between different redox forms of NO in the rat anococcygeus muscle by reducing relaxant responses to solutions of NO gas but not those produced by NO^- [54]. Therefore, it can be presumed that carboxy-PTIO selectively targets free radical NO over other redox forms of NO.

The reaction between heme iron and NO was briefly mentioned earlier. Heme proteins, such as circulating HbO, are likely to be endogenous scavengers of NO. As such the relaxant actions of NO are reduced in the presence of HbO. For example, endothelium-dependent relaxations were greatly reduced by HbO, as well as those produced by solutions of NO gas in rat aorta [133]. Many studies have used HbO to reduce endothelium-dependent relaxations as well as to authentic NO and other NO donors [100,133–135], thus supporting for a role of NO in mediating these relaxations. The scavenging actions of HbO are not exclusive for authentic NO. Reports have shown it to reduce relaxations to the RSNOs nitrosocysteine and SNAP in rat aorta [136], sodium nitroprusside and glyceryl trinitrate in guinea pig and canine arteries [116] and induce metHbO formation following treatment of red blood cells with NO donor compounds [137].

Hydroxocobalamin contains a cobalt-corrin core that bears a close resemblance to the heme porphyrin ring in HbO, and like HbO, has a high affinity for NO [138], which it combines with to form nitrosocobalamin [139]. As such, there are parallels between the effects of these two substances on NO-mediated responses. Like HbO, hydroxocobalamin reduced relaxations to both endothelium-derived NO and authentic NO in the rat aorta [133,138].

Sulfhydryl Inhibitors. When evaluating the contribution of RSNOs in eliciting relaxations, the bond formed between NO and the thiol can be prevented by using agents that modify this interaction. Examples include agents that promote the oxidation of the sulfur–nitrogen bond, that inhibit the synthesis of endogenous thiols and those that cause the depletion of sulfhydryl groups. N-ethylmaleimide, is an irreversible thiol-alkylating agent, and has been shown to inhibit endothelium-dependent cGMP accumulation in cultured rabbit pulmonary arterial endothelium cells [140]. However, it has also been reported to cause loss of sulfhydryl groups on NOS, which diminished enzyme activity in porcine aortic endothelial cells [141], and thus may confound any conclusions that could be made on the involvement of RSNOs. In addition, the selenoorganic compound ebselen has been noted as an inhibitor of NOS through an action on iron–sulfur groups found in proteins such as NOS [142,143]. Its inhibitory actions on NOS can be reversed or prevented by performing experiments in the presence of reduced thiols [144,145]. Another sulfhydryl inhibitor is ethacrynic acid, which can alkylate thiols and thus prevent nitrosothiol formation. Pretreatment of rat anococcygeus muscles with ethacrynic acid non-specifically reduced relaxations induced by nerve stimulation, NO and sodium nitroprusside [146], and the authors concluded that sulfhydryl groups participate in mediating the relaxation of these mediators. Similarly, Rapoport and Murad, demonstrated that guanylate cyclase-independent relaxations to the cGMP analogue, 8-bromo cGMP were inhibited by ethacrynic acid, inferring that sulfhydryl groups may also be central to mediating the relaxant signaling pathway beyond guanylate cyclase [147]. Therefore, studies which use agents that alter sulfhydryl interactions to define the role of RSNOs should be interpreted carefully to avoid any confounding effects on other thiol-dependent cellular functions.

NOS Inhibitors

The most often used approach to modulate the actions of NO involves the inhibition of NOS activity. Substrate or L-arginine analogues are extremely effective inhibitors of NOS and their use has been instrumental in establishing the involvement of NO in various biological processes such as macrophage-mediated cytotoxicity [148], neurotransmission [149], and vasorelaxation [150]. Because these compounds resemble the endogenous substrate, they can compete with either the interaction of L-arginine on the NOS active site or with the uptake of L-arginine thus reducing the availability of substrate for NOS catalysis [151]. The inhibition of NOS by substrate analogues can be reversed with an excess of L-arginine but not with its D-isomer [152], suggesting stereospecific binding on the active site of NOS.

Traditionally, NOS inhibitors were thought to be very effective in inhibiting NOS when used at their appropriate concentrations. For example, L-NNA is typically used at concentrations of around 100 μM, yet in pig isolated coronary arteries, maximal effects of L-NNA were observed at concentrations of less than 10 μM [153]. However, in recent years it has become apparent that either, NOS inhibitors may not be able to completely inhibit the eNOS-mediated synthesis of NO, which appears unlikely in light of the above study [153], or alternatively, that there may be a NOS inhibitor resistant source or synthesis of NO.

NO Stores in the Vasculature?

The first evidence that in the presence of NOS inhibitors endothelial cells may still be able to release NO came from a study by Kemp and Cocks [154]. Their study found that in precontracted human isolated small coronary arteries the maximum relaxation to BK was reduced, but not abolished in the presence of L-NNA (300 μM). When L-NNA was combined with HbO (20 μM) a further reduction in the maximum relaxation and an 11-fold shift to the right in the concentration-response curve to BK was observed. These findings indicated that L-NNA alone did not completely inhibit eNOS in this tissue and it was speculated that the release of NO may either be from a non-NOS source, due to an excess of L-arginine, which could overcome the effects of L-NNA, or, possibly, due to an increased metabolism or impaired uptake of L-NNA.

Cohen and colleagues [155] also reported that, in precontracted rabbit carotid arteries, ACh caused vasorelaxation and hyperpolarization, which was reduced but not inhibited in the presence of L-NAME (30 μM). With the addition of a very high concentration of L-NNA (300 μM) in conjunction with L-NAME (30 μM) small amounts of vasorelaxation and hyperpolarization were still observed. In the same study, the authors also used a porphorynic sensor to measure the release of NO and chemiluminescence techniques to measure nitrite. Indeed, NO release was detected in the presence of the NOS inhibitors and it was concluded that no other endothelium-derived relaxing factors were involved, that is NO accounted for all the vasorelaxations and hyperpolarizations in this tissue.

In addition, in rat superior mesenteric arteries in the presence of L-NNA (100 μM), ACh-induced NO release, which was associated with vasorelaxation, was detected with an NO sensor [156]. Both the L-NNA-resistant NO release and vasorelaxation were reversed by HbO (10 μM) leading the authors to suggest that NO mediated the L-NNA resistant relaxation. Recent studies have identified L-NNA-resistant vasorelaxations and NO release in response to BK in pig coronary arteries and reported that the NO release but not the vasorelaxations were inhibited following the inclusion of HbO (10 μM) [157]. In addition, an L-NAME-resistant generation of NO in response to KCl (70 mM) in the perfused mesenteric bed of the rat [158] has been identified. Although these studies put forward a strong argument for the presence of a non L-arginine source of NO several studies also argue against it in a number of tissues. For example in rat isolated aorta L-NNA (200 μM), completely inhibited relaxations induced by ACh [159]

Complicating matters are the findings that NO itself, can cause hyperpolarization. In guinea-pig uterine arteries ACh–induced hyperpolarizations were reduced by L-NMMA (20 μM) [160] and in rat mesenteric arteries at resting tension, but not under active force, both solutions of NO gas and acidified $NaNO_2$, induced smooth muscle hyperpolarization [161]. Results from several studies have suggested that these hyperpolarizations were the result of the activation of potassium (K^+) channels as the ATP sensitive K^+ (K_{ATP}) channel inhibitor, glibenclamide (10 μM), reduced hyperpolarizations to NO, but not those to ACh, in rat and rabbit mesenteric arteries [161,162]. Additional studies have also shown that NO can directly activate calcium activated K^+ (K_{Ca}) channels in rabbit cerebral arteries and rat pulmonary arteries in the presence of cGMP [163,164] and in rabbit aorta, in the absence of cGMP [165]. However, tissue and species differences are apparent as NO does not activate BK_{Ca} in cultured human endothelial cells [166] or K_{ATP} channels in rat isolated aorta [167]. In addition, it is not clear if NO can activate K^+ channels directly or indirectly through cGMP. Studies utilising the rat tail artery reported that NO could, in addition to guanylate cyclase activation, activate K_{Ca} channels directly to cause vasorelaxation [168].

In both the Kemp and Cocks study and the study by Ge and colleagues, the BK-induced relaxations [154] and hyperpolarizations [157] were not completely abolished by NOS inhibitors, or even in the combined presence of HbO and L-NNA, indicating another non-NO hyperpolarizing factor. As the relaxations were sensitive to high extracellular K^+ (30 mM) [154] and K_{Ca} channel inhibitors [157] these relaxations were attributed to EDHF. Moreover, electrophysiological studies investigating the hyperpolarizing properties of NO donors and exogenous NO in rat isolated aorta, have found NO hyperpolarizes the tissues by only small amounts (2–4 mV) [159]. Similar results have been reported in precontracted rabbit basilar arteries where exogenous NO had little effect on membrane potential while ACh-induced hyperpolarizations of around 8 mV [169]. However, ACh-induced relaxations were not affected by KCl (65 mM) or glibenclamide (10 μM) suggesting hyperpolarization did not play an important role in the relaxations [169]. Studies by Plane and colleagues may provide a possible explanation for the differences observed between the effects of endothelium-derived NO and NO donors. They reported that, in isolated precontracted carotid arteries from rabbits, both endothelium derived NO and exogenous NO induced substantial hyperpolarizations ($\sim$35 mV) which appeared to be mediated by cGMP-dependent and -independent mechanisms. However, even though the NO donors induced similar levels of hyperpolarizations ($\sim$35 mV) it appeared that the hyperpolarizations to SIN-1 and SNAP were mediated completely by the cGMP pathway [170].

Of course it is possible that there is another enzyme responsible for the production of 'residual

NO'. As mentioned previously, cytochrome P_{450} due to its structural similarity to NOS, can also produce NO [171]. However, the studies by Jia and colleagues [171] were performed in tracheal smooth muscle and to date it is not known if cytochrome P_{450} can produce NO in endothelial cells. In addition, it is difficult to assess the production of NO by cytochrome P_{450} in the vasculature as many of the inhibitors of cytochrome P_{450} have been shown to have non-specific actions, particularly on K^+ channels [172–175]. There is also evidence that NOS inhibitors themselves may be able to non-enzymatically synthesise NO [176].

The conclusions to be drawn from these studies are that it appears possible that, in some tissues, there may be a NOS inhibitor-resistant source of NO present in endothelial cells that is distinct from EDHF. In addition, when attempting to study non-NO mediated relaxations in the vasculature, it is important that a NO scavenger such as HbO (10–20 μM) be used in conjunction with NOS inhibitors and indomethacin. The next logical questions are where is this NOS-inhibitor resistant NO store and what mediates its release? Is its release mediated by the same activators of NOS synthesised NO or are there separate activation pathways which may be targeted to mediate release NO independently of NOS? And, finally, is there a similar store of NO in other cells such as VSMCs? A phenomenon that was first described many years ago is that of photorelaxation [177] and may reflect the release of NO from a non-NOS derived tissue store.

Photorelaxation

In 1955, Furchgott reported that precontracted rabbit aorta exhibited a reversible relaxation when exposed to UV light [177]. Many years later this phenomenon was revisited and studies determined that this photorelaxation was associated with an increase in cGMP levels [178] and reduced by NO scavengers such as HbO and methylene blue [179]. In addition, photorelaxation is also enhanced by agents that are decomposed by light to release NO, including L-NNA [180] and 3-nitro-1,4-dihydropyridines [181]. Given these results, it is not surprising that in response to UV light NO release from rat aortic rings has been detected with a porphorynic sensor [182]. However, studies have shown that photorelaxation is not inhibited by NOS inhibitors such as L-NMMA [179] suggesting that photorelaxation is not directly dependent upon activation of NOS [183]. There was no significant difference in the photorelaxation response in rat aorta of stroke-prone spontaneously hypertensive rats when compared to control, suggesting little change in the photorelaxation store in disease states that typically impair NO production. In addition, photorelaxation in both strains was enhanced by L-NNA (300 μM), and reduced by KCl (40 mM), tetraethylammonium (20 mM) and ouabain (1 mM) [184]. This enhancement of photorelaxation by L-NNA (100–500 μM) was also observed in studies utilising rabbit vascular and non-vascular smooth muscle [180] and rat isolated aorta and trachealis [183]. However, this enhancement may be the result of NO production from L-NNA as evidenced by studies which demonstrated a concentration dependent increase in NO from L-NNA when exposed to both room light and UV light [185,186].

Initial studies with endothelium denuded rabbit aortic tissue demonstrated that tissues relaxed to UV light and the conclusion was that photorelaxation was endothelium independent [179]. However, in 1993, Venturini and colleagues found that in the same tissue photorelaxation was progressively reduced upon repeated exposures to light leading them to hypothesize that photorelaxation is mediated by the release of NO from a 'finite' molecular store in vascular smooth muscle that can be depleted [187]. In contrast, results from experiments utilizing rat aortic rings indicated that the NO/cGMP pathway or K^+ channels were not involved in photorelaxation after methylene blue and KCl had no significant effect on photorelaxation responses [188]. However, the findings of Megson and colleagues [189,190] supported those of Venturini and colleagues [191], as they found the photorelaxation store in endothelium-denuded rat aorta could be depleted by repeated exposures to UV light. In addition, they also found that endothelial intact rat aorta, depleted of their photorelaxation stores, recovered slowly in the dark whereas those treated with L-NNA did not, leading them to the postulation that eNOS was essential for repriming the NO utilized by photorelaxation. Megson and colleagues also postulated that thiols react with oxides of nitrogen originating from eNOS to form RSNOs.

The studies of Megson and colleagues also investigated the possibility that the release of NO from RSNOs may mediate photorelaxation [189,190]. They found that the thiol alkylating agent ethacrynic acid reduced photorelaxation and postulated that NO was released from a photosensitive RSNO store in the vascular smooth muscle of rat isolated aorta. Then again, as mentioned previously, ethacrynic acid is fairly non-specific agent as it targets many different cellular proteins with thiol groups. However, the findings of Lovren and Triggle [192] indicated that a number of RSNO depleting compounds also reduced photorelaxation in rat isolated aorta supported the view that RSNOs are indeed involved in photorelaxation.

In recent studies carried out in our laboratory we wished to further investigate the role of NOS in photorelaxation by examining photorelaxation in aortic rings from nNOS deficient (−/−), iNOS −/− and eNOS −/−mice and their background control strains (B6;129SF2 & C57BL/6j). We determined that photorelaxation in mouse aorta appears to involve the release of NO, derived from RSNOs, evidently a stable form of NO, which mediates a guanylate cyclase activation of K_V channels that results in a reversible vasorelaxation [193]. Photorelaxation was observed in all strains of mice examined, suggesting there is no dependence on the expression of any particular NOS isoform.

However, the photorelaxation response in endothelium intact and eNOS −/− mice appears to be different (Fig. 3). Since, after rundown of the store utilized by photorelaxation in the eNOS −/− mouse aorta, there was no recovery of the photorelaxation response after 1 hour of rest in the dark. In contrast, the C57BL/6j control endothelium intact mouse aorta shows ~30% recovery of the photorelaxation response after 1 hour indicating some regeneration of the photorelaxation store. These findings suggest that eNOS does appear to be important *in vitro* in the regeneration of photorelaxation stores. However, eNOS does not seem to be necessary to generate photorelaxation stores *in vivo* as there is a photorelaxation response in the eNOS −/− mouse, suggesting that NO can be derived from another NOS isoform or even a non-NOS enzymatic source. We speculate that circulating blood, containing NO bound in the form of RSNOs may be an additional source of NO *in vivo* (Fig. 4). This additional source may be used together with NO generated from NOS, to regenerate the photorelaxation store of RSNOs in VSMCs. This theory is supported by the recent findings that the liberation of NO from RSNOs outside the cell may be catalysed by cell surface protein disulfide isomerase [86,87]. It is postulated that the released NO, accumulates in the membrane where is reacts with O_2 to produce dinitrogen trioxide (N_2O_3). Intracellular thiols are then nitrosated by the N_2O_3 at the membrane cytosol interface to form RSNOs inside the cell. It has been reported that the oxidation of NO by O_2 to produce N_2O_3 was slow and insignificant *in vivo* [194]. However, more recent findings have shown that the reaction between NO and O_2 is 300 fold more rapid in cell membranes [17] and thus, the theory involving protein disulfide isomerase [86,87] is a plausible one.

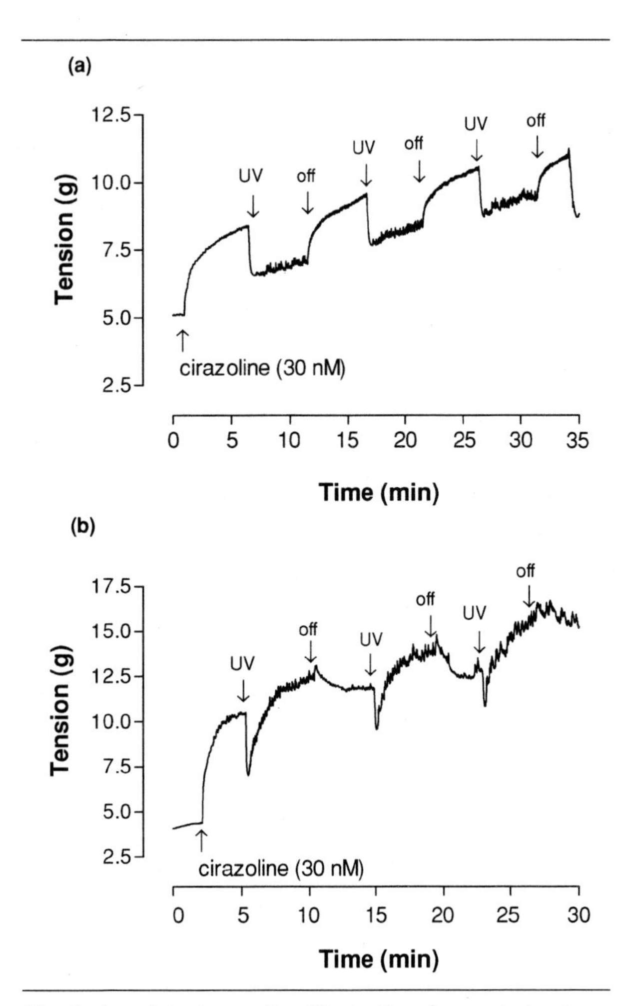

Fig. 3. *An original recording illustrating changes in tension (mN) in cirazoline precontracted aorta from (a) eNOS (−/−) and (b) C57BL / 6j mice in response to exposure to UV light (UV) and the effect of turning off it off (off).*

Physiological Implications

However, fundamental questions still remain. 1. What is the functional relevance of a molecular store of NO in vascular smooth muscle? 2. Can this store of NO be manipulated through pharmacological intervention and thus be a potential novel treatment for many different pathophysiological conditions? In particular, it would be useful in conditions in which endothelial dysfunction is involved and increased NO production would be beneficial. It has been suggested there could be beneficial effects from the photochemotherapeutic application of UV light in disease states such as leukaemia [74]. Although the use of UV light appears to be convenient tool for the assay of RSNOs in vascular smooth muscle, it is unlikely that UV light is the physiological stimulus *in vivo* that induces the release of NO from this store. Which leads to the question of what is the physiological stimulus? Further studies are required to address this question. However, there are several theories as to how NO is released from RSNOs *in vivo*, but it must be pointed out that these findings are largely from studies carried out *in vitro*. The enzyme, γ-glutamyl transpeptidase has

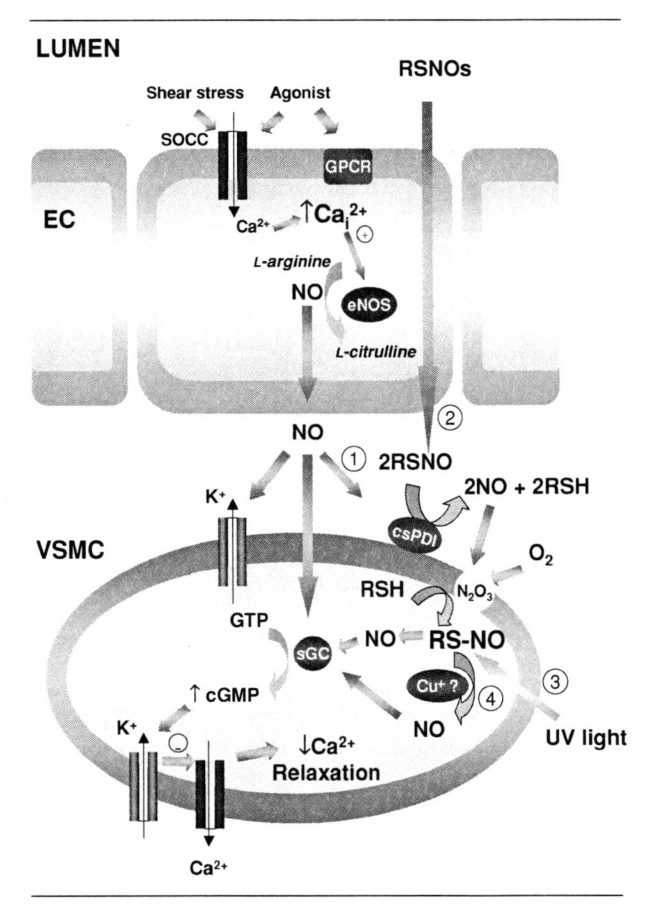

Fig. 4. *Possible pathways involved in the formation, storage and release of RSNOs in vascular smooth muscle. Activation of GPCR by agonists and activation of SOCC from the shear stress of blood flow in the membrane of EC increases Ca_i^{2+} levels, which in turn activates eNOS. The activation of eNOS results in the conversion of L-arginine to L-citrulline and the liberation of NO, which diffuses across to the VSMC and activates sGC causing an increase in cGMP. By a number of possible mechanisms, K^+ channels open, Ca^{2+} channels close and a decrease in Ca_i^{2+} results which is followed by vasorelaxation. The NO derived from this pathway (1) may also form RSNOs inside the cell after a reaction with RSH under aerobic conditions. Alternatively, NO liberated from RSNOs outside the cell (possibly from the circulating blood) by csPDI may provide an additional mechanism (2) for the formation of RSNOs inside the cell. The liberated NO reacts with O_2 in the membrane to form N_2O_3, which reacts with intracellular thiols at the cell cytosol interface to produce RSNOs. NO can be liberated from RSNOs by irradiation with UV light (3) which homolytically cleaves the S–N bond of RSNOs to release NO. Alternatively, a Cu^+ containing intracellular enzyme may catalyse the release of NO from RSNOs (4).*

Abbreviations: Ca_i^{2+}, intracellular calcium; csPDI, cell surface protein disulfide isomerase; Cu^+, copper; EC, endothelial cell; eNOS, endothelial nitric oxide synthase; GPCR, G-protein coupled receptor; NO, nitric oxide; N_2O_3, dinitrogen trioxide; O_2, oxygen; RSNO, S-nitrosothiols; RSH, thiols; S-N, sulfur–nitrogen; SOCC, stretch operated Ca^{2+} channels; sGC, soluble guanylate cyclase; VSMC, vascular smooth muscle cell.

been implicated in the decomposition of GSNO to S-nitrosocysteinylglycine, which is more susceptible to transition metal decomposition than GSNO [85]. However, studies that investigated the anti-platelet action of GSNO indicated that another enzyme was involved after the γ-glutamyl transpeptidase inhibitor, acivicin (1 mM) did not affect the rate of GSNO composition [77]. Yet, it has been shown that GSNO is a substrate for γ-glutamyl transpeptidase in the kidney but not the heart [195] indicating possible tissue-dependent differences for the expression of this enzyme. More recent studies by Lipton and colleagues have reported a possible physiological stimulus for RSNOs in the plasma [72]. They hypothesise that RSNOs perform a signalling role in the ventilatory response to hypoxia. Furthermore, they also propose a role for γ-glutamyl transpeptidase since acivicin attenuated the response to GSNO in control mice and impaired responses to GSNO were observed in γ-glutamyl transpeptidase knockout mice. After several findings implicated transition metals in the decomposition of RSNOs [75,196], Jourd'heuil and co-workers investigated the effect of CuZn-SOD on RSNOs [78]. Their results suggested that in the presence of physiological concentrations of GSH, the CuZn-SOD associated Cu^{2+} is reduced to Cu^+, which in turn decomposes RSNOs to release NO. Around the same time Singh and colleagues, reported similar findings and in addition, they also implicated the glutamate moiety in the regulation of the decomposition of GSNO as without it GSNO is readily reduced to release NO [197].

Other theories on how NO is released from RSNOs implicate the thioredoxin system [198], the putative enzyme "GSNO lysase" [199] and xanthine oxidase [200]. However, perhaps the most promising finding to date is the identification of glutathione-dependent formaldehyde dehydrogenase as a highly specific modulator of GSNO, controlling intracellular levels of both GSNO and other RSNOs [201,202]. Moreover, when the gene for this enzyme is deleted in mice, increased levels of GSNO and RSNOs were detected together with decreased release of NO from GSNO [202].

Conclusions

NO produced by the vascular endothelium acts on adjacent VSMCs to signal relaxation, and this process represents a major contributor to the humoral control of the vasculature. Owing to its reasonably reactive and free radical nature, the chemistry of NO is a large determinant of its cellular actions, distribution and metabolic fate. For example, the propensity for NO to bind with iron would suggest that it has a strong affinity for heme proteins, such as the heme moiety in soluble guanylate cyclase, of which it is known to activate to signal VSMC relaxation. Among the substances that NO may react with on route to activating soluble guanylate cyclase include oxygen, protein thiols, other free radicals and transition metal ions. The autooxidation of NO produces a substance that has a strong tendency to incorporate a NO group onto nucleophilic cellular constituents, ultimately giving rise to the formation of the NO carrier molecules, RSNOs. Many useful pharmacological tools have been developed which take advantage of the reactivity and chemical properties of NO and have been used extensively to establish the involvement of NO in physiological functions. Their use has also been instrumental in identifying a potential non-NOS mediated source of NO present in endothelial cells that is distinct from EDHF. However, more studies are needed to identify the source and cellular location of this NO store and what physiological stimuli mediate its release. In addition, studies have reported a store of RSNOs in vascular smooth cells, which under appropriate conditions can release NO. Direct exposure to UV light is one such stimulus, however, *in vivo* it appears unlikely that this would be the physiological stimulus to mediate the release of NO from this store. Therefore, although the physiological stimulus and mediators for the release of RSNOs stored within the VSMCs remain elusive, RSNOs stores in the vasculature are an exciting prospect for the targeted release of NO in the search for a novel therapeutic approach for the treatment of many diseases.

Acknowledgment

Data from the laboratory of C.T. that have been quoted in this review were supported by grants from the Canadian Institutes of Health (CIHR) and the Heart and Stroke Foundation of Alberta.

References

1. Bredt DS, Hwang PM, Glatt CE, Lowenstein C, Reed RR, Snyder SH. Cloned and expressed nitric oxide synthase structurally resembles cytochrome P-450 reductase. *Nature* 1991;351:714–718.
2. Bredt DS, Snyder SH. Nitric oxide: A physiologic messenger molecule. *Annu Rev Biochem* 1994;63:175–195.
3. McGuire JJ, Ding H, Triggle CR. Endothelium-derived relaxing factors: A focus on endothelium-derived hyperpolarizing factor(s). *Can J Physiol Pharmacol* 2001;79:443–470.
4. Feelisch M. The biochemical pathways fo nitric oxide formation from nitrovasodilators: Appropriate choice of exogenous NO donors and aspects of preparation and handling of aqueous NO solutions. *J Cardiovasc Pharmacol* 1991;17:S25–S33.
5. Malinski T, Taha Z, Grunfeld S, Patton S, Kapturczak M, Tomboulian P. Diffusion of nitric oxide in the aorta wall monitored *in situ* by porphyrinic microsensors. *Biochem Biophys Res Commun* 1993;193:1076–1082.
6. Pries AR, Secomb TW, Gaehtgens P. The endothelial surface layer. *Pflugers Arch* 2000;440:653–666.
7. Lancaster JR, Jr. A tutorial on the diffusibility and reactivity of free nitric oxide. *Nitric Oxide* 1997;1:18–30.
8. Butler AR, Flitney FW, Williams DL. NO, nitrosonium ions, nitroxide ions, nitrosothiols and iron-nitrosyls in biology: A chemist's perspective. *Trends Pharmacol Sci* 1995;16:18–22.
9. Bonner FT, Stedman G. *The Chemistry of Nitric Oxide and Redox-Related Species*, New York: John Wiley & Sons Ltd, 1996:3–27.
10. Fukuto JM. Chemistry of nitric oxide: Biologically relevant aspects. *Adv Pharmacol* 1995;34:1–15.
11. Kelm M, Schrader J. Control of coronary vascular tone by nitric oxide. *Circ Res* 1990;66:1561–1575.
12. Feelisch M, te PM, Zamora R, Deussen A, Moncada S. Understanding the controversy over the identity of EDRF. *Nature* 1994;368:62–65.
13. Moncada S, Palmer RM, Higgs EA. Biosynthesis of nitric oxide from L-arginine. A pathway for the regulation of cell function and communication. *Biochem Pharmacol* 1989;38:1709–1715.
14. Wood J, Garthwaite J. Models of the diffusional spread of nitric oxide: Implications for neural nitric oxide signalling and its pharmacological properties. *Neuropharmacology* 1994;33:1235–1244.
15. Kharitonov VG, Sundquist AR, Sharma VS. Kinetics of nitric oxide autoxidation in aqueous solution. *J Biol Chem* 1994;269:5881–5883.
16. Ford PC, Wink DA, Stanbury DM. Autoxidation kinetics of aqueous nitric oxide. *FEBS Lett* 1993;326:1–3.
17. Liu X, Miller MJ, Joshi MS, Thomas DD, Lancaster JR, Jr. Accelerated reaction of nitric oxide with O_2 within the hydrophobic interior of biological membranes. *Proc Natl Acad Sci USA* 1998;95:2175–2179.
18. Hughes MN. Relationships between nitric oxide, nitroxyl ion, nitrosonium cation and peroxynitrite. *Biochim Biophys Acta* 1999;1411:263–272.
19. Hogg N. The biochemistry and physiology of s-nitrosothiols. *Annu Rev Pharmacol Toxicol* 2002;42:585–600.
20. Arnelle DR, Stamler JS. NO^+, NO, and NO^- donation by S-nitrosothiols: Implications for regulation of physiological functions by S-nitrosylation and acceleration of disulfide formation. *Arch Biochem Biophys* 1995;318:279–285.
21. Butler AR, Rhodes P. Chemistry, analysis, and biological roles of S-nitrosothiols. *Anal Biochem* 1997;249:1–9.
22. Kowaluk EA, Fung HL. Spontaneous liberation of nitric oxide cannot account for *in vitro* vascular relaxation by S-nitrosothiols. *J Pharmacol Exp Ther* 1990;255:1256–1264.
23. Furchgott RF, Jothianandan D. Transnitrosation reactions involved in the sensitisation of rabbit aorta for relaxations by thiols. *Pharmacol Toxicol* 1998;83:44–46.
24. Pino RZ, Feelisch M. Bioassay discrimination between nitric oxide (NO.) and nitroxyl (NO^-) using L-cysteine. *Biochem Biophys Res Commun* 1994;201:54–62.
25. Zamora R, Grzesiok A, Weber H, Feelisch M. Oxidative release of nitric oxide accounts for guanylyl cyclase stimulating, vasodilator and anti-platelet activity of Piloty's

acid: A comparison with Angeli's salt. *Biochem J* 1995;312 (Pt 2):333–339.
26. Doyle MP, Mahapatro SN, Broene RD, Guy JK. Oxidation and reduction of hemoproteins by trixododitirate(II). The roel of nitrosyl hydride and nitrite. *J Am Chem Soc* 1988;110:593–599.
27. Khan AA, Schuler MM, Coppock RW. Inhibitory effects of various sulfur compounds on the activity of bovine erythrocyte enzymes. *J Toxicol Environ Health* 1987;22:481–490.
28. Beckman JS, Beckman TW, Chen J, Marshall PA, Freeman BA. Apparent hydroxyl radical production by peroxynitrite: Implications for endothelial injury from nitric oxide and superoxide. *Proc Natl Acad Sci USA* 1990;87:1620–1624.
29. Huie RE, Padmaja S. The reaction of no with superoxide. *Free Radic Res Commun* 1993;18:195–199.
30. Kirsch M, de Groot H. Formation of peroxynitrite from reaction of nitroxyl anion with molecular oxygen. *J Biol Chem* 2002;277:13379–13388.
31. Liu S, Beckman JS, Ku DD. Peroxynitrite, a product of superoxide and nitric oxide, produces coronary vasorelaxation in dogs. *J Pharmacol Exp Ther* 1994;268:1114–1121.
32. Moro MA, Darley-Usmar VM, Goodwin DA, Read NG, Zamora-Pino R, Feelisch M, Radomski MW, Moncada S. Paradoxical fate and biological action of peroxynitrite on human platelets. *Proc Natl Acad Sci USA* 1994;91:6702–6706.
33. Moro MA, Darley-Usmar VM, Lizasoain I, Su Y, Knowles RG, Radomski MW, Moncada S. The formation of nitric oxide donors from peroxynitrite. *Br J Pharmacol* 1995;116:1999–2004.
34. Mayer B, Schrammel A, Klatt P, Koesling D, Schmidt K. Peroxynitrite-induced accumulation of cyclic GMP in endothelial cells and stimulation of purified soluble guanylyl cyclase. Dependence on glutathione and possible role of S-nitrosation. *J Biol Chem* 1995;270:17355–17360.
35. Iesaki T, Gupte SA, Kaminski PM, Wolin MS. Inhibition of guanylate cyclase stimulation by NO and bovine arterial relaxation to peroxynitrite and H_2O_2. *Am J Physiol* 1999;277:H978–H985.
36. Kooy NW, Royall JA. Agonist-induced peroxynitrite production from endothelial cells. *Arch Biochem Biophys* 1994;310:352–359.
37. Marin J, Rodriguez-Martinez MA. Nitric oxide, oxygen-derived free radicals and vascular endothelium. *J Auton Pharmacol* 1995;15:279–307.
38. Pagano PJ, Ito Y, Tornheim K, Gallop PM, Tauber AI, Cohen RA. An NADPH oxidase superoxide-generating system in the rabbit aorta. *Am J Physiol* 1995;268:H2274–H2280.
39. Suzuki H, DeLano FA, Parks DA, Jamshidi N, Granger DN, Ishii H, Suematsu M, Zweifach BW, Schmid-Schonbein GW. Xanthine oxidase activity associated with arterial blood pressure in spontaneously hypertensive rats. *Proc Natl Acad Sci USA* 1998;95:4754–4759.
40. Fleming I, Michaelis UR, Bredenkotter D, Fisslthaler B, Dehghani F, Brandes RP, Busse R. Endothelium-derived hyperpolarizing factor synthase (Cytochrome P450 2C9) is a functionally significant source of reactive oxygen species in coronary arteries. *Circ Res* 2001;88:44–51.
41. Cosentino F, Sill JC, Katusic ZS. Role of superoxide anions in the mediation of endothelium-dependent contractions. *Hypertension* 1994;23:229–235.
42. Xia Y, Tsai AL, Berka V, Zweier JL. Superoxide generation from endothelial nitric-oxide synthase. A Ca^{2+}/calmodulin-dependent and tetrahydrobiopterin regulatory process. *J Biol Chem* 1998;273:25804–25808.
43. Ignarro LJ, Fukuto JM, Griscavage JM, Rogers NE, Byrns RE. Oxidation of nitric oxide in aqueous solution to nitrite but not nitrate: Comparison with enzymatically formed nitric oxide from L-arginine. *Proc Natl Acad Sci USA* 1993;90:8103–8107.
44. Benko B, Yu NT. Resonance Raman studies of nitric oxide binding to ferric and ferrous hemoproteins: Detection of Fe(III)—NO stretching, Fe(III)—N—O bending, and Fe(II)—N—O bending vibrations. *Proc Natl Acad Sci USA* 1983;80:7042–7046.
45. Gow AJ, Stamler JS. Reactions between nitric oxide and haemoglobin under physiological conditions. *Nature* 1998;391:169–173.
46. Gow AJ, Luchsinger BP, Pawloski JR, Singel DJ, Stamler JS. The oxyhemoglobin reaction of nitric oxide. *Proc Natl Acad Sci USA* 1999;96:9027–9032.
47. Gorren AC, de Boer E, Wever R. The reaction of nitric oxide with copper proteins and the photodissociation of copper-NO complexes. *Biochim Biophys Acta* 1987;916:38–47.
48. Rochelle LG, Morana SJ, Kruszyna H, Russell MA, Wilcox DE, Smith RP. Interactions between hydroxocobalamin and nitric oxide (NO): Evidence for a redox reaction between NO and reduced cobalamin and reversible NO binding to oxidized cobalamin. *J Pharmacol Exp Ther* 1995;275:48–52.
49. Murphy ME, Sies H. Reversible conversion of nitroxyl anion to nitric oxide by superoxide dismutase. *Proc Natl Acad Sci USA* 1991;88:10860–10864.
50. Fukuto JM, Chiang K, Hszieh R, Wong P, Chaudhuri G. The pharmacological activity of nitroxyl: A potent vasodilator with activity similar to nitric oxide and/or endothelium-derived relaxing factor. *J Pharmacol Exp Ther* 1992;263:546–551.
51. Fukuto JM, Gulati P, Nagasawa HT. Involvement of nitroxyl (HNO) in the cyanamide-induced vasorelaxation of rabbit aorta. *Biochem Pharmacol* 1994;47:922–924.
52. Hobbs AJ, Fukuto JM, Ignarro LJ. Formation of free nitric oxide from L-arginine by nitric oxide synthase: Direct enhancement of generation by superoxide dismutase. *Proc Natl Acad Sci USA* 1994;91:10992–10996.
53. Schmidt HH, Hofmann H, Schindler U, Shutenko ZS, Cunningham DD, Feelisch M. No .NO from NO synthase. *Proc Natl Acad Sci USA* 1996;93:14492–14497.
54. Li CG, Karagiannis J, Rand MJ. Comparison of the redox forms of nitrogen monoxide with the nitrergic transmitter in the rat anococcygeus muscle. *Br J Pharmacol* 1999;127:826–834.
55. Ellis A, Li CG, Rand MJ. Differential actions of L-cysteine on responses to nitric oxide, nitroxyl anions and EDRF in the rat aorta. *Br J Pharmacol* 2000;129:315–322.
56. Adak S, Wang Q, Stuehr DJ. Arginine conversion to nitroxide by tetrahydrobiopterin-free neuronal nitric-oxide synthase. Implications for mechanism. *J Biol Chem* 2000;275:33554–33561.

57. Rusche KM, Spiering MM, Marletta MA. Reactions catalyzed by tetrahydrobiopterin-free nitric oxide synthase. *Biochemistry* 1998;37:15503–15512.
58. Sharpe MA, Cooper CE. Reactions of nitric oxide with mitochondrial cytochrome c: A novel mechanism for the formation of nitroxyl anion and peroxynitrite. *Biochem J* 1998;332(Pt 1):9–19.
59. Wong PS, Hyun J, Fukuto JM, Shirota FN, DeMaster EG, Shoeman DW, Nagasawa HT. Reaction between S-nitrosothiols and thiols: Generation of nitroxyl (HNO) and subsequent chemistry. *Biochemistry* 1998;37:5362–5371.
60. Lang D, Hussain SA, Lewis MJ. Homocysteine inhibits endothelium-dependent relaxation in isolated rabbit aortic rings. *Br J Pharmacol* 1997;145P.
61. Quere I, Hillaire-Buys D, Brunschwig C, Chapal J, Janbon C, Blayac JP, Petit P, Loubatieres-Mariani MM. Effects of homocysteine on acetylcho. *Br J Pharmacol* 1997;122:351–357.
62. Wanstall JC, Jeffery TK, Gambino A, Lovren F, Triggle CR. Vascular smooth muscle relaxation mediated by nitric oxide donors: A comparison with acetylcholine, nitric oxide and nitroxyl ion. *Br J Pharmacol* 2001;134:463–472.
63. Wink DA, Feelisch M, Fukuto J, Chistodoulou D, Jourd'heuil D, Grisham MB, Vodovotz Y, Cook JA, Krishna M, DeGraff WG, Kim S, Gamson J, Mitchell JB. The cytotoxicity of nitroxyl: Possible implications for the pathophysiological role of NO. *Arch Biochem Biophys* 1998;351:66–74.
64. Ohshima H, Gilibert I, Bianchini F. Induction of DNA strand breakage and base oxidation by nitroxyl anion through hydroxyl radical production. *Free Radic Biol Med* 1999;26:1305–1313.
65. Miranda KM, Espey MG, Yamada K, Krishna M, Ludwick N, Kim S, Jourd'heuil D, Grisham MB, Feelisch M, Fukuto JM, Wink DA. Unique oxidative mechanisms for the reactive nitrogen oxide species, nitroxyl anion. *J Biol Chem* 2001;276:1720–1727.
66. Ma XL, Gao F, Liu GL, Lopez BL, Christopher TA, Fukuto JM, Wink DA, Feelisch M. Opposite effects of nitric oxide and nitroxyl on postischemic myocardial injury. *Proc Natl Acad Sci USA* 1999;96:14617–14622.
67. De Vriese AS, Verbeuren TJ, Van d, V, Lameire NH, Vanhoutte PM. Endothelial dysfunction in diabetes. *Br J Pharmacol* 2000;130:963–974.
68. Katusic ZS. Vascular endothelial dysfunction: Does tetrahydrobiopterin play a role? *Am J Physiol Heart Circ Physiol* 2001;281:H981–H986.
69. Cosentino F, Barker JE, Brand MP, Heales SJ, Werner ER, Tippins JR, West N, Channon KM, Volpe M, Luscher TF. Reactive oxygen species mediate endothelium-dependent relaxations in tetrahydrobiopterin-deficient mice. *Arterioscler Thromb Vasc Biol* 2001;21:496–502.
70. Reif A, Zecca L, Riederer P, Feelisch M, Schmidt HH. Nitroxyl oxidizes NADPH in a superoxide dismutase inhibitable manner. *Free Radic Biol Med* 2001;30:803–808.
71. Stamler JS, Singel DJ, Loscalzo J. Biochemistry of nitric oxide and its redox-activated forms. *Science* 1992;258:1898–1902.
72. Lipton AJ, Johnson MA, Macdonald T, Lieberman MW, Gozal D, Gaston B. S-nitrosothiols signal the ventilatory response to hypoxia. *Nature* 2001;413:171–174.
73. Williams DL. S-nitrosation and the reaction of S-nitrso compounds. *Chem Soc Rev* 1985;14:171–196.
74. Sexton DJ, Muruganandam A, McKenney DJ, Mutus B. Visible light photochemical release of nitric oxide from S-nitrosoglutathione: Potential photochemotherapeutic applications. *Photochem Photobiol* 1994;59:463–467.
75. Dicks AP, Williams DL. Generation of nitric oxide from S-nitrosothiols using protein-bound Cu^{2+} sources. *Chem Biol* 1996;3:655–659.
76. Al Sa'doni HH, Megson IL, Bisland S, Butler AR, Flitney FW. Neocuproine, a selective Cu(I) chelator, and the relaxation of rat vascular smooth muscle by S-nitrosothiols. *Br J Pharmacol* 1997;121:1047–1050.
77. Gordge MP, Meyer DJ, Hothersall J, Neild GH, Payne NN, Noronha-Dutra A. Copper chelation-induced reduction of the biological activity of S-nitrosothiols. *Br J Pharmacol* 1995;114:1083–1089.
78. Jourd'heuil D, Laroux FS, Miles AM, Wink DA, Grisham MB. Effect of superoxide dismutase on the stability of S-nitrosothiols. *Arch Biochem Biophys* 1999;361:323–330.
79. Myers PR, Minor RL, Jr., Guerra R, Jr., Bates JN, Harrison DG. Vasorelaxant properties of the endothelium-derived relaxing factor more closely resemble S-nitrosocysteine than nitric oxide. *Nature* 1990;345:161–163.
80. Keaney JF Jr., Simon DI, Stamler JS, Jaraki O, Scharfstein J, Vita JA, Loscalzo J. NO forms an adduct with serum albumin that has endothelium-derived relaxing factor-like properties. *J Clin Invest* 1993;91:1582–1589.
81. Jia L, Furchgott RF. Inhibition by sulfhydryl compounds of vascular relaxation induced by nitric oxide and endothelium-derived relaxing factor. *J Pharmacol Exp Ther* 1993;267:371–378.
82. Creager MA, Roddy MA, Boles K, Stamler JS. N-acetylcysteine does not influence the activity of endothelium-derived relaxing factor *in vivo*. *Hypertension* 1997;29:668–672.
83. Marley R, Patel RP, Orie N, Ceaser E, Darley-Usmar V, Moore K. Formation of nanomolar concentrations of S-nitroso-albumin in human plasma by nitric oxide. *Free Radic Biol Med* 2001;31:688–696.
84. Gaston B, Reilly J, Drazen JM, Fackler J, Ramdev P, Arnelle D, Mullins ME, Sugarbaker DJ, Chee C, Singel DJ. Endogenous nitrogen oxides and bronchodilator S-nitrosothiols in human airways. *Proc Natl Acad Sci USA* 1993;90:10957–10961.
85. Askew SC, Butler AR, Flitney FW, Kemp GD, Megson IL. Chemical mechanisms underlying the vasodilator and platelet anti-aggregating properties of S-nitroso-N-acetyl-DL-penicillamine and S-nitrosoglutathione. *Bioorg Med Chem* 1995;3:1–9.
86. Zai A, Rudd MA, Scribner AW, Loscalzo J. Cell-surface protein disulfide isomerase catalyzes transnitrosation and regulates intracellular transfer of nitric oxide. *J Clin Invest* 1999;103:393–399.
87. Ramachandran N, Root P, Jiang XM, Hogg PJ, Mutus B. Mechanism of transfer of NO from extracellular S-nitrosothiols into the cytosol by cell-surface protein disulfide isomerase. *Proc Natl Acad Sci USA* 2001;98:9539–9544.
88. Mulsch A, Lurie DJ, Seimenis I, Fichtlscherer B, Foster MA. Detection of nitrosyl-iron complexes by proton-electron-double-resonance imaging. *Free Radic Biol Med* 1999;27:636–646.

89. Ueno T, Suzuki Y, Fujii S, Vanin AF, Yoshimura T. *In vivo* nitric oxide transfer of a physiological NO carrier, dinitrosyl dithiolato iron complex, to target complex. *Biochem Pharmacol* 2002;63:485–493.
90. Vanin AF, Stukan RA, Manukhina EB. Physical properties of dinitrosyl iron complexes with thiol-containing ligands in relation with their vasodilator activity. *Biochim Biophys Acta* 1996;1295:5–12.
91. Mulsch A, Mordvintcev PI, Vanin AF, Busse R. Formation and release of dinitrosyl iron complexes by endothelial cells. *Biochem Biophys Res Commun* 1993;196:1303–1308.
92. Vanin AF. Endothelium-derived relaxing factor is a nitrosyl iron complex with thiol ligands. *FEBS Lett* 1991;289:1–3.
93. Muller B, Kleschyov AL, Stoclet JC. Evidence for N-acetylcysteine-sensitive nitric oxide storage as dinitrosyl-iron complexes in lipopolysaccharide-treated rat aorta. *Br J Pharmacol* 1996;119:1281–1285.
94. Lundberg JO, Weitzberg E, Lundberg JM, Alving K. Intragastric nitric oxide production in humans: Measurements in expelled air. *Gut* 1994;35:1543–1546.
95. Modin A, Bjorne H, Herulf M, Alving K, Weitzberg E, Lundberg JO. Nitrite-derived nitric oxide: A possible mediator of 'acidic-metabolic' vasodilation. *Acta Physiol Scand* 2001;171:9–16.
96. Cocks TM, Angus JA. Comparison of relaxation responses of vascular and non-vascular smooth muscle to endothelium-derived relaxing factor (EDRF), acidified sodium nitrite (NO) and sodium nitroprusside. *Naunyn Schmiedebergs Arch Pharmacol* 1990;341:364–372.
97. Boucher JL, Genet A, Vadon S, Delaforge M, Henry Y, Mansuy D. Cytochrome P450 catalyzes the oxidation of N omega-hydroxy-L-arginine by NADPH and O2 to nitric oxide and citrulline. *Biochem Biophys Res Commun* 1992;187:880–886.
98. Jia Y, Zacour M, Tolloczko B, Martin JG. Nitric oxide synthesis by tracheal smooth muscle cells by a nitric oxide synthase-independent pathway. *Am J Physiol* 1998;275:L895–L901.
99. Hobbs AJ. Soluble guanylate cyclase: The forgotten sibling. *Trends Pharmacol Sci* 1997;18:484–491.
100. Martin W, Villani GM, Jothianandan D, Furchgott RF. Selective blockade of endothelium-dependent and glyceryl trinitrate-induced relaxation by hemoglobin and by methylene blue in the rabbit aorta. *J Pharmacol Exp Ther* 1985;232:708–716.
101. Wolin MS, Cherry PD, Rodenburg JM, Messina EJ, Kaley G. Methylene blue inhibits vasodilation of skeletal muscle arterioles to acetylcholine and nitric oxide via the extracellular generation of superoxide anion. *J Pharmacol Exp Ther* 1990;254:872–876.
102. Marczin N, Ryan US, Catravas JD. Methylene blue inhibits nitrovasodil. *J Pharmacol Exp Ther* 1992;263:170–179.
103. Mayer B, Brunner F, Schmidt K. Inhibition of nitric oxide synthesis by methylene blue. *Biochem Pharmacol* 1993;45:367–374.
104. Luo D, Das S, Vincent SR. Effects of methylene blue and LY83583 on neuronal nitric oxide synthase and NADPH-diaphorase. *Eur J Pharmacol* 1995;290:247–251.
105. Garthwaite J, Southam E, Boulton CL, Nielsen EB, Schmidt K, Mayer B. Potent and selective inhibition of nitric oxide-sensitive guanylyl cyclase by 1H-[1,2,4]oxadiazolo[4,3-a]quinoxalin-1-one. *Mol Pharmacol* 1995;48:184–188.
106. Feelisch M, Kotsonis P, Siebe J, Clement B, Schmidt HH. The soluble guanylyl cyclase inhibitor 1H-[1,2,4]oxadiazolo[4,3,-a] quinoxalin-1-one is a nonselective heme protein inhibitor of nitric oxide synthase and other cytochrome P-450 enzymes involved in nitric oxide donor bioactivation. *Mol Pharmacol* 1999;56:243–253.
107. Zhao Y, Brandish PE, DiValentin M, Schelvis JP, Babcock GT, Marletta MA. Inhibition of soluble guanylate cyclase by ODQ. *Biochemistry* 2000;39:10848–10854.
108. Gryglewski RJ, Palmer RM, Moncada S. Superoxide anion is involved in the breakdown of endothelium-derived vascular relaxing factor. *Nature* 1986;320:454–456.
109. Moncada S, Palmer RM, Gryglewski RJ. Mechanism of action of some inhibitors of endothelium-derived relaxing factor. *Proc Natl Acad Sci USA* 1986;83:9164–9168.
110. Rubanyi GM, Vanhoutte PM. Oxygen-derived free radicals, endothelium, and responsiveness of vascular smooth muscle. *Am J Physiol* 1986;250:H815–H821.
111. Palmer RM, Ferrige AG, Moncada S. Nitric oxide release accounts for the biological activity of endothelium-derived relaxing factor. *Nature* 1987;327:524–526.
112. Mian KB, Martin W. Differential sensitivity of basal and acetylcholine-stimulated activity of nitric oxide to destruction by superoxide anion in rat aorta. *Br J Pharmacol* 1995;115:993–1000.
113. Ignarro LJ, Byrns RE, Buga GM, Wood KS, Chaudhuri G. Pharmacological evidence that endothelium-derived relaxing factor is nitric oxide: Use of pyrogallol and superoxide dismutase to study endothelium-dependent and nitric oxide-elicited vascular smooth muscle relaxation. *J Pharmacol Exp Ther* 1988;244:181–189.
114. Mittra S, Singh M. Possible mechanism of captopril induced endothelium-dependent relaxation in isolated rabbit aorta. *Mol Cell Biochem* 1998;183:63–67.
115. Srivastava P, Hegde LG, Patnaik GK, Dikshit M. Role of endothelial-derived reactive oxygen species and nitric oxide in norepinephrine-induced rat aortic ring contractions. *Pharmacol Res* 1998;38:265–274.
116. Inoue M, Watanabe N, Matsuno K, Sasaki J, Tanaka Y, Hatanaka H, Amachi T. Expression of a hybrid Cu/Zn-type superoxide dismutase which has high affinity for heparin-like proteoglycans on vascular endothelial cells. *J Biol Chem* 1991;266:16409–16414.
117. Abrahamsson T, Brandt U, Marklund SL, Sjoqvist PO. Vascular bound recombinant extracellular superoxide dismutase type C protects against the detrimental effects of superoxide radicals on endothelium-dependent arterial relaxation. *Circ Res* 1992;70:264–271.
118. Wambi-Kiesse CO, Katusic ZS. Inhibition of copper/zinc superoxide dismutase impairs NO.-mediated endothelium-dependent relaxations. *Am J Physiol* 1999;276:H1043–H1048.
119. Halliwell B, Gutteridge JMC. The Chemistry of Free Radicals and Related Reactive Species, Oxford: Oxford University Press, 1999:110–185.
120. Akpaffiong MJ, Taylor AA. Antihypertensive and vasodilator actions of antioxidants in spontaneously hypertensive rats. *Am J Hypertens* 1998;11:1450–1460.
121. Fontana L, McNeill KL, Ritter JM, Chowienczyk PJ. Effects of vitamin C and of a cell permeable superoxide dismutase mimetic on acute lipoprotein induced

endothelial dysfunction in rabbit aortic rings. *Br J Pharmacol* 1999;126:730–734.
122. Jackson TS, Xu A, Vita JA, Keaney JF Jr. Ascorbate prevents the interaction of superoxide and nitric oxide only at very high physiological concentrations. *Circ Res* 1998;83:916–922.
123. Aleryani S, Milo E, Rose Y, Kostka P. Superoxide-mediated decomposition of biological S-nitrosothiols. *J Biol Chem* 1998;273:6041–6045.
124. Scorza G, Pietraforte D, Minetti M. Role of ascorbate and protein thiols in the release of nitric oxide from S-nitroso-albumin and S-nitroso-glutathione in human plasma. *Free Radic Biol Med* 1997;22:633–642.
125. Gorren AC, Schrammel A, Schmidt K, Mayer B. Decomposition of S-nitrosoglutathione in the presence of copper ions and glutathione. *Arch Biochem Biophys* 1996;330:219–228.
126. De Man JG, Moreels TG, De Winter BY, Herman AG, Pelckmans PA. Neocuproine potentiates the activity of the nitrergic neurotransmitter but inhibits that of S-nitrosothiols. *Eur J Pharmacol* 1999;381:151–159.
127. Hobbs AJ, Tucker JF, Gibson A. Differentiation by hydroquinone of relaxations induced by exogenous and endogenous nitrates in non-vascular smooth muscle: Role of superoxide anions. *Br J Pharmacol* 1991;104:645–650.
128. Akaike T, Yoshida M, Miyamoto Y, Sato K, Kohno M, Sasamoto K, Miyazaki K, Ueda S, Maeda H. Antagonistic action of imidazolineoxyl N-oxides against endothelium-derived relaxing factor/.NO through a radical reaction. *Biochemistry* 1993;32:827–832.
129. Yoshida M, Akaike T, Goto S, Takahashi W, Inadome A, Yono M, Seshita H, Maeda H, Ueda S. Effect of the NO scavenger carboxy-ptio on endothelium-dependent vasorelaxation of various blood vessels from rabbits. *Life Sci* 1998;62:203–211.
130. Rand MJ, Li CG. Discrimination by the NO-trapping agent, carboxy-PTIO, between NO and the nitrergic transmitter but not between NO and EDRF. *Br J Pharmacol* 1995;116:1906–1910.
131. Pieper GM, Siebeneich W. Use of a nitronyl nitroxide to discriminate the contribution of nitric oxide radical in endothelium-dependent relaxation of control and diabetic blood vessels. *J Pharmacol Exp Ther* 1997;283:138–147.
132. Pfeiffer S, Leopold E, Hemmens B, Schmidt K, Werner ER, Mayer B. Interference of carboxy-PTIO with nitric o. *Free Radic Biol Med* 1997;22:787–794.
133. La M, Li CG, Rand MJ. Comparison of the effects of hydroxocobalamin and oxyhaemoglobin on responses to NO, EDRF and the nitrergic transmitter. *Br J Pharmacol* 1996;117:805–810.
134. Gillespie JS, Sheng H. Influence of haemoglobin and erythrocytes on the effects of EDRF, a smooth muscle inhibitory factor, and nitric oxide on vascular and non-vascular smooth muscle. *Br J Pharmacol* 1988;95:1151–1156.
135. Khan MT, Jothianandan D, Matsunaga K, Furchgott RF. Vasodilation induced by acetylcholine and by glyceryl trinitrate in rat aortic and mesenteric vasculature. *J Vasc Res* 1992;29:20–28.
136. Rand MJ, Li CG. Differential effects of hydroxocobalamin on relaxations induced by nitrosothiols in rat aorta and anococcygeus muscle. *Eur J Pharmacol* 1993;241:249–254.
137. Hrinczenko BW, Alayash AI, Wink DA, Gladwin MT, Rodgers GP, Schechter AN. Effect of nitric oxide and nitric oxide donors on red blood cell oxygen transport. *Br J Haematol* 2000;110:412–419.
138. Rajanayagam MA, Li CG, Rand MJ. Differential effects of hydroxocobalamin on NO-mediated relaxations in rat aorta and anococcygeus muscle. *Br J Pharmacol* 1993;108:3–5.
139. Kaczka EE, Wolf DE, Kuehl FA, Folkers K. Vitamin B12. Modifications of cyano-cobalamin. *J Am Chem Soc* 1951;73:3569–3572.
140. Marczin N, Ryan US, Catravas JD. Sulfhydryl-depleting agents, but not deferoxamine, modulate EDRF action in cultured pulmonary arterial cells. *Am J Physiol* 1993;265:L220–L227.
141. Patel JM, Block ER. Sulfhydryl-disulfide modulation and the role of disulfide oxidoreductases in regulation of the catalytic activity of nitric oxide synthase in pulmonary artery endothelial cells. *Am J Respir Cell Mol Biol* 1995;13:352–359.
142. Hatchett RJ, Gryglewski RJ, Mlochowski J, Zembowicz A, Radziszewski W. Carboxyebselen a potent and selective inhibitor of endothelial nitric oxide synthase. *J Physiol Pharmacol* 1994;45:55–67.
143. Hattori R, Yui Y, Shinoda E, Inoue R, Aoyama T, Masayasu H, Kawai C, Sasayama S. Effect of ebselen on bovine and rat nitric oxide synthase activity is modified by thiols. *Jpn J Pharmacol* 1996;72:191–193.
144. Zembowicz A, Hatchett RJ, Radziszewski W, Gryglewski RJ. Inhibition of endothelial nitric oxide synthase by ebselen. Prevention by thiols suggests the inactivation by ebselen of a critical thiol essential for the catalytic activity of nitric oxide synthase. *J Pharmacol Exp Ther* 1993;267:1112–1118.
145. Kim HR, Kim JW, Park JY, Je HD, Lee SY, Huh IH, Sohn UD. The effects of thiol compounds and ebselen on nitric oxide activity in rat aortic vascular responses. *J Auton Pharmacol* 2001;21:23–28.
146. Li CG, Brosch SF, Rand MJ. Inhibition by ethacrynic acid of NO-mediated relaxations of the rat anococcygeus muscle. *Clin Exp Pharmacol Physiol* 1994;21:293–299.
147. Rapoport RM, Murad F. Effects of ethacrynic acid and cystamine on sodium nitroprusside-induced relaxation, cyclic GMP levels and guanylate cyclase activity in rat aorta. *Gen Pharmacol* 1988;19:61–65.
148. Hibbs JB, Jr., Vavrin Z, Taintor RR. L-arginine is required for expression of the activated macrophage effector mechanism causing selective metabolic inhibition in target cells. *J Immunol* 1987;138:550–565.
149. Li CG, Rand MJ. Evidence for a role of nitric oxide in the neurotransmitter system mediating relaxation of the rat anococcygeus muscle. *Clin Exp Pharmacol Physiol* 1989;16:933–938.
150. Rees DD, Palmer RM, Schulz R, Hodson HF, Moncada S. Characterization of three inhibitors of endothelial nitric oxide synthase *in vitro* and *in vivo*. *Br J Pharmacol* 1990;101:746–752.
151. Bogle RG, MacAllister RJ, Whitley GS, Vallance P. Induction of NG-monomethyl-L-arginine uptake: A mechanism for differential inhibition of NO synthases? *Am J Physiol* 1995;269:C750–C756.
152. Moore PK, al Swayeh OA, Chong NW, Evans RA, Gibson A. L-NG-nitro arginine (L-NOARG), a novel, L-arginine-reversible inhibitor of endothelium-dependent

vasodilatation *in vitro*. *Br J Pharmacol* 1990;99:408–412.
153. Kilpatrick EV, Cocks TM. Evidence for differential roles of nitric oxide (NO) and hyperpolarization in endothelium-dependent relaxation of pig isolated coronary artery. *Br J Pharmacol* 1994;112:557–565.
154. Kemp BK, Cocks TM. Evidence that mechanisms dependent and independent of nitric oxide mediate endothelium-dependent relaxation to bradykinin in human small resistance-like coronary arteries. *Br J Pharmacol* 1997;120:757–762.
155. Cohen RA, Plane F, Najibi S, Huk I, Malinski T, Garland CJ. Nitric oxide is the mediator of both endothelium-dependent relaxation and hyperpolarization of the rabbit carotid artery. *Proc Natl Acad Sci USA* 1997;94:4193–4198.
156. Simonsen U, Wadsworth RM, Buus NH, Mulvany MJ. *In vitro* simultaneous measurements of relaxation and nitric oxide concentration in rat superior mesenteric artery. *J Physiol* 1999;516(Pt 1):271–282.
157. Ge ZD, Zhang XH, Fung PC, He GW. Endothelium-dependent hyperpolarization and relaxation resistance to N(G)-nitro-L-arginine and indomethacin in coronary circulation. *Cardiovasc Res* 2000;46:547–556.
158. Mendizabal VE, Poblete I, Lomniczi A, Rettori V, Huidobro-Toro JP, Adler-Graschinsky E. Nitric oxide synthase-independent release of nitric oxide induced by KCl in the perfused mesenteric bed of the rat. *Eur J Pharmacol* 2000;409:85–91.
159. Vanheel B, Van d, V, Leusen I. Contribution of nitric oxide to the endothelium-dependent hyperpolarization in rat aorta. *J Physiol* 1994;475:277–284.
160. Tare M, Parkington HC, Coleman HA, Neild TO, Dusting GJ. Hyperpolarization and relaxation of arterial smooth muscle caused by nitric oxide derived from the endothelium. *Nature* 1990;346:69–71.
161. Garland CJ, McPherson GA. Evidence that nitric oxide does not mediate the hyperpolarization and relaxation to acetylcholine in the rat small mesenteric artery. *Br J Pharmacol* 1992;105:429–435.
162. Murphy ME, Brayden JE. Nitric oxide hyperpolarizes rabbit mesenteric arteries via ATP-sensitive potassium channels. *J Physiol* 1995;486(Pt 1):47–58.
163. Robertson BE, Schubert R, Hescheler J, Nelson MT. cGMP-dependent protein kinase activates Ca-activated K channels in cerebral artery smooth muscle cells. *Am J Physiol* 1993;265:C299–C303.
164. Archer SL, Huang JM, Hampl V, Nelson DP, Shultz PJ, Weir EK. Nitric oxide and cGMP cause vasorelaxation by activation of a charybdotoxin-sensitive K channel by cGMP-dependent protein kinase. *Proc Natl Acad Sci USA* 1994;91:7583–7587.
165. Bolotina VM, Najibi S, Palacino JJ, Pagano PJ, Cohen RA. Nitric oxide directly activates calcium-dependent potassium channels in vascular smooth muscle. *Nature* 1994;368:850–853.
166. Haburcak M, Wei L, Viana F, Prenen J, Droogmans G, Nilius B. Calcium-activated potassium channels in cultured human endothelial cells are not directly modulated by nitric oxide. *Cell Calcium* 1997;21:291–300.
167. Vanheel B, Van d, V. Nitric oxide induced membrane hyperpolarization in the rat aorta is not mediated by glibenclamide-sensitive potassium channels. *Can J Physiol Pharmacol* 1997;75:1387–1392.
168. Goud C, DiPiero A, Lockette WE, Webb RC, Charpie JR. Cyclic GMP-independent mechanisms of nitric oxide-induced vasodilation. *Gen Pharmacol* 1999;32:51–55.
169. Plane F, Garland CJ. Differential effects of acetylcholine, nitric oxide and levcromakalim on smooth muscle membrane potential and tone in the rabbit basilar artery. *Br J Pharmacol* 1993;110:651–656.
170. Plane F, Wiley KE, Jeremy JY, Cohen RA, Garland CJ. Evidenc that different mechanisms underlie smooth muscle relaxation to nitric oxide and nitric oxide donors in the rabbit isolated carotid artery. *Br J Pharmacol* 1998;123:1351–1358.
171. Jia Y, Zacour M, Tolloczko B, Martin JG. Nitric oxide synthesis by tracheal smooth muscle cells by a nitric oxide synthase-independent pathway. *Am J Physiol* 1998;275:L895–L901.
172. Yuan XJ, Tod ML, Rubin LJ, Blaustein MP. Inhibition of cytochrome P-450 reduces voltage-gated K^+ currents in pulmonary arterial myocytes. *Am J Physiol* 1995;268:C259–C270.
173. Edwards G, Zygmunt PM, Hogestatt ED, Weston AH. Effects of cytochrome P450 inhibitors on potassium currents and mechanical activity in rat portal vein. *Br J Pharmacol* 1996;119:691–701.
174. Vanheel B, Calders P, Van dB, I, Van d, V. Influence of some phospholipase A2 and cytochrome P450 inhibitors on rat arterial smooth muscle K^+ currents. *Can J Physiol Pharmacol* 1999;77:481–489.
175. Iftinca M, Waldron GJ, Triggle CR, Cole WC. State-dependent block of rabbit vascular smooth muscle delayed rectifier and Kv1.5 channels by inhibitors of cytochrome P450-dependent enzymes. *J Pharmacol Exp Ther* 2001;298:718–728.
176. Moroz LL, Norby SW, Cruz L, Sweedler JV, Gillette R, Clarkson RB. Non-enzymatic production of nitric oxide (NO) from NO synthase inhibitors. *Biochem Biophys Res Commun* 1998;253:571–576.
177. Furchgott RF, Sleator WJ, McCaman MW, Elchlepp J. Relaxation of arterial strips by light and the influence of drugs on this photodynamic effect. *J Pharmacol Exp Ther* 1955;113:122.
178. Karlsson JO, Axelsson KL, Andersson RG. Effects of ultraviolet radiation on the tension and the cyclic GMP level of bovine mesenteric arteries. *Life Sci* 1984;34:1555–1563.
179. Matsunaga K, Furchgott RF. Interactions of light and sodium nitrite in producing relaxation of rabbit aorta. *J Pharmacol Exp Ther* 1989;248:687–695.
180. Chen X, Gillis CN. Enhanced photorelaxation in aorta, pulmonary artery and corpus cavernosum produced by BAY K 8644 or N-nitro-L-arginine. *Biochem Biophys Res Commun* 1992;186:1522–1527.
181. Lovren F, O'Neill SK, Bieger D, Igbal N, Knaus EE, Triggle CR. Nitric oxide, a possible mediator of 1,4-dihydropyridine-induced photorelaxation of vascular smooth muscle. *Br J Pharmacol* 1996;118:879–884.
182. Kubaszewski E, Peters A, McClain S, Bohr D, Malinski T. Light-activated release of nitric oxide from vascular smooth muscle of normotensive and hypertensive rats. *Biochem Biophys Res Commun* 1994;200:213–218.
183. Chang KC, Chong WS, Park BW, Seung BW, Chun GW, Lee IJ, Park PS. NO- and NO2-carrying molecules potentiate photorelaxation in rat trachea and aorta. *Biochem Biophys Res Commun* 1993;191:509–514.

184. Charpie JR, Peters A, Webb RC. A photoactivable source of relaxing factor in genetic hypertension. *Hypertension* 1994;23:894–898.
185. O'Neill SK, Triggle CR. *Unpublished observations* 1994.
186. Bauer JA, Fung HL. Photochemical generation of nitric oxide from nitro-containing compounds: Possible relation to vascular photorelaxation phenomena. *Life Sci* 1994;54:L1–L4.
187. Venturini CM, Palmer RM, Moncada S. Vascular smooth muscle contains a depletable store of a vasodilator which is light-activated and restored by donors of nitric oxide. *J Pharmacol Exp Ther* 1993;266:1497–1500.
188. Goud C, Watts SW, Webb RC. Photorelaxation is not attenuated by inhibition of the nitric oxide-cGMP pathway. *J Vasc Res* 1996;33:299–307.
189. Megson IL, Flitney FW, Bates J, Webster R. Repriming of vascular smooth muscle photorelaxation is dependent upon endothlium-derived nitric oxide. *Endothelium* 1995;3:39–46.
190. Megson IL, Holmes SA, Magid KS, Pritchard RJ, Flitney FW. Selective modifiers of glutathione biosynthesis and 'repriming' of vascular smooth muscle photorelaxation. *Br J Pharmacol* 2000;130:1575–1580.
191. Venturini CM, Palmer RM, Moncada S. Vascular smooth muscle contains a depletable store of a vasodilator which is light-activated and restored by donors of nitric oxide. *J Pharmacol Exp Ther* 1993;266:1497–1500.
192. Lovren F, Triggle CR. Involvement of nitrosothiols, nitric oxide and voltage-gated K^+ channels in photorelaxation of vascular smooth muscle. *Eur J Pharmacol* 1998;347:215–221.
193. Andrews KL, McGuire JJ, Triggle CR. Characterization of vascular smooth muscle photorelaxation in thoracic aorta from NOS knockout mice. *The Pharmacologist* 2002;44: A214, 130.12.
194. Hogg N, Singh RJ, Kalyanaraman B. The role of glutathione in the transport and catabolism of nitric oxide. *FEBS Lett* 1996;382:223–228.
195. Hogg N, Singh RJ, Konorev E, Joseph J, Kalyanaraman B. S-Nitrosoglutathione as a substrate for gamma-glutamyl transpeptidase. *Biochem J* 1997;323(Pt 2):477–481.
196. Singh RJ, Hogg N, Joseph J, Kalyanaraman B. Mechanism of nitric oxide release from S-nitrosothiols. *J Biol Chem* 1996;271:18596–18603.
197. Singh RJ, Hogg N, Goss SP, Antholine WE, Kalyanaraman B. Mechanism of superoxide dismutase/H(2)O(2)-mediated nitric oxide release from S-nitrosoglutathione–role of glutamate. *Arch Biochem Biophys* 1999;372:8–15.
198. Nikitovic D, Holmgren A. S-nitrosoglutathione is cleaved by the thioredoxin system with liberation of glutathione and redox regulating nitric oxide. *J Biol Chem* 1996;271:19180–19185.
199. Gordge MP, Addis P, Noronha-Dutra AA, Hothersall JS. Cell-mediated biotransformation of S-nitrosoglutathione. *Biochem Pharmacol* 1998;55:657–665.
200. Trujillo M, Alvarez MN, Peluffo G, Freeman BA, Radi R. Xanthine oxidase-mediated decomposition of S-nitrosothiols. *J Biol Chem* 1998;273:7828–7834.
201. Jensen DE, Belka GK, Du Bois GC. S-Nitrosoglutathione is a substrate for rat alcohol dehydrogenase class III isoenzyme. *Biochem J* 1998;331(Pt 2):659–668.
202. Keseru GM, Volk B, Balogh GT. Cytochrome P450 catalyzed nitric oxide synthesis: A theoretical study. *J Biomol Struct Dyn* 2000;17:759–767.
203. Dubbin PN, Zambetis M, Dusting GJ. Inhibition of endothelial nitric oxide biosynthesis by N-nitro-L-arginine. *Clin Exp Pharmacol Physiol* 1990;17:281–286.
204. Matsumoto T, Kinoshita M, Toda N. Mechanisms of endothelium-dependent responses to vasoactive agents in isolated porcine coronary arteries. *J Cardiovasc Pharmacol* 1993;21:228–234.
205. Wang YX, Poon CI, Pang CC. Vascular pharmacodynamics of NG-nitro-L-arginine methyl ester *in vitro* and *in vivo*. *J Pharmacol Exp Ther* 1993;267:1091–1099.
206. Pfeiffer S, Leopold E, Schmidt K, Brunner F, Mayer B. Inhibition of nitric oxide synthesis by NG-nitro-L-arginine methyl ester (L-NAME): Requirement for bioactivation to the free acid, NG-nitro-L-arginine. *Br J Pharmacol* 1996;118:1433–1440.
207. Dusting GJ, Read MA, Stewart AG. Endothelium-derived relaxing factor released from cultured cells: Differentiation from nitric oxide. *Clin Exp Pharmacol Physiol* 1988;15:83–92.
208. Guilmard C, Auguet M, Chabrier PE. Comparison between endothelial and neuronal nitric oxide pathways in rat aorta and gastric fundus. *Nitric Oxide* 1998;2:147–154.
209. Sobey CG, Faraci FM. Effects of a novel inhibitor of guanylyl cyclase on dilator responses of mouse cerebral arterioles. *Stroke* 1997;28:837–842.
210. Brunner F, Schmidt K, Nielsen EB, Mayer B. Novel guanylyl cyclase inhibitor potently inhibits cyclic GMP accumulation in endothelial cells and relaxation of bovine pulmonary artery. *J Pharmacol Exp Ther* 1996;277:48–53.
211. MacKenzie A, Martin W. Loss of endothelium-derived nitric oxide in rabbit aorta by oxidant stress: Restoration by superoxide dismutase mimetics. *Br J Pharmacol* 1998;124:719–728.
212. Dowell FJ, Hamilton CA, McMurray J, Reid JL. Effects of a xanthine oxidase/hypoxanthine free radical and reactive oxygen species generating system on endothelial function in New Zealand white rabbit aortic rings. *J Cardiovasc Pharmacol* 1993;22:792–797.
213. Girard P, Sercombe R, Sercombe C, Le Lem G, Seylaz J, Potier P. A new synthetic flavonoid protects endothelium-derived relaxing factor-induced relaxation in rabbit arteries *in vitro*: Evidence for superoxide scavenging. *Biochem Pharmacol* 1995;49:1533–1539.
214. Cherry PD, Omar HA, Farrell KA, Stuart JS, Wolin MS. Superoxide anion inhibits cGMP-associated bovine pulmonary arterial relaxation. *Am J Physiol* 1990;259:H1056–H1062.
215. MacKenzie A, Filippini S, Martin W. Effects of superoxide dismutase mimetics on the activity of nitric oxide in rat aorta. *Br J Pharmacol* 1999;127:1159–1164.
216. Hansen K, Nedergaard OA. Methodologic aspects of acetylcholine-evoked relaxation of rabbit aorta. *J Pharmacol Toxicol Methods* 1999;41:153–159.
217. Dudgeon S, Benson DP, MacKenzie A, Paisley-Zyszkiewicz K, Martin W. Recovery by ascorbate of impaired nitric oxide-dependent relaxation resulting from oxidant stress in rat aorta. *Br J Pharmacol* 1998;125:782–786.

218. Schnackenberg CG, Wilcox CS. The SOD mimetic tempol restores vasodilation in afferent arterioles of experimental diabetes. *Kidney Int* 2001;59:1859–1864.
219. Danser AH, Tom B, de Vries R, Saxena PR. L-NAME-resistant bradykinin-induced relaxation in porcine coronary arteries is NO-dependent: Effect of ACE inhibition. *Br J Pharmacol* 2000;131:195–202.
220. Emsley AM, Jeremy JY, Gomes GN, Angelini GD, Plane F. Investigation of the inhibitory effects of homocysteine and copper on nitric oxide-mediated relaxation of rat isolated aorta. *Br J Pharmacol* 1999;126:1034–1040.
221. Simon BC, Cohen RA. EDTA influences reactivity of isolated aorta from hypercholesterolemic rabbits. *Am J Physiol* 1992;262:H1606–H1610.
222. Plane F, Wigmore S, Angelini GD, Jeremy JY. Effect of copper on nitric oxide synthase and guanylyl cyclase activity in the rat isolated aorta. *Br J Pharmacol* 1997;121:345–350.
223. Mugge A, Elwell JH, Peterson TE, Harrison DG. Release of intact endothelium-derived relaxing factor depends on endothelial superoxide dismutase activity. *Am J Physiol* 1991;260:C219–C225.
224. Didion SP, Hathaway CA, Faraci FM. Superoxide levels and function of cerebral blood vessels after inhibition of CuZn-SOD. *Am J Physiol Heart Circ Physiol* 2001;281:H1697–H1703.
225. Omar HA, Cherry PD, Mortelliti MP, Burke-Wolin T, Wolin MS. Inhibition of coronary artery superoxide dismutase attenuates endothelium-dependent and -independent nitrovasodilator relaxation. *Circ Res* 1991;69:601–608.
226. Saiag B, Shacoori V, Bodin P, Pape D, Allain H, Burnstock G. Free radical involvement in endothelium-dependent responses of the rat thoracic aorta in moderate hypoxic conditions. *Eur J Pharmacol* 1999;372:57–63.
227. Karasu C. Time course of changes in endothelium-dependent and -independent relaxation of chronically diabetic aorta: Role of reactive oxygen species. *Eur J Pharmacol* 2000;392:163–173.
228. Andriambeloson E, Stoclet JC, Andriantsitohaina R. Mechanism of endothelial nitric oxide-dependent vasorelaxation induced by wine polyphenols in rat thoracic aorta. *J Cardiovasc Pharmacol* 1999;33:248–254.
229. Adachi T, Cohen RA. Decreased aortic glutathione levels may contribute to impaired nitric oxide-induced relaxation in hypercholesterolaemia. *Br J Pharmacol* 2000;129:1014–1020.
230. Crack P, Cocks T. Thimerosal blocks stimulated but not basal release of endothelium-derived relaxing factor (EDRF) in dog isolated coronary artery. *Br J Pharmacol* 1992;107:566–572.
231. Hutcheson IR, Chaytor AT, Evans WH, Griffith TM. Nitric oxide-independent relaxations to acetylcholine and A23187 involve different routes of heterocellular communication. Role of Gap junctions and phospholipase A2. *Circ Res* 1999;84:53–63.

Myocardial Contractile Effects of Nitric Oxide

Walter J. Paulus, MD, PhD, FESC[1] *and Jean G.F. Bronzwaer, MD*[2]

[1]*Cardiovascular Center, O.L.V. Ziekenhuis, Aalst, Belgium;*
[2]*Cardiology, VU Medical Center, Amsterdam, The Netherlands*

Abstract. **Recent experimental and clinical research solved some of the controversies surrounding the myocardial contractile effects of NO. These controversies were: (1) does NO exert a contractile effect at baseline? (2) is NO a positive or a negative inotrope? (3) Are the contractile effects of NO similar when NO is derived from NO-donors or from the different isoforms of NO synthases (NOS)? (4) Does NO exert the same effects in hypertrophied, failing or ischemic myocardium?**

Transgenic mice with cardioselective overexpression of NOS revealed NO to produce a small reduction in basal developed LV pressure and a LV relaxation-hastening effect mainly through myofilamentary desensitization. Similar findings had previously been reported during intracoronary infusions of NO-donors in isolated rodent hearts and in humans. The LV relaxation hastening effect was accompanied by increased diastolic LV distensibility, which augmented LV preload reserve especially in heart failure patients. This beneficial effect on diastolic LV function always overrode the small NO-induced attenuation in LV developed pressure in terms of overall LV performance. In most experimental and clinical conditions, contractile effects of NO were similar when NO was derived from NO-donors or produced by the different isoforms of NOS. Because expression of inducible NOS (NOS2) is frequently accompanied by elevated oxidative stress, NO produced by NOS2 can lead to peroxynitrite-induced contractile impairment as observed in ischemic or septic myocardium. Finally, shifts in isoforms or in concentrations of myofilaments can affect NO-mediated myofilamentary desensitization and alter the myocardial contractile effects of NO in hypertrophied or failing myocardium.

Key Words. **nitric oxide, myocardium, contractility, diastole, heart failure, nitric oxide synthase**

Introduction

In vascular tissue, the contractile effects of nitric oxide (NO) are easily appreciated consisting of relaxation of isolated vascular strips or vasodilation of perfused organs. In cardiac tissue, the contractile effects of NO are far less obvious and have been the subject of considerable confusion and debate. Some of the most important controversies surrounding the myocardial contractile effects of NO were: (1) Does NO exert a myocardial contractile effect under baseline conditions or only following adrenergic or cholinergic stimulation? (2) Can the contractile effects of NO be labelled as positively or negatively inotropic? (3) Are the contractile effects of NO similar for NO derived from NO-donor substances, for NO produced by NOS3 (eNOS, "endothelial-constitutive" NOS) or for NO produced by NOS2 (iNOS, "cytokine-inducible" NOS)? (4) Does hypertrophy, aging, transplantation, heart failure, ischemia or sepsis alter the observed contractile effects of NO?

Some of the controversies arose from comparison of data derived from different experimental set-ups ranging from isolated cardiomyocytes, papillary muscle strips, perfused rodent hearts, anaesthetized open-chest dogs to humans at the time of cardiac catheterization. In all these set-ups, myocardial contractility was assessed in an unequal and often incomplete way. Isolated cardiomyocytes are contracting at low load (i.e. friction on a glass surface). Extent of muscle shortening (Δl) or velocity of shortening ($\Delta l/dt$) recorded in isolated cardiomyocytes is therefore hard to compare with peak force (F) or speed of force development (dF/dt) in isometrically contracting papillary muscles. Effects on diastolic LV function can easily be appreciated in perfused hearts, open-chest dogs or humans but are difficult to detect in isolated cardiomyocytes or muscle strips because of the controlled preloading conditions. Interference by NO in parasympathetic or sympathetic outflow to the heart is absent in isolated preparations but unavoidable in humans.

The present review summarizes the current evidence for myocardial contractile effects of NO thereby highlighting the aforementioned controversies and shortcomings of the different experimental settings.

Address for correspondence: Dr. Walter J. Paulus, MD, PhD, FESC, Cardiovascular Center, O.L.V. Ziekenhuis, Moorselbaan 164, B 9300 Aalst, Belgium. Tel.: 32-53-724433; Fax: 32-53-724587; E-mail: walter.paulus@pi.be

Contractile Effects of NO in Normal Myocardium

Systolic Function

In isolated cardiomyocytes, administration of exogenous NO was first reported to be associated at baseline with a small negative inotropic effect, evident from a reduction in extent and velocity of shortening [1]. In the same preparation, inhibition of endogenous NOS3 had no baseline effect but reduced extent and velocity of shortening following adrenergic stimulation [2]. In isolated isometrically contracting cardiac muscls strips, NO produced by NOS3 of adjacent endocardial cells or derived from high, but not low, doses of NO-donors exerted a similar small negative inotropic effect evident from a reduction in peak isometric force development [3–5] without change in rate of rise of force development (dF/dt_{max}) (Fig. 1). This reduction in peak isometric force development resulted from an earlier onset of isometric relaxation, which abbreviated duration of contraction or time left for isometric tension build-up. This unique pattern of effect of NO on isometric force development revealed the underlying mechanism: a NO-induced reduction in myofilamentary calcium sensitivity because of phosphorylation of troponin I by cGMP dependent protein kinase.

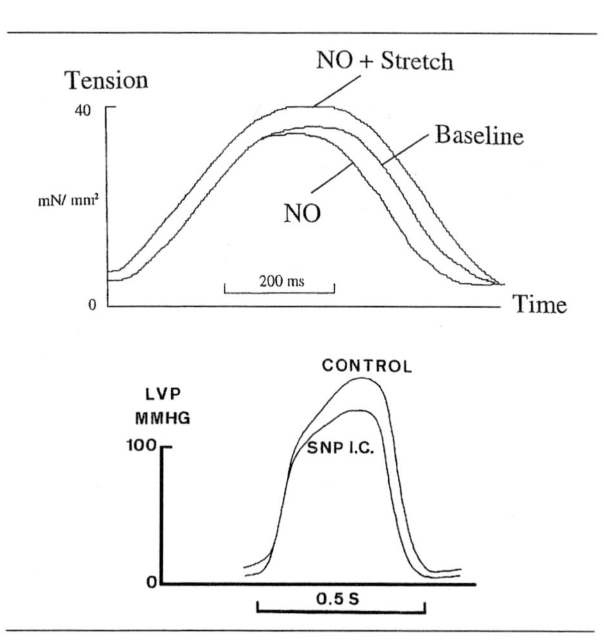

Fig. 1. Top: Contractile effects of NO on isometric tension development in an isolated cat papillary muscle. NO slightly reduces peak tension because of earlier onset of isometric relaxation and fails to alter the rate of rise of tension. The contractile effects of NO are abolished by a simultaneous rise in muscle preload (NO + stretch). Bottom: The contractile effects of NO derived from an intracoronary NO-donor infusion (sodium nitroprusside; SNP-IC) in a normal human subject. In analogy to the contractile effects in a cat papillary muscle, there is a reduction in peak LV pressure (LVP), an earlier onset of isovolumic relaxation and no effect on the rate of rise of LV pressure. There is also a drop in diastolic LV pressures during SNP-IC.

A similar relaxation hastening effect was also observed in experimental preparations using whole heart models. In isolated ejecting guinea-pig hearts perfused with buffer solution and contracting with constant preload, afterload and heart rate, both NO derived from an NO-donor [6] and NO released from the coronary endothelium following stimulation with bradykinin or substance P [7] induced an earlier onset and faster rate of LV relaxation and a small reduction in peak LV pressure without change in LV dP/dt_{max}. These changes could not be reproduced with the cGMP-independent vasodilator nicardipine and were therefore unrelated to coronary vasodilation or to coronary perfusion. These effects were also observed after addition to the perfusate of angiotensin converting enzyme inhibitors probably because of slower bradykinin breakdown and were abolished by addition of haemoglobin, which inactivates NO [8].

Experiments [9] in transgenic mice with cardioselective overexpression of NOS3 and a 60-fold increase in NOS activity (^{3}H-L-Citrulline production) confirmed NO to produce a 30% reduction in basal LV developed pressure mainly through myofilamentary desensitization and to a lesser extent through alterations in sarcolemmal calcium fluxes [10,11] or sarcoplasmic calcium handling [12–15]. A similar 10% reduction in basal LV developed pressure was also observed in transgenic mice with cardioselective overexpression of NOS2 and a 20-fold increase in ^{3}H-L-Citrulline production [16]. In these mice, addition of L-arginine to the perfusion corrected the substrate deficit for NOS2 and induced a further 10% drop in LV developed pressure. A similar drop in LV developed pressure during addition of L-arginine to an intracoronary substance P infusion had previously been reported in the human allograft [17] (Fig. 2).

A NO-induced and myofilamentary desensitization-mediated LV relaxation hastening effect was also observed in the human heart during intracoronary infusions of NO-donor substances or of substance P, which releases NO from the coronary endothelium [18–20]. In analogy to the experimental preparations, this relaxation-hastening effect abbreviated LV contraction, reduced the magnitude of the late-systolic reflected LV pressure wave and caused a ±10 mmHg drop in LV end-systolic pressure at unaltered LV end-systolic volume (Fig. 1). There were no changes in the rate of rise of LV pressure (LV dP/dt_{max}). The small drop in LV end-systolic pressure at unaltered LV end-systolic volume implied a rightward and downward shift

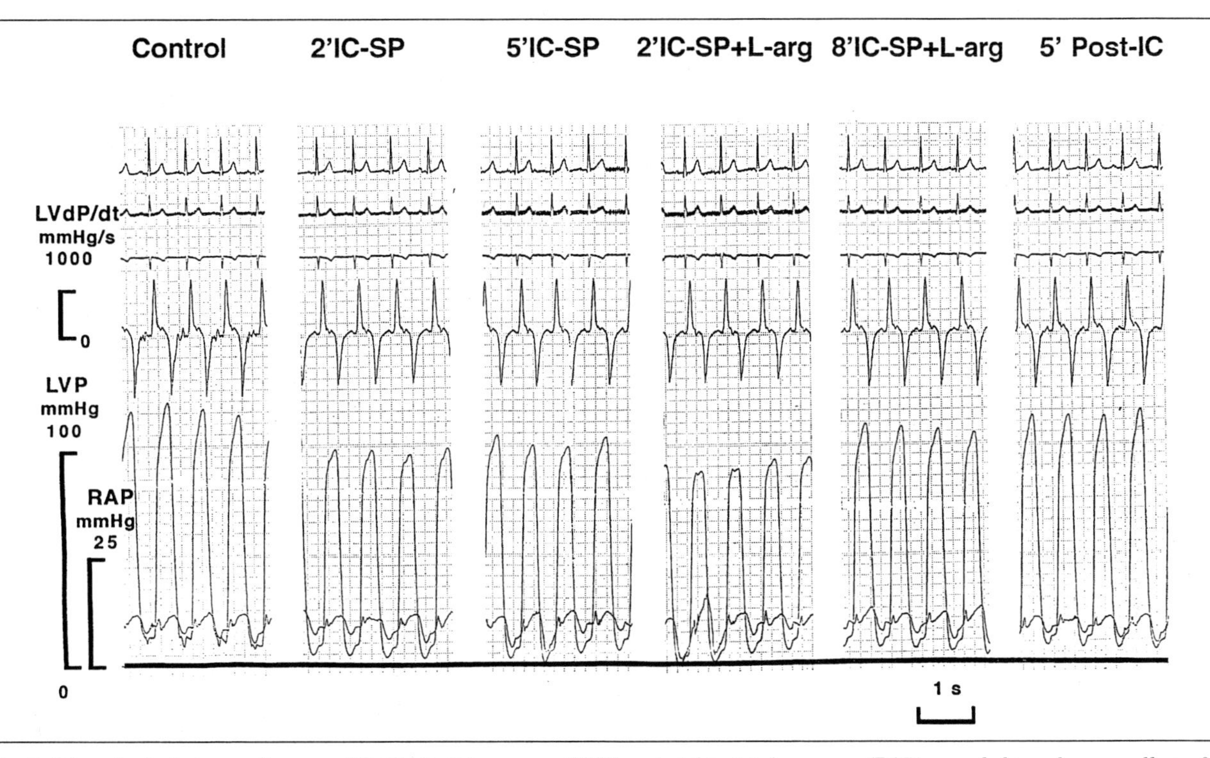

Fig. 2. *Three lead electrocardiogram, LV dP/dt, LV pressure (LVP) and right atrial pressure (RAP) recorded in a human allograft in control conditions, after 2 and 5 min of intracoronary (IC) substance P (SP) infusion, after 2 and 8 min of intracoronary (IC) substance P + L-arginine infusion (SP + L-arg) and 5 min after cessation of IC infusion. SP releases NO from the coronary endothelium, which reduces peak LV pressure without changing LV dP/dt_{max} or RAP. Addition of L-arg to the SP infusion potentiates these effects. Reproduced with permission from [17].*

of the LV end-systolic pressure-volume relation and was therefore theoretically consistent with a small negative inotropic effect. The unaltered LV dP/dt_{max} and the simultaneous fall in LV end-diastolic pressure however argue against significant hemodynamic deterioration as a result of these NO-induced negative inotropic effects on LV end-systolic performance. In fact, the NO-induced effects on LV end-systolic performance could well be beneficial for the ejecting left ventricle as they reduced mechanical work wasted in contracting against late-systolic reflected arterial pressure waves.

When isolated cardiomyocytes [21] or isometrically contracting papillary muscle strips [5] were exposed to low doses of NO-donors, a positive inotropic effect was observed probably resulting from cGMP-induced inhibition of phosphodiesterase 3 and a concomitant rise in cAMP or from direct nitrosylation of sarcolemmal or sarcoplasmic proteins. Similar positive inotropic effects of NO have been reported in anaesthetized dogs [22] and in the normal human heart, in which an intracoronary infusion of L-NMMA induced a modest (14%) drop in LV dP/dt_{max} [23]. The dose-dependent transition from a positive to a negative inotropic effect occurred at lower doses of the NO-donor in the presence of an intact endocardium and in the presence of cholinergic or ß-adrenergic stimulation [5] (Fig. 3). Larger negative inotropic effects in the presence of a functioning endocardium implies additive effects of NO derived from NO-donors and of NO produced by NOS3 of adjacent endothelial cells. Larger negative inotropic effects following cholinergic stimulation implies additive effects at the level of the second messenger cGMP, whose intracellular concentration rises both following NO-donor administration and cholinergic stimulation. The larger negative inotropic effect following ß-adrenergic stimulation suggests additive effects also at the level of the myofilaments because of simultaneous phosphorylation of troponin I by cGMP dependent protein kinase and by cAMP dependent protein kinase. The latter observation has far reaching consequences and explains why many experimental and clinical studies detected no or only small effects of NO under baseline conditions but reported large changes following ß-adrenergic stimulation. In the human heart, an intracoronary infusion of substance P following pretreatment with dobutamine resulted in a fall in LV end-systolic pressure, which was twice as large as the fall observed in control conditions and accompanied by

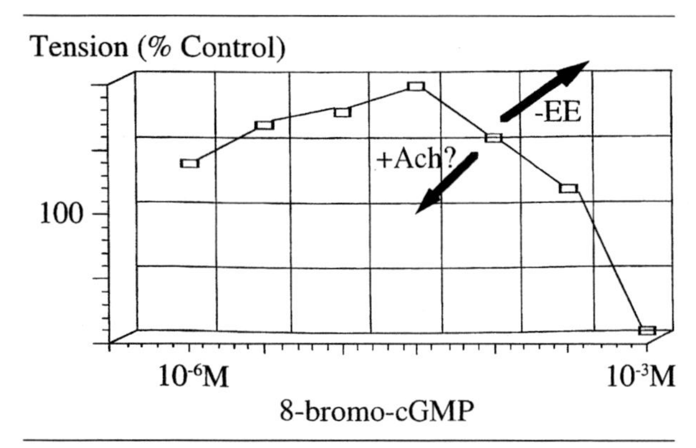

Fig. 3. *Dose response curve of peak isometric tension of an isolated cat papillary muscle versus concentration of an analogue of cGMP (8-bromo-cGMP), the second messsenger of NO. At low doses of cGMP or of NO there is a positive inotropic effect, at high doses of cGMP of of NO there is a negative inotropic effect. Removal of the endocardial endothelium (−EE) shifts the descending limb of the dose response curve to the right, addition of acetylcholine (+Ach) shifts it to the left. This implies additive effects of exogenously administered cGMP to the endogenously produced cGMP by the endothelium or following Ach administration.*

a 6% fall in LV dP/dt_{max} [24] (Fig. 4). A similar augmentation by dobutamine of NO-related myocardial contractile effects was also observed during intracoronary infusion of L-NMMA, which potentiated the dobutamine-induced rise in LV dP/dt_{max} by 51% [25] but failed to have any effect in the same patient population in control conditions [23]. Larger NO-related negative inotropic effects following ß-adrenergic stimulation also highlight the importance of the contraction mode of the experimental preparation. Differences in pre- or afterload can indeed affect calcium sensitivity of crossbridges and can counteract both cGMP- or cAMP-induced myofilamentary desensitization and concomitant changes in timing of myocardial contraction-relaxation sequence [26,27] (Fig. 1).

Diastolic Function

NO related phosphorylation of troponin I also increases myocardial diastolic distensibility through prevention of calcium-independent diastolic crossbridge cycling. Such an increase in diastolic distensibility, evident from a longer diastolic cell length at constant preload, has been observed in isolated cardiomyocytes following exposure to cGMP, the second messenger of NO or following exposure to sodium nitroprusside [28,29]. The increase in diastolic cell length and the concomitant reduction in cell shortening following exposure to sodium nitroprusside have been attributed not only to cGMP-mediated myofilamentary desensitization but also to myofilamentary desensitization related to a fall in intracellular pH. This fall in intracellular pH suggested NO, via cGMP, to disable forward sodium-proton exchange [29]. The reverse mechanism, intracellular alkalization, has been observed following administration of angiotensin II or endothelin [30] and following cardiac muscle stretch, which leads to autocrine production of angiotensin II and endothelin [31]. As discussed earlier, cardiac muscle stretch can indeed effectively counteract the NO-induced effects on timing of myocardial contraction-relaxation sequence [27] (Fig. 1). Moreover, in a beating rabbit heart, a sudden increase in LV diastolic stretch because of release of a caval occluder, leads to an instantaneous increase in diastolic intramyocardial NO concentrations measured by a porphyrinic sensor inserted in the LV wall of the beating rabbit heart [32]. Hence, a rise in LV preload triggers not only myocardial autocrine production of angiotensin II but also of NO, which can reverse the angiotensin II-induced changes in intracellular pH and in myofilamentary calcium sensitivity.

In isolated ejecting guinea-pig hearts, a perfusate containing the specific NOS inhibitor N^G-monomethyl-L-arginine, raised LV end-diastolic pressure and reduced preload recruitable LV stroke volume probably because of an acute left and upward shift of the diastolic LV pressure-volume relation [33]. In the same preparation, an increase in LV preload, caused a rise in the NO concentration of the coronary effluent [34] confirming the preload triggered enhancement of myocardial NO production recorded from porphyrinic sensors inserted in the wall of the beating rabbit heart [32]. NO-related changes in diastolic LV distensibility have been observed not only following acute administration of a NOS inhibitor but also following chronic NOS blockade. In rats receiving eight weeks of treatment with the NOS inhibitor L-NAME, the diastolic LV pressure-volume relation shifted upward with a significant reduction in LV unstressed volume and no increase in LV mass despite the elevated blood pressure [35].

In the human heart, the relaxation-hastening effect of NO was also accompanied by an increase in LV diastolic distensibility. During intracoronary infusions of NO-donors or of substance P, the LV diastolic pressure-volume relation shifted right- and downward consistent with increased LV diastolic distensibility [18–20]. Similar downward shifts of the diastolic LV pressure-volume relation had previously been observed during intravenous administration of NO-donors and had always been attributed to a sequence of events consisting of venodilation, right ventricular unloading and biventricular interaction [36,37]. The downward shifts of the diastolic LV

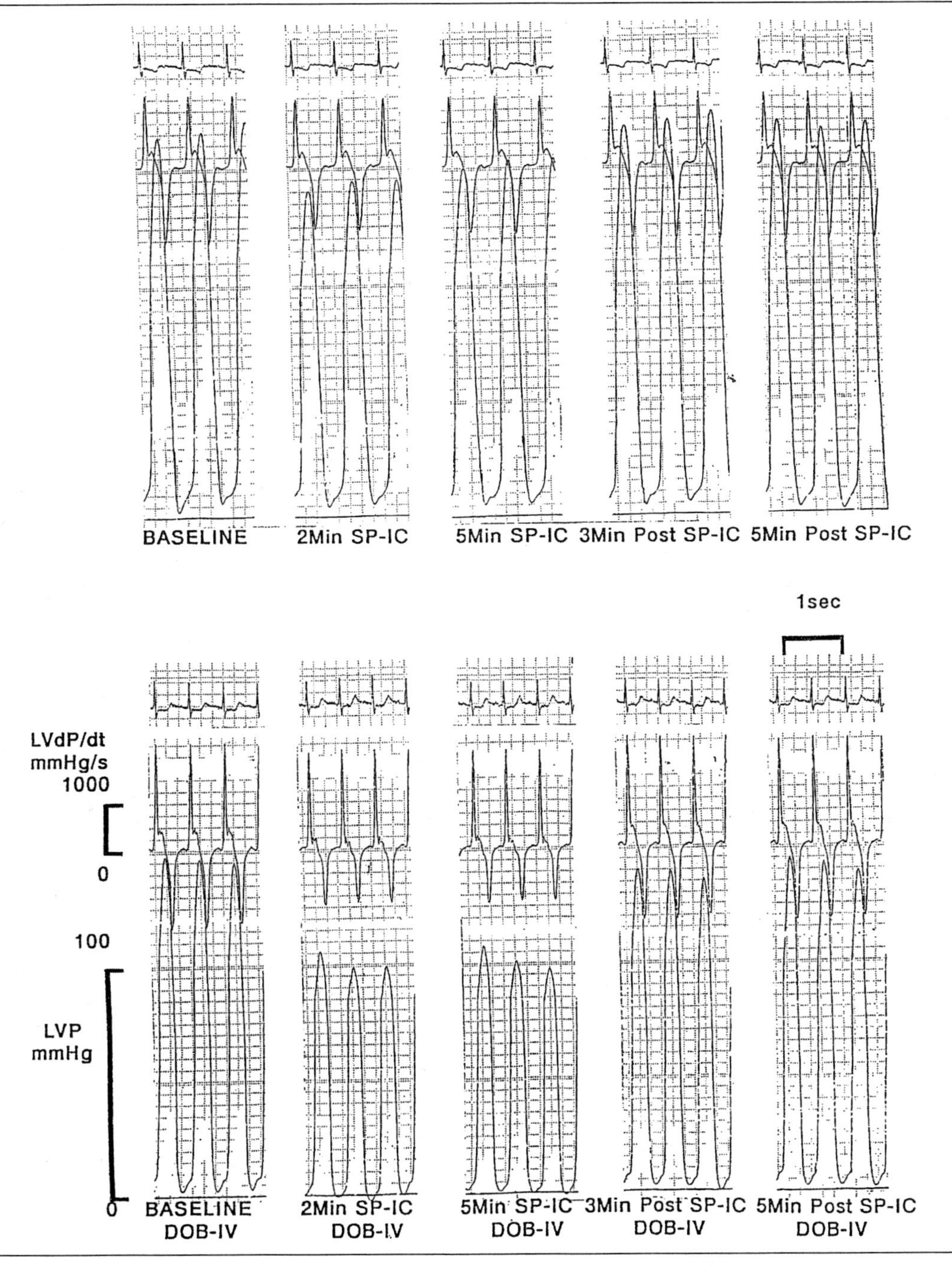

Fig. 4. *Single lead electrocardiogram, LV dP/dt and LV pressure (LVP) recorded in a human allograft at baseline, after 2 and 5 minutes of intracoronary (IC) substance P (SP) infusion and 3 and 5 minutes following infusion in the absence (top panel) and in the presence of an intravenous infusion of dobutamine (DOB-IV) (bottom panel). During intravenous infusion of dobutamine, an IC-SP infusion results in a larger drop in LV pressure than before (50 mmHg vs. 15 mmHg). Reproduced with permission from [98].*

pressure-volume relation observed during intracoronary infusion of NO-donors did however not result from such a sequence of events but from a direct myocardial effect of NO (probably phosphorylation of troponin I and reduced calcium-independent diastolic crossbridge cycling) for the following reasons: (1) the doses of NO-donors infused intracoronarily were too low to elicit significant systemic vasodilator effects; (2) a right atrial infusion of the same dose of NO-donor failed to reproduce the observed shift in the diastolic LV pressure-volume relation; (3) the shift in the diastolic LV pressure-volume relation persisted even after transforming the measured LV diastolic pressure to transmural LV diastolic pressure and (4) the fall in LV end-diastolic pressure was accompanied by an increase in LV end-diastolic volume and a fall in heart rate, both of which were incompatible with peripheral vasodilator effects.

Contractile Effects of NO in Hypertrophied and Senescent Myocardium

In hypertrophied cardiomyocytes isolated from aortic-banded rats, exposure to the NO-donor sodium nitroprusside failed to alter diastolic cell length, to reduce the amplitude of cell shortening or to lower intracellular pH, all of which were observed in cardiomyocytes from normal rats following exposure to sodium nitroprusside [29]. In a guinea-pig model of compensated LV hypertrophy induced by aortic banding, the LV relaxant effects of endothelially released NO were significantly blunted despite a preserved response to exogenous NO [38]. This defective contractile response was unrelated to alterations in coronary flow but reflected the impact on LV performance of an hypertrophy-induced endothelial dysfunction and concomitant reduction of NO bioactivity. Reduced NO bioactivity in the hypertrophied heart could have resulted from a decrease of NOS3 expression or from inactivation of NO by reactive oxygen species (ROS) such as superoxide anion. In the guinea-pig model of aortic banding [38] and in the spontaneously hypertensive rat [39], NOS3 expression was unaltered or even upregulated. This made increased ROS production the most likely cause for the reduction of NO bioactivity and for the defective LV relaxant response to coronary endothelial stimulation. Increased ROS production could have resulted from dysfunctional NOS because of deficiencies in cofactor or substrate (i.e. tetrahydrobiopterin or L-arginine) or from upregulation of other enzymes such as NADH/NADPH oxidases [40]. In the same guinea pig model of compensated LV hypertrophy produced by aortic banding, neither tetrahydrobiopterin nor L-arginine corrected the defective LV relaxant response to coronary endothelial stimulation. The latter was however restored by vitamin C, deferroxamine or superoxide dismutase and in line with these observations, protein expression of the NADPH oxidase subunits (gp91-phox and p67-phox) and myocardial NADPH oxidase activity were shown to be significantly increased in the banded animals [41].

In patients with LV hypertrophy of aortic stenosis, intracoronary administration of a NO-donor caused a marked fall in LV end-diastolic pressure and in LV end-diastolic chamber stiffness [19]. The fall in LV end-diastolic pressure observed in the aortic stenosis patients was larger than the fall observed in normal controls (−39 vs. −21%) but the fall in peak and end-systolic LV pressures was smaller than in control subjects. The larger fall in LV end-diastolic pressure observed in the aortic stenosis patients could have resulted from their higher baseline LV end-diastolic pressure and not from a higher myocardial sensitivity to the distensibility-increasing effect of NO. When the left ventricle is operating on the steeper portion of its diastolic pressure-volume relation, a NO-induced rightward shift of the diastolic pressure-volume relation will result in a larger reduction in LV end-diastolic pressure than when the left ventricle is operating on the flat portion of its diastolic pressure-volume relation. The smaller fall in peak and end-systolic LV pressures was consistent with a lower sensitivity of hypertrophied human myocardium to the relaxation-hastening effect of NO and confirmed the aforementioned reduction in contractile response to NO observed in cardiomyocytes of aortic banded rats [29]. Lower sensitivity of hypertrophied myocardium to the relaxation-hastening effect of NO could possibly be explained by lower concentrations in hypertrophied myocardium of troponin I, whose breakdown is accelerated whenever the left ventricle is chronically subjected to elevated LV filling pressures [42]. Other mechanisms for a lower sensitivity of hypertrophied myocardium to the contractile effects of NO include a lower baseline cGMP concentration in hypertrophied myocardium, which would make it operate on the flat portion of its biphasic dose-reponse to NO [5,43] (Fig. 3), or blunted downstream signaling in hypertrophied myocardium from cGMP to sodium-proton exchange [29]. Both the lower myocardial sensitivity to NO and the lower myocardial bioactivity of NO could be important mechanisms for the diastolic dysfunction of the hypertrophied LV.

The effects of aging on myocardial NOS activity are controversial. In one study, myocardial NOS3 activity was increased in the aging rat heart [44]. Despite increased myocardial NOS3 activity, left ventricular diastolic distensibility was reduced as evident from an elevation of left ventricular filling pressures. Similar to the human allograft [17], the reduction in diastolic left ventricular distensibility of the senescent rat heart reacted favourably to administration of L-arginine. Both observations warrant further investigations on the clinical use of L-arginine for the treatment of aging-induced diastolic left ventricular dysfunction. Other studies documented an aging-induced reduction of NOS3 activity [45] and upregulation of NOS2 activity [46].

Contractile Effects of NO in Transplanted Myocardium

In transplanted myocardium, myocardial contractile effects of NO were the subject of intense research because of the early demonstration of NOS2 gene expression in transplanted myocardium [47] and because of the numerous experimental studies, which linked NOS2 gene expression to contractile dysfunction in experimental preparations [48–52]. In transplant recipients free of rejection or graft vasculopathy, a Doppler echocardiographic index of LV performance revealed a significant association between systolic, diastolic or combined LV dysfunction and intensity of NOS2 gene expression in simultaneously procured LV biopsies [47]. An invasive study in transplant recipients obtained microtip LV pressure recordings and simultaneous LV angiograms at the time of annual coronary angiography but found similar indices of baseline LV function in transplant recipients with low or high myocardial NOS2 mRNA [53]. This study however observed a larger relaxation-hastening effect and a larger reduction in peak and end-systolic LV pressure in transplant recipients with high myocardial NOS2 mRNA following intravenous infusion of the ß-agonist dobutamine. These findings resembled the contractile effects of NOS3-derived NO. The latter were also studied in transplant recipients using intracoronary infusions of substance P, which releases NO from the coronary endothelium. The relaxation hastening effect and the concomitant fall in peak and end-systolic LV pressure of NOS3-derived NO were small under baseline conditions but rose drastically following pretreatment with dobutamine (Fig. 4) [24]. The similar contractile effects during dobutamine infusion of NOS2- or NOS3-derived NO also supports substantial bioavailability of NOS2-derived NO. Bioavailability of NOS2-derived NO has been questioned because of the simultaneous presence in transplanted myocardium of rejection-related oxidants, which could combine with NO to generate peroxynitrite [54] and which could thereby prevent NO from exerting its direct effects on the cardiomyocytes. The myocardial bioavailability of NOS2-derived NO was also indirectly demonstrated in transplant recipients by higher myocardial cGMP concentrations [47] and more recently by higher transcardiac nitrite/nitrate production [55] in the presence of upregulated NOS2 gene expression.

Rejection related oxidants or lower baseline myocardial concentrations of cGMP or of cAMP could explain the smaller LV relaxation-hastening effect of intracoronary NO-donor infusions in transplant recipients compared to a normal control population subjected to a similar infusion protocol [56]. In the cardiac allograft, baseline diastolic LV dysfunction predicted a poor NO-induced LV relaxation-hastening effect. The inverse correlation between the NO-induced LV relaxation-hastening effect and baseline diastolic LV dysfunction could be related to varying degrees of oxidative stress, which could induce diastolic LV dysfunction because of sarcoplasmic membrane damage and a reduction of the LV-relaxation hastening effect because of lower NO bioavailability. Low baseline cGMP concentration could result from reduced coronary NOS3 release because of graft vascular disease and could explain the characteristically shrunken appearance of the transplanted LV [57] in analogy to the small LV cavity size of rat hearts following chronic treatment with NOS inhibitors [35]. Low baseline cAMP concentration in the cardiac allograft could result from cardiac denervation and could act in concert with the presence of oxidants and low baseline cGMP concentration to blunt the NO-donor induced LV contractile effects in the cardiac allograft.

Contractile Effects of NO in Failing Myocardium

In analogy to transplanted myocardium, the early reports of increased myocardial NOS2 [58–64] and NOS3 [65] expression in failing human hearts stimulated research on the myocardial contractile effects of NO in failing myocardium because of the widely proposed but unfortunately unproven paradigm of excess NO production contributing to contractile depression.

In dogs with pacing-induced heart failure, myocardial NOS activity was significantly increased compared to control dogs. Administration of a NOS inhibitor to cardiomyocytes isolated from these hearts had no effect on baseline myocardial performance but augmented the inotropic response to ß-adrenergic stimulation [66]. Similar results were obtained in muscle strips of explanted human cardiomyopathic hearts: L-NMMA administration had no effect on baseline isometric force development but increased the dobutamine-induced rise in contractile performance [62]. Other investigators however failed to confirm these results and observed no changes in the contractile response to ß-adrenoreceptor stimulation following administration of a NOS inhibitor to cardiomyocytes isolated from cardiomyopathic human hearts [67]. Moreover, in the pacing-induced heart failure dog model, the fall in LV stroke work and the rise in LV end-diastolic pressure observed after 4 weeks of pacing was accompanied by a fall in cardiac NO production as evident from a smaller transcardiac nitrite/nitrate gradient [68]. In this same model, dampening of the myocardial ß-adrenergic response has recently been attributed to increased cGMP levels resulting not from excess NO-induced guanylyl cyclase activity but from reduced phosphodiesterase 5-mediated cGMP breakdown [69].

In patients with dilated cardiomyopathy and biopsy proven endomyocardial NOS2 gene expression, intracoronary infusion of L-NMMA had no effect on LV contractile performance [23]. In the same patient population, an intracoronary infusion of L-NMMA had previously been shown to potentiate the dobutamine-induced rise in LV dP/dt_{max} [25]. In the pacing-induced heart failure dog model, the latter finding was confirmed and related to abundance of caveolin protein in failing myocardium [70]. Caveolin protein abundance in failing myocardium reduces NOS3 activity and interferes with translocation of receptor-agonist complexes from the sarcolemma. The beneficial outcome of statins in heart failure could, apart from their effects on rho proteins, also relate to reduction of caveolin and concomitant enhancement of NOS3 activity [71]. In dilated cardiomyopathy patients with elevated LV filling pressures, intracoronary infusion of substance P enhanced myocardial NOS2 activity, slightly reduced LV end-systolic pressure but improved overall hemodynamic status as evident from higher LV stroke volume and LV stroke work [64]. These beneficial NO-related contractile effects resulted from a simultaneous NO-induced increase in diastolic LV distensibility. Low intensity of LV endomyocardial NOS2 and NOS3 gene expression was recently demonstrated to coincide with a hemodynamic phenotype of dilated cardiomyopathy patients characterized by elevated LV diastolic stiffness and reduced LV stroke work [72]. In contrast to this hemodynamic phenotype with poor prognosis, dilated cardiomyopathy patients with low LV diastolic stiffness and preserved LV stroke work had higher intensity of NOS2 and NOS3 gene expression, which attained the level observed in athlete's heart [72] (Fig. 5). Low LV diastolic stiffness, reversible right- and downward displacement of the diastolic LV pressure-volume relation and high LV preload reserve are typical features of athlete's heart and could be NO-mediated because of the documented upregulation of NOS activity and expression following regular physical exercise [73,74]. Simultaneous improvement in diastolic LV distensibility also overrode the NO-induced attenuation of the LV contractile response to dobutamine in terms of overall hemodynamic status of dilated cardiomyopathy patients. In two studies [24,75], agonist-induced coronary endothelial release of NO during intravenous administration of dobutamine resulted in a small reduction of LV contractility indices such as LV dP/dt_{max} and LV elastance but without

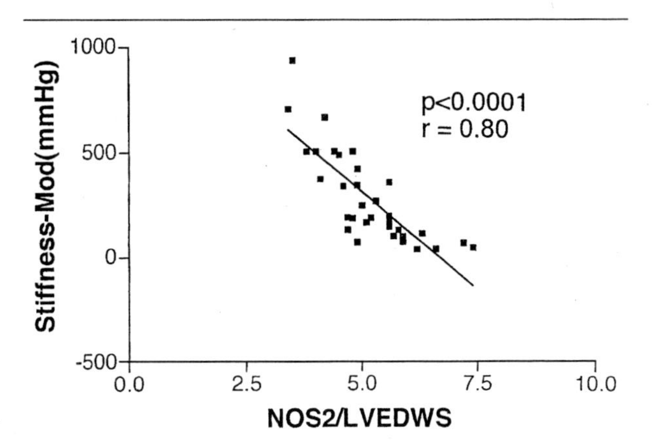

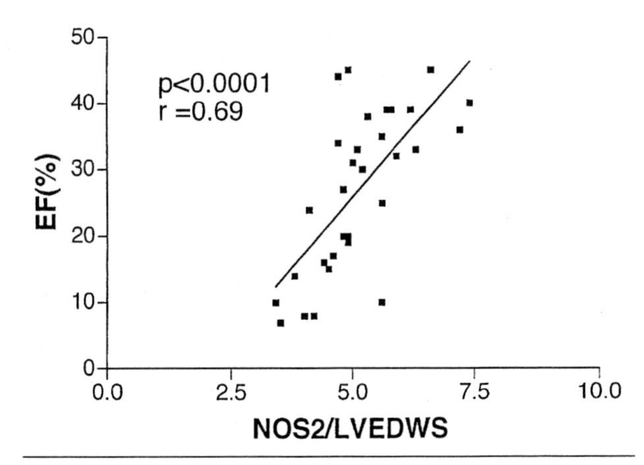

Fig. 5. *In patients with non-ischaemic dilated cardiomyopathy, an inverse relation was observed between LV diastolic stiffness (stiffness-Mod) and intensity of endomyocardial NOS2 gene expression normalised for LV end-diastolic wall stress (LVEDWS) (Top). A similar relation was also observed for LV ejection fraction (EF) [72]. Both relations imply better hemodynamic status in patients with dilated cardiomyopathy when myocardial NOS2 gene expression is elevated.*

hemodynamic deterioration as evident from unaltered LV stroke volume and reduced LV end-diastolic pressure. The fall in LV end-diastolic pressure was accompanied by an unchanged LV end-diastolic volume and was therefore consistent with an improvement in LV diastolic distensibility. NO-related modulation of diastolic LV distensibility was also observed in the pacing-induced heart failure dog model [68]. In this model, a fall in myocardial NO production occurred after 4 weeks of pacing. This fall was accompanied by a steep rise in LV end-diastolic presssure because of reduced diastolic LV distensibility and a drop in LV stroke volume. From these observations in patients with dilated cardiomyopathy and in dogs with pacing-induced heart failure, it seems justified to conclude that there is no objective evidence for NO, whether derived from NOS2 or NOS3, to account for hemodynamic deterioration of failing myocardium, either at baseline or during ß-adrenoreceptor stimulation [43,76].

Contractile Effects of NO in Ischemic and Septic Myocardium

NO exerts marked protective effects on isolated reoxygenated cardiomyocytes mostly through prevention of reoxygenation-induced hypercontracture [77]. These effects are cGMP dependent because addition of methylene blue, an inhibitor of soluble guanylyl cyclase, reduces NO-induced protection. Although a specific negative contractile effect of NO can not be excluded, cGMP-induced phosphorylation of troponin I could certainly contribute through prevention of calcium-independent diastolic crossbridge cycling [78]. Other mechanisms include reduction of osmotic fragility of the sarcolemma damaged by increased oxidative stress because of radical scavenging properties of NO [79].

In opposition to the uniform findings in isolated cardiomyocytes, reports on the outcome of inhibition of NO synthase in ischaemic-reperfused hearts are controversial. Especially under acidotic conditions, as present during low-flow ischemia, peroxynitrite anions, which result from reaction of NO with superoxide anion, get protonated and subsequently decay rapidly to generate hydoxyl radicals [80], which are highly toxic for sarcolemmal or sarcoplasmic membranes. Consistent with this sequence of events, inhibition of NO synthesis has been reported to reduce reoxygenation injury in anaesthetized piglets [81] and to reduce infarct size in rabbits [82]. In contrast to these initial results, subsequent studies reported endogenous release of NO to protect against ischemia-reperfusion. Both in the ischemic-reperfused rat [83] and rabbit [84] heart, inhibition of endogenous NO synthesis by L-nitro arginine increased the ischemia related contractile deficit possibly because of a NO-mediated scavanging effect of free radicals with a concomitant reduction in ischemic injury.

In the search for the contractile effects of endogenous NO during ischemia-reperfusion, pharmacological inhibition of NO was replaced by the use of transgenic NOS2 or NOS3 knock-out mice. Here again, the results were nonuniform: some investigators found exacerbated ischemia-reperfusion injury in NOS knock-out animals while other investigators observed a protective effect of endogenous NO production [85,86]. Using an isolated mouse heart preparation, an interesting crosstalk between NOS2 and NOS3 during ischemia-reperfusion was recently reported [87]: in NOS3 knock-out mice subjected to 30 minutes of global ischemia followed by reperfusion, a paradoxical increase of NO production was observed because of superinduction of NOS2. This increased NO production, blunted the hyperdynamic response during early reperfusion and protected the heart against myocardial injury as evident from a reduction in infarct size. Similar findings were observed in wild-type mice treated with the NO-donor S-nitroso-N-acetylpenicillamine (SNAP). In these experiments, NO-induced myofilamentary desensitization because of cGMP-induced phosphorylation of troponin I served as a brake for the myocardial hypercontractile response to early reperfusion thereby reducing energy demand and ATP deficit.

Apart from affecting the initial contractile response to ischemia-reperfusion injury, NO could also influence long-term left ventricular remodeling after myocardial infarction [88]. In the early-postinfarction period, left ventricular contractile performance in wild-type and NOS2 knock-out mice was similar, but 4 months after infarction, NOS2 knock-out mice had superior hemodynamics and better survival because of reduced apoptotic cell death and myocyte loss in the noninfarcted zone [89]. Similar findings were reported by other investigators, who noticed an earlier increase in left ventricular end-diastolic diameter and an earlier decrease in fractional shortening at 1 month post-infarction in wild type mice compared to NOS2 knock-outs [90]. In these end-stage mice-infarct models, circulating levels of free radicals become very elevated [91] and the observed late deterioration of left ventricular performance could therefore result not from NO itself but from elevated oxidative stress, which triggers peroxynitrite production and hydroxyl-related membrane damage in the presence of NO [92,93]. The superior long-term outcome following myocardial infarction in wild-type mice compared to NOS3 knock-out mice also argues against direct NO-related deleterious effects on long-term post-infarct LV remodeling [94]. Moreover, in endothelial cells, NO confers protection against apoptotic cell death via s-nitrosylation of caspases. In these cells, aging-induced loss of NOS3 activity increases sensitivity to oxidized LDL or tumor necrosis factor α induced apoptosis [95].

A detrimental role of peroxynitrite rather than NO was also demonstrated in septic or endotoxemic hearts. Once established, the depressed cardiac function after endotoxin administration was not reversed by NOS inhibition [96] and myocardial function of the endotoxemic heart remained depressed even after return of NOS2 protein levels to control value [97]. The large quantities of NO produced by NOS2 during the initial episode of endotoxemia could however have promoted the formation of peroxynitrite, which could have irreversibly impaired contractile performance through long-lasting modifications of proteins such as actin, sarcoplasmic Ca-ATPase or sarcoplasmic Ca-release channel (ryanodine receptor).

Conclusions

Recent experimental and clinical research clarified some of the controversies surrounding the myocardial contractile effects of NO. In transgenic mice with cardioselective overexpression of NOS3 [9] or of NOS2 [16] and with a 60- to 20-fold increase in myocardial NOS activity, NO produced a small reduction in basal LV developed pressure and a LV relaxation-hastening effect mainly through myofilamentary desensitization and to a lesser extent through alterations in calcium handling. Similar findings had previously been reported during intracoronary infusions of NO-donors and of substance P, which releases NO from the coronary endothelium, in isolated rodent hearts [6,7], in normal control subjects [18,20] and in patients with aortic stenosis [19]. The LV relaxation hastening effect was always accompanied by increased diastolic LV distensibility, which augmented LV preload reserve especially in dilated cardiomyopathy patients [64]. In this patient population, low endomyocardial NOS expression was accompanied by an unfavourable hemodynamic phenotype characterized by elevated LV diastolic stiffness and reduced LV stroke work [72]. The beneficial effect of NO on diastolic LV function always overrode the small NO-induced attenuation in LV developed pressure in terms of overall LV performance [43]. In most experimental and clinical conditions, contractile effects of NO were similar when NO was derived from NO-donors or produced by the different isoforms of NOS [24,53]. Because expression of inducible NOS (NOS2) is frequently accompanied by elevated oxidative stress, NO produced by NOS2 can lead to peroxynitrite formation and irreversible contractile impairment as observed in post-infarct LV remodeling [89,90] or sepsis [97]. Finally, presence of ROS and shifts of myofilaments to isoforms with reduced susceptibility to cGMP-mediated desensitization can alter the myocardial contractile effects of NO in hypertrophied or failing myocardium [29,38].

References

1. Brady AJ, Warren JB, Poole-Wilson PA, Williams TJ, Harding SE. Nitric oxide attenuates cardiac myocyte contraction. *Am J Physiol* 1993;265:H176–H182.
2. Balligand JL, Kelly RA, Marsden PA, Smith TW, Michel T. Control of cardiac muscle cell function by an endogenous nitric oxide signalling system. *Proc Natl Acad Sci USA* 1993;90:347–351.
3. Brutsaert DL, Meulemans AL, Sipido KR, Sys SU. Effects of damaging endocardial surface on the mechanical performance of isolated cardiac muscle. *Circ Res* 1988;62:358–366.
4. Smith JA, Shah AM, Lewis MJ. Factors released from endothelium of the ferret and pig modulate myocardial contraction. *J Physiol* 1991;439:1–14.
5. Mohan P, Brutsaert DL, Paulus WJ, Sys SU. Myocardial contractile response to nitric oxide and cGMP. *Circulation* 1996;93:1223–1229.
6. Grocott-Mason R, Fort S, Lewis MJ, Shah AM. Myocardial relaxant effect of exogenous nitric oxide in the isolated ejecting heart. *Am J Physiol* 1994;266:H1699–H1705.
7. Grocott-Mason R, Anning P, Evans H, Lewis MJ, Shah AM. Modulation of left ventricular relaxation in the isolated ejecting heart by endogenous nitric oxide. *Am J Physiol* 1994;267:H1804–H1813.
8. Anning PB, Grocott-Mason R, Lewis MJ, Shah AM. Enhancement of left ventricular relaxation in the isolated rat heart by an angiotensin-converting enzyme inhibitor. *Circulation* 1995;92:2660–2665.
9. Brunner F, Andrew P, Wölkart G, Zechner R, Mayer B. Myocardial contractile function and heart rate in mice with myocyte-specific overexpression of endothelial nitric oxide synthase. *Circulation* 2001;104:3097–3102.
10. Mery PF, Pavoine C, Belhassen L, Pecker F, Fischmeister R. Nitric oxide regulates cardiac Ca^{++} current. Involvment of cGMP-inhibited and cGMP-stimulated phosphodiesterases through guanylyl cyclase activation. *J Biol Chem* 1993;268:26286–26295.
11. Campbell DL, Stamler JS, Strauss HC. Redox modulation of L-type calcium channels in ferret ventricular myocytes. Dual mechanism regulation by nitric oxide and S-nitrosothiols. *J Gen Physiol* 1996;108:277–293.
12. Finkel MS, Oddis CV, Mayer OH, Hattler BG, Simmons RL. Nitric oxide synthase inhibitor alters papillary muscle force-frequency relationship. *J Pharmacol Exp Ther* 1995;272:945–952.
13. Kaye DM, Wiviott SD, Balligand JL, Simmons WW, Smith TW, Kelly RA. Frequency-dependent activation of a constitutive nitric oxide synthase and regulation of contractile function in adult rat ventricular myocytes. *Circ Res* 1996;78:217–224.
14. Stoyanovsky D, Murphy T, Anno PR, Kim Y-M, Salama G. Nitric oxide activates skeletal and cardiac ryanodine receptors. *Cell Calcium* 1997;21:19–29.
15. Ziolo MT, Katoh H, Bers DM. Expression of inducible nitric oxide synthase depresses ß-adrenergic-stimulated calcium release from the sarcoplasmic reticulum in intact ventricular myocytes. *Circulation* 2001;104:2961–2966.
16. Heger J, Gödecke A, Flögel U, Merx MW, Molojavyi A, Kühn-Velten WN, Schrader J. Cardiac-specific overexpression of inducible nitric oxide synthase does not result in severe cardiac dysfunction. *Circ Res* 2002;90:93–99.
17. Paulus WJ, Kästner S, Vanderheyden M, Shah AM, Drexler H. Myocardial contractile effects of L-arginine in the human allograft. *J Am Coll Card* 1997;29:1332–1338.
18. Paulus WJ, Vantrimpont PJ, Shah AM. Acute effects of nitric oxide on left ventricular relaxation and diastolic distensibility in humans. *Circulation* 1994;89:2070–2078.

19. Matter CM, Mandinov L, Kaufmann PA, Vassalli G, Jiang Z, Hess OM. Effects of NO-donors on LV diastolic function in patients with severe pressure-overload hypertrophy. *Circulation* 1999;99:2396–2401.
20. Paulus WJ, Vantrimpont PJ, Shah AM. Paracrine coronary endothelial control of left ventricular function in humans. *Circulation* 1995;92:2119–2126.
21. Kojda G, Kottenberg K, Nix P, Schulter KD, Piper HM, Noack E. Low increase in cGMP induced by organic nitrates and nitrovasodilators improves contractile response of rat ventricular myocytes. *Circ Res* 1996;78:91–101.
22. Preckel B, Kojda G, Schlack W, Ebel D, Kottenberg K, Noack E, Thämer V. Inotropic effects of glyceryl trinitrate and spontaneous NO donors in the dog heart. *Circulation* 1997;96:2675–2682.
23. Cotton JM, Kearney MT, Mac Carthy PA, Grocott-Mason RM, McClean DR, Heymes C, Richardson PJ, Shah AM. Effects of nitric oxide synthase inhibition on basal function and the force-frequency relationship in the normal and failing human heart *in vivo*. *Circulation* 2001;104:2318–2323.
24. Bartunek J, Shah AM, Vanderheyden M, Paulus WJ. Dobutamine enhances cardiodepressant effects of receptor-mediated coronary endothelial stimulation. *Circulation* 1997;95:90–96.
25. Hare JM, Givertz MM, Creager MA, Colucci WS. Increased sensitivity to nitric oxide synthase inhibition in patients with heart failure: Potentiation of ß-adrenergic inotropic responsiveness. *Circulation* 1998;97:161–166.
26. Coudray N, Beregi JP, Lecarpentier Y, Chemla D. Effects of isoproterenol on myocardial relaxation rate: Influence of the level of load. *Am J Physiol* 1993;265:H1645–H1653.
27. Mohan P, Paulus WJ, S.U.Sys. Cardiac muscle length increase abolishes early relaxation induced by cardiac agents. *Eur Heart J* 1994;15:145 (Abstr Suppl).
28. Shah AM, Spurgeon HA, Sollott SJ, Talo A, Lakatta EG. 8-Bromo-cGMP reduces the myofilament response to Ca^{2+} in intact cardiac myocytes. *Circ Res* 1994;74:970–978.
29. Ito N, Bartunek J, Spitzer KW, Lorell BH. Effects of the nitric oxide donor sodium nitroprusside on intercellular pH and contraction in hypertrophied myocytes. *Circulation* 1997;95:2303–2311.
30. Ito N, Kagaya Y, Weinberg EO, Barry WH, Lorell BH. Differing effects of endothelin on contraction, intracellular pH and intracellular Ca^{2+} in hypertrophied and normal rat myocytes. *J Clin Invest* 1997;99:125–135.
31. Cingolani HE, Alvarez BV, Ennis IL, Camilion de Hurtado MC. Stretch-induced alkalinization of feline papillary muscle: An autocrine-paracrine system. *Circ Res* 1998;33:775–780.
32. Pinsky DJ, Patton S, Mesaros S, Brovkovych V, Kubaszewski E, Grunfeld S, Malinski T. Mechanical transduction of nitric oxide synthesis in the beating heart. *Circ Res* 1997;81:372–379.
33. Prendergast BD, Sagach VF, Shah AM. Basal release of nitric oxide augments the Frank-Starling response in the isolated heart. *Circulation* 1997;96:1320–1329.
34. Sagach VF, Shimanskaya TV, Sagach VV, Bogomolets AA. Coronary endothelium dysfunction and heart failure. *J Heart Fail* 1998;5:79 (Abstr.).
35. Matsubara BB, Matsubara LS, Zornoff AM, Franco M, Janicki JS. Left ventricular adaptation to chronic pressure overload induced by inhibition of nitric oxide synthase in rats. *Basic Res Cardiol* 1998;93:173–181.
36. Carroll JD, Lang RM, Neumann AL, Borow KM, Rajfer SI. The differential effects of positive inotropic and vasodilator therapy on diastolic properties in patients with congestive cardiomyopathy. *Cicrulation* 1986;74:815–825.
37. Kingma L, Smiseth OA, Belenkie I, Knudtson ML, Mc Donald RPR, Tyberg JV. A mechanism for the nitroglycerin-induced downward shift of the left ventricular diastolic pressure-diameter relationship of patients. *Am J Cardiol* 1986;57:673–677.
38. MacCarthy PA, Shah AM. Impaired endothelium-dependent regulation of ventricular relaxation in pressure-overload cardiac hypertrophy. *Circulation* 2000;101:1854–1860.
39. Nava E, Noll G, Lüscher TF. Increased activity of constitutive nitric oxide synthase in cardiac endothelium in spontaneous hypertension. *Circulation* 1995;91:2310–2313.
40. Lang D, Mosfer SI, Shakesby A, Donaldson F, Lewis MJ. Coronary microvascular endothelial cell redox state in left ventricular hypertrophy. The role of angiotensin II. *Circ Res* 2000;86:463–469.
41. McCarthy PA, Grieve DJ, Li JM, Dunster C, Kelly FJ, Shah AM. Impaired endothelial regulation of ventricular relaxation in cardiac hypertrophy. Role of reactive oxygen species and NADPH oxidase. *Circulation* 2001;104:2967–2972.
42. Feng J, Schaus BJ, Fallavollita JA, Lee TC, Canty JM, Jr. Preload induces troponin I degradation independently of myocardial ischemia. *Circulation* 2001;103:2035–2037.
43. Paulus WJ, Frantz S, Kelly RA. Nitric oxide and cardiac contractility in human heart failure: Time for reappraisal. *Circulation* 2001;104:2260–2262.
44. Zieman SJ, Gerstenblith G, Lakatta EG, Rosas GO, Vandegaer K, Ricker KM, Hare JM. Upregulation of nitric oxide-cGMP pathway in aged myocardium: Physiological response to l-arginine. *Circ Res* 2001;88:97–102.
45. Gao F, Christopher TA, Lopez BL, Friedman E, Cai G, Ma XL. Mechanism of decreased adenosine protection in reperfusion injury of aging rats. *Am J Physiol* 2000;279:H329–H338.
46. Yang B, Larson DF, Shi J, Gorman M, Watson RR. Modulation of iNOS activity in senescent cardiac dysfunction. *Circulation* 2001;104(II):435 (Abstract).
47. Lewis NP, Tsao PS, Rickenbacher PR, Xue C, Johns RA, Haywood GA, von der Leyen H, Trindade PT, Cooke JP, Hunt SA, Billingham ME, Valantine HA, Fowler MB. Induction of nitric oxide synthase in the human allograft is associated with contractile dysfunction of the left ventricle. *Circulation* 1996;93:720–729.
48. Finkel MS, Odis CV, Jacob TD, Watkins SC, Hattler BG, Simmons RL. Negative inotropic effects of cytokines on the heart mediated by nitric oxide. *Science* 1992;257:387–389.
49. Brady AJB, Poole-Wilson PA, Harding SE, Warren JB. Nitric oxide production within cardiac myocytes reduces their contractility in endotoxemia. *Am J Physiol* 1992;26:H1963–H1996.
50. Balligand JL, Ungureanu D, Kelly RA, Kobzik L, Pimental D, Michel T, Smith TW. Abnormal contractile function due to induction of nitric oxide synthesis in rat cardiac myocytes follows exposure to activated macrophage-conditioned medium. *J Clin Invest* 1993;91:2314–2319.
51. Evans HG, Lewis MJ, Shah AM. Interleukin 1-ß modulates myocardial contraction via dexamethasone-sensitive production of nitric oxide. *Cardiovasc Res* 1993;27:1486–1490.

52. Kinugawa K, Takahashi T, Kohmoto O, Yao A, Aoyagi T, Momomura S, Hirota Y, Serizawa T. Nitric oxide-mediated effects of interleukin-6 on $[Ca^{2+}]_i$ and cell contraction in cultured chick ventricular myocytes. *Circ Res* 1994;75:285–295.
53. Paulus WJ, Kästner S, Pujadas P, Shah AM, Drexler H, Vanderheyden M. Left ventricular contractile effects of inducible nitric oxide synthase in the human allograft. *Circulation* 1997;96:3436–3442.
54. Ronson RS, Nakamura M, Vinten-Johansen J. The cardiovascular effects and implications of peroxynitrite. *Cardiovasc Res* 1999;44:47–59.
55. Wildhirt SM, Weis M, Schulze C, Conrad N, Pehlivanli S, Rieder G, Enders G, von Scheidt W, Reichart B. Expression of endomyocardial nitric oxide synthase and coronary endothelial function in human cardiac allografts. *Circulation* 2001;104(Suppl I):336–343.
56. Paulus WJ. Nitric oxide and cardiac contraction: Clinical studies. In: Lewis MJ and Shah AM, eds. *Endothelial Modulation of Cardiac Function*. Amsterdam: Harwood Academic Publishers 1997;35–51.
57. Paulus WJ, Bronzwaer JGF, Felice H, Kishan N, Wellens F. Deficient acceleration of left ventricular relaxation during exercise after heart transplantation. *Circulation* 1992;86:1175–1185.
58. de Belder AJ, Radomski M, Why H, Richardson PJ, Bucknall CA, Salas E, Martin JF. Nitric oxide synthase activities in human myocardium. *Lancet* 1993;341:84–85.
59. Haywood GA, Tsao PS, von der Leyen HE, Mann MJ, Keeling PJ, Trindade PT, Lewis NP, Byrne CD, Rickenbacher PR, Bishopric NH, Cooke JP, McKenna WJ, Fowler MB. Expression of inducible nitric oxide synthase in human heart failure. *Circulation* 1996;93:1087–1094.
60. Habib FM, Springall DR, Davies GJ, Oakley CM, Yacoub MH, Polak JM. Tumour necrosis factor and inducible nitric oxide synthase in dilated cardiomyopathy. *Lancet* 1996;347:1151–1155.
61. Satoh M, Nakamura M, Tamura G, Makita S, Segawa I, Tashiro A, Satodate R, Hiramori K. Inducible nitric oxide synthase and tumor necrosis factor-alpha in myocardium in human dilated cardiomyopathy. *J Am Coll Cardiol* 1997;29:716–724.
62. Drexler H, Kästner S, Strobel A, Studer R, Brodde OE, Hasenfuss G. Expression, activity and functional significance of inducible nitric oxide synthase in the failing human heart. *J Am Coll Cardiol* 1998;32:955–963.
63. Fukuchi M, Hussain SNA, Giaid A. Heterogeneous expression and activity of endothelial and inducible nitric oxide synthases in end-stage human heart failure. Their relation to lesion site and ß-adrenergic receptor therapy. *Circulation* 1998;98:132–139.
64. Heymes C, Vanderheyden M, Bronzwaer JGF, Shah AM, Paulus WJ. Endomyocardial nitric oxide synthase and left ventricular preload reserve in dilated cardiomyopathy. *Circulation* 1999;99:3009–3016.
65. Stein B, Eschenhagen T, Rüdiger J, Scholz H, Förstermann U, Gath I. Increased expression of constitutive nitric oxide synthase III, but not inducible nitric oxide synthase II, in human heart failure. *J Am Coll Cardiol* 1998;32:1179–1186.
66. Yamamoto S, Tsutsui H, Tagawa H, Saito K, Takahashi M, Tada H, Yamamoto M, Katoh M, Egashira K, Takeshita A. Role of myocyte nitric oxide in ß-adrenergic hyporesponsiveness in heart failure. *Circulation* 1997;95:1111–1114.
67. Harding SE, Davies CH, Money-Kyrle AM, Poole-Wilson PA. An inhibitor of nitric oxide synthase does not increase contraction or ß-adrenoreceptor sensitivity of ventricular myocytes from failing human heart. *Cardiovasc Res* 1998;40:523–529.
68. Recchia FA, McConnell PL, Bernstein RD, Vogel TR, Xu X, Hintze TH. Reduced nitric oxide production and altered myocardial metabolism during the decompensation of pacing-induced heart failure in the conscious dog. *Circ Res* 1998;83:969–979.
69. Senzaki H, Smith CJ, Juang GJ, Isoda T, Mayer SP, Ohler A, Paolocci N, Tomaselli GF, Hare JM, Kass DA. Cardiac phosphodiesterase 5 (cGMP-specific) modulates beta-adrenergic signaling *in vivo* and is downregulated in heart failure. *FASEB J* 2001;15:1718–1726.
70. Hare JM, Lofthouse RA, Juang GJ, Colman L, Ricker KM, Kim B, Senzaki H, Cao S, Tunin RS, Kass DA. Contribution of caveolin protein abundance to augmented nitric oxide signaling in conscious dogs with pacing-induced heart failure. *Circ Res* 2000;86:1085–1092.
71. Bauersachs J, Galuppo P, Fraccarollo D, Christ M, Ertl G. Improvement of left ventricular remodeling and function by hydroxymethylglutaryl coenzyme A reductase inhibition with cerivastatin in rats with heart failure after myocardial infarction. *Circulation* 2001;104:982–985.
72. Bronzwaer JGF, Zeitz C, Visser CA, Paulus WJ. Endomyocardial nitric oxide synthase and the hemodynamic phenotypes of human dilated cardiomyopathy and of athlete's heart. *Cardiovasc Res* 2002;55:270–278.
73. Sessa WC, Pritchard K, Seyedi N, Wang J, Hintze TH. Chronic exercise in dogs increases coronary vascular nitric oxide production and endothelial cell nitric oxide synthase gene expression. *Circ Res* 1994;74:349–353.
74. Bernstein RD, Ochoa FY, Xu X, Forfia P, Shen W, Thompson CI, Hintze TH. Function and production of nitric oxide in the coronary circulation of the conscious dog during exercise. *Circ Res* 1996;79:840–848.
75. Wittstein IS, Kass DA, Pak PH, Maughan WL, Fetics B, Hare JM. Cardiac nitric oxide production due to angiotensin-converting enzyme inhibition decreases beta-adrenergic myocardial contractility in patients with dilated cardiomyopathy. *J Am Coll Cardiol* 2001;38:429–435.
76. Hare JM, Stamler JS. NOS: Modulator, not mediator of cardiac performance. *Nat Med* 1999;5:273–274.
77. Schlüter KD, Weber M, Schraven E, Piper HM. NO donor SIN-1 protects against reoxygenation-induced cardiomyocyte injury by a dual mechanism. *Am J Physiol* 1994;267:H1461–H1466.
78. Shah AM, Silverman HS, Griffiths EJ, Spurgeon HA, Lakatta EG. cGMP prevents delayed relaxation at reoxygenation after brief hypoxia in isolated cardiac myocytes. *Am J Physiol* 1995;268:H2396–H2404.
79. Schlüter KD, Jacob G, Ruiz-Meana M, Garcia-Dorado D, Piper HM. Protection of reoxygenated cardiomyocytes against osmotic fragility by NO donors. *Am J Physiol* 1996;271:H428–H434.
80. Beckman JS, Beckman TW, Chen J, Marshall PA, Freeman BA. Apparent hydroxyl radical production by peroxynitrite: Implications for endothelial injury from nitric oxide and superoxide. *PNAS* 1990;87:1620–1624.
81. Matheis G, Sherman MP, Buckberg GD, Haybron DM, Young HH, Ignarro LJ. Role of L-arginine-nitric oxide pathway in myocardial reoxygenation injury. *Am J Physiol* 1992;262:H616–H620.

82. Patel VC, Yellon DM, Singh KJ, Neild GH, Woolfson RG. Inhibition of nitric oxide limits infarct size in the *in situ* rabbit heart. *Biochem Biophys Res Comm* 1993;194: 234–238.
83. Beresewicz A, Karwatowska-Prokopczuk E, Lewartowski B, Cedro-Ceremuzynska K. A protective rol of nitric oxide in isolated ischaemic/reperfused rat heart. *Cardiovasc Res* 1995;30:1001–1008.
84. Williams MW, Taft CS, Ramnauth S, Zhao ZQ, Vinten-Johansen J. Endogenous nitric oxide (NO) protects against ischaemia-reperfusion injury in the rabbit. *Cardiovasc Res* 1995;30:79–85.
85. Jones SP, Girod WG, Palazzo AJ, Granger DN, Grisham MB, Jourd'Heuil D, Huang PL, Lefer DJ. Myocardial ischemia-reperfusion injury is exacerbated in absence of endothelial nitric oxide synthase. *Am J Physiol* 1999;276:H1567–H1573.
86. Igarashi J, Nishida M, Hoshida S, Yamashita N, Kosaka H, Hori M, Kuzuya T, Tada M. Inducible nitric oxide synthase augments injury elicited by oxidative stress in rat cardiac myocytes. *Am J Physiol* 1998;274:C245–C252.
87. Kanno S, Lee PC, Zhang Y, Ho C, Griffith BP, Shears LL II, Billiar TR. Attenuation of myocardial ischemia/reperfusion injury by superinduction of inducible nitric oxide synthase. *Circulation* 2000;101:2742–2748.
88. Jugdutt BI. Effect of nitrates on myocardial remodeling after acute myocardial infarction. *Am J Cardiol* 1996;77:17C–23C.
89. Sam F, Sawyer DB, Xie Z, Chang DL, Ngoy S, Brenner DA, Siwik DA, Singh K, Apstein CS, Colucci WS. Mice lacking inducible nitric oxide synthase have improved left ventricular contractile function and reduced apoptotic cell death late after myocardial infarction. *Circ Res* 2001;89:351–356.
90. Feng Q, Lu X, Jones DL, Shen J, Arnold JM. Increased inducible nitric oxide synthase expression contributes to myocardial dysfunction and higher mortality after myocardial infarction in mice. *Circulation* 2001;104:700–704.
91. Tsutsui H, Ide T, Hayashidani S, Suematsu N, Shiomi T, Wen J, Nakamura K, Ichikawa K, Utsumi H, Takeshita A. Enhanced generation of reactive oxygen species in the limb skeletal muscles from a murine infarct model of heart failure. *Circulation* 2001;104:134.
92. Yasmin W, Strynadka KD, Schulz R. Generation of peroxynitrite contributes to ischemia-reperfusion injury in isolated rat hearts. *Cardiovasc Res* 1997;33:422–432.
93. Wang W, Sawicki G, Schulz R. Peroxynitrite-induced myocardial injury is mediated through matrix metalloproteinase-2. *Cardiovasc Res* 2002;53:165–174.
94. Scherrer-Crosbie M, Ullrich R, Bloch KD, Nakajima H, Nasseri B, Aretz T, Lindsey ML, Vançon AC, Huang PL, Lee RT, Zapol WM, Picard MH. Endothelial nitric oxide synthase limits left ventricular remodeling after myocardial infarction in mice. *Circulation* 2001;104:1286–1291.
95. Hoffmann J, Haendeler J, Aicher A, Rossig L, Vasa M, Zeiher A, Dimmeler S. Aging enhances the sensitivity of endothelial cells toward apoptotic stimuli. *Circ Res* 2001;89:709.
96. Decking U, Flesche C, Godecke A, Schrader J. Endotoxin-induced contractile dysfunction in guinea pig hearts is not mediated by nitric oxide. *Am J Physiol* 1995;268:H2460–H2465.
97. Mebazaa A, De Keulenaer GW, Paqueron X, Andries LJ, Ratajczak P, Lanone S, Frelin C, Longrois D, Payen D, Brutsaert DL, Sys SU. Activation of cardiac endothelium as a compensatory component in endotoxin-induced cardiomyopathy: Role of endothelin, prostaglandins and nitric oxide. *Circulation* 2001;104:3137–3144.
98. Paulus WJ, Shah AM. NO and cardiac diastolic function. *Cardiovasc Res* 1999;43:595–606.

Nitric Oxide, Platelet Function, Myocardial Infarction and Reperfusion Therapies

David Alonso and Marek W. Radomski
Department of Integrative Biology and Pharmacology, University of Texas-Houston, USA

Abstract. **Platelets play an important role in physiologic hemostasis and pathologic thrombosis that complicate the course of vascular disorders. A number of platelet functions including adhesion, aggregation and recruitment are controlled by nitric oxide (NO) generated by platelets and the endothelial cells. Derangements in this generation may contribute to the pathogenesis of thrombotic complications of vascular disorders. The pharmacologic supplementation of the diseased vasculature with drugs releasing NO may help to restore the hemostatic balance.**

Key Words. **nitric oxide, platelet function, myocardial infarction, reperfusion**

Introduction

Catastrophic events leading to myocardial ischemia and infarction result from atherosclerotic plaque rupture and are mediated to large extent by formation of occlusive platelet thrombus at the site of injury [1]. The objective of this article is to review the actions of nitric oxide (NO) in regulation of platelet function in physiology and during the coronary artery disease. We will also attempt to review the pharmacological effects of NO donors on the diseased coronary circulation.

A Brief Overview of Platelet Function

Blood platelets play a vital role in vascular hemostasis. Following an accidental injury, platelets adhere to the damaged portions of the vessel wall and then aggregate forming the hemostatic plug in order to seal the vessel and contain blood loss. This ability of the vasculature must be carefully balanced against the necessity to maintain the fluid state of blood in order to provide oxygen and nutrients to the cells and tissues. A failure to maintain blood in its fluid state leads to thrombosis, one of serious manifestations of vascular dysfunction. In order to maintain tight control over the process of platelet activation, regulatory mechanisms operate as an interactive system of agents that stimulate or inhibit the process of platelet activation. While thromboxane A_2, adenosine diphosphate and matrix metalloproteinase-2 mediate the activator platelet reactions [2], prostacyclin, ADPase and NO provide the platelet inhibitor pathways of regulation [3]. These labile mediators control the platelet surface abundance of major platelet receptors including P-selectin, glycoprotein (GP) Ib and GP IIb/IIIa [4–6]. The activation of these receptors leads to platelet adhesion to the components of the extracellular matrix and aggregation.

Regulation of Platelet Hemostasis by NO

Nitric oxide (NO) is a gaseous biological mediator first identified as the endothelium-derived relaxing factor (EDRF) [7–9]. It is generated from the guanidino-nitrogen of L-arginine yielding citrulline (for review see [10]) and plays a prominent role in controlling a variety of functions in the cardiovascular, immune, reproductive, and nervous systems [11–13]. The production of NO is catalyzed by three major isoforms of the enzyme NO synthase (NOS): neuronal (nNOS), inducible (iNOS), and endothelial (eNOS) [14–16]. The nNOS and eNOS are constitutively expressed and activated by calcium entry into cells [17,18], whereas iNOS is calcium-independent, and its synthesis is induced in inflammatory and other cell types by stimuli such as endotoxin and proinflammatory cytokines [19–21].

In 1986–1987, we analyzed the non-eicosanoid, vasodilator and platelet-inhibitory properties of endothelial cells and found that these could be accounted for by the release of EDRF [8,22,23]. Moreover, comparison of pharmacological properties of EDRF with those of NO gas, and the measurement of NO metabolites (nitrite and nitrate) released during these reactions clearly showed

Address for correspondence: Dr. Marek W. Radomski, University of Texas-Houston, 2121 W. Holcombe Blvd., Houston TX, 77030. E-mail: MAREK.RADOMSKI@UTH.TMC.EDU

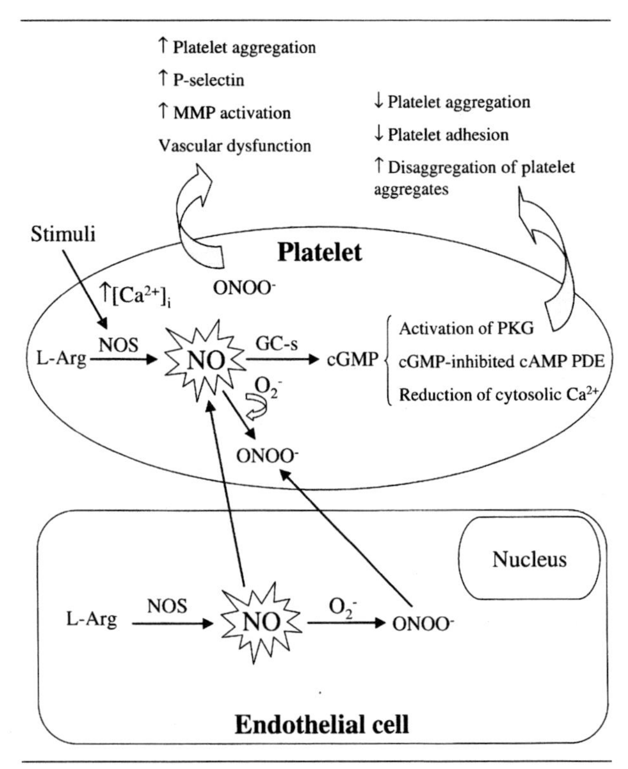

Fig. 1. *Overview of the role of NO in platelet function. Nitric oxide generated from L-arginine (L-arg) by the endothelial cells and platelets activates the soluble gunaylate cyclase (GC-s) to increase the levels of cyclic GMP that control the intracellular enzymes including protein kinase G (PKG), cGMP-inhibited cAMP phosphodiesterase (PDE), and the function of ion channels regulating calcium flux. Nitric oxide can also react with superoxide anion (O_2^-) to form peroxynitrite ($ONOO^-$).*

that generation of NO could account for the platelet-inhibitory activities of NO on platelet adhesion and aggregation [8,24–27].

In 1990–1993, we provided both biochemical and pharmacological characteristics of platelet NOS [20,28] and measured the release of NO from stimulated platelets using a selective porphyrinic microsensor [29]. During the past few years, several groups have described molecular characteristics of platelet NOS [30–33]. In addition to inhibition of platelet adhesion and aggregation, platelet-derived NO is involved in down-regulation of platelet recruitment to the aggregate [34] (Fig. 1).

Platelets in Myocardial Ischemia and Infarction

There is substantial clinical and experimental evidence implicating the interactions between platelets and vascular endothelium in the pathogenesis of myocardial ischemia. The clinical evidence for a role of platelets in myocardial ischemia includes studies indicating alterations in platelet behavior [35] and the therapeutic benefit of some antiplatelet drugs including aspirin [36] and the antagonists of GP IIb/IIIa [37]. Experimental evidence points to reductions in transmyocardial platelet number under conditions of coronary ischemia consistent with formation of coronary thrombi [38], the cyclical reductions in coronary blood flow that result from intermittent intravascular platelet aggregation [39], and the relationship between induced platelet aggregation and cardiac arrhythmias [40]. Shimada and colleagues suggested the importance of platelets in reperfusion injury showing that the depletion of both leukocytes and platelets in canine myocardial ischemia-reperfusion injury effectively protects against reperfusion injury [41].

Coronary artery occlusion or thrombus formation produces myocardial ischemia that can ultimately result in myocardial cell injury and necrosis. Atherosclerotic plaque disruption is the predominant pathogenetic mechanism underlying the acute coronary syndromes [42]. The pathogenesis of atherosclerosis appears to depend on a precise sequence of critical events based on the interaction of blood elements and lipids with the arterial wall. The pioneering work by Fuster's group showed the role of platelets in the development of atherosclerotic disease [43]. They have shown that plaque rupture leads to the exposure of collagen and vessel media, resulting in activation of platelets and the coagulation cascade that lead to occlusive thrombus formation. Willerson's group [44] speculated that the abrupt conversion from chronic stable to unstable angina and the continuum to acute myocardial infarction might result from myocardial ischemia caused by progressive platelet aggregation and dynamic vasoconstriction. Platelet aggregation and dynamic coronary artery vasoconstriction probably result from the local accumulation of several mediators including thromboxane A_2, serotonin, adenosine diphosphate, platelet activating factor, oxygen-derived free radicals thrombin, and tissue factor. This is accompanied by relative decreases in the local concentrations of endothelium-derived vasodilators and inhibitors of platelet aggregation such as prostacyclin, NO, and tissue plasminogen activator [44–46]. This helps to create a prothrombotic and vasoconstrictive environment. With severe reductions in coronary blood flow caused by these mechanisms, platelet aggregates may increase, and an occlusive thrombus composed of platelets and white and red blood cells in a fibrin mesh may develop. When coronary arteries are occluded or narrowed for a sufficient period of time by these mechanisms, myocardial necrosis, electrical instability, or sudden death may occur [47].

Nitric Oxide in the Pathogenesis of Coronary Artery Disease and Reperfusion Injury

The vasodilator and platelet-regulatory functions of endothelium are impaired during the course of vascular disorders including atherosclerosis, the coronary artery disease, essential hypertension, diabetes mellitus and preeclampsia [48]; however, the reasons underlying this impairment are not clear. A number of researchers correlated the changes in the endothelial function with the generation of NO. The endothelial dysfunction was both ascribed to decreased and enhanced generation of NO. To explain this discrepancy it was proposed that these changes in NO generation are often accompanied by reduced bioactivity of NO [49]. It is likely that the metabolism of NO and the interactions of NO with reactive oxygen species account for this reduced bioactivity of NO. In 1990, Beckman and associates first found that the reaction of NO with superoxide could take place under physiological conditions and lead to the formation of peroxynitrite ($ONOO^-$) (for refs. [50]). Peroxynitrite is a highly reactive oxidant that can oxidize various biomolecules in the cellular microenvironment. In 1994, we found that $ONOO^-$ can decrease the vasodilator and platelet-inhibitory activity of NO and prostacyclin [51,52]. Thiols and glucose [51,53] attenuated these detrimental effects of $ONOO^-$. The reaction of $ONOO^-$ with thiols in cell membranes and glucose in the extracellular fluid results in synthesis of NO donors that counteract the vasoconstrictor and platelet-aggregatory activities of the parent oxidant [51,54]. Interestingly, there is now evidence that small amounts of $ONOO^-$ may be generated during aggregation of normal platelets [55]. Thus, it is likely that under physiological conditions $ONOO^-$, when generated by platelets, is rapidly detoxified and converted to NO donors following reactions with platelet membrane thiols [54]. The oxidizing stress could decrease the efficiency of this regulating mechanism and precipitate platelet dysfunction and damage.

The precise role of NO in the reperfused heart is controversial. A number of studies have shown that NO augments postischemic arrhythmias, contractile dysfunction, and infarction [56–59]. On the other hand, a protective role has also been reported [60–62]. Moreover, pharmacological administration of L-arginine to patients suffering from coronary artery disease and peripheral arterial obstructive disease alleviated the symptoms of arterial insufficiency [63,64].

Antiplatelet Therapies, NO and NO Donors in the Treatment of Coronary Artery Disease

The major aim in the management of myocardial infarction is myocardial reperfusion therapy, i.e. complete and timely restoration of coronary blood flow. Current clinical strategies for the treatment of acute myocardial ischemia include coronary angioplasty, directional coronary atherectomy, and pharmacologic therapy to restore blood flow to the ischemic myocardium.

Previous fibrinolytic regimens for reperfusion fail to fully restore coronary blood flow in slightly less than 50% of patients with acute myocardial infarction [65,66]. Since platelets play a crucial role in the pathophysiology of myocardial infarction, platelet activation and aggregation is likely to be responsible for a large proportion of these therapeutic failures. While drugs that interfere with platelet activation and function have been available for years, more powerful agents with novel mechanisms of action are being developed. Classical platelet inhibitors such as aspirin, sulfinpyrazone, dipyridamole and ticlopidine have undergone extensive clinical testing in patients with cardiovascular disease [36,42]. The combination of aspirin and a thrombolytic agent produces maximal benefit. Recently, it has been shown that the antiaggregating effect of aspirin could be explained through its action on neutrophils. Interestingly, aspirin stimulates NO production by neutrophils inhibiting the aggregating effects of thrombin, ADP or epinephrine on platelets [67].

The final common pathway of platelet activation and aggregation is the activation of the GP IIb/IIIa receptor. Experimental and early clinical evidence suggested that GP IIb/IIIa antagonists, such as abciximab, eptifibatide, or tirofiban, might enhance reperfusion when combined with reduced doses of thrombolytic agents [68–70]. In the past decade, several strategies targeting the use of GP IIb/IIIa inhibitors have been evaluated. When GP IIb/IIIa inhibitors are used with full-dose fibrinolytics, early studies have suggested a trend toward more rapid and more complete reperfusion in an acute myocardial infarction. Later trials have examined the use of GP IIb/IIIa inhibitors in conjunction with reduced-dose fibrinolytics [71]. The results from Global Use of Strategies to Open occluded arteries-V (GUSTO-V) large-scale trial supports the use of combination therapy with reduced dose of a plasminogen activator, reteplase, and the GP IIb/IIIa inhibitor abciximab [71].

Primary coronary angioplasty has become the preferred method of reperfusion in patients with acute myocardial infarction. Still, the procedure is limited by the substantial rates of restenosis and reocclusion. A recent meta-analysis of randomized trials that compared thrombolytic therapy and primary angioplasty showed a lower incidence of death, non-fatal reinfarction, stroke, and intracranial bleeding in patients treated with primary angioplasty than in patients treated with thrombolysis [72]. Santoro and Bolognese reviewed published studies comparing angioplasty with stenting, with or without treatment with the GP IIb/IIIa inhibitor abciximab [73]. In a recent trial comparing angioplasty with stenting, Stone and colleagues reported that stenting (alone or with abciximab) was superior to standard balloon angioplasty (alone or with abciximab) in reducing the need for repeated revascularization and they recommended stent as the routine reperfusion strategy at experienced centers [74]. The main problem is that stenting is available only in specialized centers.

Studies using NO gas *in vitro* showed that NO was a potent, but short-acting (the biological half-life <4 min) inhibitor of platelet adhesion, aggregation and stimulator of platelet disaggregation [24–27]. Interestingly, animal experiments showed that inhaled NO might inhibit the development of coronary thrombosis [75]. The pharmacological significance of these findings remains to be investigated.

The effectiveness of organic nitrates as antithrombotics increases with the extent of vascular injury [76]. Furthermore, short- and long-lasting administration of nitroglycerin and isosorbide dinitrate to patients suffering from coronary artery disease and acute myocardial infarction resulted in a significant inhibition of platelet adhesion and aggregation [77,78]. A meta-analysis found significant reduction in mortality when intravenous glyceryl trinitrate or nitroprusside were used during acute course of myocardial infarction [79]. Moreover, when combined with N-acetylcysteine, glyceryl trinitrate substantially reduced myocardial infarction in unstable angina, an effect compatible with an anti-platelet effect of glyceryl trinitrate [80]. Surprisingly, GISSI III [81] and ISIS-4 [82] studies failed to show clinically beneficial effect of organic nitrates on mortality after myocardial infarction. However, further analysis of GISSI III suggests that the apparent additive effect of glyceryl trinitrate and lisinopril could be attributed to anti-platelet effects of this NO donor [83]. In addition, it is possible that nitrates may act by reducing the infarct size in small rather than large infarcts so that the neutral results of GISSI-3 and ISIS-4 may be explained by the heterogeneity of effect. Interestingly aspirin, a cyclooxygenase inhibitor, blocks only thromboxane-mediated platelet aggregation (for refs. see [84]) leaving the remaining pathways of adhesion and aggregation unopposed. In contrast, NO inhibits the activation cascade of mediators generated by all known pathways of platelet aggregation [85], including recently described matrix metalloproteinase-2-dependent platelet aggregation [2], and some pathways of platelet adhesion to sub-endothelium [86].

Because of its powerful vasodilator action, sodium nitroprusside is often used to treat vascular emergencies associated with hypertensive crisis. Since this compound shows some anti-platelet activity both *in vitro* and *in vivo* [87,88] its acute clinical effects may also be mediated, in part, through inhibition of platelet function. Recently, sodium nitroprusside was administered intrapericardially to treat experimentally induced coronary thrombosis in dogs [89]. As this route of administration of sodium nitroprusside produced less vasodilatation than systemic one, thus localized administration of this drug may offer new therapeutic possibilities for the treatment of the coronary thrombosis.

Molsidomine and its active metabolite SIN-1 inhibit experimental thrombosis and platelet aggregation in healthy volunteers and in patients suffering from acute myocardial infarction [90]. Interestingly, SIN-1 in addition to NO generates superoxide and $ONOO^-$ [91]. Since $ONOO^-$ causes platelet aggregation and counteracts the platelet inhibitory activity of NO [51], the formation of this radical may offset the anti-platelet activity of NO released from SIN-1.

The platelet-inhibitory actions of organic nitrates cannot be separated from their effects on vascular wall. The concept of platelet-selective NO donors has arisen from our experiments with S-nitrosoglutathione (GSNO) [92]. S-Nitrosoglutathione is a tripeptide S-nitrosothiol that is formed by S-nitrosylation of glutathione, the most abundant intracellular thiol. We have found that the intravenous administration of GSNO into conscious rat inhibits platelet aggregation at doses that have only small effect on the blood pressure [92]. Moreover, similar platelet/vascular differentiation is detected following intraarterial administration of GSNO into the circulation of human forearm [48]. Finally, we have infused GSNO into patients undergoing balloon angioplasty and found that this NO donor effectively protected platelets from activation at the site of angioplastic injury without altering blood pressure [93]. Interestingly, the exposure of human neutrophils to NO led to depletion of glutathione stores, activation of the hexose monophosphate shunt, synthesis of endogenous GSNO and inhibition of superoxide generation by neutrophils. Synthetic GSNO resulted in similar effects [94]. Moreover, the administration of GSNO inhibited leukocyte activation,

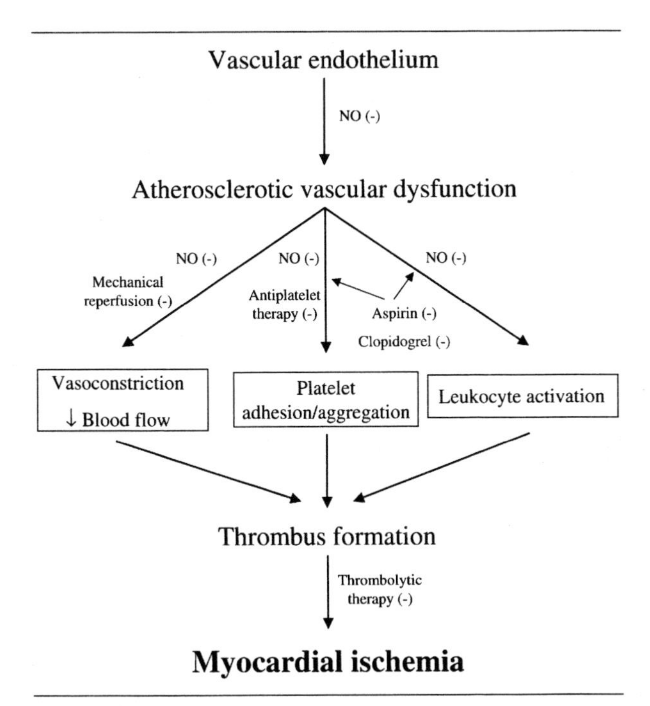

Fig. 2. Pharmacological actions of NO in myocardial ischemia. (-) denotes inhibition.

expression of iNOS and bypass-induced myocardial lesion in dogs [95]. These observations show that GSNO is a potent regulator of platelet and neutrophil functions and it may be a prototype for the development of blood cell-selective NO donors.

Recently, a new family of nitroderivatives of conventional nonsteroidal anti-inflammatory drugs has been synthesized. Among them, NCX 4016, a nitro-ester of aspirin, has been shown to exert the antiaggregatory and antithrombotic activity by a dual mechanism of action involving inhibition of cyclooxygenase and release of NO in both platelets and vascular smooth muscle cells [96,97]. A recent report has shown that NCX 4016 is cardioprotective in ischemia, producing marked improvement of postischemic ventricular dysfunction in the rabbit heart [62]. The possible pharmacological effects of NO on the sequence of events leading to the myocardial ischemia are shown in Figure 2.

Conclusions

Over the past decades, numerous studies have focused on the role of platelets in pathologic thrombosis that complicates the course of vascular disorders. There is now convincing evidence that a number of platelet functions including adhesion, aggregation and recruitment are controlled by NO generated by platelets and the endothelial cells. The findings reviewed in this article indicate that impairment of NO production and bioavailability play a role in the pathophysiology of myocardial ischemia. NO protects vascular endothelium by maintaining basal vascular tone, but it also inhibits platelet and leukocyte activation. Therefore, the pharmacological supplementation of the diseased vasculature with drugs releasing NO may help to restore the hemostatic balance and improve myocardial reperfusion. Furthermore, recent clinical trials showed that the antiplatelet therapy, aimed at blocking the final common pathway of platelet aggregation is of high therapeutic relevance to the treatment of the coronary ischemia. The combined efforts of mechanical procedures (coronary angioplasty, stenting) and a better understanding of platelet physiology and pathology have increased the therapeutic success in patients suffering from coronary ischemia.

Acknowledgments

This work is supported by the Canadian Institutes of Health Research (CIHR) and the Secretaria de Estado de Educacion y Universidades, cofunded by the European Social Fund. DA is a post-doctoral fellow of Spanish Ministry of Education, MWR is a CIHR scientist. We thank Ms. Clara Vega for her help in preparation of this manuscript.

References

1. Stein B, Badimon L, Israel DH, Badimon JJ, Fuster V. Thrombosis/platelets and other blood factors in acute coronary syndromes. *Cardiovascular Clinics* 1989;20:105–129.
2. Sawicki G, Salas E, Murat J, Miszta_Lane H, Radomski MW. Release of gelatinase A during platelet activation mediates aggregation. *Nature* 1997;386:616–619.
3. Radomski M, Radomski A. Regulation of blood cell function by the endothelial cells. In: Vallance P, Webb D, eds. *Vascular Endothelium in Human Physiology and Pathophysiology*. Harwood Academic Publishers, 2000:95–106.
4. Murohara T, Parkinson SJ, Waldman SA, Lefer AM. Inhibition of nitric oxide biosynthesis promotes P-selectin expression in platelets. Role of protein kinase C. *Arterioscler Thromb Vasc Biol* 1995;15:2068–2075.
5. Radomski A, Stewart MW, Jurasz P, Radomski MW. Pharmacological characteristics of solid-phase von Willebrand factor in human platelets. *Br J Pharmacol* 2001;134:1013–1020.
6. Jurasz P, Stewart MW, Radomski A, Khadour F, Duszyk M, Radomski MW. Role of von Willebrand factor in tumour cell-induced platelet aggregation: Differential regulation by NO and prostacyclin. *Br J Pharmacol* 2001;134:1104–1112.
7. Furchgott RF, Zawadzki JV. The obligatory role of endothelial cells in the relaxation of arterial smooth muscle by acetylcholine. *Nature* 1980;288:373–376.
8. Palmer RM, Ferrige AG, Moncada S. Nitric oxide release accounts for the biological activity of endothelium-derived relaxing factor. *Nature* 1987;327:524–526.
9. Moncada S, Radomski MW, Palmer RM. Endothelium-derived relaxing factor. Identification as nitric oxide and

role in the control of vascular tone and platelet function. *Biochem Pharmacol* 1988;37:2495–2501.
10. Marletta MA. Nitric oxide synthase: Aspects concerning structure and catalysis. *Cell* 1994;78:927–930.
11. Knowles RG, Palacios M, Palmer RM, Moncada S. Formation of nitric oxide from L-arginine in the central nervous system: A transduction mechanism for stimulation of the soluble guanylate cyclase. *Proc Natl Acad Sci USA* 1989;86:5159–5162.
12. Bredt DS, Snyder SH. Isolation of nitric oxide synthetase, a calmodulin-requiring enzyme. *Proc Natl Acad Sci USA* 1990;87:682–685.
13. Moncada S, Higgs EA. Endogenous nitric oxide: Physiology, pathology and clinical relevance. *Eur J Clin Invest* 1991;21:361–374.
14. Forstermann U, Schmidt HH, Pollock JS, Sheng H, Mitchell JA, Warner TD, Nakane M, Murad F. Isoforms of nitric oxide synthase. Characterization and purification from different cell types. *Biochem Pharmacol* 1991;42:1849–1857.
15. Moncada S, Palmer RM, Higgs EA. Nitric oxide: Physiology, pathophysiology, and pharmacology. *Pharmacol Rev* 1991;43:109–142.
16. Pollock JS, Forstermann U, Mitchell JA, Warner TD, Schmidt HH, Nakane M, Murad F. Purification and characterization of particulate endothelium-derived relaxing factor synthase from cultured and native bovine aortic endothelial cells. *Proc Natl Acad Sci USA* 1991;88:10480–10484.
17. Ignarro LJ, Buga GM, Wood KS, Byrns RE, Chaudhuri G. Endothelium-derived relaxing factor produced and released from artery and vein is nitric oxide. *Proc Natl Acad Sci USA* 1987;84:9265–9269.
18. Palmer RM, Ashton DS, Moncada S. Vascular endothelial cells synthesize nitric oxide from L-arginine. *Nature* 1988;333:664–666.
19. Marletta MA, Yoon PS, Iyengar R, Leaf CD, Wishnok JS. Macrophage oxidation of L-arginine to nitrite and nitrate: Nitric oxide is an intermediate. *Biochemistry* 1988;27:8706–8711.
20. Radomski MW, Palmer RM, Moncada S. An L-arginine/nitric oxide pathway present in human platelets regulates aggregation. *Proc Natl Acad Sci USA* 1990;87:5193–5197.
21. Szabo C, Thiemermann C. Regulation of the expression of the inducible isoform of nitric oxide synthase. *Adv Pharmacol* 1995;34:113–153.
22. Gryglewski RJ, Moncada S, Palmer RM. Bioassay of prostacyclin and endothelium-derived relaxing factor (EDRF) from porcine aortic endothelial cells. *Br J Pharmacol* 1986;87:685–694.
23. Moncada S, Palmer RM, Gryglewski RJ. Mechanism of action of some inhibitors of endothelium-derived relaxing factor. *Proc Natl Acad Sci USA* 1986;83:9164–9168.
24. Radomski MW, Palmer RM, Moncada S. Comparative pharmacology of endothelium-derived relaxing factor, nitric oxide and prostacyclin in platelets. *Br J Pharmacol* 1987;92:181–187.
25. Radomski MW, Palmer RM, Moncada S. Endogenous nitric oxide inhibits human platelet adhesion to vascular endothelium. *Lancet* 1987;2:1057–1058.
26. Radomski MW, Palmer RM, Moncada S. The role of nitric oxide and cGMP in platelet adhesion to vascular endothelium. *Biochem Biophys Res Commun* 1987;148:1482–1489.
27. Radomski MW, Palmer RM, Moncada S. The anti-aggregating properties of vascular endothelium: Interactions between prostacyclin and nitric oxide. *Br J Pharmacol* 1987;92:639–646.
28. Radomski MW, Palmer RM, Moncada S. Characterization of the L-arginine: Nitric oxide pathway in human platelets. *Br J Pharmacol* 1990;101:325–328.
29. Malinski T, Radomski MW, Taha Z, Moncada S. Direct electrochemical measurement of nitric oxide released from human platelets. *Biochem Biophys Res Commun* 1993;194:960–965.
30. Muruganandam A, Mutus B. Isolation of nitric oxide synthase from human platelets. *Biochim Biophys Acta* 1994;1200:1–6.
31. Chen LY, Mehta JL. Further evidence of the presence of constitutive and inducible nitric oxide synthase isoforms in human platelets. *J Cardiovasc Pharmacol* 1996;27:154–158.
32. Wallerath T, Gath I, Aulitzky WE, Pollock JS, Kleinert H, Forstermann U. Identification of the NO synthase isoforms expressed in human neutrophil granulocytes, megakaryocytes and platelets. *Thromb Haemost* 1997;77:163–167.
33. Berkels R, Stockklauser K, Rosen P, Rosen R. Current status of platelet NO synthases. *Thrombosis Research* 1997;87:51–55.
34. Freedman JE, Loscalzo J, Barnard MR, Alpert C, Keaney JF, Michelson AD. Nitric oxide released from activated platelets inhibits platelet recruitment. *J Clin Invest* 1997;100:350–356.
35. Mathur A, Robinson MS, Cotton J, Martin JF, Erusalimsky JD. Platelet reactivity in acute coronary syndromes: Evidence for differences in platelet behaviour between unstable angina and myocardial infarction. *Thromb Haemost* 2001;85:989–994.
36. Mehta JL, Conti CR. Aspirin in myocardial ischemia: Why, when, and how much? *Clin Cardiol* 1989;12:179–184.
37. Harrington RA. Overview of clinical trials of glycoprotein IIb-IIIa inhibitors in acute coronary syndromes. *Am Heart J* 1999;138:276–286.
38. Parratt JR. Pathophysiology of myocardial ischemia: Importance of platelet-vessel wall interactions. *Cardiovasc Drugs Ther* 1988;2:35–40.
39. Folts JD, Gallagher K, Rowe GG. Blood flow reductions in stenosed canine coronary arteries: Vasospasm or platelet aggregation? *Circulation* 1982;65:248–255.
40. Kowey PR, Verrier RL, Lown B, Handin RI. Influence of intracoronary platelet aggregation on ventricular electrical properties during partial coronary artery stenosis. *The American Journal of Cardiology* 1983;51:596–602.
41. Shimada Y, Kutsumi Y, Nishio H, Asazuma K, Tada H, Hayashi T, Nakai T, Morioka K. Role of platelets in myocardial ischemia-reperfusion injury in dogs. *Jpn Circ J* 1997;61:241–248.
42. Stein B, Fuster V. Role of platelet inhibitor therapy in myocardial infarction. *Cardiovasc Drugs Ther* 1989;3:797–813.
43. Fuster V. Role of platelets in the development of atherosclerotic disease and possible interference with platelet inhibitor drugs. *Scand J Haematol Suppl* 1981;38:1–38.
44. Willerson JT, Campbell WB, Winniford MD, Schmitz J, Apprill P, Firth BG, Ashton J, Smitherman T, Bush L, Buja LM. Conversion from chronic to acute coronary artery disease: Speculation regarding mechanisms. *The American Journal of Cardiology* 1984;54:1349–1354.

45. Willerson JT, Hillis LD, Winniford M, Buja LM. Speculation regarding mechanisms responsible for acute ischemic heart disease syndromes. *J Am Coll Cardiol* 1986;8:245–250.
46. Willerson JT. Conversion from chronic to acute coronary heart disease syndromes. Role of platelets and platelet products. *Tex Heart Inst J* 1995;22:13–19.
47. Willerson JT, Golino P, Eidt J, Campbell WB, Buja LM. Specific platelet mediators and unstable coronary artery lesions. Experimental evidence and potential clinical implications. *Circulation* 1989;80:198–205.
48. de_Belder AJ, MacAllister R, Radomski MW, Moncada S, Vallance PJ. Effects of S-nitroso-glutathione in the human forearm circulation: Evidence for selective inhibition of platelet activation. *Cardiovasc Res* 1994;28:691–694.
49. Radomski MW, Salas E. Nitric oxide–biological mediator, modulator and factor of injury: Its role in the pathogenesis of atherosclerosis. *Atherosclerosis* 1995;118 Suppl:S69–80.
50. Beckman J, Tsai J. Reactions and diffusion of nitric oxide and peroxynitrite. *The Biochemist* 1994;16:8–10.
51. Moro MA, Darley_Usmar VM, Goodwin DA, Read NG, Zamora_Pino R, Feelisch M, Radomski MW, Moncada S. Paradoxical fate and biological action of peroxynitrite on human platelets. *Proc Natl Acad Sci USA* 1994;91:6702–6706.
52. Villa LM, Salas E, Darley_Usmar VM, Radomski MW, Moncada S. Peroxynitrite induces both vasodilatation and impaired vascular relaxation in the isolated perfused rat heart. *Proc Natl Acad Sci USA* 1994;91:12383–12387.
53. Moro MA, Darley_Usmar VM, Lizasoain I, Su Y, Knowles RG, Radomski MW, Moncada S. The formation of nitric oxide donors from peroxynitrite. *Br J Pharmacol* 1995;116:1999–2004.
54. Brown AS, Moro MA, Masse JM, Cramer EM, Radomski M, Darley_Usmar V. Nitric oxide-dependent and independent effects on human platelets treated with peroxynitrite. *Cardiovasc Res* 1998;40:380–388.
55. Naseem KM, Low SY, Sabetkar M, Bradley NJ, Khan J, Jacobs M, Bruckdorfer KR. The nitration of platelet cytosolic proteins during agonist-induced activation of platelets. *FEBS Lett* 2000;473:119–122.
56. Patel VC, Yellon DM, Singh KJ, Neild GH, Woolfson RG. Inhibition of nitric oxide limits infarct size in the *in situ* rabbit heart. *Biochem Biophys Res Commun* 1993;194:234–238.
57. Woolfson RG, Patel VC, Neild GH, Yellon DM. Inhibition of nitric oxide synthesis reduces infarct size by an adenosine-dependent mechanism. *Circulation* 1995;91:1545–1551.
58. Curtis MJ, Pabla R. Nitric oxide supplementation or synthesis block—which is the better approach to treatment of heart disease? *Trends Pharmacol Sci* 1997;18:239–244.
59. Schulz R, Wambolt R. Inhibition of nitric oxide synthesis protects the isolated working rabbit heart from ischaemia-reperfusion injury. *Cardiovasc Res* 1995;30:432–439.
60. Siegfried MR, Erhardt J, Rider T, Ma XL, Lefer AM. Cardioprotection and attenuation of endothelial dysfunction by organic nitric oxide donors in myocardial ischemia-reperfusion. *The Journal of Pharmacology and Experimental Therapeutics* 1992;260:668–675.
61. Lefer DJ, Nakanishi K, Johnston WE, Vinten_Johansen J. Antineutrophil and myocardial protecting actions of a novel nitric oxide donor after acute myocardial ischemia and reperfusion of dogs. *Circulation* 1993;88:2337–2350.
62. Rossoni G, Berti M, Colonna VD, Bernareggi M, Del_Soldato P, Berti F. Myocardial protection by the nitroderivative of aspirin, NCX 4016: *In vitro* and *in vivo* experiments in the rabbit. *Ital Heart J* 2000;1:146–155.
63. Slawinski M, Grodzinska L, Kostka_Trabka E, Bieron K, Goszcz A, Gryglewski RJ. L-arginine—substrate for no synthesis—its beneficial effects in therapy of patients with peripheral arterial disease: Comparison with placebo-preliminary results. *Acta Physiol Hung* 1996;84:457–458.
64. Ceremuzynski L, Chamiec T, Herbaczynska-Cedro K. Effect of supplemental oral L-arginine on exercise capacity in patients with stable angina pectoris. *Am J Cardiol* 1997;80:331–333.
65. Conde_Pozzi I, Kleiman N. Platelet activation in acute myocardial infarction and the rationale for combination therapy. 2000;2:378–385.
66. Kleiman NS, Califf RM. Results from late-breaking clinical trials sessions at ACCIS 2000 and ACC 2000. American College of Cardiology. *J Am Coll Cardiol* 2000;36:310–325.
67. Lopez_Farre A, Riesco A, Digiuni E, Mosquera JR, Caramelo C, de_Miguel LS, Millas I, de_Frutos T, Cernadas MR, Monton M, Alonso J, and Casado S. Aspirin-stimulated nitric oxide production by neutrophils after acute myocardial ischemia in rabbits. *Circulation* 1996;94:83–87.
68. Coller BS. Platelet GP IIb/IIIa antagonists: The first anti-integrin receptor therapeutics. *J Clin Invest* 1997;99:1467–1471.
69. Coller BS. GP IIb/IIIa antagonists: Pathophysiologic and therapeutic insights from studies of c7E3 Fab. *Thromb Haemost* 1997;78:730–735.
70. Ghaffari S, Kereiakes DJ, Lincoff AM, Kelly TA, Timmis GC, Kleiman NS, Ferguson JJ, Miller DP, Califf RA, Topol EJ. Platelet glycoprotein IIb/IIIa receptor blockade with abciximab reduces ischemic complications in patients undergoing directional coronary atherectomy. EPILOG Investigators. Evaluation of PTCA to Improve Long-term Outcome by c7E3 GP IIb/IIIa Receptor Blockade. *Am J Cardiol* 1998;82:7–12.
71. The GUSTO V Investigators*. Reperfusion therapy for acute myocardial infarction with fibrinolytic therapy or combination reduced fibrinolytic therapy and platelet glycoprotein IIb/IIIa inhibition: The GUSTO V randomised trial. *Lancet* 2001;357:1905–1914.
72. Weaver WD, Simes RJ, Betriu A, Grines CL, Zijlstra F, Garcia E, Grinfeld L, Gibbons RJ, Ribeiro EE, DeWood MA, Ribichini F. Comparison of primary coronary angioplasty and intravenous thrombolytic therapy for acute myocardial infarction: A quantitative review. *JAMA* 1997;278:2093–2098.
73. Santoro G, Bolognese L. Coronary stenting and platelet glycoprotein IIb/IIIa receptor blockade in acute myocardial infarction. *Am Heart J* 2001;141:26–35.
74. Stone GW, Grines CL, Cox DA, Garcia E, Tcheng JE, Griffin JJ, Guagliumi G, Stuckey T, Turco M, Carroll JD, Rutherford BD, Lansky AJ. Comparison of angioplasty with stenting, with or without abciximab, in acute myocardial infarction. *N Engl J Med* 2002;346:957–966.
75. Adrie C, Bloch KD, Moreno PR, Hurford WE, Guerrero JL, Holt R, Zapol WM, Gold HK, Semigran MJ. Inhaled nitric oxide increases coronary artery patency after thrombolysis. *Circulation* 1996;94:1919–1926.
76. Lam JY, Chesebro JH, Fuster V. Platelets, vasoconstriction, and nitroglycerin during arterial wall injury. A

new antithrombotic role for an old drug. *Circulation* 1988;78:712–716.

77. Diodati J, Theroux P, Latour JG, Lacoste L, Lam JY, Waters D. Effects of nitroglycerin at therapeutic doses on platelet aggregation in unstable angina pectoris and acute myocardial infarction. *The American Journal of Cardiology* 1990;66:683–688.
78. Sinzinger H, Virgolini I, O_Grady J, Rauscha F, Fitscha P. Modification of platelet function by isosorbide dinitrate in patients with coronary artery disease. *Thrombosis Research* 1992;65:323–335.
79. Yusuf S, Collins R, MacMahon S, Peto R. Effect of intravenous nitrates on mortality in acute myocardial infarction: An overview of the randomised trials. *Lancet* 1988;1:1088–1092.
80. Horowitz JD, Henry CA, Syrjanen ML, Louis WJ, Fish RD, Smith TW, Antman EM. Combined use of nitroglycerin and N-acetylcysteine in the management of unstable angina pectoris. *Circulation* 1988;77:787–794.
81. GISSI-3 GISSI III study group. Effects of lisinopril and transdermal glyceryl trinitrate singly and together on 6-week mortality and ventricular function after acute myocardial infarction. *The Lancet* 1994;343:1115–1122.
82. ISIS-4 (1993) ISIS collaborative group, Oxford, U.K.: Randomised study of oral isosorbide mononitrate in over 50,000 patients with suspected acute myocardial infarction. *Circulation* 88:I–394.
83. Andrews R, May JA, Vickers J, Heptinstall S. Inhibition of platelet aggregation by transdermal glyceryl trinitrate. *Br Heart J* 1994;72:575–579.
84. Patrono C. Aspirin and human platelets: From clinical trials to acetylation of cyclooxygenase and back. *Trends Pharmacol Sci* 1989;10:453–458.
85. Salas E, Miszta-Lane H, Radomski M. Regulation of platelet function by nitric oxide and other nitrogen- and -oxygen-derived species. In: Von Bruchhausen F, Walter U, eds. *Handbook of Experimental Pharmacology*. Springer-Verlag, 1997:371–397.
86. Shahbazi T, Jones N, Radomski MW, Moro MA, Gingell D. Nitric oxide donors inhibit platelet spreading on surfaces coated with fibrinogen but not with fibronectin. *Thrombosis Research* 1994;75:631–642.
87. Levin RI, Weksler BB, Jaffe EA. The interaction of sodium nitroprusside with human endothelial cells and platelets: Nitroprusside and prostacyclin synergistically inhibit platelet function. *Circulation* 1982;66:1299–1307.
88. Hines R, Barash PG. Infusion of sodium nitroprusside induces platelet dysfunction *in vitro*. *Anesthesiology* 1989;70:611–615.
89. Willerson JT, Igo SR, Yao SK, Ober JC, Macris MP, Ferguson JJ. Localized administration of sodium nitroprusside enhances its protection against platelet aggregation in stenosed and injured coronary arteries. *Texas Heart Institute Journal* 1996;23:1–8.
90. Wautier JL, Weill D, Kadeva H, Maclouf J, Soria C. Modulation of platelet function by SIN-1A, a metabolite of molsidomine. *J Cardiovasc Pharmacol* 1989;14(suppl 11):S111–114.
91. Hogg N, Darley_Usmar VM, Wilson MT, Moncada S. The oxidation of alpha-tocopherol in human low-density lipoprotein by the simultaneous generation of superoxide and nitric oxide. *FEBS Lett* 1993;326:199–203.
92. Radomski MW, Rees DD, Dutra A, Moncada S. S-nitrosoglutathione inhibits platelet activation *in vitro* and *in vivo*. *Br J Pharmacol* 1992;107:745–749.
93. Langford EJ, Brown AS, Wainwright RJ, de_Belder AJ, Thomas MR, Smith RE, Radomski MW, Martin JF, Moncada S. Inhibition of platelet activity by S-nitrosoglutathione during coronary angioplasty. *Lancet* 1994;344:1458–1460.
94. Clancy RM, Levartovsky D, Leszczynska_Piziak J, Yegudin J, Abramson SB. Nitric oxide reacts with intracellular glutathione and activates the hexose monophosphate shunt in human neutrophils: Evidence for S-nitrosoglutathione as a bioactive intermediary. *Proc Natl Acad Sci USA* 1994;91:3680–3684.
95. Mayers I, Salas E, Hurst T, Johnson D, Radomski MW. Increased nitric oxide synthase activity after canine cardiopulmonary bypass is suppressed by s-nitrosoglutathione. *J Thorac Cardiovasc Surg* 1999;117: 1009–1016.
96. Minuz P, Zuliani V, Gaino S, Tommasoli R, Lechi A. NO-aspirins: Antithrombotic activity of derivatives of acetyl salicylic acid releasing nitric oxide. *Cardiovasc Drug Rev* 1998;16:31–47.
97. Wallace JL, Muscara MN, McKnight W, Dicay M, Del Soldato P, Cirino G. *In vivo* antithrombotic effects of a nitric oxide-releasing aspirin derivative, NCX-4016. *Thrombosis Research* 1999;93:43–50.

Nitric Oxide, Atherosclerosis and the Clinical Relevance of Endothelial Dysfunction

Todd J. Anderson, MD
Department of Medicine, University of Calgary, Calgary, AB, Canada

Abstract. **The endothelium plays a key role in vascular homeostasis through the release of a variety of autocrine and paracrine substances, the best characterized being nitric oxide. A healthy endothelium acts to prevent atherosclerosis development and its complications through a complex and favorable effect on vasomotion, platelet and leukocyte adhesion and plaque stabilization. The assessment of endothelial function in humans has generally involved the description of vasomotor responses, but more widely includes physiological, biochemical and genetic markers that characterize the interaction of the endothelium with platelets, leukocytes and the coagulation system. Stable markers of inflammation such as high sensitivity C-reactive protein are indirect and potentially useful measures of endothelial function for example.**

Attenuation of the effect of nitric oxide accounts for the majority of what is described as endothelial dysfunction. This occurs in response to atherosclerosis or its risk factors. Much remains to be learned about the molecular and genetic pathophysiological mechanisms of endothelial cell abnormalities. However, pharmacological intervention with a growing list of medications can favorably modify endothelial function, paralleling beneficial effects on cardiovascular morbidity and mortality. In addition, several small studies have provided tantalizing evidence that measures of endothelial health might provide prognostic information about an individual patient's risk of subsequent events. As such, the sum of this evidence makes the clinical assessment of endothelial function an attractive surrogate marker of atherosclerosis disease activity. The review will focus on the role of nitric oxide in atherosclerosis and the clinical relevance of these findings.

Key Words. **endothelium, nitric oxide, atherosclerosis, coronary artery disease, risk**

What is Endothelial Function?

The endothelium is the single-cell lining that covers the surface of blood vessels and numerous other structures. The strategic location that it occupies allows it to act as both a sensor and modulator within the vessel. While the term was first described by His in 1865, little was thought of its importance until the past 25 years with the key discovery of endothelium-derived autocrine factors such as prostacyclin and nitric oxide [1,2]. However, a review of older literature would suggest that researchers have questioned the importance of the endothelium for some time. In a monograph on the subject Altschul stated that "one is as old as one's endothelium" in the mid 1950s without a lot of data to support his claim [3].

Vasomotion

Our knowledge of the function of the endothelium has increased exponentially in the past decade. Broadly speaking, endothelial function refers to a physiological observation that is the result of *stimulation* of vasoactive substances released by or that interact with the vascular endothelium. More recently, *basal* endothelial function has been assessed by blockade of vasoactive substances (L-NMMA to block nitric oxide synthase). To date the majority of studies have reported endothelium-dependent vasomotion as the endpoint of interest. *In vitro* studies of arterial tone utilize pre-constricted tissue mounted on wire myographs or organ chambers where the vessel is stimulated with increasing concentrations of endothelium-dependent stimuli such as acetylcholine [2]. In humans, the measurement of vessel diameter or regional blood flow reflect vasomotion of conduit or resistance vessels respectively [4–7]. The measurement of endothelial function in humans has been reviewed in detail [8,9].

Platelet and Leukocytes

Changes in endothelial vasodilator function are likely paralleled by changes in the other properties of the endothelium (such as platelet and leukocyte adhesion) to cause a pro-atherosclerotic milieu [10]. Studies of classical platelet function have been reported to a limited degree [11]. More recently, flow cytometry has been used to evaluate the expression of a variety of integrins on both platelets and leukocytes [12–15]. The interaction of platelets and leukocytes generates circulating microparticles that can be measured and these

Address for correspondence: TJ Anderson, MD, FRCPC, Division of Cardiology, Foothills Hospital, 1432, 29th Street, NW, Calgary, AB, T2N 2T9, Canada. Tel.: 403 670-1020; Fax: 403 670-1592; E-mail: todd.anderson@calgaryhealthregion.ca have recently been demonstrated to impair endothelial function [16].

In addition, circulating levels of leukocyte adhesion molecules can be measured in the blood and have been associated with atherosclerosis and cardiac outcomes [17–20]. It was recently shown both that sICAM-1 and P-selectin was predictive of future myocardial infarction [21,22].

C-Reactive Protein

C-reactive protein (CRP) is an acute phase reactant marker produced by the liver in response to systemic inflammation. It is frequently elevated in patients with acute ischemia and myocardial infarction [23–26]. In addition, Ridker and colleagues have shown that an elevated level is a predictor for future MI or stroke [27]. It has prognostic value in a number of other conditions as well [27–34]. WHO standards have been set for high-sensitivity assessment of C-reactive protein providing consistent and reproducible results [35]. It now seems clear that CRP also plays an active role in the atherosclerosis process [36,37]. CRP should be regarded as an indirect, albeit important measure of endothelial function.

Fibrinolytic Balance

The main determinants of fibrinolysis are the balance between tissue plasminogen activator (t-PA) and plasminogen activator inhibitor (PAI-1), both of which are derived from the endothelium. Vaughan and colleagues have advanced the concept of the importance of PAI-1 as an important measure of vascular health [38–40]. Other measures of endothelial function are listed in Table 1.

Endothelium-Dependent Vasoactive Substances

While the review concentrates mainly on the role of nitric oxide in humans, it is important to note that other vasoactive substances control vascular tone in concert with NO. Two other endothelial-derived vasodilators, prostacyclin and endothelium-derived hyperpolarizing factor (EDHF) contribute significantly to arterial relaxation [1,41–44]. Duffy and colleagues have demonstrated that prostaglandins mediate both basal and metabolic vasomotion in the coronary circulation and resting blood flow in the forearm [43,44]. While the role of EDHF in humans is not clear [45], a recent human study suggested that in the human forearm, a cytochrome P_{450} product may function as an EDHF [46]. Animal studies have established the importance of K^+_{ATP} channels in the control of coronary blood flow [47,48]. In humans, blockade of K^+_{ATP} channels with glibenclamide decreases basal coronary blood flow [49].

Opposing these, is the potent vasoconstrictor endothelin-1 [50]. Basal release of ET-1 occurs in humans [51–53]. ET-1 acts on ET_A and ET_B receptors on vascular smooth muscle cells producing potent vasoconstriction [54]. ET_B receptors on endothelial cells mediate vasodilation through release of NO and PGI_2 [55]. Whereas circulating levels can be measured, ET-1 is predominantly a paracrine hormone, acting ablumինally. In a catheterization laboratory study, Kinlay et al. recently demonstrated that ET-1 activity is upregulated in atherosclerotic humans [56], demonstrating the importance of this hormone in coronary control. Other vasoconstrictors include angiotensin II and oxygen free radicals [57,58].

Nitric Oxide in Atherosclerosis

Basal release of NO has been demonstrated in the human coronary circulation [59]. Acetylcholine or serotonin cause endothelium-dependent vasodilation of epicardial coronary vessels, [6,60,61] an effect blocked in part by NO synthase inhibition

Table 1. *Measures of endothelial function*

Vasomotor	Platelet and leukocyte indices
Coronary vasomotion—QCA	Classic platelet aggregation
Coronary blood flow–doppler	Platelet activation—flow cytometry
Coronary blood flow—PET	Leukocyte adhesion—flow cytometry
Brachial ultrasound—flow-mediated vasodilation	Circulating leukocyte adhesion molecules sICAM, sVCAM, selectins
Forearm impedance plethysmography	Circulating cytokines—Interleukin –6
Skin laser doppler + iontophoresis	Circulating microparticles
Dorsal hand vein—Linear variable differential transducer	Circulating progenitor endothelial cells
Arterial compliance, pulse wave velocity and augmentation index	C-reactive protein
Pulse pressure evaluation	

QCA—quantitative coronary angiography; PET—positron emission tomography; sICAM—soluble intracellular adhesion molecule; sVCAM—soluble vascular cell adhesion molecule.

(L-NMMA) [62]. The remaining effect is due to acetylcholine-mediated release of EDHF and prostacyclin. Resistance vessel function can be determined *in vivo* by measuring blood flow [63,64]. Endothelium-dependent agonists result in an increase in blood flow in the human coronary circulation [7]. Whereas this effect has also been shown to be NO dependent [62,65], other mediators may also be involved in resistance vessel dilation and the control of CBF, particularly in response to metabolic stimuli [66,67].

Endothelial injury with resulting dysfunction is the initiating event in atherosclerosis [68], and plays an important role in the ischemic manifestations of coronary disease [69,70]. Atherosclerosis results in paradoxical epicardial vasoconstriction and attenuated increases in blood flow to endothelium-dependent stimuli [6,7,71]. In addition, Quyyumi et al. have demonstrated that basal NO activity (as reflected by decreases in CBF to the NOS inhibitor L-NMMA) is reduced in subjects with atherosclerosis [62,72,73]. However, studies from our laboratory have demonstrated preserved basal NO activity in subjects with up to moderate atherosclerosis [66].

Endothelial dysfunction is a systemic problem, with impairment of NO mediated vasodilation being demonstrated in the peripheral circulation of subjects with atherosclerosis or its risk factors [74]. Patients with coronary disease have impaired brachial artery flow-mediated vasodilation and a correlation exists between FMD and acetylcholine-mediated vasomotion in the coronary circulation [75]. Attenuation of resistance vessel function is also noted in studied utilizing impedance plethysmography [76]. Nitric oxide has been shown to mediate the effect of both flow and acetylcholine in the forearm [77,78].

Nitric oxide plays a key role in platelet and leukocyte function in humans. The infusion of the nitric oxide synthase inhibitor, L-NMMA decreases cGMP content in platelets obtained from the coronary sinus [79]. In addition, platelets from subjects with unstable angina release less nitric oxide than in those with stable angina [80]. The process of leukocyte adhesion is complex and beyond the scope of this review but it is clear that NO can inhibit leukocyte adhesion through a variety of mechanisms [81–83].

Endothelial Function Post Coronary Intervention

Spontaneous vasoconstriction (up to 40%) of the conduit vessel distal to the angioplasty site occurs within the first 30 minutes following intervention [84] and can be prevented with alpha-adrenergic stimulation [85,86]. Previous studies had demonstrated that this is a powerful predictor of restenosis in balloon angioplasty [87]. The relationship of this response to restenosis or adverse events in the stent era is not clear.

Prolonged endothelial dysfunction (up to 6 months) has been demonstrated in the distal epicardial artery particularly in patients undergoing coronary stenting [88]. We recently assessed coronary blood flow responses to acetylcholine early after coronary stenting. Attenuation of microvascular endothelial function was demonstrated and this was improved with concomitant treatment with the glycoprotein IIb/IIIa receptor blocker abciximab [89].

Platelet and neutrophil adhesion occurs following balloon injury in animal models [90]. This has also been demonstrated in coronary sinus blood from patients with stable angina undergoing coronary intervention [91]. Activation of platelets and neutrophils may be particularly marked with stent implantation contributing to endothelial dysfunction [13,92]. Platelets play an important role in neutrophil adhesion in this situation [93]. Activation of platelets and neutrophils contribute to vessel plugging and vasospasm accounting for ischemic events post stenting [94,95].

In addition, systemic elevation of leukocyte adhesion molecules has been observed following angioplasty in an acute myocardial infarction setting and in those without infarction [19]. It has been suggested that adhesion molecule expression may also be related to restenosis development [96,97], and acute ischemic events following angioplasty [98]. Liuzzo and colleagues reported that patients with unstable angina who have elevated baseline levels of CRP, had a significant further increase following coronary intervention. This was detected in peripheral blood 6 and 24 hours after angioplasty [26]. The same group also more recently reported on 121 patients undergoing single vessel angioplasty. Increasing tertiles of CRP were predictive of adverse events and clinical restenosis following balloon angioplasty [24]. More recently, larger studies have demonstrated a predictive role of CRP post coronary intervention [81–83,99,100].

Mechanisms of Impaired Nitric Oxide Activity

Nitric Oxide Synthase and Tetrahydrobiopterin

While there are many alterations in the mediators that control vascular function, abnormalities in NO and nitric oxide synthase (NOS) have been extensively studied. Nitric oxide activity is the net result of a balance between its production by NOS and its inactivation by oxygen free radicals [101]. NOS activity may be attenuated

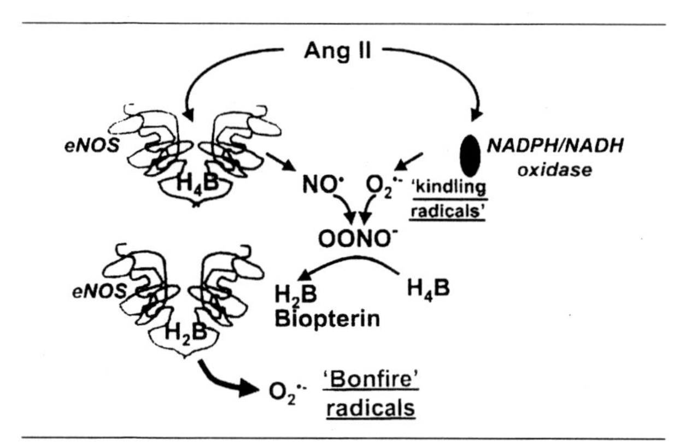

Fig. 1. Mechanisms whereby eNOS may become "uncoupled" leading to the production of superoxide instead of NO. Reproduced with permission [202].

in atherosclerotic vessels [102], but many studies have demonstrated an increase in the production of NO by NOS in the setting of cholesterol induced atherosclerosis [103]. It appears that both NO and superoxide production are increased by NOS with the net balance being a decrease in NO activity. This has lead to the concept of "uncoupling of NOS" as a major cause of endothelial dysfunction (Fig. 1) [104–106].

NO is synthesized from l-arginine in mammalian cells by a family of three NO synthases. NOS III or eNOS (endothelial NOS) is constitutively produced by endothelial cells and encoded by genes on chromosome 7 [107]. BH_4 is an essential cofactor for the proper flow of electrons to oxidize l-arginine [108]. At suboptimal concentrations of BH_4, NOS acts as an NADPH oxidase in which molecular oxygen rather than arginine becomes an electron acceptor. This leads to the production of superoxide ($\cdot O_2^-$) and hydrogen peroxide (H_2O_2) instead of NO [104,109–113]. It is probable that NOS III is continuously producing superoxide, but at suboptimal BH_4 concentrations this effect predominates.

Tetrahydrobiopterin. Thus, biopterin metabolism is critical for the regulation of NOS activity. In mammalian cells, BH_4 is synthesized through 2 distinct pathways: one is a de novo synthesis that uses GTP as a precursor via GTP cyclohydrolase I, and the other is the regeneration of BH_4 from a quinonoid form of BH_2 through a pterin salvage pathway (involving the active form of folate, 5 methyltetrahydrofolate). These pathways depend on a normal cellular redox state with oxidative stress impairing the recycling of BH_4 and are influenced by the actions of insulin [114–116]. Interestingly, it has been shown that ascorbic acid enhances NO synthase activity by increasing the concentration and stability of BH_4 [114–117].

Folic Acid and Nitric Oxide Synthase. The importance of BH_4 is further strengthened by recent studies that have demonstrated a beneficial effect of folate on endothelial function [118,119]. BH_4 is closely related in structure to other pterins including folate. In addition methylated folate plays a role in the regeneration of BH_4 through the salvage pathway. A recently published study by our group demonstrated that 5 methyltetrahydrofolate improved endothelial function and attenuated superoxide production by NOS (cell culture) to a similar degree as BH_4. Based on computer modeling of the NOS structure, it is possible that folates directly affect NOS function like BH_4 [120]. However, Stroes et al. have demonstrated that 5-MTHF was unable to alter NOS activity in enzyme that is completely devoid of pterin [121]. Further work is needed in this area to clarify the mechanisms of benefit with folates.

Genetic Abnormalities of NOS. Recent investigations have demonstrated that mutations affecting the eNOS gene may lead to impaired NO release. The Glu298Asp mutation is associated with an increased risk of coronary vasospasm and myocardial infarction in Japanese populations [122]. In a population of healthy Caucasian males we recently reported an association with the T-786-C mutation and hypertension [123]. Studies are underway to determine the relationship between these single nucleotide polymorphisms and endothelial function.

Oxidative Stress

In addition to NOS, there are many other sources of free radicals in the vessel wall. Steinberg and others have advanced the concept that oxidative modification of low-density lipoprotein (LDL) is central to the development of atherosclerosis and recent work has suggested that oxidized LDL plays an important role in abnormal endothelial vasorelaxation [124]. The detrimental effects of oxidized LDL on endothelial function are likely modulated through lysophosphatidylcholine [125,126], protein kinase C [127], and G-proteins [128]. In addition oxygen free radicals impair endothelial function through direct inactivation of NO [101]. Oxidized LDL also increases the production of endothelin in cultured cells and intact blood vessels [129].

Hypercholesterolemia induces a number of changes on vascular homeostasis, including a decrease in NO bioactivity, an increase in superoxide production, an increase in endothelin immunoreactivity [130], an increase in adhesion molecules [17], and attenuation of endothelium-dependent vasodilation [5]. It appears that cholesterol-induced endothelial dysfunction is related to the degree of LDL oxidation and not LDL concentration itself [131,132]. Acute elevations of free fatty acids and triglycerides can attenuate vasodilator responses over the course of a number of hours [133,134]. Acute hyperglycemia (6 hours) can also impair vasomotor responses [135], adding evidence to the belief that it is the milieu of the artery that determines vascular function and this can be modulated over minutes to hours.

Inflammation

Inflammatory cells are a potent source of oxygen free radicals. Microparticles generated from activated leukocytes and platelets stimulate endothelial cells to produce cytokines and adhesion molecules [136,137]. Boulanger et al. isolated circulating microparticles from patients with myocardial infarction and demonstrated impairment of endothelial NO transduction and endothelial dysfunction [16]. No effect was seen in non-ischemic patients.

As previously discussed, CRP is emerging as an important marker of atherosclerotic risk [138]. Recent studies suggest that CRP is not merely a nonspecific marker, but actively participates in lesion formation through induction of endothelial dysfunction and leukocyte activation [139–141]. Verma and colleagues recently demonstrated in cell culture experiments that human CRP upregulates adhesion molecule expression and MCP-1 production. These effect could be attenuated by endothelin and IL-6 blockade [36]. Ongoing work is looking at the effect of CRP on NO mediated events (Verma, personal communication 2002).

Treatment of Endothelial Dysfunction

This subject has been recently reviewed in detail [8]. Many therapeutic interventions that have targeted the endothelium have been evaluated in the recent past. Most have an impact on the balance between nitric oxide activity and oxidative stress. The parallel benefits on endothelial function and clinical events with several of the modalities argue strongly for the important role of endothelial health.

Lipid Lowering Therapy

The relationship between cholesterol and its subsequent lowering with statin therapy is well known. Ohara et al. [142,143] demonstrated that cholesterol feeding increases the production of superoxide anion from the endothelium of rabbit aorta, and that dietary lowering of cholesterol improves endothelial function and attenuates superoxide formation. Since then at least 20 studies have demonstrated the beneficial effects of lipid lowering on endothelial function in humans (Fig. 2) [144–146]. Whereas early studies treated patients for 6 months or more, Tamai [147]

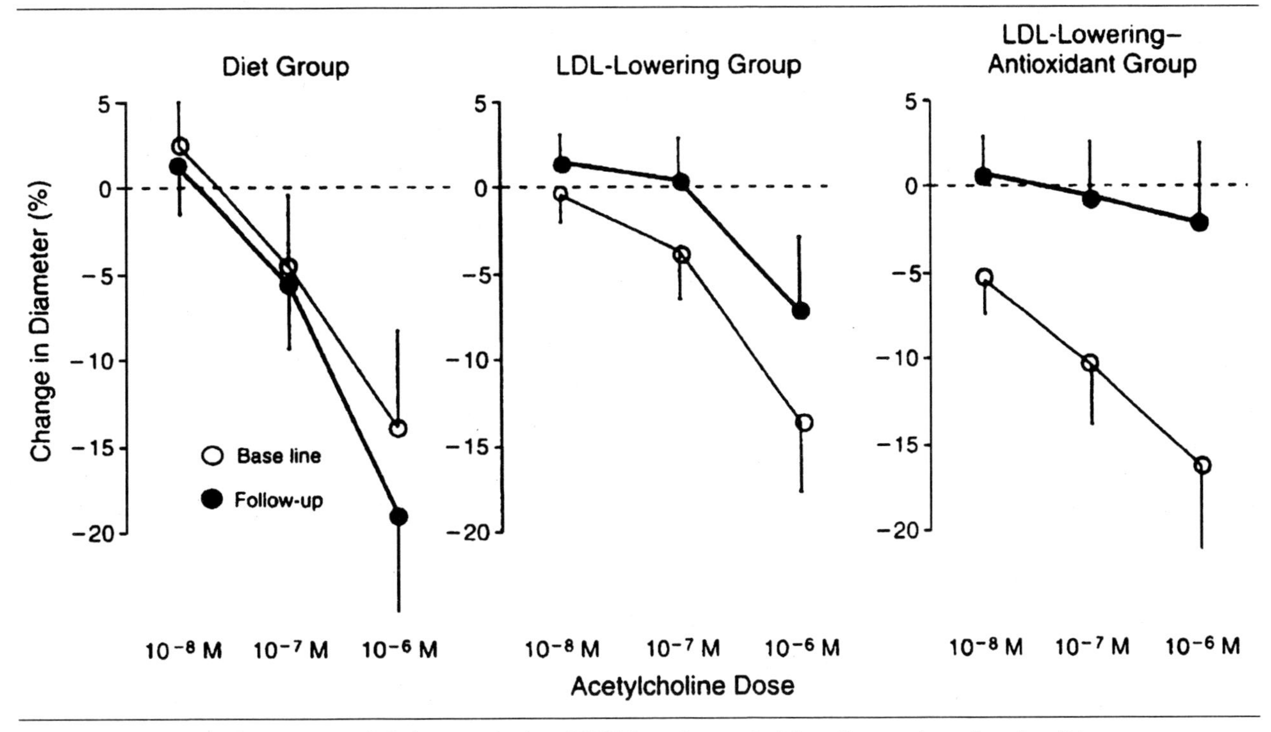

Fig. 2. *LDL lowering (lovastatin and cholestyramine) and LDL lowering-antioxidant (lovastatin and probucol) improve endothelium-dependent vasomotion after one year of therapy. Reproduced with permission [144].*

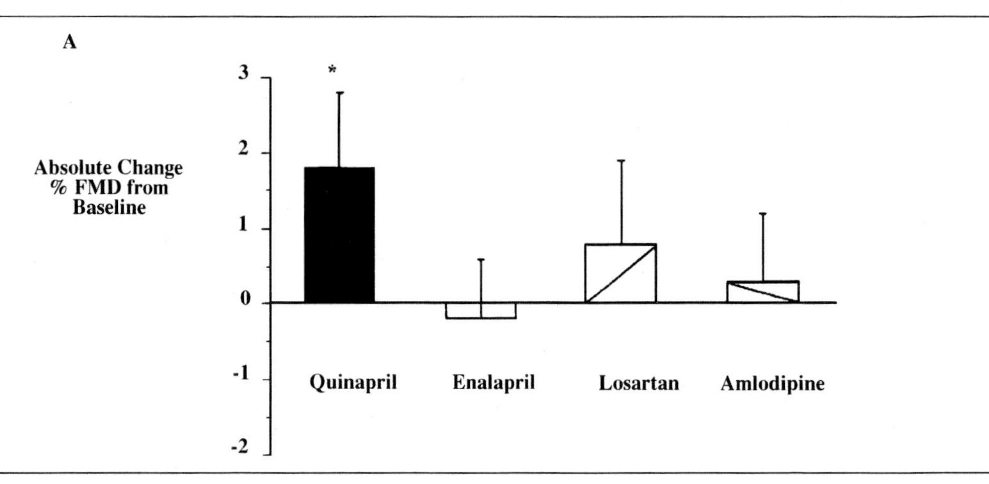

Fig. 3. 8 weeks of treatment with quinapril improves flow-mediated vasodilation in patients with coronary artery disease. Reproduced with permission [162].

modulated forearm endothelial function within hours of LDL pheresis, demonstrating the dynamic nature of the process. Whereas LDL cholesterol lowering with agents other than statins [148] improve endothelial function, there are clear pleiotrophic effects of the HMG Co-A reductase inhibitor that play an important role. These effects include but are not restricted to the following: direct effect on NOS [149], a decrease in caveolin [149], a favorable effect on inflammatory mediators within the plaque [150], mobilization of endothelial progenitor cells [151], a decrease in diabetes progression [152], and a decrease in platelet aggregation [11]. Statin therapy also decreases CRP in both primary and secondary prevention settings [153–155].

Spieker et al. infused rHDL into the forearm of healthy men and measured endothelium-dependent vasodilation to acetylcholine. The infusion of HDL resulted in improved endothelial function through an increase in the bioactivity of NO [156].

ACE-Inhibition

The HOPE trial firmly established ACE-inhibitors as an important anti-atherogenic therapeutic strategy [157]. Angiotensin II has been shown to increase superoxide production via membrane-bound NADH/NADPH [158]. ACE inhibitors could potentially improve endothelium-dependent vasodilator responses through decreased levels of angiotensin II, increased levels of bradykinin and NO.

In humans, acute administration of ACE-inhibitors augmented endothelium-dependent vasodilation in both the coronary and peripheral circulation [159,160]. Mancini and colleagues reported that the tissue specific ACE inhibitor quinapril attenuated coronary endothelial dysfunction in patients with coronary artery disease [161]. We extended these observations with quinapril to the peripheral circulation in the BANFF study [162] and suggested potential differences amongst ACE-inhibitors with respect to their effect on the endothelium (Fig. 3). Many other studies have demonstrated similar improvements with a wide range of risk factors and agents studied.

Angiotensin Receptor Blockers

While the data for ARBs is not as established as ACE-inhibitors, recent clinical trials have demonstrated the important effect of these agents on cardiorenal end-points in patients with hypertension and diabetes [163–165]. In parallel with these, there is some data to suggest that ARBs improve endothelium-dependent vasomotion. We were unable to show improved flow-mediated vasodilation in patients with coronary disease with losartan [162], however others have demonstrated a benefit with a higher dose [166]. Several studies have demonstrated attenuation of diabetic endothelial dysfunction with angiotensin receptor antagonist [167,168].

Modulation of Insulin Resistance

Obesity, insulin resistance and type II diabetes mellitus are dramatically increasing in Western society and are a huge risk for cardiovascular disease. Data from the UKPDS study suggest that in obese individuals insulin modulation with metformin might provide superior vascular protection to insulin or sulfonylureas [169], however much work needs to be done in this area. It is clear that both diabetes and insulin resistance are associated with impaired endothelium-dependent vasodilation [76,170]. Mather et al. has recently

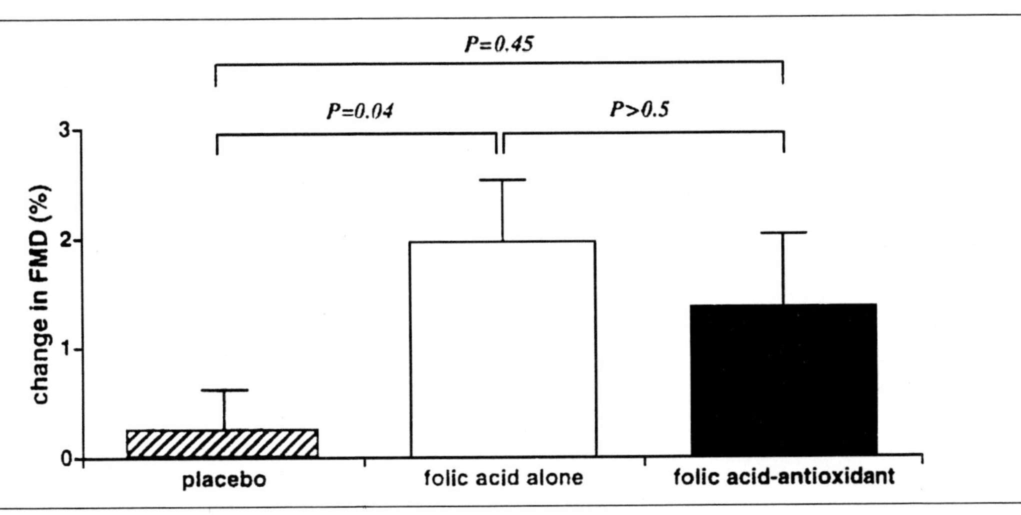

Fig. 4. *Folic acid alone improves flow-mediated vasodilation in patients with coronary artery disease. Reproduced with permission [182].*

demonstrated that metformin therapy in uncomplicated type II diabetic subjects markedly improves forearm blood flow as assessed by plethysmography. Acetycholine responses were related to insulin-resistance [171]. Further work is underway with insulin-sensitizing agents, particularly the thiazoladenediones.

Pterins

There is no data yet to demonstrate that folates reduce cardiovascular risk. The studies are currently being performed. Recent evidence suggests that BH_4 deficiency plays an important role in endothelial dysfunction, with emerging data in human. Our group has performed both *in vitro* and *in vivo* studies with BH_4. Acute incubation with BH_4 of internal thoracic arteries and saphenous veins from patients undergoing bypass surgery results in improvement in vasorelaxation to acetylcholine [172].

Only two previous studies have assessed the effect of BH_4 on endothelial function acutely in the catheterization laboratory. Setoguchi et al. demonstrated an improvement in acetylcholine-mediated CBF in subjects without coronary disease, but with endothelial dysfunction [173]. A second study demonstrated improvement in epicardial vasomotion to acetylcholine in subjects with coronary disease [174].

Recently, studies have been performed in the peripheral circulation. Stroes and colleagues measured forearm blood flow with impedance plethysmography in response to serotonin [175]. Coadministration of BH_4 improved basal and stimulated endothelial responses in subjects with hypercholesterolemia. Similar studies have been completed in cigarette smoker [176,177] and subjects with diabetes [178].

It is well known that folate improves endothelial function in patients with hyperhomocyst(e)inemia [179–183]. However, acute studies with 5-methyltetrahydrofolate (5-MTHF) suggest that the improvement in vascular function is independent of homocysteine and likely related to a direct effect of folic acid on eNOS and a reduction in superoxide production [121,184,185]. Folic acid will improve endothelial function in subjects with coronary artery disease despite relatively normal levels of homocysteine (Fig. 4) [182].

Others

A list of other endothelial modulating approaches is listed in Table 2.

Prognostic Relevance of Endothelial Function

The clinical manifestations of coronary artery disease depend on a multitude of interrelated pathophysiological processes of which endothelial dysfunction is only one. Inflammation and pertubations of vasoreactivity are key elements in the process leading to plaque instability and unstable coronary syndromes [69,186,187]. However, the relationship between our current markers of endothelial health and the status of the plaque are purely implied and I feel unproven. Abnormalities of vasomotion certainly contribute to stable angina in patients with coronary disease [188] and probably play a role in the symptoms of patients with microvascular angina and coronary vasospasm [189]. However, the burning question remains: What is the clinical relevance of markers of endothelial health?

Studies of Vasomotion

We demonstrated that endothelial dysfunction was associated with the development of

Table 2. Treatment associated with improvement of endothelial dysfunction in humans

Acute	Chronic
LDL Pheresis, fibrates	LDL lowering with statins
ACE-Inhibition	LDL lowering with resins
Antioxidants (Vitamin C and E + C)	ACE-Inhibition
Estrogen	Antioxidants (Probucol with lovastatin)
l-arginine, d-arginine	Estrogen
Tetrahydrobiopterin, Methyltetrahydrofolate	Estrogen + Progesterone
Deferoxamine	l-arginine
Glutathione	Exercise
Calcium-channel blockers	Folic acid
Angiotensin-receptor blockers	Angiotensin-receptor blockade
High density lipoprotein	Metformin
PKC inhibitors	
Abciximab	
TNF-alpha blockade—etancercept	
Endothelin-1 blockade	

atherosclerosis as assessed by intravascular ultrasound in patients post-cardiac transplantation [190]. Those subjects with normal vasodilator responses to acetylcholine immediately following transplantation developed atherosclerosis at a rate one third that of those with endothelial dysfunction during the first year of follow-up. Two trials have assessed the relationship between acetylcholine-mediated coronary endothelial function and clinical events [191,192]. Subjects with vasoconstrictor responses to acetylcholine were more likely to develop adverse cardiovascular events during follow-up of 5-10 years despite having minimal coronary disease at baseline. While important, the studies suffer from lack of events (10–16 events).

Perticone and colleagues studied 225 hypertensive subjects who underwent acetylcholine testing in the forearm with plethysmography [193]. After correcting for blood pressure, subjects with the lowest tertile of endothelial function had an increase in cardiovascular events over a 3 year follow-up. The most robust of the prognostic studies was recently reported. Heitzer et al. studied 281 subjects with coronary disease who underwent forearm acetylcholine studies [194]. Two key observations were made. Firstly, subjects with attenuated responses had more events. Secondly,

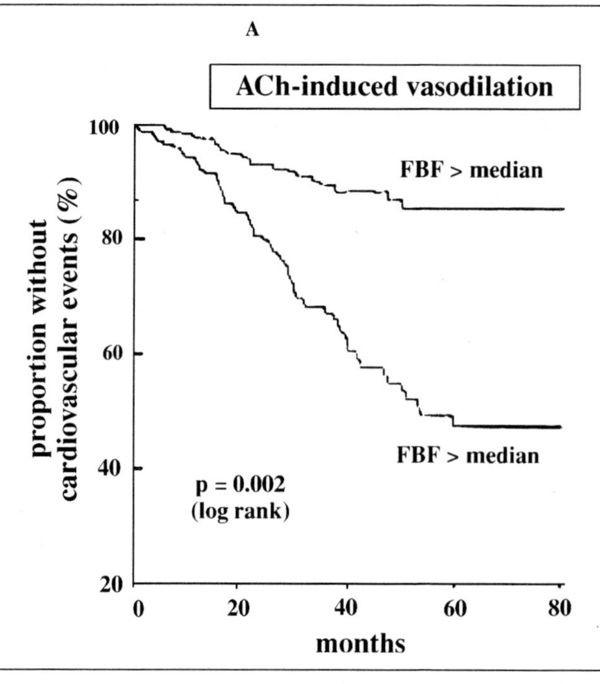

Fig. 5. Coronary artery disease subjects with impairment of acetylcholine mediated forearm blood flow (impedance plethysmography) are more likely to have an adverse event during follow-up. Reproduced with permission [194].

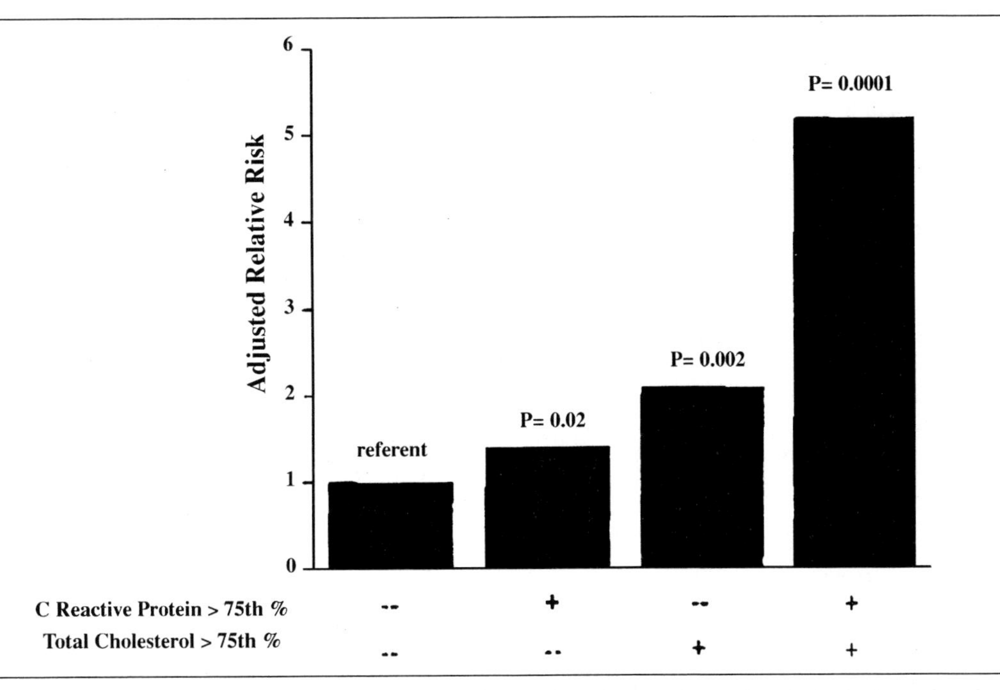

Fig. 6. *CRP is predictive of a first myocardial infarction in individuals from the Physician Health Study. Reproduced with permission [31].*

subjects with a greater acute improvement of endothelial function with vitamin C (suggesting more oxidative stress) had a worse outcome. This study was reasonably robust with 91 end-points (Fig. 5).

Prognostic studies have also suggested a role for brachial ultrasound assessment of flow-mediated vasodilation [195]. A recently reported study demonstrated that abnormalities of FMD were predictive of post-operative complications in patients undergoing non-cardiac vascular surgery [196].

All the studies done to date are too small to be definitive. Studies with thousands of subjects are underway to determine if a single measure of vasoreactivity in an *individual patient* predicts the development of atherosclerosis or its complications. (Vita 2002 personal communication) [197]. Until these are reported, measures of vasomotor responses should still be considered research tools.

Studies of Inflammation

Several systemic markers of inflammation have been assessed [198]. Studies by Ridker have demonstrated a relationship with cardiovascular outcomes with interleukin-6 [199], TNF-α [200], soluble P selectin [22], and soluble ICAM-1 [21]. However, CRP has emerged as the most promising of the inflammatory markers.

In both healthy men and women, CRP levels in the upper quartile increase risk by 2–4 fold and are at least as prognostic as lipid parameters (Fig. 6) [31]. Of great interest is the fact that subjects with elevated levels of CRP seem to gain more benefit from pharmacological therapies such as ASA or statin therapy [154,155,201]. It may be that CRP would add benefit to current risk assessment models for targeting therapy in subjects at moderate risk. Controlled trials of this type of approach are clearly warranted.

Conclusion

In the past two decades our appreciation of the role of nitric oxide in vascular homeostasis has blossomed. Endothelial function can now be readily measured in humans and is a very useful *research* tool to assess the impact of risk factors and their treatment on vascular function. A growing list of therapeutic modalities have been shown to modulate endothelial dysfunction which has important implications for the treatment of subjects at risk of developing atherosclerotic complications. In the past year, several small studies have suggested a prognostic role for measures of endothelium-dependent vasomotion. CRP is on the verge of being incorporated into clinical use to help with risk stratification. Much work remains to be done. However, it is probable that endothelial function testing will assume a prominent role in the evaluation and treatment of patients at risk of developing coronary atherosclerosis and its sequelae.

Acknowledgments

Dr. Anderson is a Scholar of the Alberta Heritage Foundation for Medical Research (Edmonton, Alberta). and the work discussed in this review has been supported in part by the Alberta Heart and Stroke Foundation (Edmonton, Alberta), the Canadian Institutes of Health Research and the Canadian Diabetes Association. The author further acknowledges the students, research staff and colleagues that have contributed to the work described in this review.

References

1. Bunting S, Gryglewski RJ, Moncada S, Vane JS. Arterial walls generate from prostaglandin endoperoxides a substance (prostaglandin X) which relaxes strips of mesenteric and coeliac arteries and inhibits its platelet aggregation. *Prostaglandins* 1976;12:897–913.
2. Furchgott RF, Zawadski JV. The obligatory role of endothelial cells in the relaxation of arterial smooth muscle by acetylcholine. *Nature* 1980;288:373–376.
3. Altschul R. *Endothelium*. New York: The MacMillan Company, 1954, pp. 1–155.
4. Celermajer DS, Sorensen KE, Gooch VM, Spiegelhalter DJ, Miller OI, Sullivan ID, Lloyd JK, Deanfield JE. Non-invasive detection of endothelial dysfunction in children and adults at risk of atherosclerosis. *Lancet* 1992;340:1111–1115.
5. Creager MA, Cooke JP, Mendelsohn ME, Gallagher SJ, Coleman SM, Loscalzo J, Dzau VJ. Impaired vasodilation of forearm resistance vessels in hypercholesterolemic humans. *J Clin Invest* 1990;86:228–234.
6. Ludmer PL, Selwyn AP, Shook TL, Wayne RR, Mudge GH, Alexander RW, Ganz P. Paradoxical vasoconstriction induced by acetylcholine in atherosclerotic coronary arteries. *N Engl J Med* 1986;315:1046–1051.
7. Zeiher AM, Drexler H, Wollschlaeger H, Just H. Endothelial dysfunction of the coronary microvasculature is associated with impaired coronary blood flow regulation in patients with early atherosclerosis. *Circulation* 1991;84:1984–1992.
8. Anderson TJ. Assessment and treatment of endothelial dysfunction in humans. *J Am Coll Cardiol* 1999;34:631–638.
9. Farouque HM, Meredith IT. The assessment of endothelial function in humans. *Coron Artery Dis* 2001;12:445–454.
10. Diodati JG, Dakak N, Gilligan DM, Quyyumi AA. Effect of atherosclerosis on endothelium-dependent inhibition of platelet activation in humans. *Circulation* 1998;98:17–24.
11. Lacoste L, Lam JYT, Hung J, Letchacovski G, Solymoss CB, Waters D. Hyperlipidemia and coronary disease: Correction of the increased thombogenic potential with cholesterol reduction. *Circulation* 1995;92:3172–3177.
12. Mickelson JK, Ali MN, Kleiman NS, Lakkis NM, Chow TC, Hughes BJ, Smith CW. Chimeric 7E3 Fab (Reopro) decreases detectabnle CD11b on neutrophils from patients undergoing coronary angioplasty. *J Am Coll Cardiol* 1999;33:97–106.
13. Gawaz M, Neumann FJ, Ott I, May A, Schomig A. Platelet activation and coronary stent implantation: Effect of antithrombotic therapy. *Circulation* 1996;94:279–285.
14. Gawaz M, Neumann FJ, Schomig A. Evaluation of platelet membrane glycoproteins in coronary artery disease. *Circulation* 1999;99:e1–e11.
15. Mickelson JK, Lakkis NM, Villarreal-Levy G, Hughes BJ, Smith CW. Leukocyte activation with platelet adhesion after coronary angioplasty: A mechanism for recurrent disease? *J Am Coll Cardiol* 1996;28:345–353.
16. Boulanger CM, Scoazec A, Ebrahimian T, Henry P, Mathieu E, Tedgui A, Mallat Z. Circulating microparticles from patients with myocardial infarction cause endothelial dysfunction. *Circulation* 2001;104:2649–2652.
17. Davi G, Romano M, Mezzetti A, Procopio A, Iacobelli S, Antiodormi T, Bucciarelli T, Alessandrini P, Cuccurullo F, Bittolo Bon G. Increased levels of soluble P-selectin in hypercholesterolemic patients. *Circulation* 1998;97:953–957.
18. Sampietro T, Tuoni M, Ferdeghini M, Ciardi A, Marraccini P, Prontera C, Sassi G, Taddei M, Bionda A. Plasma cholesterol regulates soluble cell adhesion molecule expression in familial hypercholesterolemia. *Circulation* 1997;96:1381–1385.
19. Siminiak T, Dye JF, Egdell RM, More R, Wysocki H, Sheridan DJ. The release of soluble adhesion molecules ICAM-1 and E-selectin after acute myocardial infarction and following coronary angioplasty. *International Journal of Cardiology* 1997;61:113–118.
20. Smith CW. Potential significance of circulating E-selectin [editorial; comment]. *Circulation* 1997;95:1986–1988.
21. Ridker PM, Hennekens C, Roitman-Johnson B, Stampfer MJ, Allen J. Plasma concentration of soluble intercellular adhesion molecule 1 and risks of future myocardial infarction in apparently healthy men. *Lancet* 1998;351:88–92.
22. Ridker PM, Buring JE, Rifai N. Soluble P-selectin and the risk of future cardiovascular events. *Circulation* 2001;103:491–495.
23. Biasucci LM, Liuzzo G, Grillo RL, Caligiuri G, Rebuzzi AG, Buffon A, Summaria F, Ginnetti F, Fadda G, Maseri A. Elevated levels of C-reactive protein at discharge in patients with unstable angina predict recurrent instability. *Circulation* 1999;99:855–860.
24. Buffon A, Liuzzo G, Biasucci LM, Pasqualetti P, Ramazzotti V, Rebuzzi AG, Crea F, Maseri A. Preprocedural serum levels of C-reactive protein predict early complications and late restenosis after coronary angioplasty. *J Am Coll Cardiol* 1999;34:1512–1521.
25. Liuzzo G, Biasucci LM, Gallimore JR, Caligiuri G, Buffon A, Rebuzzi AG, Pepys MB, Maseri A. Enhanced inflammatory response in patients with preinfarction unstable angina. *J Am Coll Cardiol* 1999;34:1696–1703.
26. Liuzzo G, Buffon A, Biasucci LM, Gallimore JR, Caligiuri G, Vitelli A, Altamura S, Ciliberto G, Rebuzzi AG, Crea F, Pepys MB, Maseri A. Enhanced inflamatory response to coronary angioplasty in patients with severe unstable angina. *Circulation* 1998;98:2370–2376.
27. Ridker PM, Cushman M, Stampfer MJ, Tracy R, Hennekens C. Plasma concentration of C-reactive protein and risks of developing peripheral vascular disease. *Circulation* 1997;97:425–428.
28. Ferreiros ER, Boissonnet CP, Pizarro R, Merletti PF, Corrado G, Cagide A, Bazzino OO. Independent prognostic value of elevated C-reactive protein in unstable angina. *Circulation* 1999;100:1958–1963.
29. Koenig W, Sund M, Frohlich M, Fischer HG, Lowel H, Doring A, Hutchinson WL, Pepys MB. C-reactive protein,

a sensitive marker of inflammation, predicts future risk of coronary heart disease in initially healthy middle-aged men: Results from the MONICA (Monitoring trends and determinants in cardiovascular disease) augsburg cohort study, 1984 to 1992 [In Process Citation]. *Circulation* 1999;99:237–242.
30. Morrow DA, Rifai N, Antman EM, Weiner DL, McCabe CH, Cannon CP, Braunwald E. C-reactive protein is a potent predictor of mortality independently of an in combination with troponin T in acute coronary syndromes: A TIMI 11A Substudy. *J Am Coll Cardiol* 1998;31:1460–1465.
31. Ridker PM, Glynn RJ, Hennekens C. C-reactive protein addes to the predictive value of total and HDL cholesterol in determining risk of first myocardial infarction. *Circulation* 1998;97:2007–2011.
32. Ridker PM, Rifai N, Pfeffer MA, Sacks FM, Moye LA, Goldman S, Braunwald E, for the CARE Investigators. Inflammation, pravastatin, and the risk of coronary events after myocardial infarction in patients with average cholesterol levels. *Circulation* 1998;98:839–844.
33. Ridker PM, Hennekens CH, Rifai N, Buring JE, Manson JE. Hormone replacement therapy and increased plasma concentration of C-reactive protein. *Circulation* 1999;100:713–716.
34. Roivainen M, Viik-Kajander M, Palosuo T, Toivanen P, Leinonen M, Saikku P, Tenkanen L, Manninen V, Hovi T, ntt. Infections, inflammation, and the risk of coronary heart disease. *Circulation* 2000;101:252–257.
35. Eda S, Kaufmann J, Roos W, Pohl S. Development of a new microparticle-enhanced turbidimetric assay for C-reactive protein with superior features in analytical sensitivity and dynamic range. *J Clin Lab Anal* 1998;12:137–144.
36. Verma S, Li SH, Badiwala MV, Weisel RD, Fedak PWM, Li RK, Dhillon B, Mickle DAG. Endothelin antagonism and interleukin-6 inhibition attenuate the proatherogenic effects of C-reactive protein. *Circulation* 2002;105:1890–1896.
37. Yeh ETH, Anderson HV, Pasceri V, Willerson JT. C-reactive protein: Linking inflammation to cardiovascular complications. *Circulation* 2001;104:974–975.
38. Brown NJ, Gainer JV, Murphey LJ, Vaughan DE. Bradykinin stimulates tissue plasminogen activator release from human forearm vasculature through B2 receptor-dependent, NO synthase-independent, and cyclooxygenase-independent pathway. *Circulation* 2000;102:2190–2196.
39. Vaughan DE, Lazos SA, Tong K. Angiotensin II regulates the expression of plasminogen activator inhibitor-1 in cultured endothelial cells. A potential link between the renin-angiotensin system and thrombosis. *Journal of Clinical Investigation* 1995;95:995–1001.
40. Vaughan DE. The renin-angiotensin system and fibrinolysis. [Review] [29 refs]. *American Journal of Cardiology* 1997;79:12–16.
41. Cohen RA, Vanhoutte PM. Endothelium-dependent hyperpolarization: Beyond nitric oxide and cyclic GMP. *Circulation* 1995;92:3337–3349.
42. Hui D, Waldron GJ, Galipeau D, Cole WC, Triggle CR. NO/PGIx-independent vasorelaxation and the cytochrome P450 pathway in rabbit carotid artery. *British Journal of Pharmacology* 1997;120:695–701.
43. Duffy SJ, Castle SF, Harper RW, Meredith IT. Contribution of vasodilator prostanoids and nitric oxide to resting flow, metabolic vasodilation, and flow-mediated dilation in human coronary circulation. *Circulation* 1999;100:1951–1957.
44. Duffy SJ, Tran BT, New G, Tudball RN, Esler MD, Harper RW, Meredith IT. Continuous release of vasodilator prostanoids contributes to regulation of resting forearm blood flow in humans. *American Journal of Physiology* 1998;274:t–83.
45. McGuire JJ, Ding H, Triggle CR. Endothelium-derived relaxing factors: A focus on endothelium-derived hyperpolarizing factor(s). *Can J Physiol Pharmacol* 2001;79:443–470.
46. Halcox JP, Narayanan S, Cramer-Joyce L, Mincemoyer R, Quyyumi AA. Characterization of endothelium-derived hyperpolarizing factor in the human forearm microcirculation. *Am J Physiol Heart Circ Physiol* 2001;280:H2470–H2477.
47. Duncker DJ, Van Zon NS, Altman JD, Pavek TJ, Bache RJ. Role of K + ATP channels in coronary vasodilation during exercise. *Circulation* 1993;88:1245–1253.
48. Duncker DJ, Van Zon NS, Ishibashi Y, Bache RJ. Role of K^{+}_{ATP} channels and adenosine in the regulation of coronary blood flow during exercise with normal and restricted coronary blood flow. *J Clin Invest* 1996;97:996–1009.
49. Farouque HM, Worthley SG, Meredith IT, Skyrme-Jones RA, Zhang MJ. Effect of ATP-sensitive potassium channel inhibition on resting coronary vascular responses in humans. *Circ Res* 2002;90:231–236.
50. Yanagisawa M, Kurihara H, Kimura A, Tomobe Y, Kabayashi M, Mitsui Y, Yazaki Y, Goto K, Masaki T. A novel potent vasoconstrictor peptide produced by vascular endothelial cells. *Nature* 1988;332:411–415.
51. Haynes WG, Ferro CJ, O'Kane KPJ, Somerville D, Lomax CC, Webb DJ. Systemic endothelin receptor blockade decreases peripheral vascular resistance and blood pressure in humans. *Circulation* 1996;93:1860–1870.
52. Richard V, Hogie M, Clozel M, Loffler B-M, Thuillez C. *In vivo* evidence of an endothelin-induced vasopressor tone after inhibition of nitric oxide synthesis in rats. *Circulation* 1995;91:771–775.
53. Wenzel RR, Fleisch M, Shaw S, Noll G, Kaufmann U, Schmitt R, Jones CR, Clozel M, Meier B, Luscher TF. Hemodynamic and coronary effects of the endothelin antagonist bosentan in patients with coronary artery disease. *Circulation* 1998;98:2235–2240.
54. Davenport AP, O'Reilly G, Molenaar P, Maguire JJ, Kuc RE, Sharkey A, Bacon CR, Ferro A. Human endothelin receptors characterized using reverse-transcriptase-polymerase chain reaction, *in situ* hybridization, and subtype-selective ligands BQ 123 and BQ 3020: Evidence for expression of ETb receptors in human vascular smooth muscle. *J Cardiovasc Pharmacol* 1993;22(suppl 8):S22–S25.
55. Tsukahara H, Ende H, Magazine HI, Bahou WF, Goligorsky MS. Molecular and functional characterization of the non-isopeptide-selective ETb receptor in endothelial cells: Receptor coupling to nitric oxide synthase. *J Biol Chem* 1994;269:21778–21785.
56. Kinlay S, Behrendt D, Wainstein M, Beltrame J, Fang JC, Creager MA, Selwyn AP, Ganz P. Role of endothelin-1 in

the active constriction of human atherosclerotic coronary arteries. *Circulation* 2001;104:1114–1118.
57. Dzau VJ, Burt DW, Pratt RE. Molecular biology to the renin-angiotensin system. *American Journal of Physiology* 1988;255:F563–F573.
58. Dzau VJ, Re R. Tissue angiotensin system in cardiovascular medicine: A paradigm shift. *Circulation* 1994;89:493–498.
59. Lefroy DC, Crake T, Uren NG, Davies GJ, Maseri A. Effect of inhibition of nitric oxide synthesis on epicardial coronary artery caliber and coronary blood flow in humans. *Circulation* 1993;88:43–54.
60. Golino P, Piscione F, Willerson JT, Cappelli-Bigazzi M, Focaccio A, Villari B, Indolfi C, Russolillo E, Condorelli M, Chiariello M. Divergent effects of serotonin on coronary artery dimensions and blood flow in patients with coronary atherosclerosis and control patients. *N Engl J Med* 1991;324:641–648.
61. Zeiher AM, Drexler H, Wollschlager H, Just H. Modulation of coronary vasomotor tone in humans: Progressive endothelial dysfunction with different early stages of coronary atherosclerosis. *Circulation* 1991;83:391–401.
62. Quyyumi AA, Dakak N, Andrews NP, Hussain S, Arora S, Gilligan DM, Panza JA, Cannon RO3. Nitric oxide activity in the human coronary circulation. *J Clin Invest* 1995;95:1747–1755.
63. Doucette JW, Corl PD, Payne HM, Flynn AE, Goto M, Nassi M, Segal J. Validation of a doppler guide wire for intravascular measurement of coronary artery flow velocity. *Circulation* 1992;85:1899–1911.
64. Uren NG, Marraccini P, Gistri R, de Silva R, Camici PG. Altered coronary vasodilator reserve and metabolism in myocardium subtended by normal arteries in patients with coronary artery disease. *Journal of the American College of Cardiology* 1993;22:650–658.
65. Vallance P, Collier JG, Moncada S. Effects of endothelium-derived nitric oxide on peripheral arteriolar tone in man. *Lancet* 1989;160:881–886.
66. Goodhart DM, Anderson TJ. Coronary arterial vasomotion: The role of nitric oxide and the influence of coronary atherosclerosis and its risks. *Am J Cardiol* 1998;82:1034–1039.
67. Quyyumi AA, Dakak N, Andrews NP, Gilligan DM, Panza JA, Cannon RO3. Contribution of nitric oxide to metabolic coronary vasodilation in the human heart. *Circulation* 1995;92:320–326.
68. Ross R. The pathogenesis of atherosclerosis: A perspective for the 1990's. *Nature* 1993;362:801–809.
69. Libby P. Current concepts of the pathogenesis of the acute coronary syndromes. *Circulation* 2001;104:365–372.
70. Meredith IT, Anderson TJ, Uehata A, Yeung AC, Selwyn AP, Ganz P. Role of endothelium in ischemic coronary syndromes. [Review]. *American Journal of Cardiology* 1993;72:27C-31C; discussion 31C.
71. Egashira K, Inou T, Hirooka Y, Yamada A, Maruoka Y, Kai H, Sugimachi M, Suzuki S, Takeshita A. Impaired coronary blood flow response to acetylcholine inpatients with coronary risk factors and proximal atherosclerotic lesions. *J Clin Invest* 1993;91:29–37.
72. Prasad A, Narayanan S, Waclawiw MA, Epstein N, Quyyumi AA. The insertion/deletion polymorphism of the angiotensin-converting enzyme gene determines coronary vascular tone and nitric oxide activity. *J Am Coll Cardiol* 2000;36:1579–1586.
73. Quyyumi AA, Dakak N, Mulcahy D, Andrews NP, Husain S, Panza JA, Cannon RO3. Nitric oxide activity in the atherosclerotic human coronary circulation. *J Am Coll Cardiol* 1997;29:308–317.
74. Anderson TJ, Gerhard MD, Meredith IT, Charbonneau F, Delagrange D, Creager MA, Selwyn AP, Ganz P. Systemic nature of endothelial dysfunction in atherosclerosis. [Review]. *American Journal of Cardiology* 1995;75:71B–74B.
75. Anderson TJ, Uehata A, Gerhard MD, Meredith IT, Knab S, Delagrange D, Lieberman EH, Ganz P, Creager MA, Yeung AC. Close relation of endothelial function in the human coronary and peripheral circulations. *J Am Coll Cardiol* 1995;26:1235–1241.
76. Williams SB, Cusco JA, Roddy M-A, Johnstone M, Creager MA. Impaired nitric oxide-mediated vasodilation in humans with non-insulin-dependent diabetes mellitus. *JACC* 1996;27:567–574.
77. Vallance P. Use of L-arginine and its analogs to study nitric oxide pathway in humans. *Methods Enzymol* 1996;269:453–459.
78. Mullen MJ, Kharbanda RK, Cross J, Donald AE, Taylor M, Vallance P, Deanfield JE, MacAllister RJ. Heterogenous nature of flow-mediated dilatation in human conduit arteries *in vivo*: Relevance to endothelial dysfunction in hypercholesterolemia. *Circ Res* 2001;88:145–151.
79. Andrews NP, Husain M, Dakak N, Quyyumi AA. Platelet inhibitory effect of nitric oxide in the human coronary circulation: Impact of endothelial dysfunction. *J Am Coll Cardiol* 2001;37:510–516.
80. Freedman JE, Ting B, Hankin B, Loscalzo J, Keaney JFJ, Vita JA. Impaired platelet production of nitric oxide predicts presence of acute coronary syndromes [In process citation]. *Circulation* 1998;98:1481–1486.
81. Freedman JE, Loscalzo J. Platelet-monocyte aggregates: Bridging thrombosis and inflammation. *Circulation* 2002;105:2130–2132.
82. Kubes P, Suzuki M, Granger DN. Nitric oxide: An endogenous modulator of leukocyte adhesion. *Proc Natl Acad Sci USA* 1991;88:4651–4655.
83. Kubes P. Polymorphonuclear leukocyte-endothelium interactions: A role for pro-inflammatory and anti-inflammatory molecules. *Can J Physiol Pharmacol* 1995;71:88–97.
84. Fischell TA, Bausback KN, McDonald TV. Evidence for altered epicardial coronary artery autoregulation as a cause of distal coronary vasoconstriction after successful percutaneous transluminal coronary angioplasty. *J Clin Invest* 1990;86:575–584.
85. Gregorini L, Fajadet J, Robert G, Cassagneau B, Bernis M, Marco J. Coronary vasoconstriction after percutaneous transluminal coronary angioplasty is attenuated by anti-adrenergic drugs. *Circulation* 1994;90:895–907.
86. Gregorini L, Marco J, Palombo C, Kozakova M, Anguissola JB, Cassagneau B, Bernies M, Distante A, Marco I, Fajadet J, Zanchetti A. Postischemic left ventricular dysfunction is abolished by alpha-adrenergic blocking agents. *J Am Coll Cardiol* 1998;31:992–1001.
87. Rodriguez A, Santaera O, Larribau M, Fernandez M, Sarmiento R, Perez Balino N, Newell JB, Roubin G, Palacios IF. Coronary stenting decreases restenosis in lesions with early loss in luminal diameter 24 hours after successful PTCA. *Circulation* 1995;91:1397–1402.
88. Caramori PRA, Lima VC, Seidelin PH, Newton GE, Parker JD, Adelman AG. Long-term endothelial

dysfunction after coronary artery stenting. *J Am Coll Cardiol* 1999;34:1675–1679.
89. Aymong ED, Curtis MJ, Youssef M, Graham MM, Shewchuk L, Leschuk W, Anderson TJ. Abciximab attenuates coronary microvascular endothelial dysfunction following coronary stenting. *Circulation* 2002 (In press).
90. Merhi Y, Guidoin R, Provost P, Leung TK, Lam JY. Increase of neutrophil adhesion and vasoconstriction with platelet deposition after deep arterial injury by angioplasty. *American Heart Journal* 1995;129:445–451.
91. Serrano CVJ, Ramires JA, Venturinelli M, Arie S, D'Amico E, Zweier JL, Pileggi F, da Luz PL. Coronary angioplasty results in leukocyte and platelet activation with adhesion molecule expression. Evidence of inflammatory responses in coronary angioplasty. *Journal of the American College of Cardiology* 1997;29:1276–1283.
92. van Beusekom HMM, Whelan DM, Hofma SH, Krabbendam SC, van Hinsbergh VWM, Verdouw PD, van der Giessen WJ. Long-term endothelial dysfunction is more pronounced after stenting than after balloon angioplasty in porcine coronary arteries. *J Am Coll Cardiol* 1998;32:1109–1117.
93. Merhi Y, Provost P, Guidoin R, Latour JG. Importance of platelets in neutrophil adhesion and vasoconstriction after deep carotid arterial injury by angioplasty in pigs. *Arterioscler Thromb Vasc Biol* 1997;17:1185–1191.
94. Jang Y, Lincoff MA, Plow EF, Topol EJ. Cell adhesion molecules in coronary artery disease. *J Am Coll Cardiol* 1994.
95. Topol EJ, Serruys PW. Frontiers in interventional cardiology. *Circulation* 1998;98:1802–1820.
96. Belch JJ, Shaw JW, Kirk G, McLaren M, Robb R, Maple C, Morse P. The white blood cell adhesion molecule E-selectin predicts restenosis in patients with intermittent claudication undergoing percutaneous transluminal angioplasty [see comments]. *Circulation* 1997;95:2027–2031.
97. Inoue T, Sakai Y, Fujito T, Hoshi K, Hayashi T, Takayanagi K, Morooka S. Clinical significance of neutrophil adhesion molecules expression after coronary angioplasty on the development of restenosis. *Thrombosis and Haemostasis* 1998;79:54–58.
98. Tschoepe D, Schultheib MD, Kolarov P, Schwippert B, Dannehl K, Volksw D, Nieuwenhuis HK, Kehrel B, Strauer B, Gries FA. Platelet membrane activation markers are predictive for increased risk of acute ischemic events after PTCA. *Circulation* 1993;88:37–42.
99. Heeschen C, Hamm CW, Bruemmer J, Simoons ML, for the CAPTURE Investigator. Predictive value of C-reactive protein and troponin T in patients with unstable angina: A comparative analysis. *J Am Coll Cardiol* 2000;35:1535–1542.
100. Mueller C, Buettner HJ, Hodgson JM, Marsch S, Perruchoud AP, Roskamm H, Neumann FJ. Inflammation and long-term mortality after non-ST elevation acute coronary syndrome treated with a very early invasive strategy in 1042 consecutive patients. *Circulation* 2002;105:1412–1415.
101. Rubanyi GM, Vanhoutte PM. Superoxide anions and hyperoxia inactivate endothelium-derived relaxing factor. *American Journal of Physiology* 1986;250:H822–H827.
102. Oemar BS, Tschudi MR, Godoy N, Brovkovich V, Malinkski T, Luscher T. Reduced endothelial nitric oxide synthase expression and production in human atherosclerosis. *Circulation* 1998;97:2494–2498.
103. Ohara Y, Peterson TE, Sayegh HS, Subramanian RR, Wilcox JN, Harrison DG. Dietary correction of hypercholesterolemia in the rabbit normalizes endothelial superoxide anion production. *Circulation* 1995;92:898–903.
104. Stroes ES, Hijmering M, Zandvoort M, Rabelink TJ, Faassen EE. Origin of superoxide production by nitric oxide synthase. *FEBS Lett* 1998;438:161–164.
105. Verhaar MC, Stroes E, Rabelink TJ. Folates and cardiovascular disease. *Arterioscler Thromb Vasc Biol* 2002;22:6–13.
106. Wever RMF, Lüscher TF, Cosentino F, Rabelink T. Atherosclerosis and the two faces of endothelial nitric oxide synthase. *Circulation* 1998;97:108–112.
107. Michel T, Feron O. Nitric oxide synthases: Which, where, how and why? *J Clin Invest* 1997;100:2146–2152.
108. Thony B, Auerbach G, Blau N. Tetrahydrobiopterin biosynthesis, regeneration and function. *Biochemical Journal* 2000;347:1–16.
109. Cosentino F, Katusic ZS. Tetrahydrobiopterin and dysfunction of endothelial nitric oxide synthase in coronary arteries. *Circulation* 1995;91:139–144.
110. Heinzel B, John M, Klatt P, Bohme E, Meyer B. Ca^{2+}/calmodulin-depdnent formation of hydrogen peroxide by brain nitric oxide synthase. *Biochem J* 1992;281:627–630.
111. Pritchard KA, Jr., Groszek L, Smalley DM, Sessa WC, Wu M, Villalon P, Wolin MS, Stemerman MB. Native low-density lipoprotein increases endothelial cell nitric oxide synthase generation of superoxide anion. *Circ Res* 1995;77:510–518.
112. Vasquez-Vivar J, Kalyanaraman B, Martasek P, Hogg N, Masters BS, Karoui H, Tordo P, Pritchard KAJ. Superoxide generation by endothelial nitric oxide synthase: The influence of cofactors. *Proc Natl Acad Sci USA* 2000;95:9220–9225.
113. Wermer ER, Werner-Felmayer G, Wachter H, Mayer B. Biosynthesis of nitric oxide: Dependence of pteridine metabolism. *Rev Physiol Biochem Pharmacol* 1995;127:97–135.
114. Huang A, Vita JA, Venema RC, Keaney J. Ascorbic acid enhances endothelial nitric oxide synthase activity by increasing intracellular tetrahydrobiopterin. *J Biol Chem* 2000.
115. Shinozaki K, Kashiwagi A, Nishio Y, Okamura T, Yoshida Y, Masada M, Toda N, Kikkawa R. Abnormal biopterin metabolism is a major cause of impaired endothelium-dependent relaxation through nitric oxide/O_2^- imbalance in insulin-resitant rat aorta. *Diabetes* 1999;48:2437–2445.
116. Shinozaki K, Nishio Y, Okamura T, Yoshida Y, Maegawa H, Kojima H, Masada M, Toda N, Kikkawa R, Kashiwagi A. Oral administration of tetrahydrobiopterin prevents endothelial dysfunction and vascular oxidative stress in the aortas of insulin-resistant rats. *Circ Res* 2000;87:566–573.
117. Heller R, Unbehaun A, Schellenberg B, Mayer B, Werner-Felmayer G, Wermer ER. L-ascorbic acid potentiates endothelial nitric oxide synthesis via a chemical stabilization of tetrahydrobiopterin. *Journal of Biological Chemistry* 2001;276:40–47.
118. Verhaar MC, Wever RM, Kastelein JJ, van Loon D, Milstien S, Koomans HA, Rabelink TJ. Effects of oral folic

acid supplementation on endothelial function in familial hypercholesterolemia. A randomized placebo-controlled trial. *Circulation* 1999;100:335–338.
119. Wilmink HW, Stroes ES, Erkelens WD, Gerritsen WB, Wever R, Banga JD, Rabelink TJ. Influence of folic acid on postprandial endothelial dysfunction. *Arterioscler Thromb Vasc Biol* 2000;20:185–188.
120. Hyndman ME, Verma S, Rosenfeld RJ, Anderson TJ, Parsons HG. Interaction of 5-methyltetrahydrofolate and tetrahydrobiopterin on endothelial function. *Am J Physiol Heart Circ Physiol* 2002;282:H2167–H2172.
121. Stroes ES, van Faassen EE, Yo M, Martasek P, Boer P, Govers R, Rabelink TJ. Folic acid reverts dysfunction of endothelial nitric oxide synthase. *Circ Res* 2000;86:1129–1134.
122. Yoshimura M, Yasue H, Nakayama M, Shimasaki Y, Sumida H, Sugiyama S, Kugiyama K, Ogawa H, Ogawa Y, Saito Y, Miyamoto Y, Nakao K. A missense Glu298Asp variant in the endothelial nitric oxide synthase gene is associated with coronary spasm in the Japanese. *Human Genetics* 1998;103:65–69.
123. Hyndman ME, Parsons HG, Verma S, Bridge PJ, Edworthy S, Jones C, Lonn E, Charbonneau F, Anderson TJ. The T-786–>C mutation in endothelial nitric oxide synthase is associated with hypertension. *Hypertension* 2002;39:919–922.
124. Steinberg D. Antioxidants and atherosclerosis: A current assessment. *Circulation* 1991;84:1420–1425.
125. Kugiyama K, Kerns SA, Morrisett JD, Roberts R, Henry PD. Impairment of endothelium-dependent arterial relaxation by lysolecithin in modified low-density lipoproteins. *Nature* 1990;344:160–162.
126. Kugiyama K, Ohgushi M, Sugiyama S, Murohara T, Fukunaga K, Miyamtoto E, Yasue H. Lysophosphatidylcholine inhibits surface receptor-mediated intracellular signals in endothelial cells by a pathway involving protein kinase C activation. *Circ Res* 1992;71:1422–1428.
127. Ohgushi M, Kugiyama K, Fukunaga K, Murohara T, Sugiyama S, Miyamoto E, Yasue H. Protein kinase C inhibitors prevent impairment of endothelium-dependent relaxation by oxidatively modified LDL. *Arteriosclerosis and Thrombosis* 1993;13:1525–1532.
128. Liao JK, Clark SL. Regulation of G-protein alpha i2 subunit expression by oxidized low-density lipoprotein. *J Clin Invest* 1995;95:1457–1463.
129. Boulanger CM, Tanner FC, Bea ML, Hahn AW, Werner A, Luscher TF. Oxidized low density lipoproteins induce mRNA expression and release of endothelin from human and porcine endothelium. *Circ Res* 1992;70:1191–1197.
130. Lerman A, Webster MWI, Chesebro JH, Edwards WD, Wei C-M, Fuster V, Burnett JC, Jr. Circulating and tissue endothelin immunoreactivity in hypercholesterolemic pigs. *Circulation* 1993;88:2923–2928.
131. Anderson TJ, Meredith IT, Charbonneau F, Yeung AC, Frei B, Selwyn AP, Ganz P. Endothelium-dependent coronary vasomotion relates to the susceptibility of LDL to oxidation in humans. *Circulation* 1996;93:1647–1650.
132. Heitzer T, Yla-Herttuala S, Luoma J, Kurz S, Munzel T, Just H, Olschewski M, Drexler H. Cigarette smoking potentiates endothelial dysfunction of forearm resistance vessels in patients with hypercholesterolemia: Role of oxidized LDL. *Circulation* 1996;93:1346–1353.
133. Plotnick GD, Corretti MC, Vogel RA. Effect of antioxidant vitamins on the transient impairment of endothelium-dependent brachial artery vasoactivity following a single high-fat meal. *JAMA* 1997;278:1682–1686.
134. Steinberg HO, Tarshoby M, Monestel R, Hook G, Cronin J, Johnson A, Bayazeed B, Baron AD. Elevated circulating free fatty acid levels impair endothelium-dependent vasodilation. *J Clin Invest* 1997;100:1230–1239.
135. Williams SB, Goldfine AB, Timimi FK, Ting HH, Roddy M-A, Simonson DC, Creager MA. Acute hyperglycemia attenuates endothelium-dependent vasodilation in humans *in vivo*. *Circulation* 1998;97:1695–1701.
136. Barry OP, Pratico D, Savani RC, FitzGerald GA. Modulation of monocyte-endothelial cell interactions by platelet microparticles. *J Clin Invest* 1998;102:136–144.
137. Mesri M, Altieri DC. Leukocyte microparticles stimulate endothelial cell cytokine release and tissue factor induction in a JNK1 signaling pathway. *J Biol Chem* 1999;274:23111–23118.
138. Ridker PM. High-sensitivity C-reactive protein: Potential adjunct for global risk assessment in the primary prevention of cardiovascular disease. *Circulation* 2001;103:1813–1818.
139. Pasceri V, Willerson JT, Yeh ETH. Direct proinflammatory effect of C-reactive protein on human endothelial cells. *Circulation* 2000;102:2165–2168.
140. Pasceri V, Chang J, Willerson JT, Yeh ETH. Modulation of C-reactive protein-mediated monocyte chemoattractant protein-1 induction in human endothelial cells by anti-atherosclerotic drugs. *Circulation* 2001;103:2531–2534.
141. Zwaka TP, Hombach V, Torzewski J. C-reactive protein-mediated low density lipoprotein uptake by macrophages: Implications for atherosclerosis. *Circulation* 2001;103:1194–1197.
142. Ohara Y, Peterson TE, Harrison DG. Hypercholesterolemia increases endothelial superoxide anion production. *J Clin Invest* 1993;91:2546–2551.
143. Ohara Y, Peterson TE, Zheng B, Kuo JF, Harrison DG. Lysophosphatidylcholine increases vascular superoxide anion production via protein kinase C activation. *Arteriosclerosis and Thrombosis* 1994;14:1007–1013.
144. Anderson TJ, Meredith IT, Yeung AC, Frei B, Selwyn AP, Ganz P. The effect of cholesterol-lowering and antioxidant therapy on endothelium-dependent coronary vasomotion [see comments]. *New England Journal of Medicine* 1995;332:488–493.
145. Dupuis J, Tardif JC, Cernacek P, Theroux P. Cholesterol reduction rapidly improves endothelial function after acute coronary syndromes: The RECIFE (Reduction of cholesterol in ischemia and function of the endothelium) trial. *Circulation* 1999;99:3227–3233.
146. Treasure CB, Klein L, Weintraub WS, Talley JD, Stillabower ME, Kosinski AS, Zhang J, Boccuzzi SJ, Cedarholm JC, Alexander RW. Beneficial effects of cholesterol lowering therapy on the coronary endothelium in patients with coronary artery disease. *N Engl J Med* 1995;332:481–487.
147. Tamai O, Matsuoka H, Itabe H, Wada Y, Kohno K, Imaizumi T. Single LDL apheresis improves endothelium-dependent vasodilation in hypercholesterolemic humans. *Circulation* 1997;95:76–82.
148. Leung WH, Lau CP, Wong CK. Beneficial effect of cholesterol-lowering therapy on coronary endothelium-dependent relaxation in hypercholesterolaemic patients. *Lancet* 1993;341:1496–1500.

149. Feron O, Dessy C, Desager JP, Balligand JL. Hydroxymethylglutaryl-coenzyme a reductase inhibition promotes endothelial nitric oxide synthase activation through a decrease in caveolin abundance. *Circulation* 2001;103:113–118.
150. Crisby M, Nordin-Fredriksson G, Shah PK, Yano J, Zhu J, Nilsson J. Pravastatin treatment increases collagen content and decreases lipid content, inflammation, metalloproteinases, and cell death in human carotid plaques: Implications for plaque stabilization. *Circulation* 2001;103:926–933.
151. Llevadot J, Murasawa S, Kureishi Y, Uchida S, Masuda H, Kawamoto A, Walsh K, Isner JM, Asahara T. HMG-CoA reductase inhibitor mobilizes bone marrow-derived endothelial progenitor cells. *J Clin Invest* 2001;108:399–405.
152. Freeman DJ, Norrie J, Sattar N, Neely RD, Cobbe SM, Ford I, Isles C, Lorimer AR, Macfarlane PW, McKillop JH, Packard CJ, Shepherd J, Gaw A. Pravastatin and the development of diabetes mellitus: Evidence for a protective treatment effect in the West of Scotland coronary prevention study. *Circulation* 2001;103:357–362.
153. Albert MA, Danielson E, Rifai N, Ridker PM. Effect of statin therapy on C-reactive protein levels. *JAMA* 2001;286:64–70.
154. Ridker PM, Rifai N, Pfeffer MA, Sacks F, Braunwald E. Long-term effects of pravastatin on plasma concentration of C-reactive protein. The Cholesterol and Recurrent Events (CARE) Investigators. *Circulation* 1999;100:230–235.
155. Ridker PM, Rifai N, Clearfield M, Downs JR, Weis SE, Miles JS, Gotto AM, Jr., the Air Force/Texas Coronary Atherosclerosis Prevention Study Investigators. Measurement of C-reactive protein for the targeting of statin therapy in the primary prevention of acute coronary events. *The New England Journal of Medicine* 2001;344:1959–1965.
156. Spieker LE, Sudano I, Hurlimann D, Lerch PG, Lang MG, Binggeli C, Corti R, Ruschitzka F, Luscher TF, Noll G. High-density lipoprotein restores endothelial function in hypercholesterolemic men. *Circulation* 2002;105:1399–1402.
157. HOPE Investigators: Effects of an angiotensin-converting-enzyme-inhibitor, Ramipril, on cardiovascular events in high-risk patients. *N Engl J Med* 2000;342:145–153.
158. Rajagopalan S, Kurz S, Munzel T, Tarpey M, Freeman BA, Griendling KK, Harrison DG. Angiotensin II-mediated hypertension in the rat increases vascular superoxide production via membrane NADH/NADPH oxidase activation. Contribution to alterations of vasomotor tone. *J Clin Invest* 1996;97:1916–1923.
159. Hornig B, Kohler C, Drexler H. Role of bradykinin in mediating vascular effects of angiotensin-converting enzyme inhibitors in humans. *Circulation* 1997;95:1115–1118.
160. Antony I, Lerebours G, Nitenberg A. Angiotensin-converting enzyme inhibition restores flow-dependent and cold pressor test-induced dilations in coronary arteries of hypertensive patients. *Circulation* 1996;94:3115–3122.
161. Mancini GBJ, Henry GC, Macaya C, O'Neill BJ, Pucillo AL, Carere RG, Wargovich TJ, Mudra H, Luscher TF, Klibaner MI, Haber HE, Uprichard ACG, Pepine CJ, Pitt B. Angiotensin converting enzyme inhibition with quinapril improves endothelial vasomotor dysfunction in patients with coronary artery disease: The TREND study. *Circulation* 1996;94:258–265.
162. Anderson TJ, Elstein E, Haber HE, Charbonneau F. Comparative study of ACE-inhibition, angiotensin II antagonism, and calcium channel blockade on flow-mediated vasodilation in patients with coronary disease (BANFF study). *J Am Coll Cardiol* 2000;35:60–66.
163. Brenner BM, Cooper ME, de Zeeuw D, Keane WF, Mitch WE, Parving HH, Remuzzi G, Snapinn SM, Zhang Z, Shahinfar S. Effects of losartan on renal and cardiovascular outcomes in patients with type 2 diabetes and nephropathy. *N Engl J Med* 2001;345:861–869.
164. Dahlof B, Devereux RB, Kjeldsen SE, Julius S, Beevers G, Faire U, Fyhrquist F, Ibsen H, Kristiansson K, Lederballe-Pedersen O, Lindholm LH, Nieminen MS, Omvik P, Oparil S, Wedel H. Cardiovascular morbidity and mortality in the Losartan Intervention For Endpoint reduction in hypertension study (LIFE): A randomised trial against atenolol. *Lancet* 2002;359:995–1003.
165. Lindholm LH, Ibsen H, Dahlof B, Devereux RB, Beevers G, de Faire U, Fyhrquist F, Julius S, Kjeldsen SE, Kristiansson K, Lederballe-Pedersen O, Nieminen MS, Omvik P, Oparil S, Wedel H, Aurup P, Edelman J, Snapinn S. Cardiovascular morbidity and mortality in patients with diabetes in the Losartan Intervention For Endpoint reduction in hypertension study (LIFE): A randomised trial against atenolol. *Lancet* 2002;359:1004–1010.
166. Prasad A, Tupas-Habib T, Schenke WH, Mincemoyer R, Panza JA, Waclawin MA, Ellahham S, Quyyumi AA. Acute and chronic angiotensin-1 receptor antagonism reverses endothelial dysfunction in atherosclerosis. *Circulation* 2000;101:2349–2354.
167. Cheetham C, Collis J, O'Driscoll G, Stanton K, Taylor R, Green D. Losartan, an angiotensin type I receptor antagonist,improves endothelial function in non-insulin-dependent diabetes. *J Am Coll Cardiol* 2000;36:1461–1466.
168. Cheetham C, O'Driscoll G, Stanton K et al. Losartan, an angiotensin type I receptor antagonist, improves conduit vessel endothelial function in type II diabetes. *Clinical Science* 2001;100:13–17.
169. UK Prospective Diabetes Study Group. Effect of intensive blood-glucose control with metformin on complications in overweight patients with type 2 diabetes (UKPDS 34). *Lancet* 1998;352:854–865.
170. Steinberg HO, Chaker H, Leaming R, Johnson A, Brechtel G, Baron AD. Obesity/insulin resistance is associated with endothelial dysfunction. Implications for the syndrome of insulin resistance. *J Clin Invest* 1996;97:2601–2610.
171. Mather KJ, Verma S, Anderson TJ. Improved endothelial function with metformin in diet treated type II diabetes mellitus. *J Am Coll Cardiol* 2001 (In press).
172. Verma S, Lovren F, Dumont AS, Mather KJ, Maitland A, Kieser TM, Triggle CR, Anderson TJ. Tetrahydrobiopterin improves endothelial function in human saphenous veins: A novel mechanism. *Journal of Thoracic and Cardiovascular Surgery* 2000;120:668–671.
173. Setoguchi S, Mohri M, Shimokawa H, Takeshita A. Tetrahydrobiopterin improves endothelial dysfunction in coronary microcirculation in patients without epicardial coronary artery disease. *J Am Coll Cardiol* 2001;38:493–498.

174. Maier W, Cosentino F, Lutolf RB, Fleisch M, Seiler C, Hess OM, Meier B, Luscher TF. Tetrahydrobiopterin improves endothelial function in patients with coronary artery disease. *J Cardiovasc Pharmacol* 2000;35:173–178.
175. Stroes E, Kastelein J, Erkelens W, Wever R, Koomans H, Luscher T, Rabelink T. Tetrahydrobiopterin restores endothelial function in hypercholesterolemia. *J Clin Invest* 1997;99:41–46.
176. Ueda S, Matsuoka H, Miyazaki H, Usui M, Okuda S, Imaizumi T. Tetrahydrobiopterin restores endotheial function in long-term smokers. *J Am Coll Cardiol* 2000;35:71–75.
177. Heitzer T, Brockhoff C, Mayer B, Warnholtz A, Mollnau H, Henne S, Meinertz T, Munzel T. Tetrahydrobiopterin improves endothelium-dependent vasodilation in chronic smokers. *Circ Res* 2000;86:e36–e40.
178. Heitzer T, Krohn K, Albers S, Meinertz T. Tetrahydrobiopterin improves endothelium-dependent vasodilation by increasing nitric oxide activity in patients with Type II diabetes mellitus. *Diabetologia* 2000;43:1435–1438.
179. Bellamy MF, McDowell IF, Ramsey MW, Brownlee M, Bones C, Newcombe RG, Lewis MJ. Hyperhomocysteinemia after an oral methionine load acutely impairs endothelial function in healthy adults. *Circulation* 1998;98:1848–1852.
180. Bellamy MF, McDowell IF, Ramsey MW, Brownlee M, Newcombe RG, Lewis MJ. Oral folate enhances endothelial function in hyperhomocysteinaemic subjects [see comments]. *European Journal of Clinical Investigation* 1999;29:659–662.
181. Chao CL, Chien KL, Lee YT. Effect of short-term vitamin (folic acid, vitamins B6 and B12) administration on endothelial dysfunction induced by post-methionine load hyperhomocysteinemia. *American Journal of Cardiology* 1999;84:1359–1361.
182. Title LM, Cummings PM, Giddens K, Genest JJ, Jr., Nassar BA. Effect of folic acid and antioxidant vitamins on endothelial dysfunction in patients with coronary artery disease. *J Am Coll Cardiol* 2000;36:758–765.
183. Woo KS, Chook P, Lolin YI, Sanderson JE, Metreweli C, Celermajer DS. Folic acid improves arterial endothelial function in adults with hyperhomocytinemia. *J Am Coll Cardiol* 1999;34:2002–2006.
184. Doshi SN, McDowell IF, Moat SJ, Lang D, Newcombe RG, Kredan MB, Lewis MJ, Goodfellow J. Folate improves endothelial function in coronary artery disease: An effect mediated by reduction of intracellular superoxide? *Arterioscler Thromb Vasc Biol* 2001;21:1196–1202.
185. Verhaar MC, Wever RMF, Kastelein JJP, van Dam T, Koomans HA, Rabelink TJ. 5-Methyltetrahydrofolate,the active form of folic acid restores endothelial function in familial hypercholesterolemia. *Circulation* 1998;97:237–241.
186. Libby P, Simon DI. Inflammation and thrombosis: The clot thickens. *Circulation* 2001;103:1718–1720.
187. Libby P, Ridker PM, Maseri A. Inflammation and atherosclerosis. *Circulation* 2002;105:1135–1143.
188. Hasdai D, Gibbons RJ, Holmes DR, Jr., Higano ST, Lerman A. Coronary endothelial dysfunction in humans is associated with myocardial perfusion defects. *Circulation* 1997;96:3390–3395.
189. Quyyumi AA, Cannon RO3, Panza JA, Diodati JG, Epstein SE. Endothelial dysfunction in patients with chest pain and normal coronary arteries. *Circulation* 1992;86:1864–1871.
190. Davis SF, Yeung AC, Meredith IT, Charbonneau F, Ganz P, Selwyn AP, Anderson TJ. Early endothelial dysfunction predicts the development of transplant coronary artery disease at 1 year posttransplant. *Circulation* 1996;93:457–463.
191. Schachinger V, Britten MB, Zeiher A. Impaired epicardial coronary vasoreactivity predicts for adverse cardiovascular events during long-term follow-up. *Circulation* 2000;101:1902–1907.
192. Suwaidi JA, Hamasaki S, Higano ST, Velianou JL, Araujo NA, Lerman A. Long-term follow-up of patients with mild coronary artery disease and endothelial dysfunction. *Circulation* 2000;101:948–954.
193. Perticone F, Ceravolo R, Pujia A, Ventura G, Iacopino S, Scozzafava A, Ferraro A, Chello M, Mastroroberto P, Verdecchia P, Schillaci G. Prognostic significance of endothelial dysfunction in hypertensive patients. *Circulation* 2001;104:191–196.
194. Heitzer T, Schlinzig T, Krohn K, Meinertz T, Munzel T. Endothelial dysfunction, oxidative stress, and risk of cardiovascular events in patients with coronary artery disease. *Circulation* 2001;104:2673–2678.
195. Neunteufl T, Heher S, Katzenschlager R, Wolfl G, Kostner K, Maurer G, Weidinger F. Late prognostic value of flow-mediated dilation in the brachial artery of patients with chest pain. *Am J Cardiol* 2000;86:207–210.
196. Gokce N, Keaney JF, Jr., Hunter LM, Watkins MT, Menzoian JO, Vita JA. Risk stratification for postoperative cardiovascular events via noninvasive assessment of endothelial function: A prospective study. *Circulation* 2002;105:1567–1572.
197. Anderson TJ, Robertson A, Hildebrand K, Conradson HE, Jones C, Bridge P, Edworthy S, Lonn EM, Verma S, Charbonneau F. The FATE of endothelial function testing: Rational and design of the firefighters and their endothelium (FATE) Study. *Can J Cardiol* 2002 (In press).
198. Ridker PM, Stampfer MJ, Rifai N. Novel risk factors for systemic atherosclerosis. *JAMA* 2001;285:2481–2485.
199. Ridker PM, Rifai N, Stampfer MJ, Hennekens CH. Plasma concentration of interleukin-6 and the risk of future myocardial infarction among apparently healthy Men [In process citation]. *Circulation* 2000;101:1767–1772.
200. Ridker PM, Rifai N, Pfeffer M, Sacks F, Lepage S, Braunwald E. Elevation of tumor necrosis factor-alpha and increased risk of recurrent coronary events after myocardial infarction. *Circulation* 2000;101:2149–2153.
201. Ridker PM, Cushman M, Stampfer MJ, Tracy R, Hennekens C. Inflammation, aspirin and the risk of cardiovascular disease in apparently healthy men. *New England Journal of Medicine* 1997;336:973–979.
202. Landmesser U, Merten R, Spiekermann S, Buttner K, Drexler H, Hornig B. Vascular extracellular superoxide dismutase activity in patients with coronary artery disease: Relation to endothelium-dependent vasodilation. *Circulation* 2000;101:2264–2270.

Regulation of Cardiac Remodeling by Nitric Oxide: Focus on Cardiac Myocyte Hypertrophy and Apoptosis

Kai C. Wollert, MD and Helmut Drexler, MD
Department of Cardiology and Angiology, Hannover Medical School, Hannover, Germany

Abstract. **Cardiac hypertrophy occurs in pathological conditions associated with chronic increases in hemodynamic load. Although hypertrophy can initially be viewed as a salutary response, ultimately, it often enters a phase of pathological remodeling that may lead to heart failure and premature death. A prevailing concept predicts that changes in gene expression in hypertrophied cardiac myocytes and cardiac myocyte loss by apoptosis contribute to the transition from hypertrophy to failure. In recent years, nitric oxide (NO) has emerged as an important regulator of cardiac remodeling. Specifically, NO has been recognized as a potent antihypertrophic and proapoptotic mediator in cultured cardiac myocytes. Studies in genetically engineered mice have extended these findings to the *in vivo* situation. It appears that low levels and transient release of NO by endothelial NO synthase exert beneficial effects on the remodeling process by reducing cardiac myocyte hypertrophy, cavity dilation and mortality. By contrast, high levels and sustained production of NO by inducible NO synthase seem to be maladaptive by reducing ventricular contractile function, and increasing cardiac myocyte apoptosis, and mortality. In the future, these novel insights into the role of NO in cardiac remodeling should allow the development of novel therapeutic strategies to treat cardiac remodeling and failure.**

Key Words. **nitric oxide, cGMP, hypertrophy, apoptosis**

Introduction

Nitric oxide (NO) is a free radical gas and is readily diffusible, it has a very short half-life, lasting only seconds. NO is synthesized from L-arginine by the catalytic reaction of different isoforms of nitric oxide synthases, including the neuronal type 1 isoform (nNOS), the inducible type 2 isoform (iNOS), and the endothelial, type 3 isoform (eNOS). nNOS and eNOS are constitutively expressed enzymes and are regulated predominantly at the posttranslational level, whereas in most cell types, iNOS is only expressed in response to appropriate stimuli. Small amounts of NO, produced by nNOS and eNOS, are involved in signal events that regulate neurotransmission and vascular tone. Because the activity of nNOS and eNOS is triggered, it is transient. Much larger concentrations of NO are usually provided by iNOS. Because iNOS is independent of stimulating agonists and Ca^{2+}, its activity is sustained. Both nNOS and eNOS are constitutively expressed in the myocardium. nNOS is expressed in nerve endings involved in the neurotransmission of norepinephrine, and eNOS is expressed in endothelial cells, endocardial cells, and cardiomyocytes [1]. In disease states associated with cytokine activation or inflammation, cardiac iNOS expression is induced in endothelial cells, vascular smooth muscle cells, macrophages, and cardiac myocytes [1]. The ubiquitous distribution of NO synthases within the myocardium and the versatile actions of NO make this molecule unique in the diverse cardiac effects that depend on its concentration and the spatial and temporal activity of its isoforms [2]. The versatility of NO is largely related to its capability to react with superoxide, oxygen, and transition metals. Each of the products of these reactions, peroxynitrite, NO_x, and metal-NO adducts, support additional reactions. An important physiological target of NO is the heme protein soluble guanylyl cyclase. NO activates guanylyl cyclase by interacting with its heme. Guanylyl cyclase catalyzes the formation of cGMP, which serves as a second messenger for NO and acts on a number of downstream targets, including ion channels, phosphodiesterases and kinases in a cell-type specific manner.

In recent years, a number of studies have been published which collectively indicate that nitric oxide can modulate both hypertrophy and apoptosis in the heart. Considering the pathophysiological importance of these processes, NO may play a critical role in various disease states associated with cardiac hypertrophy and failure. Specifically, in cell culture experiments, NO has been recognized as a negative regulator of the hypertrophic response and a potent proapoptotic

Address for correspondence: Kai C. Wollert, Abt. Kardiologie und Angiologie, Medizinische Hochschule Hannover, Carl-Neuberg Str.1, 30625 Hannover, Germany. Tel.: +49 (511) 532-4055; Fax: +49 (511) 532-5412; E-mail: wollert.kai@mh-hannover.de stimulus in cardiac myocytes. Recent studies in genetically manipulated mice extend these findings to the *in vivo* situation and establish that NO, produced by the different NO synthase isoforms, acts as an important regulator of the remodeling process and may even determine survival in cardiac hypertrophy and failure. In this review, we discuss the role of NO and NO synthases in cardiac remodeling. We will start with an overview of contemporary concepts of cardiac hypertrophy and apoptosis and their contribution to the remodeling process and heart failure progression.

Cardiac Hypertrophy, Remodeling and Failure

Cardiac hypertrophy occurs in response to long-term increases in hemodynamic load related to a variety of physiological and pathological conditions. The hypertrophic process is characterized by structural changes at the cardiac myocyte level that are translated into alterations in chamber size and geometry, collectively called remodeling [3,4]. In pressure-overload hypertrophy, additional sarcomeres are assembled in parallel, leading to thicker myocytes and to a concentric pattern of ventricular hypertrophy. In contrast, in volume overload, additional sarcomeres are assembled in series, leading to longer myocytes and ventricular dilatation (eccentric hypertrophy). Functional benefits of the hypertrophic response include an increase in the number of contractile elements, a lowering of wall stress through increased wall thickness in concentric hypertrophy, and increasing stroke volume by increasing end-diastolic volume in eccentric hypertrophy. Physiological forms of cardiac hypertrophy are characterized by intermittent (athlete's heart) or transitory (pregnancy) increases in workload and are considered to be purely adaptive. By contrast, pathological forms of cardiac hypertrophy observed in ischemic, hypertensive or valvular heart disease develop in response to persistent increases in hemodynamic load. Epidemiological studies have demonstrated that pathological cardiac hypertrophy is a risk factor for future cardiac events [5,6]. Indeed, pathological cardiac hypertrophy often enters a phase of pathological remodeling that leads to contractile dysfunction and ultimately to heart failure and sudden death. These observations suggest that certain components of the hypertrophic response to persistent increases in workload are maladaptive. The remainder of this discussion will focus on pathological forms of cardiac (myocyte) hypertrophy and remodeling.

Cardiac Myocyte Hypertrophy and Apoptosis and the Transition to Failure

The precise mechanism(s) responsible for the transition from adaptive hypertrophy and remodeling to maladaptive heart failure are elusive, but there are several candidate mechanisms (reviewed in [3]). Deficiencies in high-energy phosphate stores, defects in excitation-contraction coupling, excess formation of myocyte microtubules and increased production of extracellular matrix likely participate in pathological remodeling. Attenuation of ß-adrenergic signal transduction as the major means of supporting decreased myocardial performance probably contributes to the transition as well. A valuable concept that has emerged in recent years predicts that cardiac myocyte hypertrophy itself, although compensatory in the beginning, may be detrimental in the long-term [3]. Pathological cardiac myocyte hypertrophy is not simply a matter of a quantitative increase in contractile proteins and other key elements that initiate and regulate contraction, but rather, it is associated with qualitative changes in gene expression. The molecular signature of pathological cardiac myocyte hypertrophy is fetal gene induction, including changes in gene expression of contractile and calcium handling proteins. Neurohormonal systems and cytokines play a central role in promoting pathological cardiac myocyte hypertrophy in response to persistent increases in hemodynamic load. The qualitative changes in gene expression in hypertrophied myocytes lead to an impairment of myocyte contractile function which then promotes persistent activation of neurohormones and cytokines. A schematic diagram of this putative vicious circle is shown in Figure 1.

Considering the potential importance of pathological cardiac myocyte hypertrophy in the remodeling process and the transition to failure, much investigation has focused on the signaling pathways controlling the hypertrophic response at the level of the single cardiac myocyte. It has emerged that the hypertrophic response is orchestrated by neurohormonal growth factors and cytokines acting through several intracellular signaling cascades (reviewed in [7]). Signaling molecules that have been characterized as important transducers of the hypertrophic response, include $\beta1$ integrins and mechanical stretch-activated pathways, specific G protein isoforms, low-molecular-weight GTPases (Ras, RhoA, Rac), mitogen-activated protein kinase cascades, protein kinase C, calcineurin, gp130-Jak-Stat, insulin-like growth factor I receptor pathway, fibroblast growth factor and transforming growth factor-ß receptor pathways, and many others. Most, if not all, of

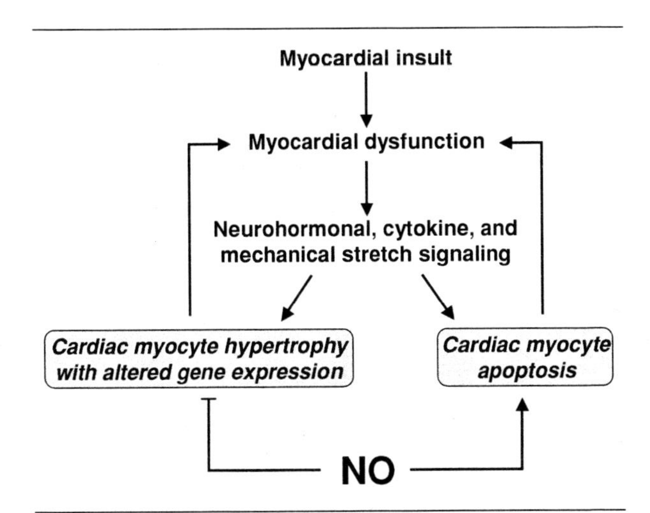

Fig. 1. Nitric oxide and its influence on two important vicious circles that promote remodeling and cardiac dysfunction in the presence of myocardial damage. Mechanical load and neurohormonal / cytokine signaling pathways induce altered gene expression and / or cell death, both of which impair myocardial function further (adapted from Braunwald and Bristow [3]). Nitric oxide influences the remodeling process, by negatively regulating the hypertrophic response and by promoting apoptosis in cardiac myocytes.

these signaling pathways produce pathological hypertrophy, i.e. hypertrophy leading to contractile dysfunction and poor clinical outcomes. Each of these signaling pathways has been implicated as a hypertrophy transducer, suggesting an almost overwhelming complexity in hypertrophy signaling cascades in cardiac myocytes. However, both experimental and clinical studies have indicated that specific inhibition of individual growth factors or signaling cascades will often diminish the activation of other interdependent signal transduction pathways and will therefore suffice to inhibit the hypertrophic response [7]. These studies support an integrated model of signal transduction in the heart such that multiple pathways are necessary for timely and effective hypertrophy.

Another recently emphasized and probably important component of the transition from hypertrophy to heart failure is cardiac myocyte apoptosis, or programmed cell death [3]. Although its role in less advanced forms of cardiac hypertrophy and failure is uncertain, cardiac myocyte apoptosis has been clearly demonstrated in end-stage failing hearts. Cardiac myocyte apoptosis is triggered by a variety of factors, including reactive oxygen species, myocyte over-stretch, hypoxia, Ca^{2+}-overload, cytokines, and neurohormones. These different factors activate a precisely orchestrated genetic program culminating in the activation of executioner caspases, the final common pathway of different proapoptotic stimuli. Caspases promote cell death by degrading critical target proteins in the nucleus, cytosol and mitochondria. A prevailing concept predicts that a progressive loss of cardiac myocytes by apoptosis will increase the hemodynamic burden on surviving myocytes and hasten their death, thereby setting up a vicious circle. Thus, as outlined in Figure 1, cardiac myocyte loss via apoptosis joins pathological cardiac myocyte hypertrophy as a process that may produce progressive myocardial dysfunction in the failing heart.

Nitric Oxide Mediates Antihypertrophic Effects

A study in spontaneously hypertensive rats (SHR) provided the first evidence that NO can promote antihypertrophic effects in the heart [8]. As shown in this study, chronic treatment with the NO precursor L-arginine attenuates cardiac hypertrophy in SHR. Importantly, L-arginine administration suppressed cardiac hypertrophy without changes in blood pressure, suggesting a direct cardiac effect. Indeed, myocardial NO_X levels were greater in L-arginine–treated SHR than in vehicle-treated controls indicating that chronic L-arginine administration enhances cardiac production of NO. L-arginine treatment not only reduced cardiac weights, but also cardiac expression of the fetal isoform skeletal α-actin, indicating that L-arginine suppresses pathological cardiac *myocyte* hypertrophy. Calderone et al. [9] demonstrated that the NO donor *S*-nitroso-*N*-acetyl-D,L-penicillamine (SNAP) causes a concentration-dependent decrease in α_1-adrenoreceptor-stimulated protein synthesis in isolated neonatal cardiac myocytes, confirming that exogenous NO exerts inhibitory effects on cardiac myocyte hypertrophy. Conversely, inhibition of endogenous NO synthesis in cardiac myocytes by *N*G-monomethyl-L-arginine potentiates α_1-adrenergic increases in protein synthesis, indicating that basal levels of endogenous NO production suppress cardiac myocyte growth [9]. Stimulation of cardiac myocytes with SNAP increases intracellular cGMP levels; moreover, the growth-suppressing effects of SNAP are mimicked by the cGMP analog 8-bromo-cGMP suggesting that the growth-inhibiting effects of NO are mediated, at least in part, by cGMP [9]. Similarly, antihypertrophic effects of L-arginine in the SHR model are associated with increased cardiac levels of cGMP [8]. Confirming and extending the results obtained in the neonatal system, antihypertrophic effects of NO via cGMP have also been demonstrated in adult cardiac myocytes [10].

What are the downstream target(s) mediating the inhibitory effects of NO and cGMP on cardiac myocyte hypertrophy? In general, cGMP effectors include cGMP-regulated phosphodiesterases, cGMP-regulated ion channels and cGMP-dependent protein kinases (PKGs) [11]. Two PKG genes have been identified in mammalian cells, encoding for PKG type I (including α and β-splice variants) and PKG type II. In cardiac myocytes, PKG I mediates negative inotropic effects of NO and cGMP [12–15]. Consistent with these previous reports, we recently demonstrated that neonatal rat ventricular cardiac myocytes express PKG type I [16], predominantly the α-splice variant (K.C.W. and S.M. Lohmann, unpublished observation). This endogenous PKG I is a downstream target for NO and cGMP in cardiac myocytes, as evidenced by the site-specific phosphorylation of vasodilator-stimulated phosphoprotein (VASP), a well-characterized PKG substrate [16]. Intriguingly, treatment of cardiac myocytes with a PKG-selective cGMP analog attenuates the hypertrophic response to α_1-adrenergic stimulation as effectively as SNAP, demonstrating that endogenous PKG I can promote antihypertrophic effects. Similar to observations in vascular smooth muscle cells [17], prolonged activation results in a downregulation of endogenous PKG I expression and function in cardiac myocytes [16]. Downregulation of PKG I likely serves as an internal brake mechanism that limits the antihypertrophic effects of NO and cGMP. Indeed, the antihypertrophic effects of NO and cGMP in cardiac myocytes can be strongly enhanced by overexpression of PKG I using adenoviral gene transfer [16]. PKG I has previously been shown to suppress the L-type Ca^{2+}-channel current and $[Ca^{2+}]_i$ transients in cardiac myocytes [12,18,19]. Ca^{2+}-influx via the L-type Ca^{2+}-channel has been implicated in the regulation of cardiac myocyte hypertrophy [20]. Recently, a number of Ca^{2+}-sensitive hypertrophic signaling pathways have been identified in cardiac myocytes, underscoring the importance of Ca^{2+} as a central regulator of the hypertrophic response [21–24]. Since the growth-inhibitory effects of NO and cGMP in cardiac myocytes can be mimicked both qualitatively and quantitatively by Ca^{2+}-channel antagonists [9], we hypothesized that antihypertrophic effects of NO and cGMP may be mediated, at least in part, via inhibition of Ca^{2+}-dependent signaling pathways by PKG I. One prominent Ca^{2+}-dependent pathway involves the Ser/Thr protein phosphatase calcineurin. Activation of calcineurin by Ca^{2+} results in the dephosphorylation and nuclear translocation of cytoplasmic latent NFAT (nuclear factor of activated T cells) transcription factors [25]. Many lines of evidence, including the antihypertrophic effects of endogenous

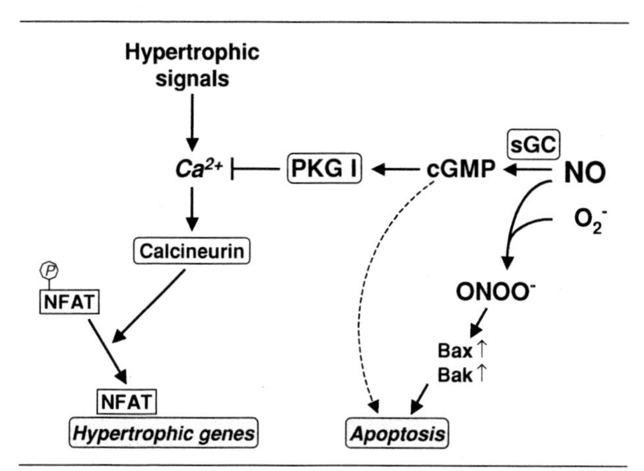

Fig. 2. Signaling pathways of NO in cardiac myocytes. Antihypertrophic effects of NO are mediated via soluble guanylyl cyclase (sGC), cGMP formation, and activation of cGMP-dependent protein kinase type I (PKG I). By contrast, proapoptotic effects of NO relate to its reaction with superoxide (O_2^-) to form peroxynitrite ($ONOO^-$). See text for details.

calcineurin inhibitory peptides [26], have contributed to the conclusion that the calcineurin-NFAT pathway plays an essential role in the hypertrophic response to growth factor stimulation. Recent data from our laboratory indicate that NO/cGMP activation of PKG I inhibits the hypertrophic calcineurin-NFAT pathway in cardiac myocytes by targeting critical, Ca^{2+}-dependent steps upstream of calcineurin. Our data indicate that inhibitory effects of PKG I upstream of calcineurin are mediated by interfering with Ca^{2+} entry via the L-type Ca^{2+}-channel (Fig. 2) [27]. It is noteworthy that in vascular smooth cells and other cell types, PKG I inhibits signaling via low-molecular-weight GTPases Ras and RhoA [28,29], mediators which have also been implicated in the regulation of cardiac myocyte hypertrophy [30–32]. Therefore, it is feasible that PKG I inhibition of such additional signaling pathways could also promote NO/cGMP antihypertrophic effects in cardiac myocytes.

Nitric Oxide Induces Apoptosis in Cardiac Myocytes

In addition to its antihypertrophic effects, NO has been shown to promote apoptosis in cardiac myocytes in a dose-dependent manner. It appears that lower concentrations of NO, which are sufficient to suppress cardiac myocyte hypertrophy, are not toxic to cardiac myocytes, and that higher concentrations of NO are required to induce caspase activation, DNA fragmentation, and cell death [16]. Treatment of neonatal cardiac myocytes with a PKG-selective cGMP analog does not promote apoptotic cell death; moreover, adenoviral overexpression of PKG I does not enhance the susceptibility of cardiomyocytes to the proapoptotic effects of NO. These data support the concept that proapoptotic effects of NO in neonatal cardiomyocytes are mediated by cGMP-independent rather than cGMP-dependent mechanisms [16]. Likewise, Arstall et al. have recently demonstrated that iNOS upregulation by interleukin-1β and interferon-γ increases apoptosis in neonatal cardiac myocytes by a process that is independent of guanylate cyclase activation and cGMP [33]. Cytokine-induced apoptosis can be blocked by iNOS inhibitors, clearly implicating iNOS in cardiac myocyte death [33]. However, concentrations of an NO donor added in similar concentrations to the amount of endogenous nitrite produced by myocytes after cytokine treatment does not stimulate cell death, suggesting that NO itself is not cytotoxic. The chemical reactivity and toxicity of NO can be greatly increased by its diffusion-limited reaction with superoxide to form peroxynitrite [34]. A role for peroxynitrite in NO mediated cardiac myocyte apoptosis was demonstrated by experiments showing that a superoxide and peroxynitrite scavenger protects myocytes from cytokine-induced apoptosis and that myocytes will rapidly undergo apoptosis after direct stimulation with peroxynitrite [33]. Several studies have shown that NO stimulated apoptosis in cardiac myocytes is associated with an altered expression pattern of apoptosis regulators of the Bcl-2 family which play a critical role in the determination of cell fate, in part, by altering mitochondrial membrane permeability [35]. NO increases the abundance of the proapoptotic Bcl-2 family members Bax and Bak in cardiac myocytes; by contrast, expression levels of the antiapoptotic Bcl-2 family members Bcl-2 and Bcl-x(L) are not affected [33,36]. It appears, therefore, that NO signaling via peroxynitrite stimulates apoptosis by producing a relative increase in proapoptotic effector proteins in cardiac myocytes. Taken together, these studies argue in favor of an oxidative, rather than a cGMP-dependent, mechanism for cytokine- and NO-mediated apoptosis in cardiac myocytes (Fig. 2) [16,33,36]. The situation may be different in adult cardiac myocytes, however, where NO has been shown to promote apoptosis in a cGMP-dependent manner [37]. Even in neonatal cells, it has been demonstrated in one study that cGMP analogs (weakly) promote apoptosis [38]. Since the degree to which NO forms peroxynitrite is dependent on the availability of oxygen free radicals, the redox state of cardiac myocytes may be an important variable in determining the apoptotic response to NO. Further studies will be required to elucidate further the contribution of cGMP-dependent mechanisms to the proapoptotic effects of NO.

Studies in Genetically Engineered Mice

Genetically engineered mice carrying gain-of-function or loss-of-function mutations of specific NOS isoforms offer the unique opportunity to address the functional significance of NO in a defined genetic background. In fact, physiological and molecular analyses of NOS-transgenic mice have allowed scientists to address some of the most fundamental questions related to NO and its putative role in cardiac remodeling [39]. The development of microsurgical techniques, and the adaptation of experimental models of cardiac hypertrophy and failure to the mouse have been critical in this regard (reviewed in [40]). To investigate the role of eNOS in post infarction cardiac remodeling, Scherrer-Crosbie et al. compared the impact of left anterior descending coronary artery ligation in wild-type and eNOS-deficient mice [41]. Although, there were no differences in infarct size between the two groups, left ventricular dilatation was more marked in eNOS deficient than in wildtype mice. Moreover, left ventricular systolic and diastolic functions were impaired to a greater extent in eNOS deficient mice. Importantly, left ventricular hypertrophy, as reflected by mass and cardiac myocyte width, was greater in eNOS deficient mice. Finally, eNOS deficiency was associated with an increased long-term mortality after myocardial infarction. eNOS deficient mice have an increased systemic blood pressure [42]. However, even after treatment with hydralazine, cardiac enlargement, contractile dysfunction, hypertrophy, and mortality were greater in eNOS deficient mice [41]. These data strongly suggest an important role for eNOS in limiting post infarction cardiac hypertrophy and dilatation. In contrast to the study by Scherrer-Crosbie et al. [41], Liu et al. recently reported that eNOS deficiency does not aggravate myocardial remodeling after coronary ligation in mice [43]. The reason(s) for the discrepant results in these two studies are not clear. Notably, however, mean infarct sizes were significantly smaller in the study by Scherrer-Crosbie et al., raising the possibility that the pathophysiological consequences of eNOS deficiency are more evident in mild forms of cardiac dysfunction and failure. Future studies should test this hypothesis.

Increased cardiac production of NO occurs in a number of disease syndromes, including myocarditis, cardiac allograft rejection, and heart failure after the induction of iNOS in the myocardium [44,45]. In the presence of adequate levels of cofactors (e.g. tetrahydrobiopterin and calmodulin) and substrates (L-arginine and oxygen), very high levels of NO production can be achieved after iNOS induction. In conditions of L-arginine or cofactor deficiency, however, iNOS reduces molecular oxygen to superoxide. As discussed above, peroxynitrite, formed by an interaction of NO with superoxide, and superoxide itself are cytotoxic to cardiac myocytes, suggesting that iNOS induction in the myocardium may exert detrimental effects. To directly investigate the role of iNOS in left ventricular remodeling, Sam et al. subjected iNOS deficient mice to coronary artery ligation [46]. iNOS deficiency did not affect mean infarct sizes after coronary ligation. However, as compared to wild-type control mice, iNOS deficient mice exhibited a less pronounced decrease in left ventricular contractile function, a reduced frequency of apoptotic myocytes in the remote myocardium, and, most importantly, reduced mortality in the chronic phase after infarction [46]. Similar results were reported by Feng et al., who also noted increased myocardial contractility and a decrease in mortality in iNOS deficient mice after infarction [47]. Mungrue et al. have used a different experimental strategy to study the pathophysiological consequences of increased iNOS expression in the myocardium and have generated transgenic mice with conditional overexpression of iNOS in cardiac myocytes [48]. iNOS overexpression was associated with a mild inflammatory cell infiltrate, cardiac fibrosis, myocyte death, increased cardiac mass, and cardiac dilatation. While a few iNOS-overexpressing mice developed overt heart failure, most animals died suddenly from heart block and asystole. Postmortem examinations of iNOS transgenic mice typically revealed normal cardiac morphology. However, a few deceased animals exhibited gross ventricular dilation and hypertrophy. These data suggest that increased myocardial iNOS activity is capable of initiating a process of cardiac remodeling that is characterized by ventricular dilatation, hypertrophy, and sudden cardiac death [48]. Remarkably, iNOS overexpression promoted an increase in cardiac mass in the study by Mungrue et al. At first glance, this observation appears to contradict previous studies demonstrating antihypertrophic effects of NO [8,9,16,41]. It is important to note, however, that upregulation of iNOS in the study by Mungrue et al. led to increased formation of superoxide and peroxynitrite in the heart with no evidence for increased production of NO_x, suggesting that iNOS overexpression led to substrate depletion and a switch from NO to superoxide production [48]. Reactive oxygen species have been recognized as prohypertrophic signaling intermediates in cardiac myocytes [49–51]. Future studies should therefore address the possibility that superoxide generation by iNOS may stimulate cardiac myocyte hypertrophy. Heger et al. recently reported that nonconditional iNOS overexpression in cardiac myocytes does not alter cardiac structure or function [52]. However, this apparent difference in phenotype likely reflects a critical difference in experimental design, with the conditional system presumably leading to higher levels of cardiac iNOS activity [48]. In future experiments, both iNOS overexpressing mouse strains will probably be subjected to hemodynamic stress; these forthcoming studies should reveal the impact of different levels of iNOS overexpression on the remodeling process.

Conclusions and Future Perspectives

Nitric oxide has emerged as an important regulator of cardiac hypertrophy, apoptosis and remodeling. The diverse cardiac effects of NO depend on its source and concentration and the activity of radical scavenging systems. It appears that low levels and transient release of NO by eNOS exert beneficial effects on the remodeling process by reducing cardiac myocyte hypertrophy, cavity dilation and mortality. By contrast, high levels and sustained production of NO by iNOS seem to be maladaptive by reducing ventricular contractile function, and increasing cardiac myocyte apoptosis, and mortality. Importantly, detrimental effects of iNOS relate, in part, to an increased formation of toxic reactive oxygen species. Hopefully, our knowledge about the biology of NO and its role in cardiac remodeling should enable us to develop novel therapeutic strategies to prevent the detrimental effects associated with cardiac hypertrophy and failure. As it appears, some "old" drugs already take advantage of this promise. For example, angiotensin-converting enzyme (ACE) inhibitors have been shown to enhance eNOS expression, and NO availability, and such a mechanism may contribute to the beneficial effects of these drugs in heart failure [53]. Indeed, the beneficial effects of ACE inhibitors (and angiotensin type 1 receptor antagonists) are severely blunted in eNOS knock-out mice with postinfarction heart failure, supporting the notion that eNOS derived NO is an important cardioprotective mediator [43]. It has recently been reported that HMG CoA reductase inhibitors (statins) and estrogens exert antihypertrophic effects *in vivo* [54–56]. Since statins and estrogens both augment eNOS expression [57,58], antihypertrophic effects of these agents may relate, in part, to an increase in NO formation by eNOS. Cell culture studies indicate that antihypertrophic and proapoptotic effects of NO are mediated via distinct signaling pathways. Inhibitory effects on cardiac myocyte hypertrophy are mediated by cGMP and its activation of cGMP-dependent protein kinase, whereas toxic effects of NO appear to be related mostly to the formation of peroxynitrite. Recently, a compound has been described that stimulates guanylyl cyclase independent from NO [59]. It has been shown that this drug lowers blood pressure and reduces mortality in a low NO, high renin rat model of hypertension [60]. Future studies should address whether this new class of drugs can be used to suppress hypertrophy without increasing the propensity to cell death. Considering the detrimental effects of increased iNOS expression in the failing heart, iNOS may be a promising therapeutic target. However, today, no therapeutic strategies have been delineated that specifically suppress iNOS activity in the failing heart. Conceptually, a number of avenues could be followed towards this goal, including the use of selective iNOS inhibitors [61], iNOS antisense oligonucleotides [62], and strategies targeting cytokines [63] or transcription factors [64] involved in the regulation of iNOS expression.

References

1. Balligand JL, Cannon PJ. Nitric oxide synthases and cardiac muscle. Autocrine and paracrine influences. *Arterioscler Thromb Vasc Biol* 1997;17:1846–1858.
2. Drexler H. Nitric oxide synthases in the failing human heart: A doubled-edged sword? *Circulation* 1999;99:2972–2975.
3. Braunwald E, Bristow MR. Congestive heart failure: Fifty years of progress. *Circulation* 2000;102(IV):14–23.
4. Lorell BH, Carabello BA. Left ventricular hypertrophy: Pathogenesis, detection, and prognosis. *Circulation* 2000;102:470–479.
5. Levy D, Garrison RJ, Savage DD, Kannel WB, Castelli WP. Prognostic implications of echocardiographically determined left ventricular mass in the Framingham Heart Study. *N Engl J Med* 1990;322:1561–1566.
6. Krumholz HM, Larson M, Levy D. Prognosis of left ventricular geometric patterns in the Framingham Heart Study. *J Am Coll Cardiol* 1995;25:879–884.
7. Molkentin JD, Dorn IG. Cytoplasmic signaling pathways that regulate cardiac hypertrophy. *Annu Rev Physiol* 2001;63:391–426.
8. Matsuoka H, Nakata M, Kohno K, Koga Y, Nomura G, Toshima H, Imaizumi T. Chronic L-arginine administration attenuates cardiac hypertrophy in spontaneously hypertensive rats. *Hypertension* 1996;27:14–18.
9. Calderone A, Thaik CM, Takahashi N, Chang DL, Colucci WS. Nitric oxide, atrial natriuretic peptide, and cyclic GMP inhibit the growth-promoting effects of norepinephrine in cardiac myocytes and fibroblasts. *J Clin Invest* 1998;101:812–818.
10. Ritchie RH, Schiebinger RJ, LaPointe MC, Marsh JD. Angiotensin II-induced hypertrophy of adult rat cardiomyocytes is blocked by nitric oxide. *Am J Physiol* 1998;275:H1370–H1374.
11. Lohmann SM, Vaandrager AB, Smolenski A, Walter U, De Jonge HR. Distinct and specific functions of cGMP-dependent protein kinases. *Trends Biochem Sci* 1997;22:307–312.
12. Mery PF, Lohmann SM, Walter U, Fischmeister R. Ca^{2+} current is regulated by cyclic GMP-dependent protein kinase in mammalian cardiac myocytes. *Proc Natl Acad Sci USA* 1991;88:1197–1201.
13. Vila-Petroff MG, Younes A, Egan J, Lakatta EG, Sollott SJ. Activation of distinct cAMP-dependent and cGMP-dependent pathways by nitric oxide in cardiac myocytes. *Circ Res* 1999;84:1020–1031.
14. Shah AM, Spurgeon HA, Sollott SJ, Talo A, Lakatta EG. 8-bromo-cGMP reduces the myofilament response to Ca^{2+} in intact cardiac myocytes. *Circ Res* 1994;74:970–978.
15. Yuasa K, Michibata H, Omori K, Yanaka N. A novel interaction of cGMP-dependent protein kinase I with troponin T. *J Biol Chem* 1999;274:37429–37434.
16. Wollert KC, Fiedler B, Gambaryan S, Smolenski A, Heineke J, Butt E, Trautwein C, Lohmann SM, Drexler H. Gene transfer of cGMP-dependent protein kinase I enhances the antihypertrophic effects of nitric oxide in cardiomyocytes. *Hypertension* 2002;39:87–92.
17. Soff GA, Cornwell TL, Cundiff DL, Gately S, Lincoln TM. Smooth muscle cell expression of type I cyclic GMP-dependent protein kinase is suppressed by continuous exposure to nitrovasodilators, theophylline, cyclic GMP, and cyclic AMP. *J Clin Invest* 1997;100:2580–2587.
18. Sumii K, Sperelakis N. cGMP-dependent protein kinase regulation of the L-type Ca^{2+} current in rat ventricular myocytes. *Circ Res* 1995;77:803–812.
19. Klein G, Drexler H, Schröder F. Protein kinase G reverses all isoproterenol induced changes of cardiac single L-type calcium channel gating. *Cardiovasc Res* 2000;48:367–374.
20. Zhang S, Hiraoka M, Hirano Y. Effects of α_1-adrenergic stimulation on L-type Ca^{2+} current in rat ventricular myocytes. *J Mol Cell Cardiol* 1998;30:1955–1965.
21. Molkentin JD, Lu JR, Antos CL, Markham B, Richardson J, Robbins J, Grant SR, Olson EN. A calcineurin-dependent transcriptional pathway for cardiac hypertrophy. *Cell* 1998;93:215–228.
22. Passier R, Zeng H, Frey N, Naya FJ, Nicol RL, McKinsey TA, Overbeek P, Richardson JA, Grant SR, Olson EN. CaM kinase signaling induces cardiac hypertrophy and activates the MEF2 transcription factor *in vivo*. *J Clin Invest* 2000;105:1395–1406.
23. Aoki H, Sadoshima J, Izumo S. Myosin light chain kinase mediates sarcomere organization during cardiac hypertrophy *in vitro*. *Nat Med* 2000;6:183–188.
24. Frey N, McKinsey TA, Olson EN. Decoding calcium signals involved in cardiac growth and function. *Nat Med* 2000;6:1221–1227.
25. Crabtree GR. Calcium, calcineurin, and the control of transcription. *J Biol Chem* 2001;276:2313–2316.
26. Taigen T, De Windt LJ, Lim HW, Molkentin JD. Targeted inhibition of calcineurin prevents agonist-induced cardiomyocyte hypertrophy. *Proc Natl Acad Sci USA* 2000;97:1196–1201.
27. Fiedler B, Lohmann SM, Smolenski A, Linnemüller S, Pieske B, Schröder F, Molkentin JD, Drexler H, Wollert KC. Inhibition of calcineurin-NFAT hypertrophy signaling by cGMP-dependent protein kinase I in cardiac myocytes. *Proc Natl Acad Sci USA* 2002;99:11363–11368.
28. Suhasini M, Li H, Lohmann SM, Boss GR, Pilz RB. Cyclic-GMP-dependent protein kinase inhibits the Ras/mitogen-activated protein kinase pathway. *Mol Cell Biol* 1998;18:6983–6994.
29. Sauzeau V, Le Jeune H, Cario-Toumaniantz C, Smolenski A, Lohmann SM, Bertoglio J, Chardin P, Pacaud P, Loirand G. Cyclic GMP-dependent protein kinase signaling

pathway inhibits RhoA-induced Ca^{2+} sensitization of contraction in vascular smooth muscle. *J Biol Chem* 2000;275:21722–21729.
30. Thorburn A, Thorburn J, Chen SY, Powers S, Shubeita HE, Feramisco JR, Chien KR. HRas-dependent pathways can activate morphological and genetic markers of cardiac muscle cell hypertrophy. *J Biol Chem* 1993;268:2244–2249.
31. Hunter JJ, Tanaka N, Rockman HA, Ross J, Chien KR. Ventricular expression of a MLC-2v-ras fusion gene induces cardiac hypertrophy and selective diastolic dysfunction in transgenic mice. *J Biol Chem* 1995;270:23173–23178.
32. Aoki H, Izumo S, Sadoshima J. Angiotensin II activates RhoA in cardiac myocytes: A critical role of RhoA in angiotensin II-induced premyofibril formation. *Circ Res* 1998;82:666–676.
33. Arstall MA, Sawyer DB, Fukazawa R, Kelly RA. Cytokine-mediated apoptosis in cardiac myocytes: The role of inducible nitric oxide synthase induction and peroxynitrite generation. *Circ Res* 1999;85:829–840.
34. Beckman JS. Parsing the effects of nitric oxide, S-nitrosothiols, and peroxynitrite on inducible nitric oxide synthase-dependent cardiac myocyte apoptosis. *Circ Res* 1999;85:870–871.
35. Haunstetter A, Izumo S. Apoptosis. Basic mechanisms and implications for cardiovascular disease. *Circ Res* 1998;82:1111–1129.
36. Ing DJ, Zang J, Dzau VJ, Webster KA, Bishopric NH. Modulation of cytokine-induced cardiac myocyte apoptosis by nitric oxide, Bak, and Bcl-x. *Circ Res* 1999;84:21–33.
37. Taimor G, Hofstaetter B, Piper HM. Apoptosis induction by nitric oxide in adult cardiomyocytes via cGMP-signaling and its impairment after simulated ischemia. *Cardiovasc Res* 2000;45:588–594.
38. Wu CF, Bishopric NH, Pratt RE. Atrial natriuretic peptide induces apoptosis in neonatal rat cardiac myocytes. *J Biol Chem* 1997;272:14860–14866.
39. Mashimo H, Goyal RK. Lessons from genetically engineered animal models. IV. Nitric oxide synthase gene knockout mice. *Am J Physiol* 1999;277:G745–G750.
40. Christensen G, Wang Y, Chien KR. Physiological assessment of complex cardiac phenotypes in genetically engineered mice. *Am J Physiol* 1997;272:H2513–H2524.
41. Scherrer-Crosbie M, Ullrich R, Bloch KD, Nakajima H, Nasseri B, Aretz HT, Lindsey ML, Vancon A-C, Huang PL, Lee RT, Zapol WM, Picard MH. Endothelial nitric oxide synthase limits left ventricular remodeling after myocardial infarction in mice. *Circulation* 2001;104:1286–1291.
42. Shesely EG, Maeda N, Kim HS, Desai KM, Krege JH, Laubach VE, Sherman PA, Sessa WC, Smithies O. Elevated blood pressures in mice lacking endothelial nitric oxide synthase. *Proc Natl Acad Sci USA* 1996;93:13176–13181.
43. Liu YH, Xu J, Yang XP, Yang F, Shesely E, Carretero OA. Effect of ACE inhibitors and angiotensin II type 1 receptor antagonists on endothelial NO synthase knockout mice with heart failure. *Hypertension* 2002;39:375–381.
44. Haywood GA, Tsao PS, von der Leyen HE, Mann MJ, Keeling PJ, Trindade PT, Lewis NP, Byrne CD, Rickenbacher PR, Bishopric NH, Cooke JP, McKenna WJ, Fowler MB. Expression of inducible nitric oxide synthase in human heart failure. *Circulation* 1996;93:1087–1094.
45. Drexler H, Kästner S, Strobel A, Studer R, Brodde OE, Hasenfuss G. Expression, activity and functional significance of inducible nitric oxide synthase in the failing human heart. *J Am Coll Cardiol* 1998;32:955–963.
46. Sam F, Sawyer DB, Xie Z, Chang DL, Ngoy S, Brenner DA, Siwik DA, Singh K, Apstein CS, Colucci WS. Mice lacking inducible nitric oxide synthase have improved left ventricular contractile function and reduced apoptotic cell death late after myocardial infarction. *Circ Res* 2001;89:351–356.
47. Feng Q, Lu X, Jones DL, Shen J, Arnold JM. Increased inducible nitric oxide synthase expression contributes to myocardial dysfunction and higher mortality after myocardial infarction in mice. *Circulation* 2001;104:700–704.
48. Mungrue IN, Gros R, You X, Pirani A, Azad A, Csont T, Schulz R, Butany J, Stewart DJ, Husain M. Cardiomyocyte overexpression of iNOS in mice results in peroxynitrite generation, heart block, and sudden death. *J Clin Invest* 2002;109:735–743.
49. Nakamura K, Fushimi K, Kouchi H, Mihara K, Miyazaki M, Ohe T, Namba M. Inhibitory effects of antioxidants on neonatal rat cardiac myocyte hypertrophy induced by tumor necrosis factor-α and angiotensin II. *Circulation* 1998;98:794–799.
50. Pimentel DR, Amin JK, Xiao L, Miller T, Viereck J, Oliver-Krasinski J, Baliga R, Wang J, Siwik DA, Singh K, Pagano P, Colucci WS, Sawyer DB. Reactive oxygen species mediate amplitude-dependent hypertrophic and apoptotic responses to mechanical stretch in cardiac myocytes. *Circ Res* 2001;89:453–460.
51. Xiao L, Pimentel DR, Wang J, Singh K, Colucci WS, Sawyer DB. Role of reactive oxygen species and NAD(P)H oxidase in α_1-adrenoceptor signaling in adult rat cardiac myocytes. *Am J Physiol Cell Physiol* 2002;282:C926–C934.
52. Heger J, Godecke A, Flogel U, Merx MW, Molojavyi A, Kuhn-Velten WN, Schrader J. Cardiac-specific overexpression of inducible nitric oxide synthase does not result in severe cardiac dysfunction. *Circ Res* 2002;90:93–99.
53. Linz W, Wohlfart P, Schölkens BA, Malinski T, Wiemer G. Interactions among ACE, kinins and NO. *Cardiovasc Res* 1999;43:549–561.
54. Hayashidani S, Tsutsui H, Shiomi T, Suematsu N, Kinugawa S, Ide T, Wen J, Takeshita A. Fluvastatin, a 3-hydroxy-3-methylglutaryl coenzyme a reductase inhibitor, attenuates left ventricular remodeling and failure after experimental myocardial infarction. *Circulation* 2002;105:868–873.
55. Bauersachs J, Galuppo P, Fraccarollo D, Christ M, Ertl G. Improvement of left ventricular remodeling and function by hydroxymethylglutaryl coenzyme a reductase inhibition with cerivastatin in rats with heart failure after myocardial infarction. *Circulation* 2001;104:982–985.
56. van Eickels M, Grohe C, Cleutjens JP, Janssen BJ, Wellens HJ, Doevendans PA. 17beta-estradiol attenuates the development of pressure-overload hypertrophy. *Circulation* 2001;104:1419–1423.
57. Hernandez-Perera O, Perez-Sala D, Navarro-Antolin J, Sanchez-Pascuala R, Hernandez G, Diaz C, Lamas S. Effects of the 3-hydroxy-3-methylglutaryl-CoA reductase inhibitors, atorvastatin and simvastatin, on the expression of endothelin-1 and endothelial nitric oxide synthase in vascular endothelial cells. *J Clin Invest* 1998;101:2711–2719.
58. Nuedling S, Kahlert S, Loebbert K, Doevendans PA, Meyer R, Vetter H, Grohe C. 17 Beta-estradiol stimulates expression of endothelial and inducible NO synthase in

rat myocardium *in vitro* and *in vivo*. *Cardiovasc Res* 1999;43:666–674.
59. Stasch JP, Alonso-Alija C, Apeler H, Dembowsky K, Feurer A, Minuth T, Perzborn E, Schramm M, Straub A. Pharmacological actions of a novel NO-independent guanylyl cyclase stimulator, BAY 41-8543: *In vitro* studies. *Br J Pharmacol* 2002;135:333–343.
60. Stasch JP, Dembowsky K, Perzborn E, Stahl E, Schramm M. Cardiovascular actions of a novel NO-independent guanylyl cyclase stimulator, BAY 41-8543: *In vivo* studies. *Br J Pharmacol* 2002;135:344–355.
61. Tunctan B, Uludag O, Altug S, Abacioglu N. Effects of nitric oxide synthase inhibition in lipopolysaccharide-induced sepsis in mice. *Pharmacol Res* 1998;38:405–411.
62. von der Leyen HE, Dzau VJ. Therapeutic potential of nitric oxide synthase gene manipulation. *Circulation* 2001;103:2760–2765.
63. Bozkurt B, Torre-Amione G, Warren MS, Whitmore J, Soran OZ, Feldman AM, Mann DL. Results of targeted anti-tumor necrosis factor therapy with etanercept (ENBREL) in patients with advanced heart failure. *Circulation* 2001;103:1044–1047.
64. Khurana R, Martin JF, Zachary I. Gene therapy for cardiovascular disease: A case for cautious optimism. *Hypertension* 2001;38:1210–1216.

Role of Nitric Oxide in the Pathophysiology of Heart Failure

Hunter C. Champion, Michel W. Skaf, and Joshua M. Hare
Division of Cardiology, Department of Medicine, Johns Hopkins Hospital, Baltimore, MD, USA

Abstract. **Nitric oxide (NO) plays critical roles in the regulation of integrated cardiac and vascular function and homeostasis. An understanding of the physiologic role and relative contribution of the three NO synthase isoforms (neuronal—NOS1, inducible—NOS2, and endothelial—NOS3) is imperative to comprehend derangements of the NO signaling pathway in the failing cardiovascular system. Several theories of NO and its regulation have developed as explanations for the divergent observations from studies in health and disease states. Here we review the physiologic and pathophysiologic influence of NO on cardiac function, in a framework that considers several theories of altered NO signaling in heart failure. We discuss the notion of spatial compartmentalization of NO signaling within the myocyte in an effort to reconcile many controversies about derangements in the influences of NO in the heart and vasculature.**

Key Words. **nitric oxide, heart failure, spatial compartmentalization, myocyte**

Introduction

The discovery of the endogenous nitric oxide (NO) pathway has revolutionized concepts of endocrine-paracrine-autocrine signaling. The diatomic molecule, NO, is produced in essentially all mammalian organs and tissues by a family of NO synthases. While NO can act as a paracrine-autocrine signaling molecule (endothelial-dependent vasorelaxation as a prototype), it also has chemical properties allowing it to act with precision in subcellular microdomains [1]. Substantial work on NO biology within the heart has better clarified how this molecule mediates/modulates essentially all key pathways involved in cardiac regulation. In this review, we outline the mechanisms for NO regulation of the heart in health and disease, with particular emphasis on heart failure (HF).

The mechanism(s) by which nitric oxide (NO) influences cardiac function and, importantly, dysfunction has been difficult to conceptualize in a unified manner because of the diverse and, in some cases, seemingly opposite effects attributed to this signaling molecule [2]. NO synthases (NOS) are expressed not only in cardiac endothelial cells but also in cardiac myocytes, where NO exerts various location-specific intracrine effects [3]. Endothelial NOS (NOS3) localizes to sarcolemmal caveolae [4] and to the mitochondria [5], whereas neuronal NOS (NOS1) is localized to the sarcoplasmic reticulum (SR) [6]. In addition, inducible NOS (NOS2) is expressed when the gene is activated in response to circulating inflammatory mediators [7,8]. Among the effects attributable to NO within the heart are influences upon signal transduction/β-adrenergic stimulation [9], mitochondrial respiration/myocardial energetics [10,11], and calcium cycling [3,12] (Table 1).

In heart failure most of the NO influences on cardiac function are deranged often in seemingly opposite directions (Table 1). The notion that NO acts in cellular microdomains with site-specific modes of signaling offers an explanation for these divergent effects. This is supported by the fact that NOS isoforms are spatially confined to different cellular organelles.

Theories of Dysregulation of NO in HF

Several theories have been explored in detail in an attempt to offer a comprehensive explantation of altered NO signaling in the failing circulation. These are discussed below.

NOS2 Induction

Following the initial report by Shulz and coworkers [13] that exposure of cardiac myocytes to cytokines led to the induction of biochemical Ca^{2+}-independent NOS activity, both (iNOS) mRNA and protein have been reported in the heart [14]. In several studies, NOS2 expression in cardiac myocytes or adjacent cells has been shown to result in contractile dysfunction, notably depression of β-adrenergic inotropic responses [14]. Based on these data and the observations that there are elevated circulating cytokines in HF [15,16], it has

Address for correspondence: Joshua M. Hare, M.D., Cardiology Division, Johns Hopkins Hospital, 600 N Wolfe Street, Carnegie 568, Baltimore, MD 21287, USA. Tel.: 410 614-4161; Fax: 410 955-3478; E-mail: jhare@mail.jhmi.edu

Table 1. Effect of endogenous NO on myocardial physiology and the pathophysiology of heart failure

	Effect of NO in normal heart	L-NMMA or $NOS3^{-/-}$	L-NMMA in CHF
Basal contractility	↑	↓(52,53)[a]	↔(52,53)[a]
Lusitropy	↑	↓(34,48)[a]	↓
β-Adrenergic inotropy	↓	↑(55,88)	↑↑(9,90)[a]
SW/MVO_2[b]	↑	↓(11)	↔(110)[a]
Ca^{2+} release (stimulated by β-agonist)	↑	↑↑(3)	?

The table depicts reported effects of endogenous NO on five aspects of cardiac physiology (first column), the impact of inhibition of NO signaling with NOS inhibition with L-N^G-monomethyl arginine (L-NMMA) or NOS3 gene deletion in the otherwise normal heart (second column), and the impact of L-NMMA in heart failure (third column).

[a]Observations have been extended to humans.

[b]SW/MVO_2, ratio of myocardial stroke work to oxygen consumption, a measure of cardiac mechanical energetic efficiency.

been suggested that iNOS expression *in vivo* may contribute to myocardial contractile dysfunction in HF.

Despite the attractiveness of this hypothesis, the precise roles of NOS2 in HF remain controversial [17]. It is possible to that NOS2 induction may have beneficial effects including: cytoprotection, decreased leukocyte adhesion, antiplatelet activity, reduced vascular permeability, and antioxidant activity. On the other hand, the widespread expression of iNOS, especially in noninflammatory cells, may have harmful consequences [8,18–20]. Potential subcellular mechanisms involved in these harmful effects include excessive direct (cGMP-independent) reactions of NO with a wide variety of proteins and enzymes including reactions with amino, thiol (—SH), diazo, and tyrosyl groups, and with heme and Fe^{2+} or sulfur centers. In addition, excess NO could lead to abnormal elevations in cGMP concentrations or disruption of NO signaling requiring precise modulation. Finally unregulated NO production also associated with oxidative stress can result in the generation of peroxynitrite and other reactive nitrogen species that alter protein function via nitration and oxidation reactions [18,21–25]. Moreover, the reaction of excess NO in the setting of superoxide can result in the inactivation of physiologically important NO [26]. The precise relevance of any of these pathways remains speculative and none has emerged as a dominant mechanism in HF.

NOS3 Downregulation

Endothelium-dependent coronary vasodilation is profoundly depressed in HF [27], leading to the suggestion that reduced NO release in all areas of the cardiovascular system is a dominant mechanism in HF. Studies at the enzymatic level are contradictory; In dogs with pacing-induced HF, aortic endothelial cells have decreased expression of NOS3 [28], while in some studies in end-stage human HF NOS3 expression is *increased* [29]. Fukuchi et al. reported a variably increased NOS3 only in subendocardial cardiac myocytes, which did not correlate with NOS activity [30]. Moreover it was reported that eNOS expression elsewhere in the heart (especially in coronary microvessels) was reduced [30]. Furthermore, the true activity of the enzyme *in vivo* may not be reflected in NOS activity assays performed in the presence of non-limiting concentrations of substrate and cofactors. Further studies are required to confirm these data, to define the mechanisms responsible for the variations in NOS2 and NOS3 gene expression in individual patients, and to directly establish the relationship between NOS gene expression and functional activity *in vivo*.

Spatial Confinement of NOS

Spatial confinement is an emerging concept in the arena of NO biology [2,3]. With the observations that NOS3 and NOS1 are present in the cardiac myocyte as well as NOS3 in the coronary endothelium, it could be argued that there is a significant degree of redundancy in the cardiac system. However, the observations that NOS3 and NOS1 have divergent roles in the mouse, suggest that the system is not redundant [3]. The idea of spatial confinement of NOS isoforms and their products may help explain the apparent differences in NO effects in the cardiac system. This review will discuss the NO literature as it relates to HF and demonstrate areas in which this concept of spatial confinement may reconcile apparent discrepancies.

Modulation of the NO System

Substrate Availability

The availability of L-arginine as substrate for NOS has been controversial. Although it has been suggested with the intracellular concentration of L-arginine should be adequate for NO production [31,32], several studies have shown a benefit to giving supplemental L-arginine in models of aging and disease [33–35]. If the concentration of L-arginine were sufficient for NOS signaling, supplemental substrate should not have an effect on the system. Recently it has been shown that arginase, an enzyme that competes for L-arginine, is upregulated in aging (unpublished observations) and diabetes [36], and competes with NOS. Arginase converts L-arginine to urea and L-ornithine and limits NO production. In these studies, arginase inhibitors were shown to increase NO production by increasing L-arginine availability for NOS [36].

Downstream Mechanisms of NO Stimulation

The activation of NOS results in the activation of a number of complex signaling pathways. Many of the cardiac actions of NO are believed to be mediated by the activation of soluble guanylyl cyclase and the resulting elevation in intracellular cGMP [31,37,38]. The predominant myocardial targets of the activation of soluble guanylyl cyclase are the cGMP-dependent protein kinases (PKGs). In addition, the cGMP-stimulated and cGMP-inhibited cyclic nucleotide phosphodiesterases play a role in regulating the downstream actions of cGMP through its inactivation. Other less understood pathways may involve cGMP-regulated ion channels and the activation of cGMP phosphatases. The increase in intracellular cGMP concentration in myocardium results in modulation of Ca^{2+}-influx, decreased myofilament response to Ca^{2+}, altered sarcoplasmic reticulum function, action potential changes, cell volume alteration, and a decrease in cellular oxygen consumption.

While NO can alter myocardial function through cGMP-dependent pathways, it is becoming increasingly apparent that cGMP-independent actions of NO in the heart may at least as if not more important [12,39]. In this regard, precise cellular signaling mediated by protein modification through nitrosylation of specific thiol residues [40] can profoundly alter protein function [1]. Among the many regulatory pathways of the cardiac ryanodine receptor, responsible for regulating Ca^{2+} induced Ca^{2+} release in excitation-contraction coupling, is thiol nitrosylation [12]. Nitrosylation may also play a role in regulating myocardial energetics at both mitochondrial [41] and creatinine kinase levels [42].

The precise effects of nitrosylation based signaling mechanisms may be disrupted during conditions of enhanced NO production (e.g. NOS2 induction). In addition, excess NO production in conjunction with reactive oxygen species (ROS) can lead to peroxynitrite formation as a result of a reaction between NO and O_2^- [18,21,22,38]. Peroxynitrite can, in low concentrations, stimulate guanylate cyclase. However, in higher concentrations, the protonation of peroxynitrite results in the formation of peroxynitrous acid which then acts to generate hydroxyl species [43,44]. These hydroxyl species induce toxic cellular effects by oxidation and protein nitration [1,12,45,46].

Role of NO on Basal Myocardial Contractility

Systolic Function

Physiologic Response. Conflicting results have made the interpretation of NO effect on systolic cardiac function very complex. Brady et al. first described a negative NO effect on basal contractility under normal condition as well as after induction of inducible NOS (iNOS or NOS2) in guinea pig myocytes [47]. Paulus et al. extended these finding to humans by showing that bicoronary infusion of NO donor sodium nitroprusside decreases estimated LV end-systolic elastance (Ees) [48]; in this work peak systolic pressure fell while end-systolic volume remained unchanged. Other studies have shown either biphasic [49–51] or negligible effects. Regardless of the effect reported, the magnitude of the response to NO donors is small.

A second approach to assess NO effects on basal contractility is the application of NOS inhibitors. This approach has the advantage of establishing the role of endogenously produced NO and may reflect biologically relevant findings to a greater extent than studies using NO donors. In this regard, Cotton and co-workers reported that intracoronary infusion of L-N^G monomethly arginine (L-NMMA) reduced an isovolumic index of contractility, LV dP/dt_{max}, in subjects with normal LV function, but did not affect basal contractility in HF [52]. Similarly, we reported that L-NMMA depressed basal myocardial contractility in normal and not in HF dogs [53,54]. Moreover, the $NOS3^{-/-}$ mouse [55,56] has mildly reduced basal $+dP/dt$ compared to wild type. On the other hand, similar studies in rat myocytes and guinea pig hearts failed to demonstrate an NO effect on basal contractility [57,58].

Response in Heart Failure. Similarly conflicting results are reported for the effects of NO on basal contractility in HF. While some studies showed a negative effect of NO on contractility in HF [59,60], others [52] showed that it has no effect. Both our group in dogs and humans and Cotton et al. in humans have shown that L-NMMA reduces $+dP/dt$ in normals but not in subjects with heart failure. These studies utilized direct comparison between normal and heart failure subjects using NOS inhibition and therefore strongly suggest that a stimulatory effect of NO on contractility is lost in HF (Table 1). Recently we demonstrated that caveolin-3, a peptide that inhibits basal NOS activity (see below), is elevated in HF, an observation that offers a potential mechanism for the loss of an NO effect on basal contractility.

Diastolic Function

Physiologic Response. Studies have consistently shown that NO hastens diastolic relaxation. This has been shown using NO donors, stimulators of NO release, cGMP, and cGMP stimulators in isolated papillary muscles [49,61,62], isolated hearts [63,64], isolated myocytes [38,65–68] and in humans [48,60]. In an isolated papillary muscle preparation, NO caused an earlier onset of isometric twitch relaxation and a reduction in peak tension development, in a setting of little or no change in the maximum rate of tension development [63]. This selective effect on myocardial relaxation has been reported using exogenous NO donors (Sodium nitroprusside; SNP), endothelium-derived NO (released by substance P), a cGMP analogue [38] and with atrial natriuretic peptide (which raises myocardial cGMP via stimulation of particulate guanylyl cyclase).

It is important to note that the effect of NO to enhance LV relaxation is selective and not in compensation to other hemodynamic changes as shown in isolated hearts studied at constant preload, afterload, and heart rate [63]. Stimulation of NO release or selective donors of NO induce earlier and faster LV relaxation, and a small reduction in peak LV pressure (LVP) without reducing LV dP/dt_{max}. These effects were independent of alterations in coronary flow and were not reproduced with the NO- and cGMP-independent vasodilator nicardipine.

Modulation by NO of the onset of LV relaxation and of LV diastolic properties has been extended to the human heart *in vivo* [48,69]. Low-dose coronary infusion of SNP or substance P results in earlier onset of LV relaxation and reduced peak end-systolic LVP, without altering LV dP/dt_{max}. It also resulted in a reduction in LV minimum diastolic pressure and LVEDP, with a rise in LV end-diastolic volume (LVEDV) and a down and rightward displacement of the LV diastolic pressure-volume relationship which are consistent with an NO-induced increase in LV diastolic distensibility (passive diastolic filling) [48,69].

Response in Heart Failure. Whether the ability of NO to modulate diastolic function is altered in heart failure remains controversial. Recchia et al. found that reduced NO metabolite release from the heart in HF correlates with elevations in LV end-diastolic pressure [10]. Moreover, lusitropy (active diastolic relaxation) is lengthened after L-NMMA and in NOS3$^{-/-}$ mice (unpublished observations). We suggested that NOS3 activity is reduced despite normal levels of NOS3 at baseline in dogs with heart failure secondary to upregulation of caveolin-3. Reduced baseline NO production in HF could contribute to diastolic impairment.

These findings must be balanced by studies demonstrating elevated NOS2 in cardiomyopathy. The increase in NOS2 or iNOS creates a milieu in which NO is produced in an uncontrolled manner and is only limited by substrate availability. The stimulus for NOS2 expression is uncertain, but as mentiomed above it has been suggested that inflammatory cytokines (e.g. TNF-α) are elevated in plasma of patients with heart failure and that these cytokines stimulate NOS2 production at the level of mRNA transcription.

Effects of NO on Inotropic Responses

Effect of NO on the β-Adrenergic Signaling Pathway

Physiologic Response. Sympathetic activation of the β-adrenergic pathway in the heart produces positive inotropic, chronotropic, and lusitropic responses (see [70] for review). These responses are mediated through the activation of the β_1- and β_2-adrenoceptors coupled to a stimulatory G protein (G_s), which in turn stimulates cAMP production by adenylyl cyclase (AC). cAMP, through cAMP-dependent protein kinase-A (PKA), directly augments the L-type Ca^{2+} current (I_{ca}), which enhances excitation-contraction coupling [71]. The best-understood signaling pathway opposing β-adrenergic cardiac activation is mediated by muscarinic-cholinergic stimulation of the heart [72–74]. Muscarinic receptors are coupled to an inhibitory G protein ($G_{i\alpha}$) that both inhibits AC production of cAMP and stimulates NO production via NOS3 [73]. NO, which stimulates the production of cGMP, can have both cGMP-dependent [62,65,66,75–77] and cGMP-independent [39,42,78] inotropic effects. With regard to β-adrenergic stimulation of cAMP, a clear role has been established for cGMP as a counterbalancing force [75,76]. Thus, in addition to muscarinic cholinergic pathways, the NO pathway may also subserve an inhibitory influence over β-adrenergic inotropic and chronotropic responses, a contention supported by early observations made in isolated myocytes [79] and intact animals [80].

Two observations strongly support that NO plays a negative feedback role with regard to β-adrenergic signaling. First, Kanai has shown that β-adrenergic agonists increase NO production [81]. Secondly, selective agonists of the β3-adrenoceptor cause negative inotropic effects [3,82–87] accompanied by increases in NO and cGMP production [82–84]. Using mice deficient for this receptor, we have demonstrated that it is responsible for NO-mediated negative feedback over β-adrenergic stimulated positive inotropy [88].

NO inhibition of β-adrenergic inotropy is primarily mediated by NOS3 which localizes to sarcolemmal caveolae in proximity to β-adrenergic receptors and effector proteins such as the L-type Ca^{2+} channel. NOS1 which localizes to the cardiac sarcoplasmic reticulum, appears to subserve a stimulatory effect over β-adrenergic inotropic responses as deletion of NOS1 leads to suppressed Ca^{2+} transients and myofilament contraction in response to isoproterenol [3].

Response in Heart Failure. An enhanced ability of NO to inhibit β-adrenergic inotropic responses is widely reported in both human dilated cardiomyopathy [9,89], in pacing-induced canine heart failure, a widely used and highly relevant animal model [53], and in cytokine treated rodent models [7,8]. Studies have used the approach of inhibiting endogenous NO production with NOS inhibitors as well as NOS agonists to demonstrate this response. With regard to the latter, Bartuneck and colleagues reported that dobutamine potentiates the effect of substance P on ventricular relaxation in dilated cardiomyopathy (DCM) patients [59]. Moinette et al. demonstrated that β_3-adrenoceptor proteins and $G_{i\alpha}$ proteins are upregulated in HF when compared to control subjects [86], providing a potential mechanism for increased NO release during β-adrenergic stimulation in HF. We have also previously shown that NOS inhibition by L-NMMA augments β-adrenergic contractility to a greater degree in patients with idiopathic dilated cardiomyopathy than in control subjects with normal LV function [9,90]. This response which is seemingly opposite to the findings that L-NMMA appears to lack an influence over basal contractility in HF exemplifies the difficulty at arriving at a unified concept of altered NO signaling in heart failure. As discussed below, pathways that regulate NOS or its downstream signaling may be altered in HF in such a way as to cause divergent effects on basal versus adrenergically stimulated myocardial contraction.

Effect of NO on the Force Frequency Response
The force frequency response (FFR), which describes the increase in cardiac contractility in response to increased heart rate, is a second major cardiac reserve mechanism. Changes in the amplitude, shape, and position of the myocardial FFR occurring in heart failure are characterized in terms of maximal isometric twitch tension, slope of the ascending limb (myocardial reserve), and position of the peak of the FFR on the frequency axis (optimum stimulation frequency). All three of these parameters are reduced in heart failure. Substantial data supports a sarcoplasmic reticulum Ca^{2+}-pump based mechanism accounting for progressive depression of the FFR in failing hearts. The manner in which NO influences myocardial contractility and the FFR has been extremely controversial. On the one hand, studies using NO donors and NOS inhibitors in both experimental systems and in humans have failed to detect NO influences [52,91,92] whereas others have shown an inhibitory influence with similar interventions [93,94].

SR Ca^{2+} stores are a major determinant of the FFR. NOS1$^{-/-}$ mice have a depressed FFR and reduced SR Ca^{2+} stores [95]. These data suggest that NOS1 enhances the FFR under physiologic conditions and may also play an important role in HF. Further studies, however, are needed to elucidate the role of NOS1 in modulating the FFR in humans under baseline condition as well as in HF.

Influence of NO on Myocardial Efficiency (the Hemodynamic/Energetic Aspect of HF)

The failing heart displays substantial energy inefficiency in both isolated muscle, and the intact heart [96–99]. This phenomenon can be best described as "mechanoenergetic uncoupling," given that the depression of contractile force is not matched by a concomitant depression of energy consumption. Among the proposed mechanisms is enhanced oxidative stress stemming from mitochondrial [100] and cytosolic free radical generating systems [101]. Xanthine oxidase (XO) is prominent among these enzymes, because it produces superoxide as a byproduct of the terminal two steps of purine metabolism [97]. XO is upregulated in failing myocardium of experimental animal models [97,102] and humans [103], and its inhibition by allopurinol improves the contractility and mechanical efficiency (the ratio between ventricular work performed and oxygen consumed) of intact failing hearts [11].

NO may influence energetics both at the SR [6,12,104,105] and the mitochondrial [5,41,106] levels. In the former, NOS activity has been shown to augment SR Ca^{2+}-release channel activity via protein nitrosylation in a manner regulated by oxidative stress [6,12,104,107,108]. Oxidative stress likely promotes maximal channel activity, reducing the ability of NO to exert feedback regulation of SR Ca^{2+} release. Such a situation is likely to lead to "futile" Ca^{2+} cycling, which increases ATP expenditure [109]. NO may additionally influence SR Ca^{2+}-ATPase activity; however, this remains controversial with studies demonstrating contradictory results [6,107]. On the other hand, NO might play a protective role in HF by attenuating β-adrenergic contractility and hence sparing myocardial oxygen consumption [110]. We have shown that NOS inhibition blocked the inotropic and energetic effect of allopurinol and asocorbic acid in HF, suggesting that NO may play an endogenous protective role with regard to oxidative disruption of energetics in HF [11].

Mechanisms of NO Regulation in HF

Role of Caveolin 3

Inhibition of NOS leads to an enhancement of β-adrenergic inotropic responses as previously discussed. This observation supported the hypothesis that NO signaling contributes to β-adrenergic chronotropic and inotropic hyporesponsiveness in situations such as sepsis [7,111], aging [7,34,112], and HF [9,89,113]. With regard to HF, we have previously shown that L-NMMA augments β-adrenergic contractility to a greater degree in patients with idiopathic dilated cardiomyopathy than in control subjects with normal LV function [9]. Whether or not upregulation of one or more NOS isoforms explains this response has not been consistently demonstrated as discussed above, and other mechanisms that regulate NOS have been sought.

Recently we demonstrated that caveolin-3 is elevated in dogs with pacing induced HF [53]. Caveolins are scaffolding proteins found in caveolae, plasmalemmal microdomains that participate in signal transduction by means of colocalizing membrane receptors with signal transduction effectors [114–117]. Because caveolin inhibits NOS activity by preventing calmodulin activation, it may exert dual regulation of NOS: inhibition of basal activity yet augmentation of agonist-stimulated actions by virtue of bringing NOS in proximity with activating receptors. Western analysis and electron microscopy revealed ~2-fold increase in both caveolin-3 protein abundance and myocyte sarcolemmal caveolae, respectively [53]. Confocal imaging identified caveolin-3 localized to the sarcolemma and T-tubules and colocalized with NOS3 at these sites in HF myocytes. Moreover, caveolin-3 abundance correlated with the augmentation of dobutamine contractility due to NOS inhibition in HF dogs. This increased expression of caveolin-3 has the potential to produce greater inhibition of NOS3 at baseline, while triggering more NO release after β-adrenergic stimulation.

Role of β-3 Adrenoreceptors

In the failing heart, β_1- and β_2-adrenoceptors are either downregulated or desensitized [118]. In contrast, the abundance of β_3-adrenoceptors (β-AR) appears to increase in human heart failure [86]. In the normal heart, β_3-AR participates in nitric oxide (NO)–mediated negative feedback control over contractility within the sympathetic nervous system. Stimulation of β_3-ARs with β_3-AR selective agonists produces negative inotropic effect in human donor hearts through NO signaling [82,83,85], and β_3-AR deficiency blocked NO-dependent inhibition of myocardial contractility in transgenic mice [88]. The β_3-AR is coupled selectively to NOS3 and not NOS1 as predicted by subcellular localization of these isoforms as evidenced by loss of negative inotropic effects of $\beta 3$ agonists in $NOS3^{-/-}$ but not $NOS1^{-/-}$ mice [3].

As previously mentioned, in failing human hearts, β_3-AR abundance is increased, and the balance between opposing inotropic influences of β_1-ARs and β_3-ARs is significantly altered, providing a potential mechanism for progressive deterioration in cardiac function. In a recent study in the failing canine heart, it has been shown that $\beta 3$-ARs are, like in humans, upregulated and associated with enhanced β_3-AR–mediated negative modulation of myocyte contractile response and $[Ca^{2+}]_i$ regulation [119]. Although this effect was found to be largely secondary to the effect of NO, other pathways may be involved because the myocyte response to β_3-AR activation was only partially inhibited by pretreatment of myocytes with the non-selective NO synthase inhibitor L-NAME. It has, therefore, been suggested that upregulated β_3-ARs in CHF may exacerbate dysfunctional $[Ca^{2+}]_i$ homeostasis and regulation, causing greater inhibition of cardiac contraction and relaxation and worsening heart failure. Alternatively, NO released in the myocardium following β_3-adrenoceptor stimulation could enhance diastolic relaxation (thereby increasing diastolic reserve of the failing heart) and reduce oxygen consumption [60,110]. Future studies on the effects of selective $\beta 3$-AR antagonists and agonists are needed to determine what therapeutic role the modulation of this receptor may play in heart failure.

Role of Phosphodiesterase (PDE5) Expression and Activity

The impact of nitric oxide formation and release can be regulated via second messenger systems associated with the activation of soluble guanylyl cyclase and the regulation of cGMP [80,120]. A mechanism gaining attention is augmented cGMP synthesis. cGMP, which is produced by two isoforms of guanylyl cyclase—a membrane-bound form (ANP/BNP receptors) and a cytosolic form activated by nitric oxide (NO)—enhances or opposes the action of cAMP in a concentration-dependent manner [120]. Evidence has accumulated that myocardial cGMP signaling increases in heart failure via both ANP/BNP and NO pathways, [120,121], and likely contributes to reduced β-adrenergic responsiveness. cGMP content is also regulated by catabolic enzymes, yet little is known about their role in the heart. Various phosphodiesterases regulate cGMP catabolism, including PDE5A, PDE6, PDE9A, PDE10A, and PDE11A [121] and their expression is often selective to specific tissues. Although the peripheral and coronary vascular effects of the inhibition of PDE5A have been widely studied, relatively little is known about the role of PDE5A in the regulation of cardiac function [121]. We have recently shown that PDE5A modulates systolic and diastolic β-adrenergic responsiveness in the intact *in vivo* heart [121]. In normal hearts, PDE5A inhibition potently depressed dobutamine-stimulated inotropic and lusitropic responses consistent with the effect of NO to inhibit β-adrenergic responses.

This study also provided novel evidence for a reduced physiological effect from PDE5A inhibition in failing hearts [121]. HF was accompanied by decreased protein expression and activity in total heart extracts. Cardiac myocytes however had unchanged PDE5 but a loss of PDE5 protein distribution along the Z-band of the myocyte. Interestingly, PDE5 colocalizes with caveolin-3 providing additional evidence for the importance of protein-protein interactions between molecules of the NO signaling pathway.

In terms of the functional impact of altered PDE5 distribution in HF, PDE5 inhibition failed to reduce β-adrenergic responses in HF animals. These findings represent another example whereby regulation of NO signaling is disrupted in HF; in this case cGMP degradation is reduced by loss of normal PDE5 activity and this in turn could enhance the ability of NO-cGMP pathways to offset β-adrenergic signaling in the failing heart [121].

Taken together the findings regarding caveolin-3, the β3-adrenoceptor and PDE5 strongly support the concept of disruption of spatially confined NO signaling. In this context, changes in NOS abundance may be viewed as secondary to or a bystander phenomenon to dysregulation of a signaling pathway tightly regulated by proteins that are spatially confined and linked to each other by protein-protein interactions.

Role of Compartmentalization vs Diffusion in the Explanation of the Effects of NO on Myocardial Function

Subcellular localization of nitric oxide (NO) synthases with effector molecules is an important regulatory mechanism for NO signaling [122,123]. In the heart, NO inhibits L-type Ca^{2+} channels but stimulates sarcoplasmic reticulum (SR) Ca^{2+} release, leading to variable effects on myocardial contractility. We have shown that spatial confinement of specific NO synthase isoforms regulates this process and may explain the differences observed in previous studies. Endothelial NO synthase (NOS3) localizes to caveolae, where compartmentalization with β-adrenergic receptors and L-type Ca^{2+} channels allows NO to inhibit β-adrenergic-induced inotropy. Neuronal NO synthase (NOS1), however, is targeted to cardiac SR. NO stimulation of SR Ca^{2+} release via the ryanodine receptor (RyR) *in vitro* suggests that NOS1 has an opposite, facilitative effect on contractility. Supporting this contention, NOS1-deficient mice have suppressed inotropic response, whereas NOS3-deficient mice have enhanced contractility, owing to corresponding changes in SR Ca^{2+} release. Thus, NOS1 and NOS3 mediate independent, and in some cases opposite, effects on cardiac structure and function. These findings are highly supportive of a paradigm for intracrine NO signaling, whereby spatial confinement of different NO synthase isoforms allows NO signals to have independent, and even opposite, effects on cardiac phenotype. As such, local regulation of effector molecules is a central mechanism by which NO exerts biological activity. Thus it can be envisioned that NO provides exquisite fine-tuning of organ function by recruiting different downstream NO signaling pathways (for example, S-nitrosylation or cGMP production) within distinct microdomains of the same cell.

A specific role for NO in myocardial energetics is also consistent with the model of spatial confinement. Indeed the mitochondria contains a specific NOS, alternatively described as either NOS3 or NOS1 [124], and the possibility of a unique isoform remains an active speculation. As discussed above both NO and free radical formation regulate mitochondrial function in intact heart energetics. Spatially unique signaling phenomena offer an explanation for divergent effects of NO at mitochondrial [10,11] vs. β-adrenergic signaling levels [9,89] in HF. It is possible that the decrease in mitochondrial respiration in heart failure is a result of increased reactive oxygen species such as superoxide in the mitochondria that limit the utilization of NO for respiration [100].

A model of spatial confinement of NO signaling must be reconciled with the high diffusibility of NO and the close proximity of endothelial cells to myocytes. The best rationalization is that although NO is highly diffusible it is also highly reactive with thiols, hemoproteins, and oxygen. Thus, organelles that are critically dependent on NO for optimal function (the SR and mitochondria in the heart) have local NOS sources.

Cardiac Hypertrophy

NO activity in the heart influences not only cardiac function but also its structure, exerting an anti-hypertrophic effect [125,126]. We have shown that both NOS isoforms found in the heart play protective roles in this regard. Both NOS1$^{-/-}$ and NOS3$^{-/-}$ mice develop age-related hypertrophy, although only NOS3$^{-/-}$ mice are hypertensive. NOS1/3$^{-/-}$ double knockout mice have an additive phenotype of marked ventricular remodeling [3]. The independent contributions of NO synthase isoforms to the maintenance of cardiac architecture has important implications for the pathophysiology of heart failure [3].

Summary

Since its discovery, nitric oxide has been shown to play a critical role in the regulation of cardiac and vascular function. An understanding of the physiologic role and relative contribution of NOS1, NOS2, and NOS3 under normal conditions is imperative in order to understand the mechanisms of the derangements of the NO signaling pathway with heart failure. In this review of the literature as it relates to the influence of NO on cardiac function in normal and diseased states, it is clear that there is a very tight regulation of the molecule with regard to its formation and release as well as its cellular impact. A theory of spatial compartmentalization of the NO pathway in the myocyte helps reconcile many controversies in the literature regarding differences in the effects of NO in the heart and vasculature in heart failure.

Acknowledgment

This work is supported by NIH grant HL65455 and a Paul Beeson Physician Faculty Scholars in Aging Research Award to Dr. Joshua Hare.

References

1. Stamler JS, Lamas S, Fang FC. Nitrosylation. The prototypic redox-based signaling mechanism. *Cell* 2001;106:675–683.
2. Hare JM, Stamler JS. NOS: Modulator, not mediator of cardiac performance. *Nat Med* 1999;5:273–274.
3. Barouch LA, Harrison RW, Skaf MW, Rosas GO, Cappola TP, Kobeissi ZA, Hobai IA, Lemmon CA, Burnett AL, O'Rourke B, Rodriguez ER, Huang PL, Lima JA, Berkowitz DE, Hare JM. Nitric oxide regulates the heart by spatial confinement of nitric oxide synthase isoforms. *Nature* 2002;416:337–339.
4. Michel T, Feron O. Nitric oxide synthases: Which, where, how and why? *J Clin Invest* 1997;100:2146–2152.
5. Bates TE, Loesch A, Burnstock G, Clark JB. Mitochondrial nitric oxide synthase: A ubiquitous regulator of oxidative phosphorylation? *Biochem Biophys Res Comm* 1996;218:40–44.
6. Xu KY, Huso DL, Dawson T, Bredt DS, Becker LC. NO synthase in cardiac sarcoplasmic reticulum. *Proc Natl Acad Sci USA* 1999;96:657–662.
7. Rosas GO, Zieman SJ, Donabedian M, Vandegaer K, Hare JM. Augmented age-associated innate immune responses contribute to negative inotropic and lusitropic effects of lipopolysaccharide and interferon gamma. *J Mol Cell Cardiol* 2001;33:1849–1859.
8. Funakoshi H, Kubota T, Kawamura N, Machida Y, Feldman AM, Tsutsui H, Shimokawa H, Takeshita A. Disruption of inducible nitric oxide synthase improves beta-adrenergic inotropic responsiveness but not the survival of mice with cytokine-induced cardiomyopathy. *Circ Res* 2002;90:959–965.
9. Hare JM, Givertz MM, Creager MA, Colucci WS. Increased sensitivity to nitric oxide synthase inhibition in patients with heart failure: Potentiation of β-adrenergic inotropic responsiveness. *Circulation* 1998;97:161–166.
10. Recchia FA, McConnell PI, Bernstein RD, Vogel TR, Xu X, Hintze TH. Reduced nitric oxide production and altered myocardial metabolism during the decompensation of pacing-induced heart failure in the conscious dog. *Circ Res* 1998;83:969–979.
11. Saavedra WF, Paolocci N, St John ME, Skaf MW, Stewart GC, Xie JS, Harrison RW, Zeichner J, Mudrick D, Marban E, Kass DA, Hare JM. Imbalance between xanthine oxidase and nitric oxide synthase signaling pathways underlies mechanoenergetic uncoupling in the failing heart. *Circ Res* 2002;90:297–304.
12. Xu L, Eu JP, Meissner G, Stamler JS. Activation of the cardiac calcium release channel (Ryanodine receptor) by Poly-S-Nitrosylation. *Science* 1998;279:234–237.
13. Schulz R, Nava E, Moncada S. Induction and potential biological relevance of a Ca$^{(2+)}$-independent nitric oxide synthase in the myocardium. *Br J Pharmacol* 1992;105:575–580.
14. Balligand J-L, Ungureanu-Longrois D, Simmons WW, Pimental D, Malinski TA, Kapturczak M, Taha Z, Lowenstein CJ, Davidoff AJ, Kelly RA, Smith TW, Michel T. Cytokine-inducible nitric-oxide synthase (iNOS) expression in cardiac myocytes: Characterization and regulation of iNOS expression and detection of iNOS activity in single cardiac myocytes *in vitro*. *J Biol Chem* 1994;269:27580–27588.
15. Torre-Amione G, Kapadia S, Benedict C, Oral H, Young JB, Mann DL. Proinflammatory cytokine levels in patients with depressed left ventricular ejection fraction: A report from the studies of left ventricular dysfunction (SOLVD). *J Am Coll Cardiol* 1996;27:1201–1206.

16. Torre-Amione G, Kapadia S, Lee J, Durand J-B, Bies RD, Young JB, Mann DL. Tumor necrosis factor-α and tumor necrosis factor receptors in the failing human heart. *Circulation* 1996;93:704–711.
17. Heger J, Godecke A, Flogel U, Merx MW, Molojavyi A, Kuhn-Velten WN, Schrader J. Cardiac-specific overexpression of inducible nitric oxide synthase does not result in severe cardiac dysfunction. *Circ Res* 2002;90:93–99.
18. Ferdinandy P, Danial H, Ambrus I, Rothery RA, Schulz R. Peroxynitrite is a major contributor to cytokine-induced myocardial contractile failure. *Circ Res* 2000;87:241–247.
19. Shindo T, Ikeda U, Ohkawa F, Kawahara Y, Yokoyama M, Shimada K. Nitric oxide synthesis in cardiac myocytes and fibroblasts by inflammatory cytokines. *Cardiovasc Res* 1995;29:813–819.
20. Pinsky DJ, Cai B, Yang X, Rodriguez C, Sciacca RR, Cannon PJ. The lethal effects of cytokine-induced nitric oxide on cardiac myocytes are blocked by nitric oxide synthase antagonism or transforming growth factor β. *J Clin Invest* 1995;95:677–685.
21. Ferdinandy P, Panas D, Schulz R. Peroxynitrite contributes to spontaneous loss of cardiac efficiency in isolated working rat hearts. *Am J Physiol* 1999;276:H1861–H1867.
22. Oyama J, Shimokawa H, Momii H, Cheng X, Fukuyama N, Arai Y, Egashira K, Nakazawa H, Takeshita A. Role of nitric oxide and peroxynitrite in the cytokine-induced sustained myocardial dysfunction in dogs *in vivo*. *J Clin Invest* 1998;101:2207–2214.
23. Xie Y-W, Kaminski PM, Wolin MS. Inhibition of rat cardiac muscle contraction and mitochondrial respiration by endogenous peroxynitrite formation during posthypoxic reoxygenation. *Circ Res* 1998;82:891–897.
24. Szabo C, Ferrer-Sueta G, Zingarelli B, Southan GJ, Salzman AL, Radi R. Mercaptoethylguanidine and guanidine inhibitors of nitric-oxide synthase react with peroxynitrite and protect against peroxynitrite- induced oxidative damage. *J Biol Chem* 1997;272:9030–9036.
25. Lopez BL, Liu GL, Christopher TA, Ma X. Peroxynitrite, the product of nitric oxide and superoxide, causes myocardial injury in the isolated perfused rat heart. *Coronary Artery Disease* 1997;8:149–153.
26. Harrison DG. Cellular and molecular mechanisms of endothelial cell dysfunction. *J Clin Invest* 1997;100:2153–2157.
27. Ito K, Akita H, Kanazawa K, Yamada S, Terashima M, Matsuda Y, Yokoyama M. Comparison of effects of ascorbic acid on endothelium-dependent vasodilation in patients with chronic congestive heart failure secondary to idiopathic dilated cardiomyopathy versus patients with effort angina pectoris secondary to coronary artery disease. *Am J Cardiol* 1998;82:762–767.
28. Smith CJ, Sun D, Hoegler C, Roth BS, Zhang X, Zhao G, Xu XB, Kobari Y, Pritchard KJ, Sessa WC, Hintze TH. Reduced gene expression of vascular endothelial NO synthase and cyclooxygenase-1 in heart failure. *Circ Res* 1996;78:58–64.
29. Stein B, Eschenhagen T, Rudiger J, Scholz H, Forstermann U, Gath I. Increased expression of constitutive nitric oxide synthase III, but not inducible nitric oxide synthase II, in human heart failure. *J Am Coll Cardiol* 1998;32:1179–1186.
30. Fukuchi M, Hussain SN, Giaid A. Heterogeneous expression and activity of endothelial and inducible nitric oxide synthases in end-stage human heart failure: Their relation to lesion site and beta-adrenergic receptor therapy. *Circulation* 1998;98:132–139.
31. Werner ER, Werner-Felmayer G, Mayer B. Tetrahydrobiopterin, cytokines, and nitric oxide synthesis. *Proc Soc Exp Biol Med* 1998;219:171–182.
32. Shimizu S, Ishii M, Momose K, Yamamoto T. Role of tetrahydrobiopterin in the function of nitric oxide synthase, and its cytoprotective effect (Review). *Int J Mol Med* 1998;2:533–540.
33. Cooke JP, Dzau VJ. Derangements of the nitric oxide synthase pathway, L-arginine, and cardiovascular diseases. *Circulation* 1997;96:379–382.
34. Zieman SJ, Gerstenblith G, Lakatta EG, Rosas GO, Vandegaer K, Ricker KM, Hare JM. Upregulation of the nitric oxide-cGMP pathway in aged myocardium: Physiological response to L-arginine. *Circ Res* 2001;88:97–102.
35. McNamara DB, Bedi B, Aurora H, Tena L, Ignarro LJ, Kadowitz PJ, Akers DL. L-arginine inhibits balloon catheter-induced intimal hyperplasia. *Biochem Biophys Res Commun* 1993;193:291–296.
36. Bivalacqua TJ, Hellstrom WJ, Kadowitz PJ, Champion HC. Increased expression of arginase II in human diabetic corpus cavernosum: In diabetic-associated erectile dysfunction. *Biochem Biophys Res Commun* 2001;283:923–927.
37. Balligand JL, Cannon PJ. Nitric oxide synthases and cardiac muscle. Autocrine and paracrine influences. *Arterioscler Thromb Vasc Biol* 1997;17:1846–1858.
38. Shah AM, MacCarthy PA. Paracrine and autocrine effects of nitric oxide on myocardial function. *Pharmacol Ther* 2000;86:49–86.
39. Campbell DL, Stamler JS, Strauss HC. Redox modulation of L-type calcium channels in ferret ventricular myocytes. Dual mechanism regulation by nitric oxide and S-nitrosothiols. *J Gen Physiol* 1996;108:277–293.
40. Sun J, Xin C, Eu JP, Stamler JS, Meissner G. Cysteine-3635 is responsible for skeletal muscle ryanodine receptor modulation by NO. *Proc Natl Acad Sci USA* 2001;98:11158–11162.
41. Clementi E, Brown GC, Feelisch M, Moncada S. Persistent inhibition of cell respiration by nitric oxide: Crucial role of S-nitrosylation of mitochondrial complex I and protective action of glutathione. *Proc Natl Acad Sci USA* 1998;95:7631–7636.
42. Gross WL, Bak MI, Ingwall JS, Arstall MA, Smith TW, Balligand J-L, Kelly RA. Nitric oxide inhibits creatine kinase and regulates rat heart contractile reserve. *Proc Natl Acad Sci USA* 1996;93:5604–5609.
43. Moro MA, Darley-Usmar VM, Lizasoain I, Su Y, Knowles RG, Radomski MW, Moncada S. The formation of nitric oxide donors from peroxynitrite. *Br J Pharmacol* 1995;116:1999–2004.
44. Wu M, Pritchard KA Jr, Kaminski PM, Fayngersh RP, Hintze TH, Wolin MS. Involvement of nitric oxide and nitrosothiols in relaxation of pulmonary arteries to peroxynitrite. *Am J Physiol* 1994;266:H2108–H2113.
45. Gow AJ, Chen Q, Hess DT, Day BJ, Ischiropoulos H, Stamler JS. Basal and stimulated protein S-nitrosylation in multiple cell types and tissues. *J Biol Chem* 2002;277:9637–9640.
46. Hess DT, Matsumoto A, Nudelman R, Stamler JS. S-nitrosylation: Spectrum and specificity. *Nat Cell Biol* 2001;3:E46–E49.

47. Brady AJB, Poole-Wilson PA, Harding SE, Warren JB. Nitric oxide production within cardiac myocytes reduces their contractility in endotoxemia. *Am J Physiol* 1992;263:H1963–H1966.
48. Paulus WJ, Vantrimpont PJ, Shah AM. Acute effects of nitric oxide on left ventricular relaxation and diastolic distensibility in humans. Assessment by bicoronary sodium nitroprusside infusion. *Circulation* 1994;89: 2070–2078.
49. Mohan P, Brutsaert DL, Paulus WJ. Myocardial contractile response to nitric oxide and cGMP. *Circulation* 1996;93:1223–1229.
50. Kojda G, Kottenberg K, Noack E. Inhibition of nitric oxide synthase and soluble guanylate cyclase induces cardiodepressive effects in normal rat hearts. *Eur J Pharmacol* 1997;334:181–190.
51. Kojda G, Kottenberg K. Regulation of basal myocardial function by NO. *Cardiovasc Res* 1999;41:514–523.
52. Cotton JM, Kearney MT, MacCarthy PA, Grocott-Mason RM, McClean DR, Heymes C, Richardson PJ, Shah AM. Effects of nitric oxide synthase inhibition on basal function and the force-frequency relationship in the normal and failing human heart *in vivo*. *Circulation* 2001;104:2318–2323.
53. Hare JM, Lofthouse RA, Juang GJ, Colman L, Ricker KM, Kim B, Senzaki H, Cao S, Tunin RS, Kass DA. Contribution of caveolin protein abundance to augmented nitric oxide signaling in conscious dogs with pacing-induced heart failure. *Circ Res* 2000;86:1085–1092.
54. Harrison RW, Thakkar RN, Senzaki H, Ekelund UE, Cho E, Kass DA, Hare JM. Relative contribution of preload and afterload to the reduction in cardiac output caused by nitric oxide synthase inhibition with L-N(G)-methylarginine hydrochloride 546C88. *Crit Care Med* 2000;28:1263–1268.
55. Gyurko R, Kuhlencordt P, Fishman MC, Huang PL. Modulation of mouse cardiac function *in vivo* by eNOS and ANP. *Am J Physiol Heart Circ Physiol* 2000;278:H971–H981.
56. Huang PL, Huang Z, Mashimo H, Bloch KD, Moskowitz MA, Bevan JA, Fishman MC. Hypertension in mice lacking the gene for endothelial nitric oxide synthase. *Nature* 1995;377:239–242.
57. Abi-Gerges N, Fischmeister R, Mery PF. G protein-mediated inhibitory effect of a nitric oxide donor on the L-type Ca^{2+} current in rat ventricular myocytes. *J Physiol (Lond)* 2001;531:117–130.
58. Gallo MP, Malan D, Bedendi I, Biasin C, Alloatti G, Levi RC. Regulation of cardiac calcium current by NO and cGMP-modulating agents. *Pflugers Arch* 2001;441:621–628.
59. Bartunek J, Shah AM, Vanderheyden M, Paulus WJ. Dobutamine enhances cardiodepressant effects of receptor-mediated coronary endothelial stimulation. *Circulation* 1997;95:90–96.
60. Heymes C, Vanderheyden M, Bronzwaer JG, Shah AM, Paulus WJ. Endomyocardial nitric oxide synthase and left ventricular preload reserve in dilated cardiomyopathy. *Circulation* 1999;99:3009–3016.
61. Flesch M, Kilter H, Cremers B, Lenz O, Sudkamp M, Kuhn-Regnier F, Bohm M. Acute effects of nitric oxide and cyclic GMP on human myocardial contractility. *J Pharmacol Exp Ther* 1997;281:1340–1349.
62. Smith JA, Shah AM, Lewis MJ. Factors released from endocardium of the ferret and pig modulate myocardial contraction. *J Physiol* 1991;439:1–14.
63. Grocott-Mason R, Anning P, Evans H, Lewis M, Shah A. Modulation of left ventricular relaxation in isolated ejecting heart by endogenous nitric oxide. *Am J Physiol* 1994;267:H1804–H1813.
64. Anning PB, Grocott-Mason RM, Lewis MJ, Shah AM. Enhancement of left ventricular relaxation in the isolated heart by an angiotensin-converting enzyme inhibitor. *Circulation* 1995;92:2660–2665.
65. Shah AM, Lewis MJ, Henderson AH. Effects of 8-bromo-cyclic GMP on contraction and on inotropic response of ferret cardiac muscle. *J Mol Cell Cardiol* 1991;23:55–64.
66. Shah AM, Spurgeon HA, Sollott SJ, Talo A, Lakatta EG. 8-bromo-cGMP reduces the myofilament response to Ca^{2+} in intact cardiac myocytes. *Circ Res* 1994;74:970–978.
67. Shah AM. Paracrine modulation of heart cell function by endothelial cells. *Cardiovasc Res* 1996;31:847–867.
68. Vila-Petroff MG, Younes A, Egan J, Lakatta EG, Sollott SJ. Activation of distinct cAMP-dependent and cGMP-dependent pathways by nitric oxide in cardiac myocytes. *Circ Res* 1999;84:1020–1031.
69. Paulus WJ, Vantrimpont PJ, Shah AM. Paracrine coronary endothelial control of left ventricular function in humans. *Circulation* 1995;92:2119–2126.
70. Dzimiri N. Regulation of beta-adrenoceptor signaling in cardiac function and disease. *Pharmacol Rev* 1999;51:465–501.
71. Hartzell HC, Fischmeister R. Opposite effects of cyclic GMP and cyclic AMP on Ca^{2+} current in single heart cells. *Nature* 1986;323:273–275.
72. Henning RJ, Khalil IR, Levy MN. Vagal stimulation attenuates sympathetic enhancement of left ventricular function. *Am J Physiol* 1990;258:H1470–H1475.
73. Hare JM, Kim B, Flavahan NA, Ricker KM, Peng X, Colman L, Weiss RG, Kass DA. Pertussis toxin-sensitive G proteins influence nitric oxide synthase III activity and protein levels in rat heart. *J Clin Invest* 1998;101:1424–1431.
74. Hare JM, Keaney JF Jr, Balligand JL, Loscalzo J, Smith TW, Colucci WS. Role of nitric oxide in parasympathetic modulation of beta-adrenergic myocardial contractility in normal dogs. *J Clin Invest* 1995;95(1):360–366.
75. Mery P-F, Lohmann SM, Walter U, Fischmeister R. Ca^{2+} current is regulated by cyclic GMP-dependent protein kinase in mammalian cardiac myocytes. *Proc Natl Acad Sci USA* 1991;88:1197–1201.
76. Mery P-F, Pavoine C, Belhassen L, Pecker F, Fischmeister R. Nitric oxide regulates cardiac Ca^{2+} current. Involvement of cGMP-inhibited and cGMP-stimulated phosphodiesterases through guanylyl cyclase activity. *J Biol Chem* 1993;268:26286–26295.
77. Ji GJ, Fleischmann BK, Bloch W, Feelisch M, Andressen C, Addicks K, Hescheler J. Regulation of the L-type Ca^{2+} channel during cardiomyogenesis: Switch from NO to adenylyl cyclase-mediated inhibition. *FASEB J* 1999;13:313–324.
78. Paolocci N, Ekelund UEG, Isoda T, Ozaki M, Vandegaer K, Georgakopoulos D, Harrison R, Kass DA, Hare JM. cGMP-independent inotropic effect of nitric oxide and peroxynitirite donors: Potential role for S-nitrosylation. *Am J Physiol* 2000;279:H1982–H1988.

79. Balligand J-L, Kelly RA, Marsden PA, Smith TW, Michel T. Control of cardiac muscle cell function by an endogenous nitric oxide signaling system. *Proc Natl Acad Sci USA* 1993;90:347–351.
80. Keaney JF Jr, Hare JM, Kelly RA, Loscalzo J, Smith TW, Colucci WS. Inhibition of nitric oxide synthase potentiates the positive inotropic response to β-adrenergic stimulation in normal dogs. *Am J Physiol* 1996;271:H2646–H2652.
81. Kanai AJ, Mesaros S, Finkel MS, Oddis CV, Birder LA, Malinski T. β-Adrenergic regulation of constitutive nitric oxide synthase in cardiac myocytes. *Am J Physiol* 1997;273:C1371–C1377.
82. Gauthier C, Leblais V, Kobzik L, Trochu J, Khandoudi N, Bril A, Balligand J-L, Le Marec H. The negative inotropic effect of β_3-adrenoreceptor stimulation is mediated by activation of a nitric oxide synthase pathway in human ventricle. *J Clin Invest* 1998;102:1377–1384.
83. Gauthier C, Tavernier G, Charpentier F, Langin D, Le Marec H. Functional β_3-adrenoceptor in the human heart. *J Clin Invest* 1998;98:556–562.
84. Gauthier C, Leblais V, Kobzik L, Trochu J-N, Khandoudi N, Bril A, Balligand J-L, LeMarec.H. The negative inotropic effect of β_3-adrenoceptor stimulation is mediated by activation of a nitric oxide synthase pathway in human ventricle. *J Clin Invest* 1998;102:1377–1384.
85. Gauthier C, Tavernier G, Trochu J, Leblais V, Laurent K, Langin D, Escande D, Le Marec H. Interspecies differences in the cardiac negative inotropic effects of β_3-adrenoreceptor agonists. *J Pharmacol Exp Ther* 1999;290:687–693.
86. Moniotte S, Kobzik L, Feron O, Trochu JN, Gauthier C, Balligand JL. Upregulation of beta(3)-adrenoceptors and altered contractile response to inotropic amines in human failing myocardium. *Circulation* 2001;103:1649–1655.
87. Shen Y-T, Cervoni P, Claus T, Vatner SF. Differences in β_3-adrenergic receptor cardiovascular regulation in conscious primates, rats and dogs. *J Pharmacol Exp Ther* 1996;278:1435–1443.
88. Varghese P, Harrison RW, Lofthouse RA, Georgakopoulos D, Berkowitz DE, Hare JM. Beta(3)-adrenoceptor deficiency blocks nitric oxide-dependent inhibition of myocardial contractility. *J Clin Invest* 2000;106:697–703.
89. Hare JM, Loh E, Creager MA, Colucci WS. Nitric oxide inhibits the contractile response to β-adrenergic stimulation in humans with left ventricular dysfunction. *Circulation* 1995;92:2198–2203.
90. Wittstein IS, Kass DA, Maughan WL, Pak PH, Fetics B, Hare JM. Cardiac nitric oxide production due to angiotensin converting-enzyme inhibition decreases β-adrenergic myocardial contractility in patients with dilated cardiomyopathy. *J Am Coll Cardiol* 2001;38:429–435.
91. Prabhu SD, Azimi A, Frosto T. Nitric oxide effects on myocardial function and force-interval relations: Regulation of twitch duration. *J Mol Cell Cardiol* 1999;31:2077–2085.
92. Prabhu SD, Freeman GL. Effect of tachycardia heart failure on the restitution of left ventricular function in closed-chest dogs. *Circulation* 1995;91:176–185.
93. Kaye DM, Wiviott SD, Balligand J-L, Simmons WW, Smith TW, Kelly RA. Frequency-dependent activation of a constitutive nitric oxide synthase and regulation of contractile function in adult rat ventricular myocytes. *Circ Res* 1996;78:217–224.
94. Finkel MS, Oddis CV, Mayer OH, Hattler BG, Simmons RL. Nitric oxide synthase inhibitor alters papillary muscle force-frequency relationship. *J Pharmacol Exp Ther* 1995;272:945–952.
95. Harrison RW, Skaf MW, Berkowitz DE, Shoukas AA, Hare JM. Cardiac nitric oxide synthase 1 preserves the force-frequency response in mice. *Circulation* 2001;104:II–436 (Abstract).
96. Ishihara H, Yokota M, Sobue T, Saito H. Relation between ventriculoarterial coupling and myocardial energetics in patients with idiopathic dilated cardiomyopathy. *J Am Coll Cardiol* 1994;23:406–416.
97. Ekelund UEG, Harrison RW, Shokek O, Thakkar RN, Tunin RS, Senzaki H, Kass DA, Marbán E, Hare JM. Intravenous allopurinol decreases myocardial oxygen consumption and increases mechanical efficiency in dogs with pacing-induced heart failure. *Circ Res* 1999;85:437–445.
98. Wolff MR, Buck SH, Stoker SW, Greaser ML, Mentzer RM. Myofibrillar calcium sensitivity of isometric tension is increased in human dilated cardiomyopathies: Role of altered beta-adrenergically mediated protein phosphorylation. *J Clin Invest* 1996;98:167–176.
99. Wolff MR, De Tombe PP, Harasawa Y, Burkhoff D, Bier S, Hunter WC, Gerstenblith G, Kass DA. Alterations in left ventricular mechanics, energetics, and contractile reserve in experimental heart failure. *Circ Res* 1992;70:516–529.
100. Ide T, Tsutsui H, Kinugawa S, Utsumi H, Kang D, Hattori N, Uchida K, Arimura K, Egashira K, Takeshita A. Mitochondrial electron transport complex I is a potential source of oxygen free radicals in the failing myocardium. *Circ Res* 1999;85:357–363.
101. Saugstad OD. Role of xanthine oxidase and its inhibitor in hypoxia: Reoxygenation injury. *Pediatrics* 1996;98:103–107.
102. de Jong JW. Xanthine oxidoreductase activity in perfused hearts of various species, including humans. *Circ Res* 1990;67:770–773.
103. Cappola TP, Kass DA, Nelson GS, Berger RD, Rosas GO, Kobeissi ZA, Marban E, Hare JM. Allopurinol improves myocardial efficiency in patients with idiopathic dilated cardiomyopathy. *Circulation* 2001;104:2407–2411.
104. Xu KY, Zweier JL, Becker LC. Functional coupling between glycolysis and sarcoplasmic reticulum Ca^{2+} transport. *Circ Res* 1995;77:88–97.
105. Xu L, Mann GE, Meissner G. Regulation of cardiac Ca^{2+} release channel (ryanodine receptor) by Ca^{2+}, H^{+}, Mg^{2+}, and adenine nucleotides under normal and simulated ischemic conditions. *Circ Res* 1996;79:1100–1109.
106. Clementi E, Brown GC, Foxwell N, Moncada S. On the mechanism by which vascular endothelial cells regulate their oxygen consumption. *Proc Natl Acad Sci USA* 1999;96:1559–1562.
107. Eu JP, Sun J, Xu L, Stamler JS, Meissner G. The skeletal muscle calcium release channel: Coupled O_2 sensor and NO signaling functions. *Cell* 2000;102:499–509.
108. Eu JP, Xu L, Stamler JS, Meissner G. Regulation of ryanodine receptors by reactive nitrogen species. *Biochemical Pharmacology* 1999;57:1079–1084.
109. Takasago T, Goto Y, Kawaguichi O, Hata K, Saeki A, Nishioka T, Suga H. Ryanodine wastes oxygen consumption for Ca^{2+} handling in the dog heart. *J Clin Invest* 1993;92:823–830.

110. Shinke T, Takaoka H, Takeuchi M, Hata K, Kawai H, Okubo H, Kijima Y, Murata T, Yokoyama M. Nitric oxide spares myocardial oxygen consumption through attenuation of contractile response to beta-adrenergic stimulation in patients with idiopathic dilated cardiomyopathy. *Circulation* 2000;101:1925–1930.
111. Balligand J-L, Ungureanu D, Kelly RA, Kobzik L, Pimental D, Michel T, Smith TW. Abnormal contractile function due to induction of nitric oxide synthesis in rat cardiac myocytes follows exposure to activated macrophage-conditioned medium. *J Clin Invest* 1993;91:2314–2319.
112. Lakatta EG. Cardiovascular regulatory mechanisms in advanced age. *Physiol Rev* 1993;73:413–467.
113. Hare JM, Colucci WS. Role of nitric oxide in the regulation of myocardial function. *Prog Cardiovasc Dis* 1995;38:155–166.
114. Feron O, Smith TW, Michel T, Kelly RA. Dynamic targeting of the agonist-stimulated m2 muscarinic acetylcholine receptor to caveolae in cardiac myocytes. *J Biol Chem* 1997;272:17744–17748.
115. Schwencke C, Okumura S, Yamamoto M, Geng YJ, Ishikawa Y. Colocalization of beta-adrenergic receptors and caveolin within the plasma membrane. *J Cell Biochem* 1999;75:64–72.
116. Schwencke C, Yamamoto M, Okumura S, Toya Y, Kim SJ, Ishikawa Y. Compartmentation of cyclic adenosine 3′,5′-monophosphate signaling in caveolae. *Mol Endocrinol* 1999;13:1061–1070.
117. Yamamoto M, Okumura S, Oka N, Schwencke C, Ishikawa Y. Downregulation of caveolin expression by cAMP signal. *Life Sci* 1999;64:1349–1357.
118. Brodde OE. Beta-adrenoceptors in cardiac disease. *Pharmacol Ther* 1993;60:405–430.
119. Cheng HJ, Zhang ZS, Onishi K, Ukai T, Sane DC, Cheng CP. Upregulation of functional beta(3)-adrenergic receptor in the failing canine myocardium. *Circ Res* 2001;89:599–606.
120. Hart CY, Hahn EL, Meyer DM, Burnett JC Jr, Redfield MM. Differential effects of natriuretic peptides and NO on LV function in heart failure and normal dogs. *Am J Physiol Heart Circ Physiol* 2001;281:H146–H154.
121. Senzaki H, Smith CJ, Juang GJ, Isoda T, Mayer SP, Ohler A, Paolocci N, Tomaselli GF, Hare JM, Kass DA. Cardiac phosphodiesterase 5 (cGMP-specific) modulates beta-adrenergic signaling *in vivo* and is down-regulated in heart failure. *FASEB J* 2001;15:1718–1726.
122. Brenman JE, Chao DS, Gee SH, McGee AW, Craven SE, Santillano DR, Wu Z, Huang F, Xia H, Peters MF, Froehner SC, Bredt DS. Interaction of nitric oxide synthase with the postsynaptic density protein PSD-95 and α1-syntrophin mediated by PDZ domains. *Cell* 1996;84:757–767.
123. Fang M, Jaffrey SR, Sawa A, Ye K, Luo X, Snyder SH. Dexras1: A G protein specifically coupled to neuronal nitric oxide synthase via CAPON. *Neuron* 2000;28:183–193.
124. Kanai AJ, Pearce LL, Clemens PR, Birder LA, VanBibber MM, Choi SY, de Groat WC, Peterson J. Identification of a neuronal nitric oxide synthase in isolated cardiac mitochondria using electrochemical detection.*Proc Natl Acad Sci USA* 2001;98:14126–14131.
125. Calderone A, Thaik CM, Takahashi N, Chang DLF, Colucci WS. Nitric oxide, atrial natriuretic peptide, and cyclic GMP inhibit the growth-promoting effects of norepinephrine in cardiac myocytes and fibroblasts. *J Clin Invest* 1998;101:812–818.
126. Yang XP, Liu YH, Shesely EG, Bulagannawar M, Liu F, Carretero OA. Endothelial nitric oxide gene knockout mice: Cardiac phenotypes and the effect of angiotensin-converting enzyme inhibitor on myocardial ischemia/reperfusion injury. *Hypertension* 1999;34:24–30.

Nitric Oxide, Cell Death, and Heart Failure

Jun-ichi Oyama, MD,[1] Stefan Frantz, MD,[3]
Charles Blais, Jr., PhD,[1] Ralph A. Kelly, MD,[2]
and Todd Bourcier, PhD[1]

[1] Cardiovascular Division, Brigham and Women's Hospital, Boston, MA, USA; [2] Genzyme Corporation, Framingham, MA, USA; [3] Medizinische Universitätsklinik Würzburg, Würzburg, Germany

Abstract. **Strong evidence links cardiomyocyte loss to the pathology of some forms of heart failure. Both necrotic and apoptotic modes of cell death have been invoked as the mechanism underlying progressive cardiomyocyte dropout. Nitric oxide (NO) has received particular attention as a candidate reactive oxygen intermediate that influences not only cardiac function, but also cell death elicited by both apoptotic and necrotic mechanisms. NO is produced by resident cardiac cells under stress, and is produced in large quantities by activated immune cells that infiltrate the injured heart. A review of the literature, however, reveals that the actions of NO on apoptotic cell death are complex, especially in the context of heart disease, and that the practical contribution of NO to cell death in heart disease is yet to be defined.**

Key Words. **nitric oxide, apoptosis, heart failure, necrosis**

Introduction

Progressive degeneration of left ventricular (LV) function is a characteristic of most forms of heart failure [1]. This steady decline in LV function is characterized by increased LV chamber dilatation and the activation of neurohumoral mechanisms that are persistently active. These phenomena are part of a compensatory process that maintains cardiac output in the short run [2–5]. However, if the underlying etiology of the heart's reduced function is not remedied, these previously beneficial mechanisms may progress to end-stage disease. How and why this progression occurs is a point of controversy and ongoing research. Studies over the past decade have demonstrated that myocyte death and inadequate myocyte proliferation occur in heart failure, leading to progressive LV dysfunction [6–9].

Cell Death and Heart Failure

A number of studies document the continuous loss of myocytes in aging hearts apparently free of obvious cardiac disease [10], as well as in a number of experimental animal and human cardiac pathologies, including models of cardiac overload [11], pacing [12], infarction, and reperfusion injury [12], and in human myocardial infarction [13,14], myocarditis [15], and end-stage heart failure [10,16]. A review of the literature uncovers two particularly pertinent issues: 1) *which type of cell death—apoptotic or necrotic—predominates and is of most relevance in each cardiac pathology*, and 2) *to what extent does the loss of myocytes contribute to progressive ventricular remodeling*? The reader is directed elsewhere for a more comprehensive discussion of these issues [6,8,9,17].

The relative contribution of apoptotic vs. necrotic cell death to total myocyte loss in various cardiac pathologies cannot, as yet, be quantitated with certainty. This is due in part to the limitations of current assays that measure apoptosis as opposed to cells undergoing necrotic cell death, DNA repair, replication, and phagocytic elimination of apoptotic bodies [18–22]. Also, not knowing the kinetics myocyte cell death *in vivo* precludes calculations of average rates of cell death over longer periods of time (e.g., years). The use of several distinct measures of cell death combined with innovation in biochemical methods will reduce technical error and improve estimates of cell death in heart failure [8,23]. Knowing which type of cell death is triggered in a given cardiac pathology is more than of passing interest because programmed, apoptotic cell death may be more amenable to intervention than necrotic cell death [24,25]. Thus, idiopathic or dilated cardiomyopathy may benefit more from an intervention that reduces rates of apoptotic cell death than in cases of myocardial infarction, where myocyte necrosis is typically thought to exceed myocyte apoptosis. Of note, variable rates of myocyte apoptosis have now been documented in cases of myocardial infarction and ischemia/reperfusion injury [24,25], thus raising

Address for correspondence: Ralph A. Kelly, MD, Genzyme Corporation, 15 Pleasant Street Connector, PO Box 9322, Framingham, MA 01701-9322, USA. Tel.: 508-271-3313; Fax: 508-271-2692; E-mail: ralph.kelly@genzyme.com questions about the predominance of necrosis in these pathologies.

Results from several murine models clearly show that a rapid loss of myocytes can elicit heart failure and death. The IL6 family of receptors transduce cell survival signals, and IL6 is increased following cardiac injury [26,27]. Hirota et al. [26] created a cardiac-specific knockout for gp130, the common subunit of the IL6 receptor family. In a pressure overload model of heart failure, a staggering 34% of cardiocytes were engaged in apoptosis, and over 90% of the animals rapidly died. Directly activating pro-apoptotic pathways by overexpressing caspase 8 in the heart using transgenic methods also resulted in cardiocyte apoptosis sufficient to cause rapid death of the animal [28]. Lastly, cardiac-specific overexpression of the GTP-binding protein Gsα, a model that simulates enhanced β-adrenergic activity in heart failure, resulted in cardiomyopathy in aging mice that was associated with greater degrees of baseline and isoproterenol-stimulated cardiomyocyte apoptosis [29]. Thus, dropout of a sufficient number of apoptotic myocytes indeed can cause experimental LV dysfunction. Whether current estimates of cell death in human cardiomyopathies are sufficient to impact LV dysfunction in human disease remains unproven, although it is likely that apoptosis plays a role in human heart failure.

Nitric Oxide and Cell Death

Nitric oxide is a reactive free radical gas and thus, in sufficient quantities, can be directly cytotoxic. In activated macrophages for example, NO is an important cytotoxic molecule that, as part of innate immunity, suppresses growth of pathogens as well as decreasing the rate of tumor growth in some models [30,31]. The cytotoxicity of NO and its reactive oxidant peroxynitrite is accomplished through mechanisms linked to a necrotic-type of cell death, including inhibition of DNA synthesis, decreased mitochondrial respiration, and aconitase activity [32–35]. In addition, Albina et al. [36] have demonstrated a link between NO production and apoptosis, also in macrophages. Thus, NO has the potential to affect both necrotic and apoptotic cell death. All three isoforms of the NO synthase (NOS) are responsible for producing NO from L-arginine and oxygen. Expression of the NOS isoforms in the heart is the focus of accompanying reviews by Dr. Drexler and Dr. Duncan. It is reasonable to suggest that NO produced by Type II NOS (inducible NOS or 'iNOS') is probably involved in directly causing cell death under pathological situations. As evidence of this suggestion, iNOS is expressed in a spectrum of cardiac diseases, but not in normal hearts free from disease [37–40]. Also, the amounts of NO produced by iNOS far exceed that produced by endothelial-type I NOS (eNOS), thus increasing the likelihood of achieving levels of NO that yield peroxynitrite, protein nitrosylation, and cytotoxicity. Indeed, the earliest studies linking NO to myocyte death *in vitro* and *in vivo* identified iNOS as the likely source of increased NO production [41,42]. Since these studies were published, the relation between NO, cell death, and apoptosis has become increasingly complex, as the following discussion will demonstrate.

Pleiotropic Effects of NO and Regulation of Cell Death: In Vitro Studies

A large number of *in vitro* studies ascribe pro-apoptotic and anti-apoptotic effects to NO in different cell types. An overview of the pleiotropic effects of NO on cell death is depicted in Figure 1. These pathways are described below, followed by a discussion of results derived specifically from cardiomyocytes cultured *in vitro*.

Induction of Apoptosis by NO

(1) DNA Damage. NO can block the supply of deoxyribonucleotides by inhibiting ribonucleotide reductase in tumor cell lines, thus reducing DNA

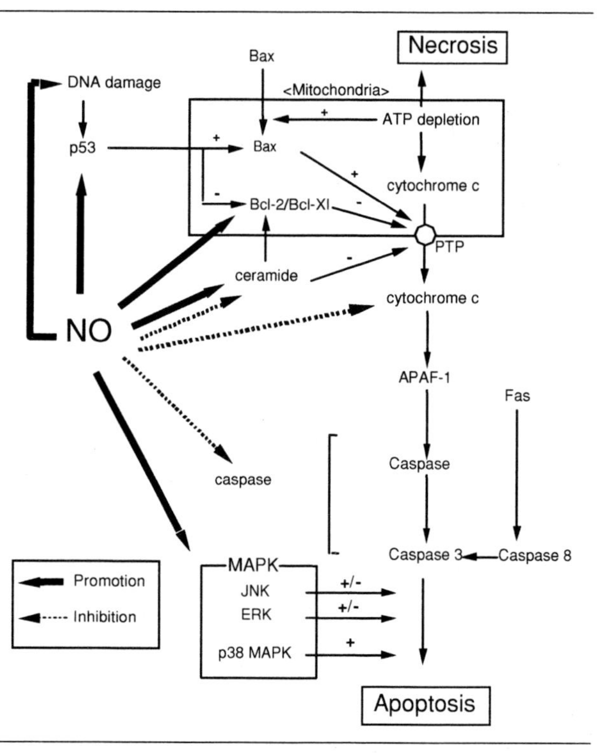

Fig. 1. Schematic representing the pleiotropic effects of nitric oxide on mechanisms of cell death. See text for details.

synthesis [43]. NO and its reactive oxidant, peroxynitrite, can also induce DNA strand breakage directly in vascular smooth muscle cells and macrophages [43,44]. Damage to DNA triggers secondary effects including upregulation of the tumor suppressor p53 [45] and activation of the DNA repair enzyme polyadenosine diphosphate ribosyl synthetase (PARS) [44]. This energy-consuming repair mechanism may eventually lead to apoptotic cell breakdown.

(2) Activation of Mitochondrial Pro-Apoptotic Pathways. An important pro-apoptotic pathway in most cell types involves the release of critical mitochondrial molecules. Of particular importance is the mitochondrial loss of cytochrome C. NO and peroxynitrite are reported to provoke release of cytochrome C from the mitochondria to the cytosol by opening of the mitochondrial permeability transition pore (PTP) [32,33,46]. Following release into the cytosol, cytochrome C can bind to Apaf-1 that, in turn, activates a caspase 9-dependent apoptotic signaling cascade [47,48]. Activation of caspases leads to a lethal breakdown of cellular target proteins and cell death [23,47,48]. Additionally, NO can indirectly increase production of superoxide and hydroxyl radical by competing with molecular oxygen for binding sites on cytochrome C oxidase in mitochondria isolated from rat hearts [49]. Thus, NO can impair mitochondrial respiration and reduce ATP synthesis [33] that can lead to apoptotic and necrotic cell death, depending on the extent of ATP depletion [89].

(3) Increase in the Tumor Suppressor Gene p53. p53 is a transcriptional transactivator protein involved in the induction of apoptosis and regulation of the cell cycle in a variety of cells that have sustained DNA damage. NO can increase the expression of p53 in various cell types, including cardiac myocytes [45,50,51], presumably through peroxynitrite production and DNA strand breakage. Induction of p53 may increase apoptosis by suppressing anti-apoptotic proteins like Bcl-xl and by increasing levels of pro-apoptotic proteins like Bax. That p53 may be an important mediator of NO-induced apoptosis was suggested by the anti-apoptotic effect of antisense oligonucleotides to p53 in cells exposed to NO. However, NO is still capable of inducing apoptosis in cell lines defective of p53 production [45], thus questioning the relevance of this pro-apoptotic pathway for NO-induced cell death. Intriguingly, p53 downregulates transcription of eNOS through a p53 binding site in the eNOS promoter. It has been suggested that the transcriptional regulation of eNOS represents a feedback mechanism to prevent overexpression of pro-apoptotic NO [52].

(4) Activation of Mitogen Activated Protein Kinases (MAPK). Nitric oxide can activate the mitogen activated protein kinase family, including JNK/SAPK (c-JUN N-terminal kinases/stress-activated protein kinases) and ERK (extracellular signal-regulated kinases), which have been linked to activation of caspase-3 [53]. Whether activation of these kinases promote or inhibit NO-induced cell death is unclear. Suppressing the activity of JNK/SAPK and ERK proteins can reduce the amount of NO-induced apoptotic cell death [54,55]. Conversely, other experiments suggest that increasing the activity of JNK/SAPK and ERK proteins protects cells from NO-induced apoptotic cell death [56] (see below).

(5) cGMP and Ceramide. NO increases cGMP production through a high affinity interaction with the heme moeity of cytosolic guanylyl cyclase [57,58]. Some reports have implicated cGMP as mediating NO-induced apoptotic cell death [59], whereas other studies excluded this pathway [60]. Thus, cGMP's role in NO-induced cell death has not yet been clarified. It should be noted, however, that cGMP is not required for NO to interact with and alter mitochondrial enzymes required for generating ATP, or for the release of cytochrome C from mitochondria [61,62]. NO can also increase cellular ceramide levels by induction of a neutral sphingomyelinase activity and inhibition of ceramidase activities [63,64]. Ceramide formation can then trigger several pro-apoptotic pathways, including the release of mitochondrial cytochrome C, caspase activation, and the suppression of Bcl-2 [65].

Inhibition of Apoptosis by NO

Nitric oxide can also exert *anti*-apoptotic effects in different cell types including hepatocytes [66], cardiac myocytes [67], and vascular endothelial cells [68]. Several mechanisms have been described.

(1) Inhibition of Caspases. NO is reported to suppress apoptosis by interrupting caspase activation [66,69,70]. Caspases harbor a reactive cysteine residue at the catalytic core that is required for proteolytic activity. NO is capable of inactivating at least 7 distinct caspases by S-nitrosylation of this cysteine residue [69]. Among these, inactivation of caspase-8 and caspase-3, which represent the upstream and downstream components of the apoptotic caspase cascade, is suggested to underlie NO-induced anti-apoptotic effects in some cell types [66,69–71].

(2) Regulation of Anti-Apoptosis Related Genes. NO can induce several anti-apoptotic genes including proteins of the cytoprotective Bcl-2 family [72]. Also, NO-induced inhibition of TNF (tumor necrosis factor) or actinomycin-induced apoptosis is suggested to occur through increased expression of heat shock protein 70 [73].

(3) NO as a Scavenger of Radical Species. There is also evidence from mesencephalic cells that nitric oxide may suppress superoxide/hydrogen peroxide-mediated cytotoxicity by acting as a scavenger of reactive oxygen species [74].

NO, Cell Death, and Cardiomyocytes In Vitro

A distinct separation between NO's pro- and anti-apoptotic effects is apparent from *in vitro* studies of cardiomyocytes in culture and from isolated perfused hearts. Application of exogenous NO in the form of NO donors to neonatal or adult rat myocytes reliably leads to rapid apoptotic and, depending on NO-donor concentration, necrotic cell death [51,75]. More importantly, endogenous production of NO by the NO synthases also promotes cardiomyocyte cell death. Elevated levels of cytokines are a hallmark feature of cardiac injury, heart failure, and myocarditis. Combinations of some of these cytokines, particularly TNFα, IL1β, and IFNγ, can kill myocytes and vascular cells *in vitro* with delayed kinetics of 48–72 hours [60,76,77]. Expression of iNOS results in copious NO production in a time course shorter than cytokine-induced apoptotic events [76]. Indeed, arginine-analog inhibitors of iNOS limit cytokine-induced NO production and cell death, thus identifying NO production as the mechanism that mediates cytokine-induced apoptosis of cardiomyocytes, at least *in vitro* [60,76,77].

The mechanism of NO-induced myocyte apoptosis remains unclear. Rabkin et al. [75] suggests that sodium nitroprusside kills cardiomyocytes indirectly through production of hydrogen peroxide, a well-known pro-apoptotic oxidant in myocytes [48]. Some investigators provide evidence that NO promotes myocyte death through a cGMP-dependent mechanism, other studies find no such evidence. As mentioned above, NO can activate JNK/SAPK and MAPK, and this effect occurs in cultured myocytes. However, blocking activation of these kinases inhibited NO-induced apoptosis in one study [55], but increased NO-induced apoptosis in another [56]. This discrepancy may lie in the pharmacological versus molecular techniques each group utilized for inhibiting the kinases. Nonetheless, the role of JNK/SAPK and MAPK in NO-induced apoptosis of cardiomyocytes remains unclear. It is worth highlighting that many studies using cardiomyocytes *in vitro* report that inhibiting caspases also prevents NO-induced apoptosis [56,59,76,78]. Identifying the critical caspases inactivated by the broad-spectrum caspase inhibitors would be a logical next step in elucidating the mechanism of NO-induced apoptosis of cardiomyocytes.

Evidence of NO Regulating Death Pathways in Heart Failure: Ex Vivo and In Vivo Studies

Direct evidence that NO production *in vivo* is sufficient to elicit myocardial cell death was provided by a study by Kawaguchi et al. [79], wherein Type III NOS (eNOS) was transfected into rat hearts by use of inactivated Sendai virus-coated liposomes. Increased production of NO caused extensive myocardial degeneration, marked by cell shrinkage and histochemical evidence of DNA strand breakage consistent with an apoptotic-type cell death [79]. It was suggested in this study that the reported morphological changes to the myocardium are similar to changes seen in acute myocarditis and ischemic injury. It is likely, however, that the pattern, time course, and extent of NO production in these latter pathologies differ from NO production in eNOS-transfected hearts. Thus, this study clearly showed that sufficient NO production from a NOS is capable of eliciting myocardial damage and cell death, but the connection between endogenous NO production and cell death in the pathologies mentioned above remains tenuous.

Ischemia/Reperfusion Injury

A cytoprotective role for NO has been found in whole-animal models of myocardial ischemia/reperfusion (I/R) injury [80]. Several investigators have performed experiments with I/R injury *ex vivo* in isolated perfused hearts, and have suggested that reduced apoptotic cell death may underlie at least part of NO's protective effect. Weiland et al. [78] reported that perfusing hearts with the NOS inhibitor N^G-mono-methyl-L-arginine (L-NMMA) increased the amount of apoptotic cell death that occurred following a round of ischemia/reperfusion. Exacerbation of apoptosis by L-NMMA coincided with increased activation of caspase-3 and no change in the level of the cytoprotective factor Bcl-2. Similar results using NOS inhibitors were reported by Czarnowska et al. [81] using isolated guinea pig hearts. Studies with cardiomyocytes in culture suggest that the NO's cytoprotective effect in ischemia/reperfusion induced cell death is due to modest depolarization and reduced calcium loading of myocardial mitochondria during the ischemic phase [82]. Thus, *ex vivo* studies suggest that mechanisms innate to the heart (i.e., independent of circulating factors) are largely responsible for the cytoprotective role of NO in I/R injury *in vivo*. It bears stressing that these experiments likely identify the eNOS as the source of cytoprotective NO, because iNOS would not be present in the experimental models employed in these studies.

Heart Transplantation

An intriguing role for NO and regulation of apoptosis is now appreciated in the pathology of rejection in *heart transplantation*: Szabolcs et al. [42] first reported on a parallelism between the time, extent of iNOS expression, and degree of apoptosis of cardiac muscle cells in a heterotopic heart transplantation model in rats. Using genetic mouse models, Koglin et al. [83] reported that iNOS knockout mice were less able to acutely reject a transplanted heart than iNOS sufficient recipients. Interestingly, greater graft rejection was noted in iNOS knockout recipients at longer time points (>55 days). Koglin et al. [84] later showed that less myocyte apoptosis occurred in hearts exhibiting less transplantation rejection in iNOS knockout recipients. Less acute transplantation rejection in iNOS knockout recipients was also reported by Szabolcs et al. [85]. However, in the latter study, protection against acute rejection was observed only with donor hearts from iNOS knockout mice. Despite this discrepancy, a case can thus be made that NO produced from iNOS, be it from infiltrating immune cells or myocardial cells, promotes tissue damage and apoptotic cell death during acute rejection of the transplanted heart. During chronic rejection, NO limits cell death, perhaps in part by limiting the extent of infiltration by inflammatory cells, and appears to be cytoprotective.

Myocardial Infarction

Recent studies have linked NO, apoptosis, and chronic cardiac remodeling *in vivo*. Indeed, three studies examined eNOS and iNOS knockout mice in an infarction model of chronic heart failure. eNOS KO mice developed greater left ventricular end diastolic dimensions and lower fractional shortening 28 days post-myocardial infarction among animals with equal infarct sizes and equal blood pressures. Mortality in eNOS KO mice was also significantly higher [86]. Unfortunately, cell loss by apoptosis or by other means was not measured in this study. Nonetheless, this data suggest a protective role for eNOS, and supports results from isolated perfused hearts subjected to ischemia/reperfusion injury [78,81,82]. In contrast, increased iNOS activity appears to be detrimental: iNOS KO mice show higher basal left ventricular $+dP/dt$ and increased response to dobutamine 30 days following myocardial infarction [87]. A second study noted higher survival rates and higher peak left ventricular developed pressure 4 months after myocardial infarction in iNOS KO mice when compared to wild type animals [88]. These early studies document that NO from iNOS contributes to an increase in apoptosis following myocardial infarction, but thus far only an association is noted between reduced survival, depressed hemodynamic function, and apoptosis.

Summary

The actions of NO on apoptotic cell death are complex, especially in the context of heart disease. NO acts not through a single specific receptor, but rather through a number of different molecules that serve as potential targets for oxidative modification. NO influences cell death elicited by both apoptotic and necrotic mechanisms. Results from *in vitro* and *in vivo* studies suggest that in large part, the net effect of NO on cell death depends on the amount of NO produced. In the heart, this is dictated by many variables, including the type of NOS responsible for producing NO, the prevailing redox state, the oxygen tension in the microenvironment, and the extent of NO production over time. The reports discussed above provide evidence for an important link between NO, cell death, and some forms of heart failure. The degree to which NO-dependent cell death contributes to cardiac disease, however, is a key question that remains thus far unanswered.

References

1. Colucci WS, Braunwald E. *Pathophysiology of Heart Failure*, 6th ed. Philadephia: W.B. Saunders Company, 2001. (Braunwald E, Zipes DP, Libby P, eds. Heart Disease: A textbook of Cardiovascular Medicine).
2. Anversa P, Olivetti G, Capasso JM. Cellular basis of ventricular remodeling after myocardial infarction. *Am J Cardiol* 1991;68(14):7D–16D.
3. Curtiss C, Cohn JN, Vrobel T, Franciosa JA. Role of the renin-angiotensin system in the systemic vasoconstriction of chronic congestive heart failure. *Circulation* 1978;58(5):763–770.
4. Levine TB, Francis GS, Goldsmith SR, Simon AB, Cohn JN. Activity of the sympathetic nervous system and renin-angiotensin system assessed by plasma hormone levels and their relation to hemodynamic abnormalities in congestive heart failure. *Am J Cardiol* 1982;49(7):1659–1666.
5. Pfeffer MA, Lamas GA, Vaughan DE, Parisi AF, Braunwald E. Effect of captopril on progressive ventricular dilatation after anterior myocardial infarction. *N Engl J Med* 1988;319(2):80–86.
6. Anversa P, Nadal-Ginard B. Myocyte renewal and ventricular remodelling. *Nature* 2002;415(6868):240–243.

7. Beltrami AP, Urbanek K, Kajstura J, et al. Evidence that human cardiac myocytes divide after myocardial infarction. *N Engl J Med* 2001;344(23):1750–1757.
8. Kang PM, Izumo S. Apoptosis and heart failure: A critical review of the literature. *Circ Res* 2000;86(11):1107–1113.
9. Sabbah HN. Apoptotic cell death in heart failure. *Cardiovasc Res* 2000;45(3):704–712.
10. Olivetti G, Abbi R, Quaini F, et al. Apoptosis in the failing human heart. *N Engl J Med* 1997;336(16):1131–1141.
11. Li Z, Bing OH, Long X, Robinson KG, Lakatta EG. Increased cardiomyocyte apoptosis during the transition to heart failure in the spontaneously hypertensive rat. *Am J Physiol* 1997;272(5 Pt 2):H2313–H2319.
12. Leri A, Liu Y, Malhotra A, et al. Pacing-induced heart failure in dogs enhances the expression of p53 and p53-dependent genes in ventricular myocytes. *Circulation* 1998;97(2):194–203.
13. Olivetti G, Quaini F, Sala R, et al. Acute myocardial infarction in humans is associated with activation of programmed myocyte cell death in the surviving portion of the heart. *J Mol Cell Cardiol* 1996;28(9):2005–2016.
14. Saraste A, Pulkki K, Kallajoki M, Henriksen K, Parvinen M, Voipio-Pulkki LM. Apoptosis in human acute myocardial infarction. *Circulation* 1997;95(2):320–323.
15. Toyozaki T, Hiroe M, Tanaka M, Nagata S, Ohwada H, Marumo F. Levels of soluble Fas ligand in myocarditis. *Am J Cardiol* 1998;82(2):246–248.
16. Narula J, Haider N, Virmani R, et al. Apoptosis in myocytes in end-stage heart failure. *N Engl J Med* 1996;335(16):1182–1189.
17. Anversa P, Kajstura J. Myocyte cell death in the diseased heart. *Circ Res* 1998;82(11):1231–1233.
18. Eastman A, Barry MA. The origins of DNA breaks: A consequence of DNA damage, DNA repair, or apoptosis? *Cancer Invest* 1992;10(3):229–240.
19. Gold R, Schmied M, Giegerich G, et al. Differentiation between cellular apoptosis and necrosis by the combined use of in situ tailing and nick translation techniques. *Lab Invest* 1994;71(2):219–225.
20. Kanoh M, Takemura G, Misao J, et al. Significance of myocytes with positive DNA in situ nick end-labeling (TUNEL) in hearts with dilated cardiomyopathy: Not apoptosis but DNA repair. *Circulation* 1999;99(21):2757–2764.
21. Ohno M, Takemura G, Ohno A, et al. "Apoptotic" myocytes in infarct area in rabbit hearts may be oncotic myocytes with DNA fragmentation: Analysis by immunogold electron microscopy combined with In situ nick end-labeling. *Circulation* 1998;98(14):1422–1430.
22. Sloop GD, Roa JC, Delgado AG, Balart JT, Hines MO, 3rd, Hill JM. Histologic sectioning produces TUNEL reactivity. A potential cause of false-positive staining. *Arch Pathol Lab Med* 1999;123(6):529–532.
23. Haunstetter A, Izumo S. Apoptosis: Basic mechanisms and implications for cardiovascular disease. *Circ Res* 1998;82(11):1111–1129.
24. Haunstetter A, Izumo S. Future perspectives and potential implications of cardiac myocyte apoptosis. *Cardiovasc Res* 2000;45(3):795–801.
25. Haunstetter A, Izumo S. Toward antiapoptosis as a new treatment modality. *Circ Res* 2000;86(4):371–376.
26. Hirota H, Chen J, Betz UA, et al. Loss of a gp130 cardiac muscle cell survival pathway is a critical event in the onset of heart failure during biomechanical stress. *Cell* 1999;97(2):189–198.
27. Pannitteri G, Marino B, Campa PP, Martucci R, Testa U, Peschle C. Interleukins 6 and 8 as mediators of acute phase response in acute myocardial infarction. *Am J Cardiol* 1997;80(5):622–625.
28. Wencker D, Nguyen K, Khine C, et al. Myocyte apoptosis is sufficient to cause dilated cardiomyopathy. *Circulation* 1999;100(Suppl I):1–17.
29. Geng YJ, Ishikawa Y, Vatner DE, et al. Apoptosis of cardiac myocytes in Gsalpha transgenic mice. *Circ Res* 1999;84(1):34–42.
30. Hibbs JB Jr, Taintor RR, Vavrin Z. Macrophage cytotoxicity: Role for L-arginine deiminase and imino nitrogen oxidation to nitrite. *Science* 1987;235(4787):473–476.
31. Hibbs JB Jr, Vavrin Z, Taintor RR. L-arginine is required for expression of the activated macrophage effector mechanism causing selective metabolic inhibition in target cells. *J Immunol* 1987;138(2):550–565.
32. Brown GC, Borutaite V. Nitric oxide, cytochrome c and mitochondria. *Biochem Soc Symp* 1999;66:17–25.
33. Brown GC. Nitric oxide and mitochondrial respiration. *Biochim Biophys Acta* 1999;1411(2–3):351–369.
34. Garg UC, Hassid A. Nitric oxide-generating vasodilators and 8-bromo-cyclic guanosine monophosphate inhibit mitogenesis and proliferation of cultured rat vascular smooth muscle cells. *J Clin Invest* 1989;83(5):1774–1777.
35. Hausladen A, Fridovich I. Superoxide and peroxynitrite inactivate aconitases, but nitric oxide does not. *J Biol Chem* 1994;269(47):29405–29408.
36. Albina JE, Cui S, Mateo RB, Reichner JS. Nitric oxide-mediated apoptosis in murine peritoneal macrophages. *J Immunol* 1993;150(11):5080–5085.
37. Balligand JL, Cannon PJ. Nitric oxide synthases and cardiac muscle. Autocrine and paracrine influences. *Arterioscler Thromb Vasc Biol* 1997;17(10):1846–1858.
38. de Belder AJ, Radomski MW, Why HJ, et al. Nitric oxide synthase activities in human myocardium. *Lancet* 1993;341(8837):84–85.
39. de Belder AJ, Radomski MW, Why HJ, Richardson PJ, Martin JF. Myocardial calcium-independent nitric oxide synthase activity is present in dilated cardiomyopathy, myocarditis, and postpartum cardiomyopathy but not in ischaemic or valvar heart disease. *Br Heart J* 1995;74(4):426–430.
40. Haywood GA, Tsao PS, von der Leyen HE, et al. Expression of inducible nitric oxide synthase in human heart failure. *Circulation* 1996;93(6):1087–1094.
41. Pinsky DJ, Cai B, Yang X, Rodriguez C, Sciacca RR, Cannon PJ. The lethal effects of cytokine-induced nitric oxide on cardiac myocytes are blocked by nitric oxide synthase antagonism or transforming growth factor beta. *J Clin Invest* 1995;95(2):677–685.
42. Szabolcs M, Michler RE, Yang X, et al. Apoptosis of cardiac myocytes during cardiac allograft rejection. Relation to induction of nitric oxide synthase. *Circulation* 1996;94(7):1665–1673.
43. Lepoivre M, Flaman JM, Bobe P, Lemaire G, Henry Y. Quenching of the tyrosyl free radical of ribonucleotide reductase by nitric oxide. Relationship to cytostasis induced in tumor cells by cytotoxic macrophages. *J Biol Chem* 1994;269(34):21891–21897.
44. Szabo C, Zingarelli B, O'Connor M, Salzman AL. DNA strand breakage, activation of poly (ADP-ribose) synthetase, and cellular energy depletion are involved in the cytotoxicity of macrophages and smooth muscle

cells exposed to peroxynitrite. *Proc Natl Acad Sci USA* 1996;93(5):1753–1758.
45. Messmer UK, Ankarcrona M, Nicotera P, Brune B. p53 expression in nitric oxide-induced apoptosis. *FEBS Lett* 1994;355(1):23–26.
46. Ghafourifar P, Bringold U, Klein SD, Richter C. Mitochondrial nitric oxide synthase, oxidative stress and apoptosis. *Biol Signals Recept* 2001;10(1–2):57–65.
47. Du C, Fang M, Li Y, Li L, Wang X. Smac, a mitochondrial protein that promotes cytochrome c-dependent caspase activation by eliminating IAP inhibition. *Cell* 2000;102(1):33–42.
48. von Harsdorf R, Li PF, Dietz R. Signaling pathways in reactive oxygen species-induced cardiomyocyte apoptosis. *Circulation* 1999;99(22):2934–2941.
49. Poderoso JJ, Carreras MC, Lisdero C, Riobo N, Schopfer F, Boveris A. Nitric oxide inhibits electron transfer and increases superoxide radical production in rat heart mitochondria and submitochondrial particles. *Arch Biochem Biophys* 1996;328(1):85–92.
50. Forrester K, Ambs S, Lupold SE, et al. Nitric oxide-induced p53 accumulation and regulation of inducible nitric oxide synthase expression by wild-type p53. *Proc Natl Acad Sci USA* 1996;93(6):2442–2447.
51. Pinsky DJ, Aji W, Szabolcs M, et al. Nitric oxide triggers programmed cell death (apoptosis) of adult rat ventricular myocytes in culture. *Am J Physiol* 1999;277(3 Pt 2):H1189–H1199.
52. Mortensen K, Skouv J, Hougaard DM, Larsson LI. Endogenous endothelial cell nitric-oxide synthase modulates apoptosis in cultured breast cancer cells and is transcriptionally regulated by p53. *J Biol Chem* 1999;274(53):37679–37684.
53. Xia Z, Dickens M, Raingeaud J, Davis RJ, Greenberg ME. Opposing effects of ERK and JNK-p38 MAP kinases on apoptosis. *Science* 1995;270(5240):1326–1331.
54. Jun CD, Pae HO, Kwak HJ, et al. Modulation of nitric oxide-induced apoptotic death of HL-60 cells by protein kinase C and protein kinase A through mitogen-activated protein kinases and CPP32-like protease pathways. *Cell Immunol* 1999;194(1):36–46.
55. Taimor G, Rakow A, Piper HM. Transcription activator protein 1 (AP-1) mediates NO-induced apoptosis of adult cardiomyocytes. *FASEB J* 2001;15(13):2518–2520.
56. Andreka P, Zang J, Dougherty C, Slepak TI, Webster KA, Bishopric NH. Cytoprotection by Jun kinase during nitric oxide-induced cardiac myocyte apoptosis. *Circ Res* 2001;88(3):305–312.
57. Ignarro LJ. Biosynthesis and metabolism of endothelium-derived nitric oxide. *Annu Rev Pharmacol Toxicol* 1990;30:535–560.
58. Moncada S, Higgs A. The L-arginine-nitric oxide pathway. *N Engl J Med* 1993;329(27):2002–2012.
59. Shimojo T, Hiroe M, Ishiyama S, Ito H, Nishikawa T, Marumo F. Nitric oxide induces apoptotic death of cardiomyocytes via a cyclic-GMP-dependent pathway. *Exp Cell Res* 1999;247(1):38–47.
60. Arstall MA, Sawyer DB, Fukazawa R, Kelly RA. Cytokine-mediated apoptosis in cardiac myocytes: The role of inducible nitric oxide synthase induction and peroxynitrite generation. *Circ Res* 1999;85(9):829–840.
61. Brookes PS, Salinas EP, Darley-Usmar K, et al. Concentration-dependent effects of nitric oxide on mitochondrial permeability transition and cytochrome c release. *J Biol Chem* 2000;275(27):20474–20479.
62. Tejedo J, Bernabe JC, Ramirez R, Sobrino F, Bedoya FJ. NO induces a cGMP-independent release of cytochrome c from mitochondria which precedes caspase 3 activation in insulin producing RINm5F cells. *FEBS Lett* 1999;459(2):238–243.
63. Huwiler A, Pfeilschifter J, van den Bosch H. Nitric oxide donors induce stress signaling via ceramide formation in rat renal mesangial cells. *J Biol Chem* 1999;274(11):7190–7195.
64. Takeda Y, Tashima M, Takahashi A, Uchiyama T, Okazaki T. Ceramide generation in nitric oxide-induced apoptosis. Activation of magnesium-dependent neutral sphingomyelinase via caspase-3. *J Biol Chem* 1999;274(15):10654–10660.
65. Di Nardo A, Benassi L, Magnoni C, Cossarizza A, Seidenari S, Giannetti A. Ceramide 2 (N-acetyl sphingosine) is associated with reduction in Bcl-2 protein levels by Western blotting and with apoptosis in cultured human keratinocytes. *Br J Dermatol* 2000;143(3):491–497.
66. Li J, Bombeck CA, Yang S, Kim YM, Billiar TR. Nitric oxide suppresses apoptosis via interrupting caspase activation and mitochondrial dysfunction in cultured hepatocytes. *J Biol Chem* 1999;274(24):17325–17333.
67. Cheng W, Li B, Kajstura J, et al. Stretch-induced programmed myocyte cell death. *J Clin Invest* 1995;96(5):2247–2259.
68. Kim YM, Bombeck CA, Billiar TR. Nitric oxide as a bifunctional regulator of apoptosis. *Circ Res* 1999;84(3):253–256.
69. Li J, Billiar TR, Talanian RV, Kim YM. Nitric oxide reversibly inhibits seven members of the caspase family via S-nitrosylation. *Biochem Biophys Res Commun* 1997;240(2):419–424.
70. Rossig L, Fichtlscherer B, Breitschopf K, et al. Nitric oxide inhibits caspase-3 by S-nitrosation *in vivo*. *J Biol Chem* 1999;274(11):6823–6826.
71. Dimmeler S, Haendeler J, Sause A, Zeiher AM. Nitric oxide inhibits APO-1/Fas-mediated cell death. *Cell Growth Differ* 1998;9(5):415–422.
72. Genaro AM, Hortelano S, Alvarez A, Martinez C, Bosca L. Splenic B lymphocyte programmed cell death is prevented by nitric oxide release through mechanisms involving sustained Bcl-2 levels. *J Clin Invest* 1995;95(4):1884–1890.
73. Kim YM, de Vera ME, Watkins SC, Billiar TR. Nitric oxide protects cultured rat hepatocytes from tumor necrosis factor-alpha-induced apoptosis by inducing heat shock protein 70 expression. *J Biol Chem* 1997;272(2):1402–1411.
74. Wink DA, Hanbauer I, Krishna MC, DeGraff W, Gamson J, Mitchell JB. Nitric oxide protects against cellular damage and cytotoxicity from reactive oxygen species. *Proc Natl Acad Sci USA* 1993;90(21):9813–9817.
75. Rabkin SW, Kong JY. Nitroprusside induces cardiomyocyte death: Interaction with hydrogen peroxide. *Am J Physiol Heart Circ Physiol* 2000;279(6):H3089–H3100.
76. Ing DJ, Zang J, Dzau VJ, Webster KA, Bishopric NH. Modulation of cytokine-induced cardiac myocyte apoptosis by nitric oxide, Bak, and Bcl-x. *Circ Res* 1999;84(1):21–33.
77. Song W, Lu X, Feng Q. Tumor necrosis factor-alpha induces apoptosis via inducible nitric oxide synthase in neonatal mouse cardiomyocytes. *Cardiovasc Res* 2000;45(3):595–602.
78. Weiland U, Haendeler J, Ihling C, et al. Inhibition of endogenous nitric oxide synthase potentiates

ischemia-reperfusion-induced myocardial apoptosis via a caspase-3 dependent pathway. *Cardiovasc Res* 2000; 45(3):671–678.

79. Kawaguchi H, Shin WS, Wang Y, et al. *In vivo* gene transfection of human endothelial cell nitric oxide synthase in cardiomyocytes causes apoptosis-like cell death. Identification using Sendai virus-coated liposomes. *Circulation* 1997;95(10):2441–2447.
80. Lefer DJ, Nakanishi K, Johnston WE, Vinten-Johansen J. Antineutrophil and myocardial protecting actions of a novel nitric oxide donor after acute myocardial ischemia and reperfusion of dogs. *Circulation* 1993;88(5 Pt 1):2337–2350.
81. Czarnowska E, Kurzelewski M, Beresewicz A, Karczmarewicz E. The role of endogenous nitric oxide in inhibition of ischemia/reperfusion-induced cardiomyocyte apoptosis. *Folia Histochem Cytobiol* 2001;39(2):179–180.
82. Rakhit RD, Mojet MH, Marber MS, Duchen MR. Mitochondria as targets for nitric oxide-induced protection during simulated ischemia and reoxygenation in isolated neonatal cardiomyocytes. *Circulation* 2001;103(21):2617–2623.
83. Koglin J, Glysing-Jensen T, Mudgett JS, Russell ME. NOS2 mediates opposing effects in models of acute and chronic cardiac rejection: Insights from NOS2-knockout mice. *Am J Pathol* 1998;153(5):1371–1376.
84. Koglin J, Granville DJ, Glysing-Jensen T, et al. Attenuated acute cardiac rejection in NOS2 −/− recipients correlates with reduced apoptosis. *Circulation* 1999;99(6):836–842.
85. Szabolcs MJ, Ma N, Athan E, et al. Acute cardiac allograft rejection in nitric oxide synthase-2(−/−) and nitric oxide synthase-2(+/+) mice: Effects of cellular chimeras on myocardial inflammation and cardiomyocyte damage and apoptosis. *Circulation* 2001;103(20):2514–2520.
86. Scherrer-Crosbie M, Ullrich R, Bloch KD, et al. Endothelial nitric oxide synthase limits left ventricular remodeling after myocardial infarction in mice. *Circulation* 2001;104(11):1286–1291.
87. Feng Q, Lu X, Jones DL, Shen J, Arnold JM. Increased inducible nitric oxide synthase expression contributes to myocardial dysfunction and higher mortality after myocardial infarction in mice. *Circulation* 2001;104(6):700–704.
88. Sam F, Sawyer DB, Xie Z, et al. Mice lacking inducible nitric oxide synthase have improved left ventricular contractile function and reduced apoptotic cell death late after myocardial infarction. *Circ Res* 2001;89(4):351–356.
89. Leist M, Single B, Naumann H, Fava E, Simon B, Kuhne S, Nicotera P. Inhibition of mitochondrial ATP generation by nitric oxide switches apoptosis to necrosis. *Exp Cell Res* 1999;249:396–403.

Acute and Chronic Endothelial Dysfunction: Implications for the Development of Heart Failure

Axel Linke, Fabio Recchia, Xiaoping Zhang, and Thomas H. Hintze

Department of Physiology, New York Medical College, Valhalla, NY 10595, USA

Abstract. **Heart failure has been characterized by a reduction in cardiac contractile function resulting in reduced cardiac output. The clinical symptoms including mild tachycardia, reduced arterial pressure, increased venous or filling pressure and exercise intolerance have conceptually, to a large degree, been attributed to cardiac myocyte dysfunction. More recently, a vascular component has been recognized to contribute to heart failure. Among the most studied vascular mechanisms that might contribute to the development of heart failure has been the reduced production of nitric oxide or the reduced bioactivity of NO associated with both basic models of heart failure and disease in patients. The still evolving concept that heart failure is a cytokine activated state has, in addition, focused attention on the possibility that the cytokine driven isoform of NO synthase (NOS), iNOS, may produce sufficient quantities of NO to actually suppress cardiac myocyte function contributing to the reduced inotropic state in the failing heart. Thus, our view of the role of NO in the development of heart failure has evolved from simply a reduction in production of NO in blood vessels, to altered substrate availability (i.e. L-arginine), to increased scavenging of NO by superoxide anion, to increased production of NO from iNOS. As these concepts develop, our approach to the therapeutics of heart failure has also progressed with the recognition of the need to develop treatments directed towards addressing one or more of these etiologies. This review will focus on these aspects of the involvement of NO in the development of heart failure and some of the treatments that have developed from our understanding of the basic biology of NO to address these pathohysiologic states.**

Key Words. **nitric oxide, eNOS, iNOS, downregulation, L-arginine, oxygen radicals, cAMP**

Introduction

Despite impressive progress in the medical treatment of chronic heart failure over the last decades, the mortality rate remains high, the quality of life impaired and its prevalence increasing [1]. Advances in research have introduced a different pathophysiological concept of the disease, recognizing chronic heart failure as a systemic rather than only a cardiac disorder involving hemodynamic, neurohumoral and peripheral vascular derangements. Enhanced peripheral vasoconstriction in response to exercise and impaired vasodilatation after stimulation with agonists are key features of endothelial dysfunction in CHF [2–4]. The disruption of endothelial cell function has been attributed to activation of the sympathetic nervous system [5], the renin-angiotensin system [6], and the pituitary-vasopressin axis [6].

Considerable attention has been focused on the endothelium itself, revealing a major role in the regulation of vascular tone [7,8]. Among other factors, endothelial cells regulate blood vessel diameter via the release of nitric oxide (NO) in response to stimulation with agonists, like acetylcholine or bradykinin [9,10] and mechanical stimuli such as changes in blood flow velocity or endothelial shear stress [11], leading to a relaxation of the vascular smooth muscle [7–11]. Three different isoforms of the nitric oxide synthase (NOS) are capable of producing NO: nNOS (type 1 isoform, [NOS1]), iNOS (type 2 isoform [NOS2]) and eNOS (type 3 isoform [NOS3]). eNOS is constitutively expressed in endothelial cells [12], endocardial cells [13], and perhaps cardiomyocytes [13], whereas nNOS was found in nerve endings that release norepinephrine [14]. In response to inflammation, particularly in end stage heart failure [15–17], iNOS is expressed and generates large amounts of nitric oxide independent of agonist-stimulation and calcium [18]. However, eNOS-derived NO is the predominant regulator of vascular tone and the importance of nNOS- and iNOS-derived NO seems to be negligible [7–11].

Downregulation of eNOS as a Contributor to Heart Failure

Early studies in animals by Elsner et al. and by Kaiser et al. indicated that inhibition of NO

Address for correspondence: Thomas H. Hintze, PhD, Department of Physiology, New York Medical College, Valhalla, NY 10595, USA. Tel.: 914-594-3633; Fax: 914-594-4108; E-mail: Thomas_Hintze@NYMC.edu synthase using L-NAME had no effect on arterial pressure, cardiac function or peripheral resistance in dogs with heart failure [19,20]. In contrast, profound hemodynamic effects after inhibtion of NO synthesis are seen in all animals [21] including man. More systematic studies indicated that pacing induced dilated cardiac myopathy and heart failure was associated with normal or increased NO production early during the development of heart failure, but a reduction in NO during cardiac decompensation [22–24]. For instance, Hintze et al. showed that flow and agonist induced large epicardial artery dilation was normal up to three weeks of pacing, i.e. compensated function- a time when LV end diastolic pressure increased to 18 mmHg, and that after that, flow and agonist induced NO dependent dilation decreased [22]. Other studies by Zhao et al. and Shen et al. indicated that reflex cholinergic-NO mediated falls in coronary vascular resistance were also markedly attenuated after cardiac decompensation [25,26]. Isolated coronary microvessels produced less nitrite, the hydration product of NO, *in vitro* after cardiac decompensation [27]. Shear-induced dilation of single perfused coronary arterioles from dogs with pacing-induced decompensated heart failure was altered such that tremendous levels of shear were needed to cause minimal increases in coronary arteriolar diameter [28].

Blood flow in the heart is critically dependent upon cardiac metabolic demand and substrate utilization because of the tight coupling of metabolism to the control of vascular resistance. Thus, during studies of alterations in the control of the coronary circulation by NO in heart failure, Recchia et al. also determined the impact of reduced NO production on cardiac metabolism [29]. Recchia et al. found that during cardiac decompensation there was a reduced production of NO*x* (nitrate and nitrite) across the heart at the time of cardiac decompensation [29]. From these studies, two important concepts have emerged: first that reduced NO production is associated with increased myocardial oxygen consumption and second that reduced NO production is associated with a shift in substrate use from fatty acids to glucose in the heart. More recent studies in the mouse heart [30] and in the dog heart [31] confirm these concepts and support the conclusion that NO production from eNOS is reduced in heart failure. The mechanism for the reduced NO production in these models of heart failure was originally shown by Smith et al. to be due to reduced mRNA and protein for eNOS in aortic endothelial cells [32]. More recently, Zhang and Hintze have shown a reduced protein for eNOS in microvessels harvested from dogs with decompensated heart failure [33]. In contrast, Wang et al., Bernstein et al., Zhao et al. and Sessa et al., respectively, have shown that exercise training results in increased large coronary artery dilation [34], NO production across the heart [35], increased NO dependent coronary dilation [36] and increased aortic endothelial cell eNOS mRNA and protein [37]; the exact opposite of heart failure.

Consequences of the Reduced Coronary eNOS Protein for the Treatment of Heart Failure

If the reduction in eNOS enzyme contributes to the development of cardiac decompensation, then therapies designed to restore eNOS activity and NO production should be effective, at least in part, in the treatment of heart failure. Figure 1 shows a schematic approach to some of these therapies. In a recent study in humans using ACE inhibitors (the HOPE trial) indicates ACE inhibition results in reduced all cause mortality in patients with heart disease [38]. ACE inhibitors were originally developed as bradykinin potentiating substances, not angiotensin converting enzyme inhibitors, although it was later realized that ACE and kininase I are the same enzyme. Thus by blocking the

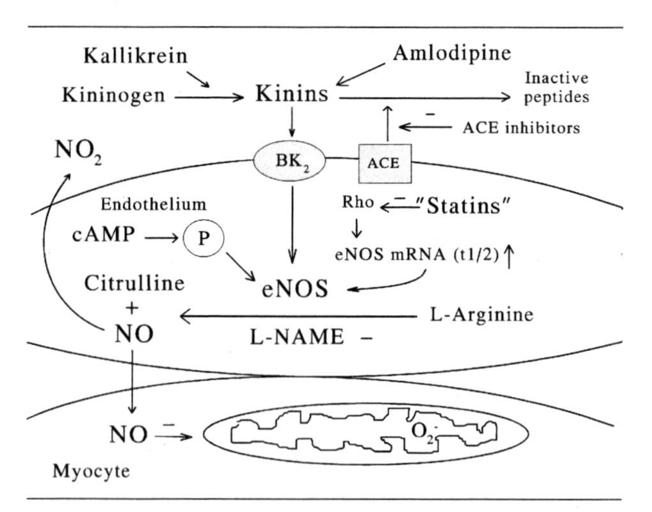

Fig. 1. *This figure shows a conceptual diagram of the potential therapeutic methods for stimulating NO production. NO is formed from the metabolism of L-arginine by eNOS. ENOS is stimulated by the bradykinin 2 receptor (BK2). ACE inhibitors increase local bradykinin levels or the half life of bradykinin by blocking kinin metabolism. Amlodipine, an L-type calcium channel blocking agent stimulates the formation of kinins locally to also activate the BK2 receptor. Statins, through a Rho kininase-dependent mechanism may increase or preserve eNOS protein. Finally, through a PKB-related mechanism, cAMP can phosporylate eNOS leading to its activation and enhanced NO production. Each of these individual mechanisms may increase NO production when eNOS is low; perhaps a combination of these therapeutic modalities would be even more effective.*

breakdown of kinins that through the B2 kinin receptor stimulate NO production, NO production should increase. Indeed, Zhang et al. have shown that ACE inhibitors can increase the production of nitrite from coronary microvessels from dogs with pacing induced heart failure [39] and Kichuk et al. have shown similar effects in coronary microvessels from explanted failing human heart [40]. Interestingly, a L-type calcium channel blocking agent, amlodipine, seems to selectively increase NO production from coronary microvessels from the failing heart through a kinin dependent mechanism, although a mechanism different from ACE inhibition i.e. stimulation of kinin formation [41,42].

Data from Laufs et al. indicates that statins may increase the mRNA for eNOS and hence the protein [43]. Recently we have found that dogs or rats treated with statins have increased bioactivity of NO, NO production and eNOS protein [44,45]. Preliminary data also indicate that treatment of dogs with chronic left ventricular pacing with statins delays the onset of decompensated heart failure, preserves NO production by coronary microvessels, and preserves eNOS protein [46]. Thus, statins may also have a beneficial effect in the treatment of heart failure by maintaining NO production from eNOS. Finally, Fulton et al. [47] and Dimmler et al. [48] have found that phosporylation of eNOS can increase the activity of the enzyme, in fact it is necessary for any activity of the enzyme at all. This increase in activity can be stimulated by drugs which activate adenyl cyclase through a PKB dependent mechanism [33]. In coronary microvessels from dogs with pacing-induced heart failure there is a reduction in eNOS protein, however, increasing the phosphorylation of the downregulated eNOS can restore NO production [33]. Thus in the future, and perhaps as part of the mechanism of action of some new beta blockers, increasing the phosporylation of eNOS may be beneficial in the treatment of heart failure.

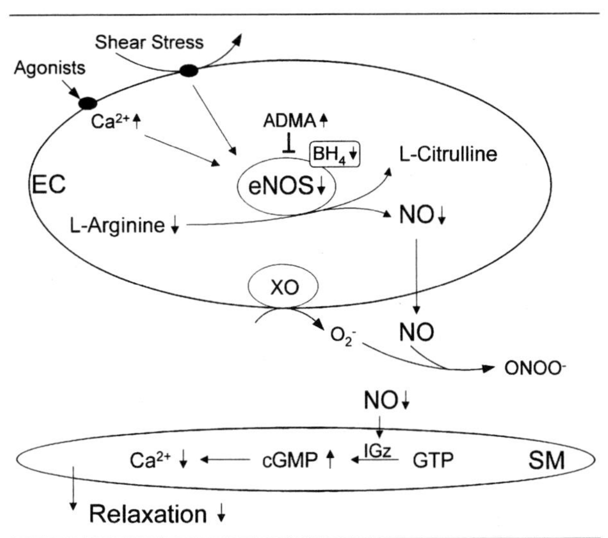

Fig. 2. *In a normal endothelial cell (EC), nitric oxide (NO) is produced by conversion of L-arginine to L-citrulline. This reaction is mediated by the endothelial isoform of the NO synthase (eNOS) in the presence of the cofactor tetrahydrobiopterin (BH_4) in response to agonist stimulation or shear stress. NO diffuses to the vascular smooth muscle cell (SM) and stimulates guanylate cyclase (IGz) to convert guanosine triphosphate (GTP) into cyclic guanosine monophosphate (cGMP). This leads to a reduction in intracellular calcium (Ca^{2+}), culminating in vasorelaxation. In patients with chronic heart failure (CHF), bioavailability of L-arginine in endothelial cells might be reduced, resulting in a impaired NO production. This is further aggravated by the reduced eNOS protein expression, the deficiency in BH_4 and the blockade of eNOS activity by asymmetric dimethyl arginine (ADMA). Large parts of the remaining NO may react with free oxygen radicals (O_2^-) produced in elevated concentrations by xanthine oxidase, NAD(P)H oxidases and mitochondrial enzymes forming peroxynitrite ($ONOO^-$). Consequently, the decrease in bioavailable NO leads to an impaired vasorelaxation. Interventions aimed to correct these derangements, like L-arginine or BH_4 administration replenishing intracellular pools, exercise training that increases eNOS protein expression, allopurinol that blocks XO activity thereby decreasing oxidative stress, or vitamin C that scavenges free radical and improves BH4-binding to eNOS, may partially restore endothelial function in patients with CHF.*

Chronic Heart Failure and L-Arginine

NO is synthesized from L-arginine, a semi-essential amino acid, by oxidation of its terminal guanidino nitrogen in the presence of the co-factor tetrahydrobiopterin, flavin mononucleotide and flavin adenine dinucleotide (Fig. 2 [49]). *In vitro* studies suggest, that L-arginine is not the rate limiting step for NO production, at least in healthy subjects. The Km for NOS was found to be in the range between 1–5 μmol/l, whereas the intracellular concentration of L-arginine normally reaches the millimolar range (0.8–2.0 mmol/L) and theoretically should saturate eNOS. However, numerous studies have shown an improvement of endothelial function after oral L-arginine administration indicating that experimental kinetics of NOS might differ from *in vivo*, especially in patients with chronic heart failure. Hirooka et al. were the first to demonstrate that L-arginine significantly augments the maximal vasodilatory response to acetylcholine and the decrease in forearm vascular resistance during reactive hyperemia in patients with chronic heart failure [50]. The therapeutic properties of L-arginine were not restricted to the forearm circulation, since L-arginine infusion also increased cardiac output

[51], enhanced lower limb vasodilatory reserve [52], and had beneficial effects on glomerular filtration rate [53]. Rector and co-workers found that oral L-arginine over a period of 6 weeks increases blood flow during exercise, improves arterial compliance and functional status of patients. Furthermore, circulating levels of the endogenous vasoconstrictor endothelin dropped during L-arginine administration [54]. In another clinical trial involving patients with CHF, the improvement in endothelium-dependent vasodilatation of the radial artery after 4 weeks of oral L-arginine supplementation was comparable to that after 4 weeks of local hand grip exercise training (a method to upregulate eNOS), but the combination of these interventions exerted additive effects on endothelial function [55]. In contrast, Chin-Dusting et al. failed to demonstrate any favorable effect of L-arginine on endothelial dysfunction in patients suffering from CHF. This might be explained by the fact that patients in each study may have different severies of heart failure, however, the administration of L-arginine in the water-soluble syrup in the latter study raised concerns due to the unknown bioavailability [56].

Numerous reasons may account for the improvement of endothelial function in response to L-arginine therapy. First, chronic heart failure is also associated with increased oxidative stress. In this regard, highly reactive oxygen species were proposed to impair endothelial function by scavenging NO thereby reducing the NO bioavailability and reacting with lipoproteins. Oxidized lipoproteins may alter the affinity of L-arginine for eNOS. Despite the absence of L-arginine eNOS transfers electrons to oxygen thereby producing oxygen radicals. A further stimulation of eNOS by agonists like acetylcholine might unmask the intracellular L-arginine deficiency [57,58]. Second, the absence of continuous physiological stimuli like shear or agonists in an endothelial cell culture might reflect one reason why the Km for L-arginine is much higher *in vivo* than *in vitro* [59]. Third, the intracellular compartimentalization of L-arginine in microdomains with lower L-arginine concentrations in direct vicinity to eNOS has been proposed as a mechanism for an impaired eNOS activity [60]. Fourth, elevated plasma levels of asymmetric dimethylarginine (ADMA), an endogenous competitive eNOS inhibitor, were found in patients with chronic heart failure and proposed to be responsible for endothelial dysfunction [61]. However, the inhibitory effects of ADMA on endothelium-dependent vasodilatation could be reversed at least in part by L-arginine in other experimental settings [62]. Fifth, the expression of the cationic amino acid precursor (CAT-1) that ensures the uptake of L-arginine, the substrate of eNOS, into endothelial cells was found to be reduced by 80% in patients with chronic heart failure. The impaired expression of CAT-1 was accompanied by a reduced forearm clearance of L-arginine and endothelial dysfunction underlining the pathophysiological importance. Given the fact that the L-arginine plasma concentration is far below the saturation of the CAT transporter, and therefore an increase in L-arginine concentrations either by intravenous infusion or by an oral administration should result in an enhanced L-arginine transport into endothelial cells and partially overcome the reduced expression of the transporter [63]. Sixth, L-arginine was shown to reverse the inhibitory effect of L-glutamine on the recycling of L-arginine from L-citrulline [64] and prevents the induction of arginase that degrades L-arginine to urea and ornithine [65]. Seventh, L-arginine may directly scavenge free radicals thereby improving endothelial function [66]. Eigth, endothelium-independent mechanisms should be considered, such as vasodilatation through insulin, glucagon and growth hormone release [67–69].

Oxidative Stress and Chronic Heart Failure

Elevated concentrations of reactive oxygen species such as superoxide anions and lipid peroxides were found in patients with chronic heart failure [70]. This increase in oxidative stress was proposed to play a key role in the pathogenesis and progression of CHF [71–74]. Moreover, reactive oxygen species may account for the impaired endothelium-dependent vasodilatation in response to agonists or flow by scavenging endothelium-released nitric oxide [75]. A number of different enzymes responsible for free radical production have been suggested, among them NADPH oxidases [76], xanthine oxidase [77,78] and mitochondrial enzymes [79]. However, the maintenance of normal endothelial function is also dependent on the balance of radical producing and radical scavenging mechanisms, the latter being reduced in chronic heart failure [80].

Correction of endothelial dysfunction has become a therapeutic target in patients with CHF and studies using antioxidants like vitamin C have been aimed at improving endothelial function [75,81–83]. Hornig et al. were the first to show that acute intra-arterial administration or oral vitamin C over a period of 4 weeks significantly improved flow-mediated vasodilatation by about 50% in patients with CHF. The portion of flow-mediated dilatation mediated by nitric oxide was found to be selectively improved and almost normalized by acute as well as chronic vitamin C treatment [75]. These results supported the notion that endothelial dysfunction in patients with CHF is, at least in part, due to an increased quenching of nitric oxide by free radicals.

Previous studies assessed the effects of vitamin C on endothelial function in patients that had developed chronic heart failure due to coronary artery disease [75,83]. Atherosclerosis itself is associated with an increased oxygen radical production and oxidative stress, therefore concerns have been raised that the improvement of endothelium-dependent vasodilatation after vitamin C is associated with a restoration of endothelial dysfunction in atherosclerosis and not in heart failure per se [83]. To address this issue, Erbs et al. assessed the effects of an intra-arterial vitamin C infusion on endothelium-dependent vasodilatation in two different patient populations: one with dilated cardiomyopathy and one with ischemic heart disease [84]. Patients with DCM and IHD had a blunted response to acetylcholine stimulation compared to healthy subjects, but intra-arterial administration of vitamin C improved acetylcholine-mediated vasodilatation to a lesser extend in DCM than in ICM. Despite the positive effects of vitamin C, endothelium-dependent vasodilatation was still attenuated in patients with CHF in comparison with healthy controls suggesting that other factors besides oxidative stress contribute to endothelial dysfunction in CHF perhaps downregulation of eNOS. In contrast, one study failed to show any improvement of endothelial function in response to vitamin C in patients with dilated cardiomyopathy [82]. It is conceivable that the intra-venous application of vitamin C used in this study was not sufficient to completely scavenge oxygen radicals in the vascular bed of the forearm [82]. However, others found reduced plasma concentrations of free radicals and enhanced flow-mediated vasodilatation after intravenous vitamin C in patients with CHF [83]. These differences might be explained by the work of Ito et al. who examined patients with very mild heart failure [82] whereas others recruited patients in more advanced stages of the disease when oxidative stress is higher [75,83]. Since oral administration does not increase arterial concentrations of vitamin C to levels necessary for radical scavenging, other factors may account for the improvement of endothelial function after long-term oral vitamin C therapy [75,83]. Recently, it is has been demonstrated that vitamin C increases tetrahydrobiopterin availability and augments the affinity of tetrahydrobiopterin for eNOS resulting in an augmented NO production [85]. Furthermore, vitamin C was shown to reduce free radical production by the inhibition of NADPH oxidases [83] and to increase eNOS expression [86]. Vitamin C also protects intracellular gluthatione from oxidation thereby enhancing endothelial NO production [87] and NO stability [88]. Finally, vitamin C promotes the regeneration of vitamin E thereby improving the intracellular antioxidant capacity [89].

Xanthine oxidase was identified as a free radical generator in chronic heart failure, producing superoxide anions and hydrogen peroxide as byproducts [77]. The enzyme is expressed as xanthine dehydrogenase in the capillary endothelium and the endothelium of small arteries and converted into XO via proteolysis or thiol oxidation [90,91]. An increase in serum uric acid levels in patients with CHF originated from xanthine oxidase activation [92] and is further accompanied by impaired peripheral blood flow, increased vascular resistance and higher NYHA functional class [93]. Recently, Doehner et al. aimed to improve endothelial function and vasodilatory capacity in patients with CHF by inhibiton of xanthine oxidase [78]. Infusion of the XO-inhibitor allopurinol improved agonist-mediated endothelium-dependent vasodilatation in uremic patients with CHF, but did not effect ACH-mediated forearm vasodilatation in CHF patients with normal serum uric acid levels. Oral allopurinol treatment for one week reduced uric acid levels in hyperurimic patients with CHF, improved flow-dependent flow in arms and legs and decreased allantoin levels, a marker of oxygen free radical formation. This change in uric acid levels due to allopurinol therapy was correlated with improvement in flow-dependent flow, underscoring the pathophysiological relevance of the free radical production by XO. Furthermore, xanthine oxidase inhibition by allopurinol seems to exert regenerative effects on the failing myocardium since it reduces myocardial oxygen consumption, improves contractility and thereby efficiency [77,94].

In contrast to allopurinol, blockade of aldosterone receptors by spironolactone, in addition to standard therapy, was already shown to improve survival in patients with chronic heart failure [95]. Besides the limitation of excessive extracellular matrix turnover due to spironolactone therapy in patients suffering form CHF [96], an improvement of endothelial function [97] was proposed to account for the beneficial effects of the aldosterone receptor blocker on mortality and morbidity in the RALES study [95]. Further evidence from experimental studies in humans and animals supports this notion. Spironolactone increases nitric oxide bioactivity, improves endothelial vasodilator dysfunction and suppresses the conversion from angiotensin I to angiotensin II in the vasculature of patients with CHF [97]. A reduction in vascular oxidative stress may contribute to the therapeutic effects of spironolactone, since it decreased the expression of the p22phox subunit of the NADPH oxidase [98]. The restoration of endothelial function with a reduction in peripheral resistance might contribute to an improvement in tissue perfusion enhancing exercise capacity, clinical symptoms and presumably survival.

Exercise Training and CHF

Since endothelial dysfunction has been recognized as a contributor to exercise intolerance in CHF, several approaches have been tested to rejuvenate the endothelium. Among them, exercise training became a promising tool to amplify basal nitric oxide release and to improve endothelium-dependent vasodilatation in patients with chronic heart failure [2,99,100]. The correlation between the restoration of endothelial function and the increment in exercise capacity suggested that peripheral hypoperfusion might limit exercise in CHF [2]. Alterations of the skeletal muscle function were also identified as potential determinants of exercise intolerance independent from peripheral endothelial dysfunction [101,102].

Most of the clinical trials tested the effects of local exercise training regimens like handgrip exercise or bicycle ergometer training on endothelial function of the trained extremity showing an improvement of endothelium-dependent vasodilatation [2,55,99]. Over the last several years, it has been extensively discussed whether there is a critical proportion of body mass that has to be trained to promote systemic effects on the endothelium. One study addressing this issue demonstrated systemic effects with an improved endothelium-dependent vasodilatation of the radial artery in patients with CHF after bicycle ergometer training [103].

Recently, the effects of an exercise training program on central hemodynamics in relation to peripheral endothelial function were assessed [104]. Six months of regular aerobic training results in a significant increase in left ventricular ejection fraction and a reduction in left ventricular dimensions in patients with CHF. Changes in total peripheral resistance were inversely correlated with stroke volume at rest and during maximal exercise, suggesting that cardiac function improved as a consequence of afterload reduction due to an augmented peripheral vasodilatation. In contrast, an attenuation of sympathetic drive or an increased vagal tone after exercise training might also account for the reduction in peripheral vascular resistance [105,106]. Furthermore, a partial correction of the adrenergic tone reflected by a reduction in plasma radiolabeled norepinephrine and a restoration of heart rate variability were attributed to the improvement of endothelial function in CHF due to regular physical exercise [107]. Exercise training not only exerts positive effects on endothelial dysfunction in CHF, it also improves exercise capacity, quality of life, and most importantly reduces mortality. Therefore, physical exercise especially when tailored to each individual should be considered as a promising tool to improve survival in patients with CHF [108]. At least a portion of this is due to enhanced NO production and bioavailability.

Molecular Changes of the Endothelium in Response to Physical Exercise

During exercise training cardiac output is intermittently elevated resulting in an increased shear stress on the vascular endothelium. Studies using cultured human and bovine endothelial cells clearly revealed an increase in eNOS gene expression in response to shear stress [109,110]. Even stronger physiological evidence is provided by animal studies showing elevated eNOS expression and NO release from coronary conduit arteries [111,112] as well as resistance vessels after regular physical exercise [113]. However, local differences in shear stress and blood pressure in response to exercise might account for the non-uniform changes in eNOS protein expression within the coronary vascular tree [114].

A rapid increase in flow e.g. after adenosine stimulation induces a flow-dependent vasodilatation. However, this flow-dependent vasodilatation occurs within seconds, orders of magnitude faster than an increase in eNOS protein expression can occur. Recently, protein kinase A dependent eNOS phosphorylation at serine 1179 was identified as a mechanism leading to an instantaneous increase in NO production in response to shear stress [115]. Caveolin is an eNOS binding protein in caveolae which inactivates eNOS. In cell culture eNOS dissociation from caveolin and binding to calmodulin after stimulation with flow were also shown to rapidly elevate NO production [116].

Moreover, shear stress augments the velocity of the endothelial L-arginine transport leading to improved substrate availability for eNOS and consequently enhanced NO production [117]. Since endothelium-derived NO also induces the expression of the radical scavenger enzyme ecSOD, exercise training might not only result in an increased NO bioavailability by an enhanced NO production but also by a reduced NO breakdown by reactive oxygen species [118].

Summary

Initial studies indicated an important role for the downregulation of eNOS in the coronary and peripheral circulations and this is still a characteristic of heart failure. Other mechanisms which impinge on the bioavailability of NO from eNOS such as reduced L-arginine (the substrate), increased oxygen radical scavenging of NO (superoxide), and the presence of cytokine induced iNOS may all contribute to the development of heart failure. Understanding the contribution of each of these mechanisms will lead to new insights into the etiology of heart failure. In addition, with that understanding comes the ability to develop new rational therapies for the treatment of the causes of heart failure.

Acknowledgments

This work was supported by PO-1 HL 43023, RO-1 HL 50142, 62573* and 61290. Axel Linke is the recipient of a fellowship from the Deutsche Forschungsgmeinschaft Grant Li 946/1-1. Xiaoping Zhang was supported by a Scientist Development Grant from the American Heart Association.

References

1. Cowie MR, Mosterd A, Wood DA, Deckers JW, Poole-Wilson PA, Sutton GC, Grobbee DE. The epidemiology of heart failure. *Eur Heart J* 1997;18:208–225.
2. Hambrecht R, Fiehn E, Weigl C, Gielen S, Hamann C, Kaiser R, Yu J, Adams V, Niebauer J, Schuler G. Regular physical exercise corrects endothelial dysfunction and improves exercise capacity in patients with chronic heart failure. *Circulation* 1998;98:2709–2715.
3. Zelis R, Flaim SF. Alterations in vasomotor tone in congestive heart failure. *Prog Cardiovasc Dis* 1982;24:437–459.
4. Kubo SH, Rector TS, Bank AJ, Williams RE, Heifetz SM. Endothelium-dependent vasodilation is attenuated in patients with heart failure. *Circulation* 1991;84:1589–1596.
5. Hasking GJ, Esler MD, Jennings GL, Burton D, Johns JA, Korner PI. Norepinephrine spillover to plasma in patients with congestive heart failure: Evidence of increased overall and cardiorenal sympathetic nervous activity. *Circulation* 1986;73:615–621.
6. Francis GS, Rector TS, Cohn JN. Sequential neurohumoral measurements in patients with congestive heart failure. *Am Heart J* 1988;116:1464–1468.
7. Palmer RM, Ferrige AG, Moncada S. Nitric oxide release accounts for the biological activity of endothelium-derived relaxing factor. *Nature* 1987;327:524–526.
8. Ignarro LJ, Buga GM, Wood KS, Byrns RE, Chaudhuri G. Endothelium-derived relaxing factor produced and released from artery and vein is nitric oxide. *Proc Natl Acad Sci USA* 1987;84:9265–9269.
9. Furchgott RF, Zawadzki JV. The obligatory role of endothelial cells in the relaxation of arterial smooth muscle by acetylcholine. *Nature* 1980;288:373–376.
10. Cherry PD, Furchgott RF, Zawadzki JV, Jothianandan D. Role of endothelial cells in relaxation of isolated arteries by bradykinin. *Proc Natl Acad Sci USA* 1982;79:2106–2110.
11. Rubanyi GM, Romero JC, Vanhoutte PM. Flow-induced release of endothelium-derived relaxing factor. *Am J Physiol* 1986;250:H1145–H1149.
12. Moncada S, Higgs A. The L-arginine-nitric oxide pathway. *N Engl J Med* 1993;329:2002–2012.
13. Kelly RA, Balligand JL, Smith TW. Nitric oxide and cardiac function. *Circ Res* 1996;79:363–380.
14. Schwarz P, Diem R, Dun NJ, Forstermann U. Endogenous and exogenous nitric oxide inhibits norepinephrine release from rat heart sympathetic nerves. *Circ Res* 1995;77:841–848.
15. Ishibashi Y, Shimada T, Murakami Y, Takahashi N, Sakane T, Sugamori T, Ohata S, Inoue S, Ohta Y, Nakamura K, Shimizu H, Katoh H, Hashimoto M. An inhibitor of inducible nitric oxide synthase decreases forearm blood flow in patients with congestive heart failure. *J Am Coll Cardiol* 2001;38:1470–1476.
16. Habib FM, Springall DR, Davies GJ, Oakley CM, Yacoub MH, Polak JM. Tumour necrosis factor and inducible nitric oxide synthase in dilated cardiomyopathy. *Lancet* 1996;347:1151–1155.
17. Haywood GA, Tsao PS, der Leyen HE, Mann MJ, Keeling PJ, Trindade PT, Lewis NP, Byrne CD, Rickenbacher PR, Bishopric NH, Cooke JP, McKenna WJ, Fowler MB. Expression of inducible nitric oxide synthase in human heart failure. *Circulation* 1996;93:1087–1094.
18. Nathan C, Xie QW. Nitric oxide synthases: Roles, tolls, and controls. *Cell* 1994;78:915–918.
19. Elsner D, Muntze A, Kromer EP, Reigger GAA. Systemic vasoconstriction induced by inhibition of nitric oxide synthesis is attenuated in conscious dogs with heart failure. *Cardiovasc Res* 1991;25:438–440.
20. Kaiser L, Spickard RC, Oliver NB. Heart failure depresses endothelium-dependent responses in canine femoral artery. *Am J Physiol* 1989;256:H962–H967.
21. Shen W, Lundborg M, Wang J, Stewart J, Xu X, Ochoa M, Hintze TH. Role of EDRF in the regulation of regional blood flow during exercise. *J Appl Physiol* 1994;77:165–172.
22. Hintze TH, Wang J, Seyedi N, Wolin M. Myocardial hypertrophy and failure: Association between alterations in the production or release of EDRF/NO and myocardial dysfunction. In: Bevan J, Kaley G, Rubayani G, eds. *Flow-Dependent Regulation of Vascular Function*. Oxford Press, 1995.
23. Larosa G, Forrester C. Coronary B-adrenoreceptor function is modified by the endothelium in heart failure. *J Vasc Res* 1996;33:62–70.
24. Redfield MM, Aarhus LL, Wright RS, Burnett JC. Cardiorenal and neurohumoral function in a canine model of early left ventricular dysfunction. *Circulation* 1993;87:2016–2022.
25. Zhao G, Shen W, XU X, Ochoa M, Bernstein R. Hintze TH. Selective impairment of vagal-mediated NO dependent coronary vasodilation in conscious dogs after pacing induced heart failure. *Circulation* 1995;91:2655–2663.
26. Shen W, Wang J, Ochoa M, XU X, Hintze TH. Role of endothelium-derived relaxing factor in parasympathetic coronary vasodilation following carotid chemoreflex activation in conscious dogs. *Am J Physiol* 1994;267:H605–H613.
27. Zhang X, Recchia F, Bernstein RD, XU X, Nasjletti, Hintze TH. Kinin-mediated coronary nitric oxide production contributes to the therapeutic actions of ACE and NEP inhibitors and amlodipine in the treatment of heart failure. *JPET* 1999;288:742–751.

28. Sun D, Huang A, Zhao G, Bernstein RD, Forfia P, XU X, Koller A, Kaley G, Hintze TH. Reduced NO-dependent arteriolar dilation during the development of cardiomyopathy. *Am J Physiol* 2000;278:H461–H468.
29. Recchia FA, McConnell PI, Bernstein RD, Vogel TR, Xu XB, Hintze TH. Reduced nitric oxide production and altered myocardial metabolism during the decompensation of pacing-induced heart failure in conscious dogs. *Circ Res* 1998;83:969–979.
30. Tada H, Thompson CI, Recchia FA, Loke KE, Ochoa M, Smith CJ, Shesely EG, Kaley G, Hintze TH. Myocardial glucose uptake is regulated by nitric oxide via endothelial nitric oxide synthase in the Langendorff Mouse heart. *Circ Res* 2000;86:270–278.
31. Recchia FA, Osorio JC, Chandler MP, Xu X, Panchal AR, Lopashuk GD, Hintze TH, Stanley WC. Reduced synthesis of NO causes marked alterations in myocardial substrate metabolism in conscious dogs. *Am J Physiol* 2002;282:E197–E206.
32. Smith CJ, Sun D, Hoegler C, Zhao G, XU X, Kobari Y, Pritchard K, Sessa WC, Hintze TH. Reduced gene expression of vascular nitric oxide synthase and cyclooxygenase-1 in heart failure. *Circ Res* 1996;78:58–64.
33. Zhang X, Tada H, Wang Z, Hintze TH. cAMP signal transduction: A potential compensatory pathway for coronary endothelial nitric oxide production after heart failure. *ATVB* 2002;22:1273–1278.
34. Wang J, Wolin MS, Hintze TH. Chronic exercise enhances endothelium-mediated dilation of epicardial coronary artery in conscious dogs. *Circ Res* 1993;73:829–838.
35. Bernstein RD, Ochoa FY, Xu X, Forfia P, Shen W, Thompson CI, Hintze TH. Function and production of nitric oxide in the coronary circulation of the conscious dog during exercise. *Circ Res* 1996;79:840–848.
36. Zhao G, Zhang X, Xu X, Ochoa M, Hintze TH. Exercise training enhances reflex cholinergic, NO dependent coronary dilation in conscious dogs. *Circ Res* 1997;80:868–876.
37. Sessa WC, Pritchard K, Seyedi N, Wang J, Hintze TH. Chronic exercise in dogs increases coronary vascular nitric oxide production and endothelial nitric oxide gene expression. *Circ Res* 1994;74:349–353.
38. Yusuf S, Sleight P, Pogue J, Bosch J, Davies R, Dagenais G. Effect of an angiotensin-converting-enzyme inhibitor, ramapril, on cardiovascular events in high-risk patients. *N Engl J Med* 2000;342:201–202.
39. Zhang X, Xie Y, Nasjletti A, Xu X, Wolin MS, Hintze TH. ACE inhibitors stimulate nitric oxide production to modulate myocardial oxygen consumption. *Circulation* 1997;95:176–182.
40. Kichuk MR, Seyedi N, X Zhang, Marboe CC, Michler RE, Addonizio LJ, Kaley G, Nasjletti A, Hintze TH. Regulation of nitric oxide production in human coronary microvessels and the contribution of local kinin formation. *Circulat* 1996;94:44–51.
41. Zhang X, Hintze TH. Amlodipine releases nitric oxide from canine coronary microvessels-an unexpected mechanism of action of a calcium-channel blocking agent. *Circulat* 1998;97:576–580.
42. Zhang X, Kichuk MR, Mital S, Oz M, Michler RE, Nasjletti A, Kaley G, Hintze TH. Amlodipine promotes kinin-mediated nitric oxide production in coronary microvessels from failing human heart. *Am J Cardiol* 1999;84:27L–33L.
43. Lauf U, LaFata V, Liao JK. Inhibition of 3-hydroxy-3-methylglutaryl (HMG)-CoA reductase blocks hypoxia-mediated downregulation of endotheial nitric oxide synthase. *J Biol Chem* 1997;272:31725–31729.
44. Mital S, X Zhang, G Zhao, Bernstein RD, Smith CJ, Fulton DL, Sessa WC, Liao JK, Hintze TH. Simvastatin upregulates coronary vascular nitric oxide synthase and nitric oxide production in conscious dogs. *Am J Physiol* 2000;279:H2649–H2657.
45. Mital S, Magneson A, Loke KE, Liao J, Forfia P, Hintze TH. Simvastatin acts synergistically with ACE inhibitors and amlodipine to decrease oxygen consumption in the rat heart. *J Cardiovas Pharm* 2000;36:248–254.
46. Trochu J-N, Mital S, Xu X, Ochoa M, Liao J, Recchia FA, Hintze TH. Preservation of NO production by stains: A new therapy for the treatment of heart failure. *FASEB J* 2001;15:A783 (abstract).
47. Fulton D, Gratto JP, McCabe TJ, Fontana J, Fujio Y, Walsh K, Franke T, Papapetropoulos A, Sessa WC. Regulation of endothelium-derived nitric oxide production by protein kinase Akt. *Nature* 1999;399:597–601.
48. Dimmler S, Fleming I, Fisslthaler B, Hermann C, Busse R, Zeiher AM. Activation of nitric oxide synthase in endothelial cells by Akt-dependent phosporylation. *Nature* 1999;399:601–605.
49. Anggard E. Nitric oxide: Mediator, murderer, and medicine. *Lancet* 1994;343:1199–1206.
50. Hirooka Y, Imaizumi T, Tagawa T, Shiramoto M, Endo T, Ando S, Takeshita A. Effects of L-arginine on impaired acetylcholine-induced and ischemic vasodilation of the forearm in patients with heart failure. *Circulation* 1994;90:658–668.
51. Koifman B, Wollman Y, Bogomolny N, Chernichowsky T, Finkelstein A, Peer G, Scherez J, Blum M, Laniado S, Iaina A. Improvement of cardiac performance by intravenous infusion of L-arginine in patients with moderate congestive heart failure. *J Am Coll Cardiol* 1995;26:1251–1256.
52. Kanaya Y, Nakamura M, Kobayashi N, Hiramori K. Effects of L-arginine on lower limb vasodilator reserve and exercise capacity in patients with chronic heart failure. *Heart* 1999;81:512–517.
53. Watanabe G, Tomiyama H, Doba N. Effects of oral administration of L-arginine on renal function in patients with heart failure. *J Hypertens* 2000;18:229–234.
54. Rector TS, Bank AJ, Mullen KA, Tschumperlin LK, Sih R, Pillai K, Kubo SH. Randomized, double-blind, placebo-controlled study of supplemental oral L-arginine in patients with heart failure. *Circulation* 1996;93:2135–2141.
55. Hambrecht R, Hilbrich L, Erbs S, Gielen S, Fiehn E, Schoene N, Schuler G. Correction of endothelial dysfunction in chronic heart failure: Additional effects of exercise training and oral L-arginine supplementation. *J Am Coll Cardiol* 2000;35:706–713.
56. Chin-Dusting JP, Kaye DM, Lefkovits J, Wong J, Bergin P, Jennings GL. Dietary supplementation with L-arginine fails to restore endothelial function in forearm resistance arteries of patients with severe heart failure. *J Am Coll Cardiol* 1996;27:1207–1213.
57. Kubes P, Kanwar S, Niu XF, Gaboury JP. Nitric oxide synthesis inhibition induces leukocyte adhesion via superoxide and mast cells. *FASEB J* 1993;7:1293–1299.

58. Ohara Y, Peterson TE, Harrison DG. Hypercholesterolemia increases endothelial superoxide anion production. *J Clin Invest* 1993;91:2546–2551.
59. Toutouzas PC, Tousoulis D, Davies GJ. Nitric oxide synthesis in atherosclerosis. *Eur Heart J* 1998;19:1504–1511.
60. McDonald KK, Zharikov S, Block ER, Kilberg MS. A caveolar complex between the cationic amino acid transporter 1 and endothelial nitric-oxide synthase may explain the "arginine paradox". *J Biol Chem* 1997;272:31213–31216.
61. Usui M, Matsuoka H, Miyazaki H, Ueda S, Okuda S, Imaizumi T. Increased endogenous nitric oxide synthase inhibitor in patients with congestive heart failure. *Life Sci* 1998;62:2425–2430.
62. Bode-Boger SM, Boger RH, Kienke S, Junker W, Frolich JC. Elevated L-arginine/dimethylarginine ratio contributes to enhanced systemic NO production by dietary L-arginine in hypercholesterolemic rabbits. *Biochem Biophys Res Commun* 1996;219:598–603.
63. Kaye DM, Ahlers BA, Autelitano DJ, Chin-Dusting JP. *In vivo* and *in vitro* evidence for impaired arginine transport in human heart failure. *Circulation* 2000;102:2707–2712.
64. Buga GM, Singh R, Pervin S, Rogers NE, Schmitz DA, Jenkinson CP, Cederbaum SD, Ignarro LJ. Arginase activity in endothelial cells: Inhibition by NG-hydroxy-L-arginine during high-output NO production. *Am J Physiol* 1996;271:H1988–H1998.
65. Arnal JF, Munzel T, Venema RC, James NL, Bai CL, Mitch WE, Harrison DG. Interactions between L-arginine and L-glutamine change endothelial NO production. An effect independent of NO synthase substrate availability. *J Clin Invest* 1995;95:2565–2572.
66. Wascher TC, Posch K, Wallner S, Hermetter A, Kostner GM, Graier WF. Vascular effects of L-arginine: Anything beyond a substrate for the NO-synthase? *Biochem Biophys Res Commun* 1997;234:35–38.
67. Pedrinelli R, Ebel M, Catapano G, Dell'Omo G, Ducci M, Del Chicca M, Clerico A. Pressor, renal and endocrine effects of L-arginine in essential hypertensives. *Eur J Clin Pharmacol* 1995;48:195–201.
68. Schmidt HH, Warner TD, Ishii K, Sheng H, Murad F. Insulin secretion from pancreatic B cells caused by L-arginine-derived nitrogen oxides. *Science* 1992;255:721–723.
69. Giugliano D, Marfella R, Verrazzo G, Acampora R, Coppola L, Cozzolino D, D'Onofrio F. The vascular effects of L-Arginine in humans. The role of endogenous insulin. *J Clin Invest* 1997;99:433–438.
70. Belch JJ, Bridges AB, Scott N, Chopra M. Oxygen free radicals and congestive heart failure. *Br Heart J* 1991;65:245–248.
71. Sobotka PA, Brottman MD, Weitz Z, Birnbaum AJ, Skosey JL, Zarling EJ. Elevated breath pentane in heart failure reduced by free radical scavenger. *Free Radic Biol Med* 1993;14:643–647.
72. Singh N, Dhalla AK, Seneviratne C, Singal PK. Oxidative stress and heart failure. *Mol Cell Biochem* 1995;147:77–81.
73. McMurray J, Chopra M, Abdullah I, Smith WE, Dargie HJ. Evidence of oxidative stress in chronic heart failure in humans. *Eur Heart J* 1993;14:1493–1498.
74. Keith M, Geranmayegan A, Sole MJ, Kurian R, Robinson A, Omran AS, Jeejeebhoy KN. Increased oxidative stress in patients with congestive heart failure. *J Am Coll Cardiol* 1998;31:1352–1356.
75. Hornig B, Arakawa N, Kohler C, Drexler H. Vitamin C improves endothelial function of conduit arteries in patients with chronic heart failure. *Circulation* 1998;97:363–368.
76. Bauersachs J, Bouloumie A, Fraccarollo D, Hu K, Busse R, Ertl G. Endothelial dysfunction in chronic myocardial infarction despite increased vascular endothelial nitric oxide synthase and soluble guanylate cyclase expression: Role of enhanced vascular superoxide production. *Circulation* 1999;100:292–298.
77. Cappola TP, Kass DA, Nelson GS, Berger RD, Rosas GO, Kobeissi ZA, Marban E, Hare JM. Allopurinol improves myocardial efficiency in patients with idiopathic dilated cardiomyopathy. *Circulation* 2001;104:2407–2411.
78. Doehner W, Schoene N, Rauchhaus M, Leyva-Leon F, Pavitt DV, Reaveley DA, Schuler G, Coats AJ, Anker SD, Hambrecht R. Effects of xanthine oxidase inhibition with allopurinol on endothelial function and peripheral blood flow in hyperuricemic patients with chronic heart failure: Results from 2 placebo-controlled studies. *Circulation* 2002;105:2619–2624.
79. Ide T, Tsutsui H, Kinugawa S, Utsumi H, Kang D, Hattori N, Uchida K, Arimura K, Egashira K, Takeshita A. Mitochondrial electron transport complex I is a potential source of oxygen free radicals in the failing myocardium. *Circ Res* 1999;85:357–363.
80. Yucel D, Aydogdu S, Cehreli S, Saydam G, Canatan H, Senes M, Cigdem TB, Nebioglu S. Increased oxidative stress in dilated cardiomyopathic heart failure. *Clin Chem* 1998;44:148–154.
81. Gokce N, Keaney JF, Jr., Frei B, Holbrook M, Olesiak M, Zachariah BJ, Leeuwenburgh C, Heinecke JW, Vita JA. Long-term ascorbic acid administration reverses endothelial vasomotor dysfunction in patients with coronary artery disease. *Circulation* 1999;99:3234–3240.
82. Ito K, Akita H, Kanazawa K, Yamada S, Terashima M, Matsuda Y, Yokoyama M. Comparison of effects of ascorbic acid on endothelium-dependent vasodilation in patients with chronic congestive heart failure secondary to idiopathic dilated cardiomyopathy versus patients with effort angina pectoris secondary to coronary artery disease. *Am J Cardiol* 1998;82:762–767.
83. Ellis GR, Anderson RA, Lang D, Blackman DJ, Morris RH, Morris-Thurgood J, McDowell IF, Jackson SK, Lewis MJ, Frenneaux MP. Neutrophil superoxide anion—generating capacity, endothelial function and oxidative stress in chronic heart failure: Effects of s. *J Am Coll Cardiol* 2000;36:1474–1482.
84. Erbs S, Mobius-Winkler S, Gielen S, Schoene N, Linke A, Schulze PC, Hambrecht R. Correction of endothelial dysfunction by vitamin C: Different effects in ischemic heart disease and dilative cardiomyopathy. *Circulation* 2000;102:II–55.
85. Heller R, Munscher-Paulig F, Grabner R, Till U. L-Ascorbic acid potentiates nitric oxide synthesis in endothelial cells. *J Biol Chem* 1999;274:8254–8260.
86. Mizutani A, Maki H, Torii Y, Hitomi K, Tsukagoshi N. Ascorbate-dependent enhancement of nitric oxide formation in activated macrophages. *Nitric Oxide* 1998;2:235–241.
87. Murphy ME, Piper HM, Watanabe H, Sies H. Nitric oxide production by cultured aortic endothelial cells in response to thiol depletion and replenishment. *J Biol Chem* 1991;266:19378–19383.

88. Stamler JS, Singel DJ, Loscalzo J. Biochemistry of nitric oxide and its redox-activated forms. *Science* 1992;258:1898–1902.
89. Packer JE, Slater TF, Willson RL. Direct observation of a free radical interaction between vitamin E and vitamin C. *Nature* 1979;278:737–738.
90. Jarasch ED, Grund C, Bruder G, Heid HW, Keenan TW, Franke WW. Localization of xanthine oxidase in mammary-gland epithelium and capillary endothelium. *Cell* 1981;25:67–82.
91. Enroth C, Eger BT, Okamoto K, Nishino T, Nishino T, Pai EF. Crystal structures of bovine milk xanthine dehydrogenase and xanthine oxidase: Structure-based mechanism of conversion. *Proc Natl Acad Sci USA* 2000;97:10723–10728.
92. Bakhtiiarov ZA. [Changes in xanthine oxidase activity in patients with circulatory failure]. *Ter Arkh* 1989;61:68–69.
93. Doehner W, Rauchhaus M, Florea VG, Sharma R, Bolger AP, Davos CH, Coats AJ, Anker SD. Uric acid in cachectic and noncachectic patients with chronic heart failure: Relationship to leg vascular resistance. *Am Heart J* 2001;141:792–799.
94. Saavedra WF, Paolocci N, St John ME, Skaf MW, Stewart GC, Xie JS, Harrison RW, Zeichner J, Mudrick D, Marban E, Kass DA, Hare JM. Imbalance between xanthine oxidase and nitric oxide synthase signaling pathways underlies mechanoenergetic uncoupling in the failing heart. *Circ Res* 2002;90:297–304.
95. Pitt B, Zannad F, Remme WJ, Cody R, Castaigne A, Perez A, Palensky J, Wittes J. The effect of spironolactone on morbidity and mortality in patients with severe heart failure. Randomized Aldactone Evaluation Study Investigators. *N Engl J Med* 1999;341:709–717.
96. Zannad F, Alla F, Dousset B, Perez A, Pitt B. Limitation of excessive extracellular matrix turnover may contribute to survival benefit of spironolactone therapy in patients with congestive heart failure: Insights from the randomized aldactone evaluation study (RALES). Rales Investigators. *Circulation* 2000;102:2700–2706.
97. Farquharson CA, Struthers AD. Spironolactone increases nitric oxide bioactivity, improves endothelial vasodilator dysfunction, and suppresses vascular angiotensin I/angiotensin II conversion in patients with chronic heart failure. *Circulation* 2000;101:594–597.
98. Bauersachs J, Heck M, Fraccarollo D, Hildemann SK, Ertl G, Wehling M, Christ M. Addition of spironolactone to angiotensin-converting enzyme inhibition in heart failure improves endothelial vasomotor dysfunction: Role of vascular superoxide anion formation and endothelial nitric oxide synthase expression. *J Am Coll Cardiol* 2002;39:351–358.
99. Hornig B, Maier V, Drexler H. Physical training improves endothelial function in patients with chronic heart failure. *Circulation* 1996;93:210–214.
100. Katz SD, Yuen J, Bijou R, LeJemtel TH. Training improves endothelium-dependent vasodilation in resistance vessels of patients with heart failure. *J Appl Physiol* 1997;82:1488–1492.
101. Hambrecht R, Schulze PC, Gielen S, Linke A, Mobius-Winkler S, Yu J, Kratzsch JJ, Baldauf G, Busse MW, Schubert A, Adams V, Schuler G. Reduction of insulin-like growth factor-I expression in the skeletal muscle of noncachectic patients with chronic heart failure. *J Am Coll Cardiol* 2002;39:1175–1181.
102. Hambrecht R, Fiehn E, Yu J, Niebauer J, Weigl C, Hilbrich L, Adams V, Riede U, Schuler G. Effects of endurance training on mitochondrial ultrastructure and fiber type distribution in skeletal muscle of patients with stable chronic heart failure. *J Am Coll Cardiol* 1997;29:1067–1073.
103. Linke A, Schoene N, Gielen S, Hofer J, Erbs S, Schuler G, Hambrecht R. Endothelial dysfunction in patients with chronic heart failure: Systemic effects of lower-limb exercise training. *J Am Coll Cardiol* 2001;37:392–397.
104. Hambrecht R, Gielen S, Linke A, Fiehn E, Yu J, Walther C, Schoene N, Schuler G. Effects of exercise training on left ventricular function and peripheral resistance in patients with chronic heart failure: A randomized trial. *JAMA* 2000;283:3095–3101.
105. Piepoli M, Clark AL, Volterrani M, Adamopoulos S, Sleight P, Coats AJ. Contribution of muscle afferents to the hemodynamic, autonomic, and ventilatory responses to exercise in patients with chronic heart failure: Effects of physical training. *Circulation* 1996;93:940–952.
106. Kiilavuori K, Toivonen L, Naveri H, Leinonen H. Reversal of autonomic derangements by physical training in chronic heart failure assessed by heart rate variability. *Eur Heart J* 1995;16:490–495.
107. Coats AJ, Adamopoulos S, Radaelli A, McCance A, Meyer TE, Bernardi L, Solda PL, Davey P, Ormerod O, Forfar C. Controlled trial of physical training in chronic heart failure. Exercise performance, hemodynamics, ventilation, and autonomic function. *Circulation* 1992;85:2119–2131.
108. Belardinelli R, Georgiou D, Cianci G, Purcaro A. Randomized, controlled trial of long-term moderate exercise training in chronic heart failure: Effects on functional capacity, quality of life, and clinical outcome. *Circulation* 1999;99:1173–1182.
109. Ranjan V, Xiao Z, Diamond SL. Constitutive NOS expression in cultured endothelial cells is elevated by fluid shear stress. *Am J Physiol* 1995;269:H550–H555.
110. Noris M, Morigi M, Donadelli R, Aiello S, Foppolo M, Todeschini M, Orisio S, Remuzzi G, Remuzzi A. Nitric oxide synthesis by cultured endothelial cells is modulated by flow conditions. *Circ Res* 1995;76:536–543.
111. Sessa WC, Pritchard K, Seyedi N, Wang J, Hintze TH. Chronic exercise in dogs increases coronary vascular nitric oxide production and endothelial cell nitric oxide synthase gene expression. *Circ Res* 1994;74:349–353.
112. Wang J, Wolin MS, Hintze TH. Chronic exercise enhances endothelium-mediated dilation of epicardial coronary artery in conscious dogs. *Circ Res* 1993;73:829–838.
113. Woodman CR, Muller JM, Laughlin MH, Price EM. Induction of nitric oxide synthase mRNA in coronary resistance arteries isolated from exercise-trained pigs. *Am J Physiol* 1997;273:H2575–H2579.
114. Laughlin MH, Pollock JS, Amann JF, Hollis ML, Woodman CR, Price EM. Training induces nonuniform increases in eNOS content along the coronary arterial tree. *J Appl Physiol* 2001;90:501–510.
115. Boo YC, Sorescu G, Boyd N, Shiojima I, Walsh K, Du J, Jo H. Shear stress stimulates phosphorylation of endothelial nitric-oxide synthase at Ser1179 by Akt-independent

mechanisms: Role of protein kinase A. *J Biol Chem* 2002;277:3388–3396.

116. Rizzo V, McIntosh DP, Oh P, Schnitzer JE. *In situ* flow activates endothelial nitric oxide synthase in luminal caveolae of endothelium with rapid caveolin dissociation and calmodulin association. *J Biol Chem* 1998;273:34724–34729.

117. Posch K, Schmidt K, Graier WF. Selective stimulation of L-arginine uptake contributes to shear stress-induced formation of nitric oxide. *Life Sci* 1999;64:663–670.

118. Fukai T, Siegfried MR, Ushio-Fukai M, Cheng Y, Kojda G, Harrison DG. Regulation of the vascular extracellular superoxide dismutase by nitric oxide and exercise training. *J Clin Invest* 2000;105:1631–1639.

The Role of NOS in Heart Failure: Lessons from Murine Genetic Models

Imran N. Mungrue,[1,3] Mansoor Husain, MD,[1,3,4] and Duncan J. Stewart, MD[2,3,4]

[1]Division of Cell & Molecular Biology, The Toronto General Hospital Research Institute, 12EN-221, 101 College St, Toronto, ON, M5G 2C4, Canada; [2]The Terrence Donnelly Heart Centre, St Michael's Hospital, 7-081 Queen, 30 Bond St, Toronto, ON, M5B 1W8, Canada; [3]Departments of Medicine, Laboratory Medicine & Pathobiology and the Heart & Stroke Richard Lewar Centre of Excellence, University of Toronto, Canada; [4]The corresponding authors have contributed equally to this work.

Abstract. **Nitric Oxide Synthases (NOSs) are a group of related proteins that produce nitric oxide (NO). In mammals, there are three known members of this gene family: nNOS (*NOS1*), iNOS (*NOS2*) and eNOS (*NOS3*). Each has been disrupted by targeted gene ablation in mice and the corresponding phenotypes examined. These mice have allowed an examination of the contribution of each NOS in a variety of experimental models and continue to provided insights into the patho-physiological role of NOS and NO. With increasing sophistication, murine transgenic approaches continue to offer a wealth of information, and invaluable tools to further study the NOS system. The focus of this review will be an examination of the tools available, and the insights gained from studies done on murine NOS genetic models in the context of heart failure.**

Key Words. **nitric oxide synthase (NOS), heart failure, knockout mouse, conditional transgenic mouse, nitric oxide (NO)**

Introduction

The prototypical Nitric Oxide Synthase (NOS) nNOS (Fig. 1) was first isolated and purified [1] in 1990, and was subsequently cloned and sequenced [2]. Two closely related enzymes, iNOS and eNOS, were isolated [3–7] and cloned shortly thereafter [8–13]. An enormous body of research now exists describing the molecular, cellular and pharmacological properties of these enzymes, as well as their contribution to several physiological and disease processes.

Given their sequence similarity, the NOS enzymes appear to have evolved from an ancestral P-450 cytochrome type enzyme, and contain a NADPH-dependent cytochrome P-450 reductase motif at the C-terminus [2]. The NOS C-terminus shuttles electrons (analogous to cytochrome P-450 reductases) from NADPH to FAD, FMN and then to a heme-coordinated iron (Fe^{3+}) within the NOS N-terminal oxygenase domain (Figs. 2 and 3) [14–16].

The oxygenase domain of NOS is functionally equivalent to the P-450 oxidases, however, their distinct sequence and crystal structure suggests convergent functional evolution from different ancestral motifs [17]. While the activities of the C and N-terminals may be functionally independent [14], the conversion of L-arginine to NO (Fig. 2) requires both domains and homodimerization through a N-terminal interface [16], requiring heme and stabilized by BH_4 (tetrahydrobiopterin), L-arginine, and Zinc [17–19]. The reaction catalyzed by the N-terminus (Fig. 3) proceeds via a stable intermediate, and thus consists of at least two steps [16]. The first step involves binding of oxygen (O_2) to the heme moiety, and oxidation of a guanido N molecule of L-arginine to form N^G-hydroxy-L-arginine [20]. A second O_2 molecule is then combined with this intermediate leading to the production of NO and citrulline [21].

The common NOS nomenclature is derived from anatomic and functional perspectives. Endothelial NOS, or eNOS (NOS3, NOS III, ecNOS), is expressed in endothelial cells and is involved in the production of vaso-active membrane permeable NO gas that participates in the regulation of blood pressure [22]. Neuronal NOS, nNOS (NOS1, NOS I, ncNOS, bNOS), originally identified in neurons, mediates synaptic signaling in a similar manner [23] and may be

Address for correspondence: Mansoor Husain and Duncan J. Stewart, The Toronto General Hospital Research Institute, 12EN-221, 200 Elizabeth St, Toronto, ON, M5G 2C4, Canada. Tel.: (416) 340-3188; Fax: 340-4021; E-mail: mansoor.husain@utoronto.ca

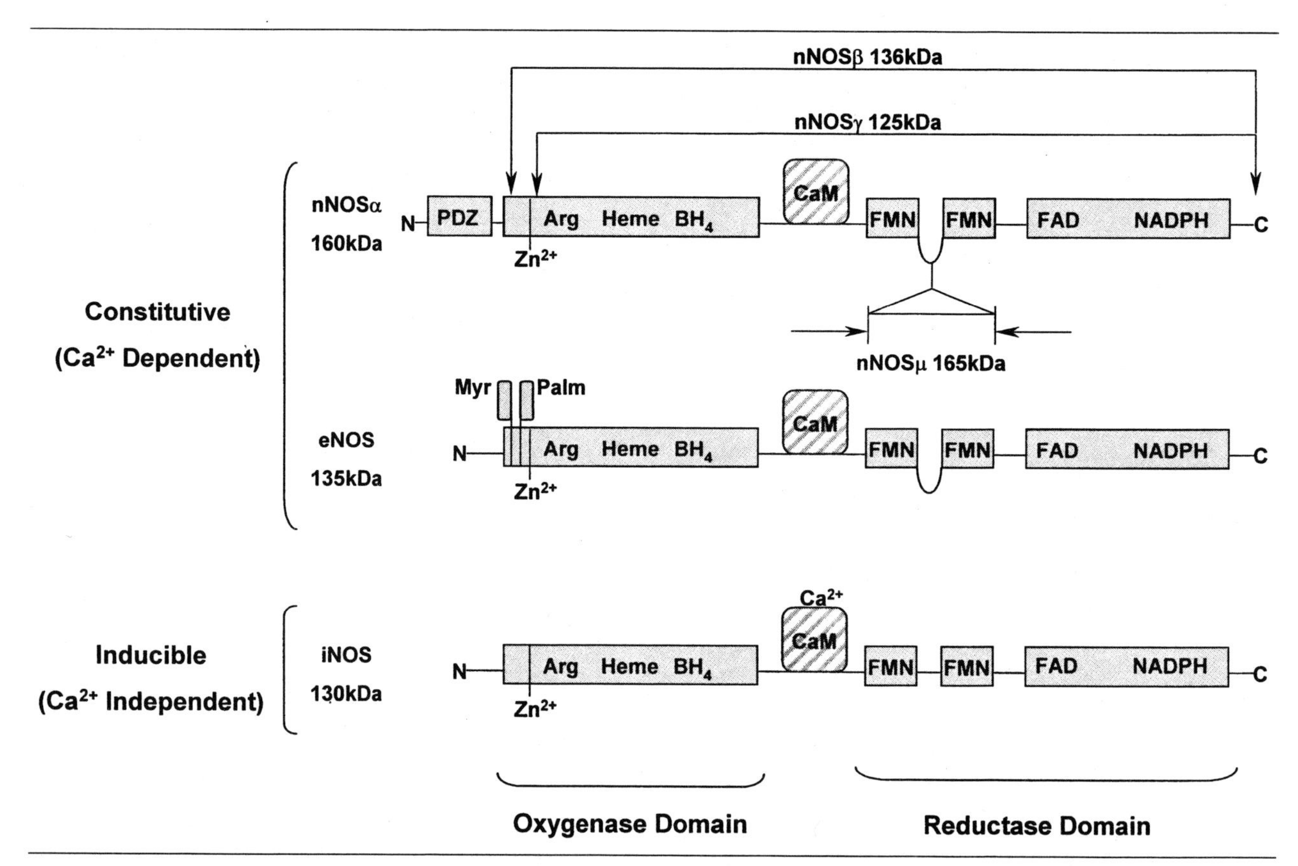

Fig. 1. Structural domains of NOS enzymes. Alignment of conserved regions of the NOS enzymes. Note calmodulin (CaM), cofactor and substrate binding regions. Isoform specific features including membrane-localizing PDZ (nNOSα and nNOSμ), myristoylation (Myr) and palmitoylation (Palm) sites (eNOS) are also shown. Arg (L-arginine), BH_4 (tetrahydrobiopterin), FAD (flavin adenine dinucleotide), FMN (flavin mononuclotide), NADPH (nicotinamide adenine dinucleotide phosphate), Zn^{2+} (zinc).

involved in memory formation [24]. The neuronal and endothelial isoforms are tightly regulated by calcium-calmodulin, and generate small amounts of NO, which have precise actions on adjacent cells. In contrast, once expressed, the cytokine-inducible iNOS (NOS2, NOS II, macNOS) produces high levels of NO independent of intracellular calcium [25]. A high concentration of NO may itself be toxic in the oxidative environments that often accompany iNOS induction [26]. Indeed, the combination of NO with other radical oxygen species may directly mediate cell toxicity, and comprises a critical component of the host immune response [26].

In addition to these well-established roles, a more promiscuous pattern of expression has become evident, and a growing number of processes have implicated NO and NOS [27–31], including apoptosis [32], angiogenesis [33], peristalsis [34], micturition [34], cutaneous wound repair [35], skeletal muscle contraction [36], sarcoplasmic reticulum Ca^{2+} release [37] and glucose matabolism [38]. This review will focus on the role of the NOS genes in the context of heart failure, with emphasis on lessons learned from mouse genetic models (Table 1). Readers are directed to the other contributions in this publication for a more comprehensive examination of the contributions of NO and NOS to other aspects of cardiac function and disease.

nNOS

Expression of neuronal NOS (nNOS) has been demonstrated in a population of developing [39] and adult neurons [40], the adventitia of a subset of neuronal blood vessels [41], and in skeletal muscle [36]. However, the finding that nNOS is also expressed in sarcoplasmic reticulum (SR) membrane vesicles of cardiac myocytes [42] suggests a functional role in the heart. Interestingly, Xu et al. demonstrated functional nNOS activity and NO production in isolated cardiac SR vesicles, which specifically inhibited ^{45}Ca uptake by the SR calcium ATPase. These data support a mechanism whereby NO produced by nNOS could modulate Ca^{2+} dynamics within the heart [42]. Unfortunately, the use of non-selective NOS inhibitors to study these pathways is complicated by their inability to distinguish between the NOS isoforms. Moreover, the physiological regulation and

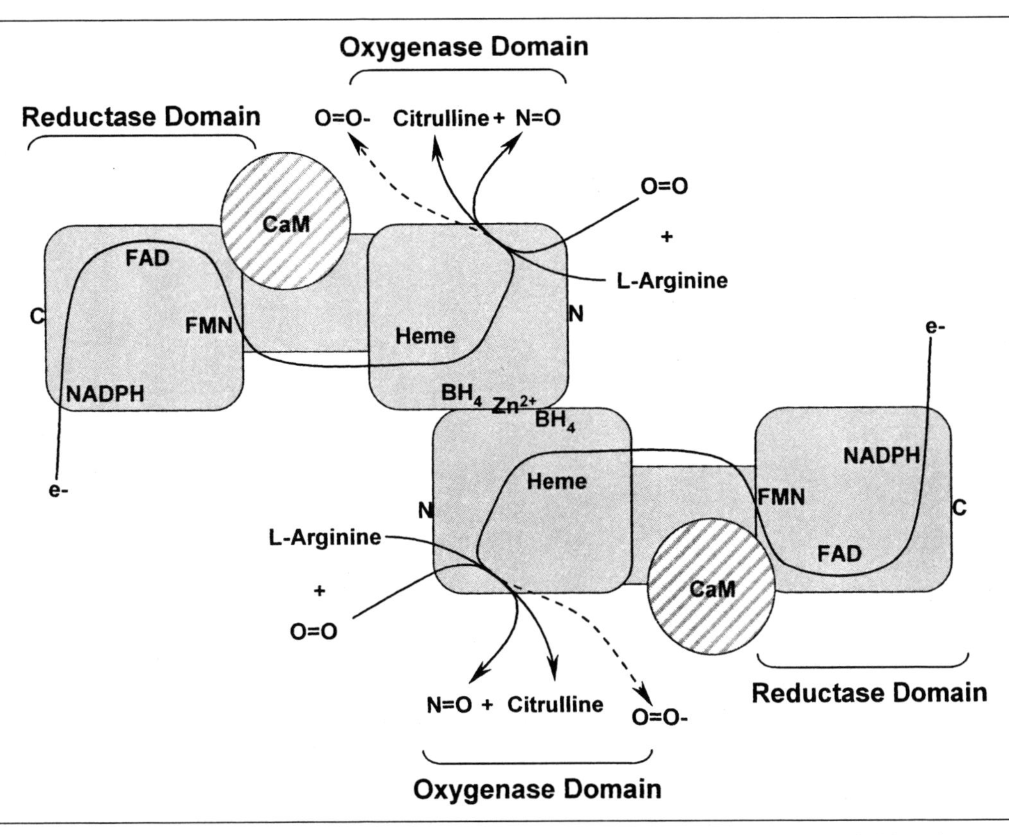

Fig. 2. *Model of the functional NOS dimer. A functional NOS dimer in association with calmodulin (CaM), cofactors and substrates. Note the overall oxo-reductase reaction catalyzed in the presence of sufficient (solid lines), or limiting (dashed lines) substrate and/or BH_4 (tetrahydrobiopterin). e- (reducing equivalent), FAD (flavin adenine dinucleotide), FMN (flavin mononuclotide), N=O (nitric oxide), NADPH (nicotinamide adenine dinucleotide phosphate), O=O (oxygen), Zn^{2+} (zinc).*

sub-cellular localization of each NOS is distinct, which suggests that the combined contribution of NOS in regulating cardiac function and pathological remodeling is complex. The use of specific gene ablation and transgenic overexpression models allows dissection of the distinct roles of each NOS isoform individually, and provides the means to test the contributions of each NOS in a myriad of physiological or pathological processes that maintain normal cardiac function or lead to disease.

nNOS Knockout Mice

nNOS was the first NOS locus to be targeted for homologous recombination in mice [43]. Huang et al. deleted the region of *NOS1* flanking exon 2, which included translation start of full-length brain-spliced nNOS (nNOSα). These mice were bred to homozygosity (nNOS$^{\Delta/\Delta}$), and did not display any reproductive or survival defects [43]. However, nNOS$^{\Delta/\Delta}$ mice had enlarged stomachs, with pyloric sphincter and circular muscle hypertrophy [43], complex behavioral perturbations [44], insulin resistance [45], and geriatric cardiac left ventricular (LV) hypertrophy [46]. Subsequent studies have highlighted that wild-type and nNOS$^{\Delta/\Delta}$ mice may produce distinct nNOS mRNA splice variants (nNOSβ and nNOSγ) by utilizing different translation start sites, which exclude the region targeted by the knockout strategy [47], or that have an extra exon inserted (nNOSμ, Fig. 1) [48]. Consequently, translation from this locus is not completely 'knocked out', rather nNOS$^{\Delta/\Delta}$ mice exhibit a low baseline level of neuronal NOS activity, but no nNOSα or nNOSμ expression [43]. While the precise identity of cardiac-expressed nNOS splice variants have yet to be described, nNOS$^{\Delta/\Delta}$ mice are not expected to have SR membrane nNOS expression, due to ablation of the PDZ targeting motif contained within exon 2 [49]. These possible limitations should be considered in any study employing this nNOS$^{\Delta/\Delta}$ model.

Involvement of nNOS in Modulation of Heart Rate (HR). While the effects of pharmacological modulations of the NOS/NO system on vagal-mediated bradycardia has produced differing results, the findings that nNOS$^{\Delta/\Delta}$ mice had higher mean HR, and lower HR variance [50] were intriguing, as similar perturbations are associated with increased mortality in humans with

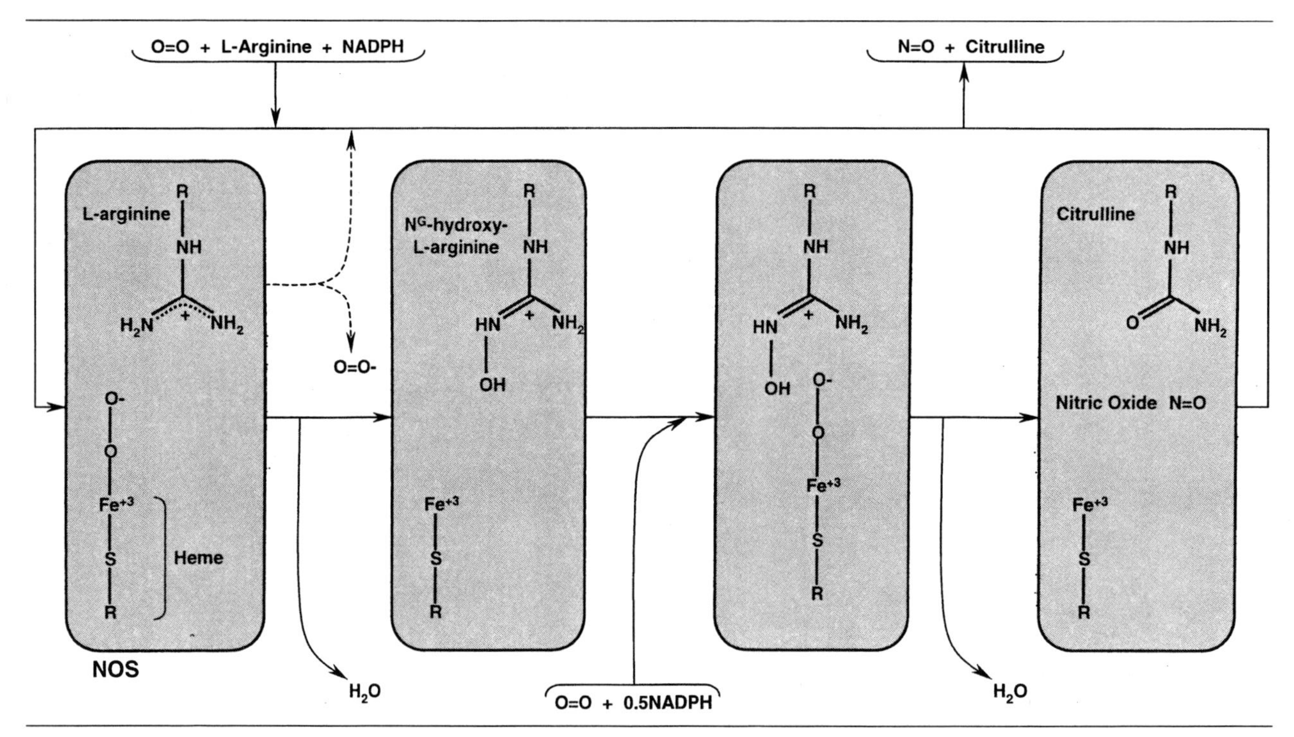

Fig. 3. Reaction at NOS oxygenase Catalytic site. Schematic of the NOS N-terminal oxygenase domain active site showing a simplified reaction mechanism in the presence of sufficient (solid lines), or limiting (dashed lines) substrate and/or critical cofactors. NADPH (nicotinamide adenine dinucleotide phosphate) O=O (oxygen).

heart failure [51]. While nNOS$^{\Delta/\Delta}$ mice also have blunted HR and HR variability in response to atropine, they display a normal response to β-adrenergic blockade with propanolol, suggesting reduced baseline parasympathetic tone [50]. Furthermore, in the absence of inhibitory cardiac G protein activity (pertussis toxin treatment), a specific role for NO in a parallel pathway mediating autonomic bradycardia was also demonstrated [50]. Subsequent work with isolated atria and intact right vagus nerve preparations from nNOS$^{\Delta/\Delta}$ mice, have shown that nNOS is involved in mediating vagal slowing of heart rate through an NO-mediated mechanism [52]. Interestingly, this was not dependent on receptor-mediated (carbamylcholine dependent) parasympathetic signaling [52].

Involvement of nNOS in Modulating Cardiac Contractility. Recently, Barouch et al. reported that nNOS$^{\Delta/\Delta}$ mice have suppressed inotropic responses to isoproterenol, and that nNOS was associated with the cardiac myocyte ryanodine receptor in wild type mice [46]. Additionally, isoproterenol treatment led to decreased $[Ca^{2+}]_i$ and sarcomere shortening in nNOS$^{\Delta/\Delta}$ mice versus controls, suggesting a mechanism whereby nNOS positively modulates the β-adrenergic-stimulated cardiac reserve [46].

Myocardial Reperfusion Injury in nNOS$^{\Delta/\Delta}$ Mice. Jones et al. compared the responses of nNOS$^{\Delta/\Delta}$ mice and controls in a myocardial reperfusion injury model [53]. While there was a significant increase in cardiac polymorphonuclear leukocyte infiltration, infarct size did not differ in nNOS$^{\Delta/\Delta}$ mice [53,54]. These studies imply that nNOS is not involved in response to ischemia and reperfusion in the heart, as opposed to the brain where a negative contribution of nNOS was previously established in the nNOS$^{\Delta/\Delta}$ model following cerebral ischemia [55].

Cardiac nNOS Genetic Models: 'To Be Continued...'

The nNOS$^{\Delta/\Delta}$ mouse has provided insights into the roles of nNOS in the neuronal, digestive, skeletal-muscular and cardiovascular systems. The role of nNOS in mediating decreases in vagal-dependent HR and HR variability, make this pathway a target for new pharmacological approaches in the management of heart failure. The production of cardiovascular-directed conditional nNOS transgenic over-expression and/or conditional knockout models would offer complementary insights into the roles of nNOS in cardiovascular patho-physiology, and represent important avenues for future research.

Table 1. *Summary murine genetic NOS models*

	Model	Baseline phenotype	Experimental insights
nNOS	$nNOS^{\Delta/\Delta}$ [43]	Enlarged stomach, pyloric sphincter stenosis, increased aggression in males, geriatric LV hypertrophy and insulin resistance.	nNOS involved in autonomic vagal bradycardia [50,52], and increases inotropy [46].
eNOS	$eNOS^{-/-}$ [61–63]	Hypertension, decreased body wt, decreased waking HR, decreased plasma renin, increased HR variability, increased cardiac prepro ANP, impaired angiogenesis, insulin resistance, decreased aggression in males, limb defects, bicuspid aortic valve, geriatric LV hypertrophy and increased vascular adventitial nNOS expression.	Role of nNOS in BP regulation [62,70–72], eNOS negatively modulates β-adrenergic stimulation of cardiac contractility [46] and ventricular relaxation [65], eNOS reduces infarct size following cardiac ischemia [54,77,78].
	$eNOS^{-/-}$; $apoE^{-/-}$ (DKO) [74,75]	Hypertension, coronary arteroslcerosis with high fat diet + ...?	Anti-atherogenic role for eNOS [74,75].
	ppET-eNOS [80]	Hypotension, decreased response to ACh, ATPγS, SNP and NTG.	eNOS activities in BP regulation [80], beneficial roles in sepsis [82] and vascular remodeling [83].
	αMHC-eNOS [84]	Decreased LVP and cardiac hypertrophy (only in highest expressing line).	Biphasic effect of eNOS/NO: positive inotropy at low and negative at high concentrations [84].
iNOS	$iNOS^{-/-}$ [94–96]	No abnormalities.	Biphasic roles in septic shock [94–96,98–101], vascular remodeling [74,76,104–106,109,110], cardiac transplant [111,112] and ischemia [113].
	$iNOS^{-/-}$; $apoE^{-/-}$ (DKO) [74,104,105]	No abnormalities.	
	αMHC-iNOS [115]	No cardiac phenotype.	
	$\alpha MtTA^{+}/iNOS^{+}$ [114]	Cardiomyopathy, sudden death, brady-arrhythmia and decreased Cx40 in AV-node and BB.	

List of NOS genetic models, known baseline phenotypes and experimental insights described within this review. LV (left ventricle), HR (heart rate), ANP (atrial natriuretic peptide), DKO (double knockout), SNP (sodium nitroprusside), ACh (acetylcholine), NTG (nitroglycerin), LVP (LV systolic pressure), ppET (prepro endothelin 1), αMHC (α-myosin heavy chain), αMtTA (α-MHC-tetracycline regulated transcriptional transactivator), AV (atrioventricular) BB (bundle branch), apoE (apolipoprotein E), and Cx40 (connexin 40).

eNOS

The endothelial constitutive NOS (eNOS, NOS3, NOS III, ecNOS) is the major NOS isoform expressed in the heart, except during cardiac pathology where cytokine-induced iNOS expression may dominate. While eNOS expression occurs mainly in vascular endothelium and plays an important role in regulation of vascular resistance and myocardial perfusion, this isoform may also be expressed in endocardial endothelium, and cardiac myocytes [56]. eNOS is regulated by complex mechanisms that include reversible calmodulin (CaM) binding, induction by shear stress, post-translational caveoli targeting, and serine, threonine and/or tyrosine phosphorylation [57–60].

eNOS Knockout Mice

Three groups independently produced mice with targeted disruption of the *NOS3* locus [61–63]. Huang et al. first demonstrated that the absence of eNOS led to a decrease in acetylcholine-dependent relaxation in intact aortic rings, and that these mice were hypertensive, supporting a role for eNOS derived NO in blood pressure homeostasis [62]. Interestingly, inhibition of residual NOS activity in these mice, with a non-selective NOS inhibitor, resulted in a paradoxical normalization of blood pressure, suggesting additional eNOS-independent hypertensive pathways [62]. Consistent with this finding was the observation that $nNOS^{\Delta/\Delta}$ mice had a tendency towards hypotension with anesthesia, suggesting a role for nNOS in BP homeostasis [55].

Using a different targeting strategy, Shesely et al. confirmed that $eNOS^{-/-}$ mice exhibited hypertension. However, this group also noted that $eNOS^{-/-}$ mice had lower body weights at 14 weeks of age, lower waking HR, and a decrease in plasma renin activity [63]. The $eNOS^{-/-}$ mice produced by Godecke et al. also displayed hypertension and decreased waking HR [61]. In addition to these initial characterizations, $eNOS^{-/-}$ mice were subsequently shown to have increased HR variability [64], increased cardiac prepro ANP (atrial natriuretic peptide) mRNA expression [65], impaired angiogenesis [66], insulin resistance [45], behavioral dysfunctions [67], age-dependent LV hypertrophy [46], and developmental limb [68] and cardiac valve abnormalities [69]. These findings point to diverse and subtle influences of eNOS-derived NO on a wide range of developmental and physiological processes. Indeed, the importance of eNOS may have been underestimated due to a strong negative selection bias, since the majority of eNOS$^{-/-}$ mice appear to succumb shortly after birth (Qingping Feng, personal communication), likely due to cardiovascular abnormalities that are incompatible with post-natal survival. As with the nNOS$^{\Delta/\Delta}$ model, the contributions of the above baseline phenotypes must be taken into account in any studies comparing experimental manipulations in eNOS$^{-/-}$ mice versus wild-type littermate controls.

Interaction Between NOS Isoforms in Regulation of Blood Pressure. The role of eNOS in the regulation of blood pressure and vascular tone has been well characterized over the last two decades. However, new evidence suggesting that nNOS may also be involved in the modulation of blood pressure has been generated in the eNOS$^{-/-}$ model. In eNOS$^{-/-}$ mice there is a compensatory up-regulation of nNOS in the arterial endothelium [70] and preserved functional responses to acetylcholine in resistance vessels such as the pial [71], mesenteric [72] and coronary arteries [70], but not in larger conduit vessels such as the carotid artery or aorta [62,72]. These data would suggest that nNOS might functionally replace or augment eNOS activity and contribute to vascular relaxation, similar to the functional redundancy between the constitutive NOS isoforms in mediating long-term potentiation in the brain [24]. This phenomenon may be important in the setting of cardiac and vascular disease, where nNOS expression may partially compensate for a down-regulation of eNOS. Indeed, constitutive NOS functional redundancy may serve a teleologically protective role, as opposed to a principally pathogenic role of immune-stimulated iNOS expression, which may also be present in the setting of cardiovascular disease.

Endothelial Function in eNOS$^{-/-}$ Mice. The production of NO by eNOS has been suggested to play a protective role in prevention of vascular endothelial dysfunction, a major contributing factor to cardiovascular disease. One mechanism for the protective effect of eNOS-derived NO is the prevention of endothelial leukocyte adhesion. Wild-type mice treated with a non-specific NOS inhibitor (L-NAME) showed increased leukocyte adhesion, suggesting a NOS-dependent mechanism. However, at baseline, eNOS$^{-/-}$ mice did not show a similar increase in vascular wall leukocyte adhesion [73]. This paradox was resolved by the observation that a nNOS selective inhibitor (7-nitroindazole) induced leukocyte adhesion in eNOS$^{-/-}$, but not wild-type mice, and that a compensatory increase in nNOS expression was observed in eNOS$^{-/-}$ mice [73]. This finding is important because it further supports the concept that the two constitutive NOS isoforms may have overlapping cardiovascular functions, with nNOS functionally compensating or augmenting eNOS deficiency.

Atherosclerosis in eNOS$^{-/-}$ Mice. Two groups have bred eNOS$^{-/-}$ mice into the apolipoproteinE knockout (apoE$^{-/-}$) background in the C57Bl/6 mouse strain [74,75], and have reported increased atherosclerotic lesions. While the degree and regional distribution of lesions differed in the two reports, this may have been due to differences in the respective mouse diets. When maintained on normal chow, 4-month-old eNOS$^{-/-}$; apoE$^{-/-}$ double knockout (eNOS/apoE DKO) mice had increased numbers and areas of aortic lesions compared to apoE$^{-/-}$ mice, and some evidence of proximal coronary artery lesions [74]. Interestingly, treatment of apoE$^{-/-}$ mice with a non-specific NOS inhibitor, led to hypertension and an increase in severity of aortic lesions compared to untreated apoE$^{-/-}$ mice [74]. Additionaly, treatment with an angiotensin-converting enzyme (ACE) inhibitor (enalapril), reduced both blood pressure and aortic lesion size in eNOS/apoE-DKO mice, which were also hypertensive compared to apoE$^{-/-}$ mice [74]. Together, these findings led the authors to suggest that hypertension was the major contributor to the etiology of atherosclerotic lesion formation in these mice [74].

A more striking distribution and level of atherosclerotic lesion formation was reported in eNOS/apoE-DKO mice fed a high-fat diet for 16 weeks [75]. These mice had severe aortic atherosclerotic lesions, with frequent aortic aneurysm and dissection, and evidence of lesions in the distal coronary tree [75]. These abnormalities were not observed in apoE$^{-/-}$ on the same diet for 16 or 24 weeks, or in the previous study with a normal diet [74,75]. These mice had impaired LV function and cardiac hypertrophy, possibly a result of chronic myocardial ischemia, resulting from coronary artery disease [75]. These mice were also hypertensive, consistent with the hypothesis linking hypertension to the severity of atherosclerosis in eNOS$^{-/-}$ mice [75]. While the absence of apoE in concert with a high-fat diet and hypertension may represent a supra-physiological condition, this is the first murine model system demonstrating distal coronary atherosclerosis that parallels human disease. Of note, Lee et al. reported increased arteriosclerosis in an aortic transplant model in

$eNOS^{-/-}$ mice compared to controls [76]. Together, these reports suggest that the eNOS/NO system plays an important anti-atherogenic role, which may be functionally depressed in coronary artery disease.

Cardiac Function in $eNOS^{-/-}$ ***Mice.*** The roles of NOS and NO in regulation of cardiac function are complex, controversial and likely multifactorial [29]. As such, the $eNOS^{-/-}$ model is an invaluable tool with which to explore the importance of these pathways in relation to cardiac function and pathology. Although baseline cardiac contractile function (dP/dt_{max}) in Langendorff preparations or *in vivo* catheterizations was not different in $eNOS^{-/-}$ mice as compared to littermate controls, $eNOS^{-/-}$ mice displayed increased contractile responses to the β-adrenergic agonist isoproterenol [24,46,65]. This response was not related to differences in afterload, as the same effect was observed when phenylephrine-treated wild-type mice, with similar blood pressure, were compared with $eNOS^{-/-}$ mice [65]. These data suggest a role for eNOS in modulating the cardiac contractile response to β-adrenergic stimulation. However, this finding suggests a negative inotropic role for eNOS-derived NO, as opposed to a positive inotropic role for nNOS-derived NO, implying that sub-cellular NOS compartmentalization leads to unique isoform-specific NO effects on cardiac physiology. Interestingly, the findings that double knockout $eNOS^{-/-}$; $nNOS^{\Delta/\Delta}$ mice displayed phenotype similar to the $nNOS^{\Delta/\Delta}$ suggested that the nNOS-dependent role, likely modulating SR-dependent Ca^{+2} release, lies downstream of the β-adrenergic stimulation of caveoli-localized eNOS [46]. These findings are intriguing in light of other data that suggest neuronal and vascular eNOS/nNOS functional overlap, and highlight the tissue-specific and subcellular complexities of the NOS/NO system. The use of conditional and tissue-specific transgenic models should provide further complementary insights into the diverse roles of NOS in cardiac physiology and disease.

In addition, while the ventricular relaxation constant (Tau, τ) is increased in wild-type mice treated with a non-selective NOS inhibitor (L-NNA), $eNOS^{-/-}$ mice have normal baseline ventricular relaxation, and are not affected by L-NNA [65]. These results suggest a role for eNOS in ventricular relaxation that is masked in the $eNOS^{-/-}$ model, possibly due to a compensatory increase in cardiac ANP expression [65].

There is considerable controversy surrounding the contribution of NOS/NO in cardiac ischemia and reperfusion. For example, infarct size was decreased in wild-type mice treated with an ACE inhibitor (ramiprilat) [77]. However, this effect was absent in $eNOS^{-/-}$ mice, implying a protective role for eNOS-derived NO in this process [77]. In addition, results from Langendorff isolated heart experiments in $eNOS^{-/-}$ mice demonstrated increased infarct size compared to controls, after 30 min global ischemia and 30 min reperfusion. These data were consistent with a protective role for eNOS-derived NO in the setting of ischemia and reperfusion [54]. Another group examined infarct size in isolated hearts from $eNOS^{-/-}$ mice and reported a decrease in the ischemic preconditioning threshold compared to control mice [78]. While four cycles of preconditioning produced no difference in infarct size in $eNOS^{-/-}$ mice, less than 4 cycles of preconditioning led to increased infarct sizes compared to control littermates [78]. However, Flogel et al. reported improved cardiac function in Langendorff preparations from $eNOS^{-/-}$ mice following ischemia-reperfusion, suggesting that eNOS contributed to myocardial injury in this model [79]. Further experimentation using the $eNOS^{-/-}$ model is needed to help define the specific role of eNOS and NO in cardiac ischemia and reperfusion injury.

Prepro Endothelin-1 (ppET) Promoter Directed eNOS Transgenic Mice. Ohashi et al. employed a 9.2 kb fragment of the murine ppET promoter to drive expression of a bovine eNOS mRNA in mice [80]. Four lines of ppET-eNOS mice were reported with bovine eNOS mRNA expressed in heart lung, aorta and uterus, with lesser expression in brain, liver, kidney and intestine [80]. In addition, bovine eNOS protein was detected in particulate, but not cytosolic, fractions of heart, lung and aorta by Western blot; and immuno-staining confirmed an increase in endothelial cell-specific eNOS expression, and a 10-fold increase in calcium-dependent NOS activity in the aorta [80]. Transgenic ppET-eNOS mice were hypotensive, and treatment with a non-selective NOS inhibitor (L-NAME) elevated BP in both wild type and ppET-eNOS mice to the same level [80]. Of note, there were no differences in awake heart rate, body weight, organ weight, or blood biochemistry, including plasma renin, or ET-1. However, ppET-eNOS mice had depressed vascular responses to acetylcholine, ATPγS, sodium nitroprusside or nitro-glycerin, but normal relaxation to forskolin or dibutyryl cAMP [80]. These findings are intriguing when compared with the $eNOS^{-/-}$ model, and support a role for eNOS-derived NO in regulating blood pressure. Indeed, these data point in particular to a role for eNOS in setting the baroreceptor point, and suggested that chronic eNOS up-regulation may lead to desensitization in response to muscarinic agonists and NO.

Subsequently, the contribution of decreased soluble guanylate cyclase, and cGMP-dependent protein kinase levels, have been proposed as mechanisms for impaired endothelium-dependent relaxation in ppET-eNOS mice [81].

Responses to endotoxin shock [82], and common carotid artery ligation [83] have also been examined in ppET-eNOS mice. Following challenge with lipopolysaccharide (LPS), ppET-eNOS mice were protected from septic shock, as demonstrated by a blunted hypotensive response and reduced pulmonary leukocyte infiltration and edema [82]. ppET-directed over-expression of eNOS was also shown to decrease both neointimal and medial thickening in a common carotid artery ligation model of vascular remodeling [83]. The vascular endothelium from ligated ppET-eNOS mice also had decreased leukocyte infiltrates, reduced intracellular adhesion molecule-1 (ICAM-1) and vascular cellular adhesion mlecule-1 (VCAM-1) expression [83]. Collectively, these reports highlight the wealth of information that may be discerned from transgenic over-expression models, especially when complementary knockout models also exist. Specifically, these data further support beneficial roles for eNOS in preventing cardiovascular septic shock, adverse vascular remodeling, regulating blood pressure, and modulating the baroreceptor response.

αMHC Promoter Driven eNOS Over-Expressing Transgenic Mice. To explore the consequences of cardiomyocyte-specific eNOS expression, the αMHC promoter was used to drive the expression of a human eNOS transgene in mice (αMHC-eNOS) [84]. Three lines of transgenic mice were produced with increased cardiac eNOS protein expression and a ∼10–100 fold increase in total cardiac NOS activity [84]. While the levels of NO generation in αMHC-eNOS mice were not reported, there was no change in superoxide generation, estimated using *Cypridina* luciferin chemiluminescence [84]. Though there was no difference in blood pressure, heart rate or body weight in any αMHC-eNOS mice, the line with highest level of transgenic protein expression had increased heart weights, and all lines displayed depressed left ventricular peak pressure (LVP) [84]. Interestingly, while a non-hypertensive dose of L-NAME (50 mg/Kg/Day) depressed LVP in wild-type mice, suggesting a net positive inotropic effect of NOS-derived NO, αMHC-eNOS mice had increased LVP following NOS blockade, suggesting a net negative inotropic effect [84]. The mechanism for this biphasic effect of NO on cardiac function is intriguing in light of the opposing inotropic roles of cardiomyocyte sarcoplasmic reticulum-localized nNOS and caveoli-localized eNOS gleaned from the $eNOS^{-/-}$, $nNOS^{-/-}$ and double knockout $eNOS^{-/-}$; $nNOS^{-/-}$ models [46]. These data suggest that the net effect of non-selective NOS inhibition, with the dose of L-NAME used, primarily blunted the positive inotropic contribution of nNOS in wild-type mice. However, in αMHC-eNOS mice the same treatment lead to reduced contractility, that was likely a result of decreased cardiomyocyte Ca^{2+}-sensitivity, possibly resulting from chronically elevated NO generation [84].

Cardiac eNOS Genetic Models: 'Complementary Perspectives'

The $eNOS^{-/-}$, ppET-eNOS and αMHC-eNOS mouse models have provided insights into the roles of eNOS in several physiological and developmental pathways. These complementary models have highlighted a role for vascular eNOS-derived NO in regulating blood pressure homeostasis, setting the baroreceptor point and suggest functional overlap with nNOS. However, the distinct cardiomyocyte sub-cellular distributions of eNOS and nNOS facilitate opposite effects on cardiac inotropy. This finding is interesting because it highlights distinct NOS/NO pathways in different sub-domains of a single cell type. Additionally, the beneficial contribution of eNOS in preventing the formation of atherosclerotic lesions and reducing infarct size after ischemia, suggests that eNOS and NO augmentation may represent a target for therapy. These insights highlight the types of studies that are possible, and implicate eNOS deficiency in hypertension, atherosclerosis, and myocardial infarction, the principal contributors to heart failure.

iNOS

As opposed to the critical calcium-dependent regulation of constitutive NOS enzymes, iNOS has been described as calcium-insensitive, likely due to its tight non-covalent interaction with calmodulin (CaM) and Ca^{2+}. While evidence for 'baseline' iNOS expression has been elusive, IRF-1 [85] and NF-κB [25,86] -dependent activation of the inducible NOS promoter supports an inflammation-mediated stimulation of this transcript. From a functional perspective, it is important to recognize that induction of the high-output iNOS usually occurs in an oxidative environment, and thus high levels of NO have the opportunity to react with superoxide leading to peroxynitrite formation and cell toxicity [25]. Moreover, generation of NO is dependent on the availability of sufficient amounts of substrate and/or co-factors, in particular BH_4. Relative deficiency of L-arginine and BH_4 leads to an "uncoupling" of NOS activity and the production of superoxide (O_2^-) anion instead of NO. Of note, the continuous high-level of Ca^{2+}/CaM-independent activity of inducible NOS may predispose to depletion of substrate and cofactors and 'convert' iNOS into a predominantly O_2^--generating enzyme (Figs. 2 and 3). While all NOS enzymes may generate O_2^- under *in vitro* conditions [87–89], iNOS may be the most likely to produce O_2^- *in vivo*. These properties may define the roles of iNOS in host immunity, enabling its participation in anti-microbial and anti-tumor activities [31].

An increasing number of reports suggest that inflammatory signals can activate the iNOS promoter in nearly any cell type. Importantly, cardiac myocytes have been shown to express iNOS in a variety of cardiac diseases [90–92], and in particular heart failure [93]. Given the potential complexities of NOS interactions, it is nearly impossible to determine the roles of increased iNOS activity in these cardiac conditions using non-specific pharmacological agents. Indeed, iNOS may be pathogenic either through direct cytotoxicity or by interfering with the beneficial activities of the constitutive NOSs. Thus, murine genetic models provide invaluable tools to ascertain both necessary and sufficient roles for iNOS in association with cardiovascular physiology and immune-mediated cardiovascular diseases, including heart failure.

iNOS Knockout Mice

Mice with *NOS2* gene disruption were produced independently by three groups [94–96]. MacMicking et al. were the first to generate iNOS$^{-/-}$ mice with no detectable baseline or lipopolysaccharide (LPS)-induced iNOS expression in heart, kidney, lung, or spleen [94]. Furthermore, peritoneal macrophage cultures from iNOS$^{-/-}$ mice treated with interferon-γ (IFNγ) and LPS, or IFNα/β and LPS, did not express iNOS, and IFNγ or IFNγ and TNFα did not elicit increased NO production [94]. By contrast, each of these treatments produced significant iNOS expression or NO production in littermate controls.

Homozygous iNOS$^{-/-}$ mice were born with expected frequency, displaying no developmental defects or mortality [94]. Adult mice were able to reproduce, and exhibited no abnormalities of histology, hematology or blood biochemistry [94]. In addition, iNOS$^{-/-}$ mice were able to mount a cellular inflammatory response following treatment with thioglycollate broth, sodium periodate or IFNγ and LPS [94]. Furthermore, activated macrophages from iNOS$^{-/-}$ mice were able to elicit a normal H_2O_2-defined oxidative burst, following activation with LPS and IFNγ, phorbol myristate, *Bacillus Calmette-Guérin*, or *Listeria monocytogenes* [94], but displayed reduced nitrotyrosine staining [97]. These data demonstrated an apparently intact cellular inflammatory response and oxidative capacity in iNOS$^{-/-}$ mice, but a lack of macrophage-inducible NO and diminished peroxynitrite generation. Two other groups independently confirmed these results [95,96].

Sepsis and Hemodynamic Shock in iNOS$^{-/-}$ Mice. While iNOS$^{-/-}$ mice were indistinguishable from controls under sterile vivarium conditions, they succumbed to a 10-fold lower dose of *Listeria monocytogenes*, and had increased histological signs of bacterial infection [94]. Additionally, in a cecal ligation and puncture-induced model of sepsis, survival was decreased in iNOS$^{-/-}$ mice as compared to controls [98]. There was also an increase in T-cell proliferation in iNOS$^{-/-}$ mice following treatment with *Leishmania major* or carrageenin [95]. These findings highlighted critical roles for iNOS in mediating antibacterial immunity, and negatively regulating T-cell accumulation. However, when treated with LPS, anesthetized iNOS$^{-/-}$ mice did not display hypotensive shock and, unlike littermate controls that succumbed to the same treatment, knockout mice did not exhibit any significant mortality [94,95]. These data suggest that iNOS plays a detrimental role during septic shock, and is necessary for LPS-induced hypotension. Finally, MacMicking et al. demonstrated that if sensitized by pre-treatment with heat-killed *Propionobacterium acnes*, the response of waking iNOS$^{-/-}$ mice to LPS was not different from littermate controls [94,96]. This finding indicates the presence of additional iNOS-independent immune pathways involved in the response to sepsis. These findings are intriguing, in that they support diverse roles for iNOS in cellular immunity, manifest as both beneficial and detrimental response mechanisms.

Vascular Contractility and Permeability in iNOS$^{-/-}$ Mice. The murine iNOS$^{-/-}$ model has provided insight into the contribution of iNOS to the increased microvascular permeability that follows endotoxin shock. Responses to norepinephrine have been examined in isolated mesenteric arteries from iNOS$^{-/-}$ and control mice, and 4 h following LPS challenge [99]. While norepinephrine-induced constriction was decreased in arteries from wild-type mice following LPS, this effect was not observed in iNOS$^{-/-}$ mice [99]. Pre-treatment of wild-type mice with the iNOS-selective inhibitor aminoguanidine was able to partially rescue the decreased sensitivity to norepinephrine following LPS, but had no effect in iNOS$^{-/-}$ mice [99]. These findings were corroborated in isolated cremasteric arterioles from mice with cecal ligation and puncture-induced sepsis [100]. In these experiments, arteries from iNOS$^{-/-}$ mice had improved norepinephrine-dependent vasoconstriction, and additionally iNOS$^{-/-}$ mice displayed decreased mortality [100]. These data suggest that iNOS plays a role in antagonizing constrictor responses in septic shock. Another study has examined the mesenteric microvascular response in iNOS$^{-/-}$ mice following endotoxin treatment using intravital microscopy [101]. Endotoxin treatment induced an acute vasoconstrictor response and increased vascular permeability in both iNOS$^{-/-}$ and control mice. While this increase in permeability was blunted in iNOS$^{-/-}$ mice after 20 min, it continued to increase in controls [101]. These reports are interesting, and highlight a biphasic role for iNOS in sustaining, but not initiating, acute micro-vascular constriction and permeability following endotoxin treatment, and decreasing norepinephrine sensitivity.

Atherosclerosis in iNOS Knockout Mice. Ihrig et al. recently reported that untreated iNOS$^{-/-}$ mice were transiently hypertensive at 3, but not 9 or 12 months, and had a two-fold increase in plasma cholesterol, which correlated with the development of aortic atherosclerosis [102]. However, a second group has reported that 8 wk old iNOS$^{-/-}$ mice do not develop increased atherosclerotic lesions, compared to control mice, after being fed a high fat diet for 15 wks, but had decreased collagen content in lesions [103]. These findings are complex, however, neither group used littermate controls and therefore the contribution of genetic drift cannot be excluded.

Recently iNOS$^{-/-}$; apoE$^{-/-}$ double knockout (iNOS/apoE DKO) mice have been produced by three groups [74,104,105]. When maintained on a normal (low-fat) diet, iNOS/apoE DKO did not display any difference in lesion formation as compared to apoE$^{-/-}$ littermate controls [74]. However, when fed a high-fat diet, aortic lesion size was decreased in iNOS/apoE DKO vs. littermate control apoE$^{-/-}$ mice [104,105]. These data are intriguing since iNOS expression is well known to occur in atherosclerotic lesions, and suggest that iNOS inhibition may represent a potential therapeutic strategy. In addition, these findings suggest that treatment of iNOS/apoE DKO with a high-fat diet may unmask the pro-atherogenic potential of iNOS, contradicting the above experiments in the wild-type (apoE$^{+/+}$) background. The production of complementary vascular-specific iNOS-overexpressing models, in wild type and pro-atherogenic backgrounds, should provide further insights into these contradictory findings.

In a heart transplant model, cardiac allografts introduced into iNOS$^{-/-}$ mice had increased micro-vascular lumen occlusion and intimal smooth muscle cell accumulation, as compared to wild-type recipients [106]. These data suggested that iNOS played a beneficial role in prevention of transplant arteriosclerosis. In yet another transplant model, aortic allografts from iNOS$^{-/-}$ or control mice were transplanted into a wild-type host, and the extent of vascular remodeling examined [76]. In this model, transplanted aortic segments from iNOS$^{-/-}$ mice were not different from wild-type allografts, suggesting that iNOS was not involved in vascular remodeling following transplant atherosclerosis [76]. However, these authors also showed that simultaneous treatment with an iNOS over-expressing viral vector could decrease transplant atherosclerosis in the eNOS$^{-/-}$ model, which had significantly increased intimal thickening [76]. Excluding the differences between the transplant models, these data suggest a biphasic role for iNOS in transplant atherogenesis. Indeed, iNOS may have beneficial effects, but also contribute to intimal hyperplasia, and have complicated interactions with constitutive NOS. A more detailed temporal and tissue-specific examination of the contributions of iNOS in these processes, awaits the production of more sophisticated regulated and tissue-specific knockout or over-expressing systems.

In an elastase-induced abdominal aortic aneurysm model leading to chronic inflammation, Lee et al. examined the contribution of iNOS induction by comparing iNOS$^{-/-}$ and control mice [97]. While iNOS$^{-/-}$ female mice had more severe aneurysms, male mice or ovariectomized females were not different from controls [97]. These data are interesting in light of the interactions between NO and estrogen [107,108], and suggest that synergistic interactions may have anti-atherogenic effects.

Restenosis in iNOS Knockout Mice. Since vascular iNOS expression has been described in restenosis, Chyu et al. examined the response of iNOS$^{-/-}$ mice implanted with a carotid periadventitial collar [109]. Neointimal accumulation and VCAM-1 expression were decreased in vascular segments of iNOS$^{-/-}$ mice, as compared to controls [109]. These data suggested that iNOS is necessary for neointima formation in this model, and that this pathway may involve VCAM-1. Interestingly, Tolbert et al. reported no difference in the response of iNOS$^{-/-}$ male or female mice in a carotid artery ligation model when compared to sex-matched control mice [110]. However, female mice had decreased neointimal lesions, regardless of genotype [110]. These reports suggest that iNOS may have differing roles depending on the model examined, probably related to the extent of associated inflammation. In addition this study suggests that the protective effects of estrogens were independent of iNOS [110], as opposed to data from an abdominal aortic aneurysm (atherosclerosis) model where estrogen-iNOS interactions were reported [97]. Thus, iNOS-estrogen interactions are complex and may have differing effects on vascular remodeling, possibly depending on the conditions of the experimental model examined.

Cardiac Patho-Physiology in iNOS Knockout Mice. iNOS$^{-/-}$ mice provide an excellent model to examine the contribution of iNOS to cardiac transplant rejection. Koglin et al. transplanted cardiac allografts into iNOS$^{-/-}$ mice and controls, and studied the histological rejection scores after 7 and 55 days [111]. An improved score was observed early in iNOS$^{-/-}$ mice, as compared to wild type controls with endogenous iNOS, but the opposite was observed in the chronic experiment [111]. These data suggest that iNOS plays a dual role in cardiac transplantation, having early detrimental effects but improving long-term survival. Additionally, expression of the pro-apoptotic markers p53 and Bcl-2 were increased, and anti-apoptotic Bax and Bcl-X_1 decreased, in the iNOS$^{-/-}$ group, suggesting a mechanism involving the iNOS-dependent activation of p53 in stimulation of apoptosis through Bax, and antagonizing the anti-apoptotic effects of Bcl-2 and Bcl-X_1 [112].

The role of iNOS in cardiac ischemia has been examined in Langendorff preparations from iNOS$^{-/-}$ mice [113]. No acute differences in cardiac contractility, heart rate, coronary hemodynamics, or leakage of intracellular creatine kinase or lactate dehydrogenase were observed, suggesting no significant role for iNOS in the early phase of cardiac ischemia [113].

αMHC Directed iNOS Over-Expressing Transgenic Mice. We have recently reported a binary transgenic mouse model with Doxycycline (DOX)-regulated, and cardiomyocyte-specific expression of human iNOS (iNOS^{+}/αMtTA^{+}) [114]. Two independent lines of iNOS^{+}/αMtTA^{+} mice displayed DOX-reversible human iNOS expression in cardiomyocytes with a 10-fold increase in total cardiac NOS activity, increased peroxynitrite generation and a 2-fold increase in L-NAME inhibited lucigenin chemiluminescence (an indicator of superoxide generation) [114]. We also reported significant cardiac hypertrophy, atrio-ventricular dilation, mild cardiac inflammation (histo-pathology) and an infrequent occurrence of heart failure in iNOS^{+}/αMtTA^{+} mice as compared to littermate or DOX-treated controls [114]. Of particular interest, these mice had a significant incidence of sudden cardiac death [114]. This phenotype was correlated with cardiac brady-arrhythmias, atrio-ventricular (AV)-node dysfunction and down-regulation of connexin 40 in the AV-node and proximal His-Purkinje fibers [114]. These findings suggest that enhanced cardiac iNOS expression, as occurs in a variety of cardiac diseases, has the potential to cause hypertrophy, dilation, cardiomyopathy, AV-node dysfunction, brady-arrhythmia, sudden cardiac death, and heart failure.

In contrast to our results, Heger et al. recently reported a non-conditional model with αMHC promoter-directed expression of human iNOS (αMHC-iNOS) [115]. These mice have increased cardiac iNOS activity but no alterations in cardiac structure or function [115]. We believe this difference in phenotype likely reflects a critical difference in experimental design, as the 'non-conditional' transgenic approach may have pre-selected lines without significant cardiac toxicity [115]. Indeed, significant embryonic lethality was observed in our model, which was fully rescued by DOX suppression of iNOS transgene expression during development. Therefore, the use of a DOX-regulatable conditional system for cardiac-selective transgene expression allowed the bypass of embryonic mortality, and prevention of developmental adaptation *in utero* [114].

Heger et al. reported S-ethylisothiourea-inhibitable iNOS activity in the αMHC-iNOS animals and these data are presented as a 260–400-fold increase over control mice exhibiting no significant iNOS activity over baseline [115]. Unfortunately, such a comparison is prone to overestimation of the absolute levels of iNOS activity, and may have little bearing on total NOS activity. Since Heger et al. did not present total NOS activity, a direct comparison with our model cannot be made [115]. Additional variables influencing the differences in phenotype may include epigenetic transgene dosing effects, or adaptive compensatory effects of other loci such as the up-regulation of genes regulating substrate (L-arginine) or cofactor (BH_4) availability, or oxidative stress prevention (SOD) in the αMHC-iNOS mice.

Cardiac iNOS Genetic Models: 'Another Time and Place...'
The iNOS$^{-/-}$, αMHC-iNOS and iNOS^{+}/αMtTA^{+} models are important tools with which to uncover insights into the roles of iNOS in a variety of relevant physiological and developmental pathways. A large body of data has been generated using iNOS$^{-/-}$ mice to determine whether this molecule is necessary in several inflammation-related

processes. Experiments focusing on cardiac pathophysiology using the $iNOS^{-/-}$ model have highlighted both survival benefit and detriment. These experiments support the notion of $iNOS^{-/-}$ as a 'double edged sword'. Indeed, iNOS represents an essential component necessary for proper immune function, but conversely may be involved in cytotoxicity, when ectopically induced or over-expressed, for example in cardiac or vascular myocytes. These insights suggest that a better understanding of iNOS biology holds promise for human therapy, and that simplistic pharmacological manipulations may be problematic.

Increasingly sophisticated and powerful murine genetic models should provide even further insights into processes where iNOS modulates pathophysiology. For example, the $iNOS^{+}/\alpha MtTA^{+}$ mice have suggested that cardiomyocyte iNOS function contributes to cardiomyopathy and is sufficient to cause lethal brady-arrhythmia. This model will also allow the design of experiments that test titratable and regulatable cardiomyocyte-specific iNOS activity in a variety of relevant cardiovascular processes that relate to heart failure.

Summary

The mouse is the most ideal genetically modifiable mammalian model presently available. Spatial expression problems have been overcome with tissue-specific promoter elements, and reversible transcriptional strategies allow precise temporal modulation of transgene expression. Additionally, gene ablation techniques are increasing in sophistication, and can provide insights complementary to over-expression models. The use of engineered LoxP sites and conditional tissue-specific expression of the LoxP-targeted *Cre* recombinase, have been used to produce 'conditional tissue-specific knockouts', that overcome limitations of traditional knockouts [116]. Moreover, the recent discovery of RNAi methodology [117], in combination with a conditional tissue specific promoter may provide 'reversible' tissue-specific gene silencing in the future.

While the production and use of both knockout and transgenic over-expression models have informed on the sufficient and necessary roles for NOS, there remains inherent complexities as to the absolute isoform-specific sources of NO and other possible reaction products of NOS (e.g. O_2^-, $ONOO^-$). While these shortcomings must be taken into consideration in any model examined, the studies using these experimental systems exemplify important methodologies in the present post-genomic era that will aid in defining cardiovascular pathophysiology at the molecular level.

Acknowledgments

Imran N. Mungrue is supported by a CIHR (Canadian Institutes for Health research)/HSFO (Heart and Stroke Foundation of Ontario) studentship. Duncan J. Stewart is supported by CIHR (MT-11620) and HSFO (NA-4789) operating grants. Mansoor Husain is a CIHR Clinician Scientist and supported by CIHR (MT-14648) and HSFO (NA-4389) operating grants.

References

1. Bredt DS, Snyder SH. Isolation of nitric oxide synthetase, a calmodulin-requiring enzyme. *Proc Natl Acad Sci USA* 1990;87:682–685.
2. Bredt DS, Hwang PM, Glatt CE, Lowenstein C, Reed RR, Snyder SH. Cloned and expressed nitric oxide synthase structurally resembles cytochrome P-450 reductase. *Nature* 1991;351:714–718.
3. Yui Y, Hattori R, Kosuga K, Eizawa H, Hiki K, Kawai C. Purification of nitric oxide synthase from rat macrophages. *J Biol Chem* 1991;266:12544–12547.
4. Stuehr DJ, Cho HJ, Kwon NS, Weise MF, Nathan CF. Purification and characterization of the cytokine-induced macrophage nitric oxide synthase: An FAD- and FMN-containing flavoprotein. *Proc Natl Acad Sci USA* 1991;88:7773–7777.
5. Hevel JM, White KA, Marletta MA. Purification of the inducible murine macrophage nitric oxide synthase. Identification as a flavoprotein. *J Biol Chem* 1991;266:22789–22791.
6. Forstermann U, Pollock JS, Schmidt HH, Heller M, Murad F. Calmodulindependent endothelium-derived relaxing factor/nitric oxide synthase activity is present in the particulate and cytosolic fractions of bovine aortic endothelial cells. *Proc Natl Acad Sci USA* 1991;88:1788–1792.
7. Pollock JS, Forstermann U, Mitchell JA, Warner TD, Schmidt HH, Nakane M, Murad F. Purification and characterization of particulate endothelium-derived relaxing factor synthase from cultured and native bovine aortic endothelial cells. *Proc Natl Acad Sci USA* 1991;88:10480–10484.
8. Xie QW, Cho HJ, Calaycay J, Mumford RA, Swiderek KM, Lee TD, Ding A, Troso T, Nathan C. Cloning and characterization of inducible nitric oxide synthase from mouse macrophages. *Science* 1992;256:225–228.
9. Janssens SP, Shimouchi A, Quertermous T, Bloch DB, Bloch KD. Cloning and expression of a cDNA encoding human endothelium-derived relaxing factor/nitric oxide synthase. *J Biol Chem* 1992;267:14519–14522.
10. Sessa WC, Harrison JK, Barber CM, Zeng D, Durieux ME, D'Angelo DD, Lynch KR, Peach MJ. Molecular cloning and expression of a cDNA encoding endothelial cell nitric oxide synthase. *J Biol Chem* 1992;267:15274–15276.
11. Nishida K, Harrison DG, Navas JP, Fisher AA, Dockery SP, Uematsu M, Nerem RM, Alexander RW, Murphy TJ. Molecular cloning and characterization of the constitutive bovine aortic endothelial cell nitric oxide synthase. *J Clin Invest* 1992;90:2092–2096.
12. Lowenstein CJ, Glatt CS, Bredt DS, Snyder SH. Cloned and expressed macrophage nitric oxide synthase contrasts with the brain enzyme. *Proc Natl Acad Sci USA* 1992;89:6711–6715.

13. Lyons CR, Orloff GJ, Cunningham JM. Molecular cloning and functional expression of an inducible nitric oxide synthase from a murine macrophage cell line. *J Biol Chem* 1992;267:6370–6374.
14. Stuehr DJ. Structure-function aspects in the nitric oxide synthases. *Annu Rev Pharmacol Toxicol* 1997;37:339–359.
15. Marletta MA. Nitric oxide synthase: Aspects concerning structure and catalysis. *Cell* 1994;78:927–930.
16. Alderton WK, Cooper CE, Knowles RG. Nitric oxide synthases: Structure, function and inhibition. *Biochem J* 2001;357:593–615.
17. Crane BR, Arvai AS, Gachhui R, Wu C, Ghosh DK, Getzoff ED, Stuehr DJ, Tainer JA. The structure of nitric oxide synthase oxygenase domain and inhibitor complexes. *Science* 1997;278:425–431.
18. Crane BR, Arvai AS, Ghosh DK, Wu C, Getzoff ED, Stuehr DJ, Tainer JA. Structure of nitric oxide synthase oxygenase dimer with pterin and substrate. *Science* 1998;279:2121–2126.
19. Raman CS, Li H, Martasek P, Kral V, Masters BS, Poulos TL. Crystal structure of constitutive endothelial nitric oxide synthase: A paradigm for pterin function involving a novel metal center. *Cell* 1998;95:939–950.
20. Abu-Soud HM, Gachhui R, Raushel FM, Stuehr DJ. The ferrous-dioxy complex of neuronal nitric oxide synthase. Divergent effects of L-arginine and tetrahydrobiopterin on its stability. *J Biol Chem* 1997;272:17349–17353.
21. Korth HG, Sustmann R, Thater C, Butler AR, Ingold KU. On the mechanism of the nitric oxide synthase-catalyzed conversion of N omega-hydroxyl-L-arginine to citrulline and nitric oxide. *J Biol Chem* 1994;269:17776–17779.
22. Busse R, Fleming I. Regulation and functional consequences of endothelial nitric oxide formation. *Ann Med* 1995;27:331–340.
23. Bredt DS, Snyder SH. Nitric oxide: A physiologic messenger molecule. *Annu Rev Biochem* 1994;63:175–195.
24. Son H, Hawkins RD, Martin K, Kiebler M, Huang PL, Fishman MC, Kandel ER. Long-term potentiation is reduced in mice that are doubly mutant in endothelial and neuronal nitric oxide synthase. *Cell* 1996;87:1015–1023.
25. Xie Q, Nathan C. The high-output nitric oxide pathway: Role and regulation. *J Leukoc Biol* 1994;56:576–582.
26. Moilanen E, Vapaatalo H. Nitric oxide in inflammation and immune response. *Ann Med* 1995;27:359–367.
27. Nava E, Noll G, Luscher TF. Nitric oxide in cardiovascular diseases. *Ann Med* 1995;27:343–351.
28. Dusting GJ, Macdonald PS. Endogenous nitric oxide in cardiovascular disease and transplantation. *Ann Med* 1995;27:395–406.
29. Hare JM, Colucci WS. Role of nitric oxide in the regulation of myocardial function. *Prog Cardiovasc Dis* 1995;38:155–166.
30. Christopherson KS, Bredt DS. Nitric oxide in excitable tissues: Physiological roles and disease. *J Clin Invest* 1997;100:2424–2429.
31. Nathan C. Inducible nitric oxide synthase: What difference does it make? *J Clin Invest* 1997;100:2417–2423.
32. Kim PK, Zamora R, Petrosko P, Billiar TR. The regulatory role of nitric oxide in apoptosis. *Int Immunopharmacol* 2001;1:1421–1441.
33. Ziche M, Morbidelli L. Nitric oxide and angiogenesis. *J Neurooncol* 2000;50:139–148.
34. Burnett AL. Nitric oxide control of lower genitourinary tract functions: A review. *Urology* 1995;45:1071–1083.
35. Frank S, Kampfer H, Wetzler C, Pfeilschifter J. Nitric oxide drives skin repair: Novel functions of an established mediator. *Kidney Int* 2002;61:882–888.
36. Kobzik L, Reid MB, Bredt DS, Stamler JS. Nitric oxide in skeletal muscle. *Nature* 1994;372:546–548.
37. Eu JP, Sun J, Xu L, Stamler JS, Meissner G. The skeletal muscle calcium release channel: Coupled O2 sensor and NO signaling functions. *Cell* 2000;102:499–509.
38. Roberts CK, Barnard RJ, Scheck SH, Balon TW. Exercise-stimulated glucose transport in skeletal muscle is nitric oxide dependent. *Am J Physiol* 1997;273:E220–E225.
39. Bredt DS, Snyder SH. Transient nitric oxide synthase neurons in embryonic cerebral cortical plate, sensory ganglia, and olfactory epithelium. *Neuron* 1994;13:301–313.
40. Cork RJ, Perrone ML, Bridges D, Wandell J, Scheiner CA, Mize RR. A web-accessible digital atlas of the distribution of nitric oxide synthase in the mouse brain. *Prog Brain Res* 1998;118:37–50.
41. Nozaki K, Moskowitz MA, Maynard KI, Koketsu N, Dawson TM, Bredt DS, Snyder SH. Possible origins and distribution of immunoreactive nitric oxide synthase-containing nerve fibers in cerebral arteries. *J Cereb Blood Flow Metab* 1993;13:70–79.
42. Xu KY, Huso DL, Dawson TM, Bredt DS, Becker LC. Nitric oxide synthase in cardiac sarcoplasmic reticulum. *Proc Natl Acad Sci USA* 1999;96:657–662.
43. Huang PL, Dawson TM, Bredt DS, Snyder SH, Fishman MC. Targeted disruption of the neuronal nitric oxide synthase gene. *Cell* 1993;75:1273–1286.
44. Nelson RJ, Demas GE, Huang PL, Fishman MC, Dawson VL, Dawson TM, Snyder SH. Behavioural abnormalities in male mice lacking neuronal nitric oxide synthase. *Nature* 1995;378:383–386.
45. Shankar RR, Wu Y, Shen HQ, Zhu JS, Baron AD. Mice with gene disruption of both endothelial and neuronal nitric oxide synthase exhibit insulin resistance. *Diabetes* 2000;49:684–687.
46. Barouch LA, Harrison RW, Skaf MW, Rosas GO, Cappola TP, Kobeissi ZA, Hobai IA, Lemmon CA, Burnett AL, O'Rourke B, et al. Nitric oxide regulates the heart by spatial confinement of nitric oxide synthase isoforms. *Nature* 2002;416:337–339.
47. Brenman JE, Chao DS, Gee SH, McGee AW, Craven SE, Santillano DR, Wu Z, Huang F, Xia H, Peters MF, et al. Interaction of nitric oxide synthase with the postsynaptic density protein PSD-95 and alpha1-syntrophin mediated by PDZ domains. *Cell* 1996;84:757–767.
48. Silvagno F, Xia H, Bredt DS. Neuronal nitric-oxide synthase-mu, an alternatively spliced isoform expressed in differentiated skeletal muscle. *J Biol Chem* 1996;271:11204–11208.
49. Brenman JE, Chao DS, Xia H, Aldape K, Bredt DS. Nitric oxide synthase complexed with dystrophin and absent from skeletal muscle sarcolemma in Duchenne muscular dystrophy. *Cell* 1995;82:743–752.
50. Jumrussirikul P, Dinerman J, Dawson TM, Dawson VL, Ekelund U, Georgakopoulos D, Schramm LP, Calkins H, Snyder SH, Hare JM, et al. Interaction between neuronal nitric oxide synthase and inhibitory G protein activity in heart rate regulation in conscious mice. *J Clin Invest* 1998;102:1279–1285.

51. Woo MA, Stevenson WG, Moser DK, Trelease RB, Harper RM. Patterns of beat-to-beat heart rate variability in advanced heart failure. *Am Heart J* 1992;123:704–710.
52. Choate JK, Danson EJ, Morris JF, Paterson DJ. Peripheral vagal control of heart rate is impaired in neuronal NOS knockout mice. *Am J Physiol Heart Circ Physiol* 2001;281:H2310–H2317.
53. Jones SP, Girod WG, Huang PL, Lefer DJ. Myocardial reperfusion injury in neuronal nitric oxide synthase deficient mice. *Coron Artery Dis* 2000;11:593–597.
54. Sumeray MS, Rees DD, Yellon DM. Infarct size and nitric oxide synthase in murine myocardium. *J Mol Cell Cardiol* 2000;32:35–42.
55. Huang Z, Huang PL, Panahian N, Dalkara T, Fishman MC, Moskowitz MA. Effects of cerebral ischemia in mice deficient in neuronal nitric oxide synthase. *Science* 1994;265:1883–1885.
56. Kelly RA, Balligand JL, Smith TW. Nitric oxide and cardiac function. *Circ Res* 1996;79:363–380.
57. Shaul PW. Regulation of endothelial nitric oxide synthase: Location, Location, Location. *Annu Rev Physiol* 2002;64:749–774.
58. Fleming I, Busse R. Signal transduction of eNOS activation. *Cardiovasc Res* 1999;43:532–541.
59. Robinson LJ, Busconi L, Michel T. Agonist-modulated palmitoylation of endothelial nitric oxide synthase. *J Biol Chem* 1995;270:995–998.
60. Robinson LJ, Michel T. Mutagenesis of palmitoylation sites in endothelial nitric oxide synthase identifies a novel motif for dual acylation and subcellular targeting. *Proc Natl Acad Sci USA* 1995;92:11776–11780.
61. Godecke A, Decking UK, Ding Z, Hirchenhain J, Bidmon HJ, Godecke S, Schrader J. Coronary hemodynamics in endothelial NO synthase knockout mice. *Circ Res* 1998;82:186–194.
62. Huang PL, Huang Z, Mashimo H, Bloch KD, Moskowitz MA, Bevan JA, Fishman MC. Hypertension in mice lacking the gene for endothelial nitric oxide synthase. *Nature* 1995;377:239–242.
63. Shesely EG, Maeda N, Kim HS, Desai KM, Krege JH, Laubach VE, Sherman PA, Sessa WC, Smithies O. Elevated blood pressures in mice lacking endothelial nitric oxide synthase. *Proc Natl Acad Sci USA* 1996;93:13176–13181.
64. Stauss HM, Godecke A, Mrowka R, Schrader J, Persson PB. Enhanced blood pressure variability in eNOS knockout mice. *Hypertension* 1999;33:1359–1363.
65. Gyurko R, Kuhlencordt P, Fishman MC, Huang PL. Modulation of mouse cardiac function *in vivo* by eNOS and ANP. *Am J Physiol Heart Circ Physiol* 2000;278:H971–H981.
66. Lee PC, Salyapongse AN, Bragdon GA, Shears LL II, Watkins SC, Edington HD, Billiar TR. Impaired wound healing and angiogenesis in eNOS-deficient mice. *Am J Physiol* 1999;277:H1600–H1608.
67. Demas GE, Kriegsfeld LJ, Blackshaw S, Huang P, Gammie SC, Nelson RJ, Snyder SH. Elimination of aggressive behavior in male mice lacking endothelial nitric oxide synthase. *J Neurosci* 1999;19:RC30.
68. Gregg AR, Schauer A, Shi O, Liu Z, Lee CG, O'Brien WE. Limb reduction defects in endothelial nitric oxide synthase-deficient mice. *Am J Physiol* 1998;275:H2319–H2324.
69. Lee TC, Zhao YD, Courtman DW, Stewart DJ. Abnormal aortic valve development in mice lacking endothelial nitric oxide synthase. *Circulation* 2000;101:2345–2348.
70. Huang A, Sun D, Shesely EG, Levee EM, Koller A, Kaley G. Neuronal NOS-dependent dilation to flow in coronary arteries of male eNOS-KO mice. *Am J Physiol Heart Circ Physiol* 2002;282:H429–H436.
71. Meng W, Ayata C, Waeber C, Huang PL, Moskowitz MA. Neuronal NOS-cGMP-dependent ACh-induced relaxation in pial arterioles of endothelial NOS knockout mice. *Am J Physiol* 1998;274:H411–H415.
72. Chataigneau T, Feletou M, Huang PL, Fishman MC, Duhault J, Vanhoutte PM. Acetylcholine-induced relaxation in blood vessels from endothelial nitric oxide synthase knockout mice. *Br J Pharmacol* 1999;126:219–226.
73. Sanz MJ, Hickey MJ, Johnston B, McCafferty DM, Raharjo E, Huang PL, Kubes P. Neuronal nitric oxide synthase (NOS) regulates leukocyte-endothelial cell interactions in endothelial NOS deficient mice. *Br J Pharmacol* 2001;134:305–312.
74. Knowles JW, Reddick RL, Jennette JC, Shesely EG, Smithies O, Maeda N. Enhanced atherosclerosis and kidney dysfunction in eNOS(−/−)Apoe(−/−) mice are ameliorated by enalapril treatment. *J Clin Invest* 2000;105:451–458.
75. Kuhlencordt PJ, Gyurko R, Han F, Scherrer-Crosbie M, Aretz TH, Hajjar R, Picard MH, Huang PL. Accelerated atherosclerosis, aortic aneurysm formation, and ischemic heart disease in apolipoprotein E/endothelial nitric oxide synthase double-knockout mice. *Circulation* 2001;104:448–454.
76. Lee PC, Wang ZL, Qian S, Watkins SC, Lizonova A, Kovesdi I, Tzeng E, Simmons RL, Billiar TR, Shears LL, II. Endothelial nitric oxide synthase protects aortic allografts from the development of transplant arteriosclerosis. *Transplantation* 2000;69:1186–1192.
77. Yang XP, Liu YH, Shesely EG, Bulagannawar M, Liu F, Carretero OA. Endothelial nitric oxide gene knockout mice: Cardiac phenotypes and the effect of angiotensin-converting enzyme inhibitor on myocardial ischemia/reperfusion injury. *Hypertension* 1999;34:24–30.
78. Bell RM, Yellon DM. The contribution of endothelial nitric oxide synthase to early ischaemic preconditioning: The lowering of the preconditioning threshold. An investigation in eNOS knockout mice. *Cardiovasc Res* 2001;52:274–280.
79. Flogel U, Decking UK, Godecke A, Schrader J. Contribution of NO to ischemia-reperfusion injury in the saline-perfused heart: A study in endothelial NO synthase knockout mice. *J Mol Cell Cardiol* 1999;31:827–836.
80. Ohashi Y, Kawashima S, Hirata K, Yamashita T, Ishida T, Inoue N, Sakoda T, Kurihara H, Yazaki Y, Yokoyama M. Hypotension and reduced nitric oxide-elicited vasorelaxation in transgenic mice overexpressing endothelial nitric oxide synthase. *J Clin Invest* 1998;102:2061–2071.
81. Yamashita T, Kawashima S, Ohashi Y, Ozaki M, Rikitake Y, Inoue N, Hirata K, Akita H, Yokoyama M. Mechanisms of reduced nitric oxide/cGMP-mediated vasorelaxation in transgenic mice overexpressing endothelial nitric oxide synthase. *Hypertension* 2000;36:97–102.
82. Yamashita T, Kawashima S, Ohashi Y, Ozaki M, Ueyama T, Ishida T, Inoue N, Hirata K, Akita H, Yokoyama M. Resistance to endotoxin shock in transgenic

mice overexpressing endothelial nitric oxide synthase. *Circulation* 2000;101:931–937.

83. Kawashima S, Yamashita T, Ozaki M, Ohashi Y, Azumi H, Inoue N, Hirata K, Hayashi Y, Itoh H, Yokoyama M. Endothelial NO synthase overexpression inhibits lesion formation in mouse model of vascular remodeling. *Arterioscler Thromb Vasc Biol* 2001;21:201–207.
84. Brunner F, Andrew P, Wolkart G, Zechner R, Mayer B. Myocardial contractile function and heart rate in mice with myocyte-specific overexpression of endothelial nitric oxide synthase. *Circulation* 2001;104:3097–3102.
85. Kamijo R, Harada H, Matsuyama T, Bosland M, Gerecitano J, Shapiro D, Le J, Koh SI, Kimura T, Green SJ, et al. Requirement for transcription factor IRF-1 in NO synthase induction in macrophages. *Science* 1994;263:1612–1615.
86. Xie QW, Kashiwabara Y, Nathan C. Role of transcription factor NF-kappa B/Rel in induction of nitric oxide synthase. *J Biol Chem* 1994;269:4705–4708.
87. Xia Y, Zweier JL. Superoxide and peroxynitrite generation from inducible nitric oxide synthase in macrophages. *Proc Natl Acad Sci USA* 1997;94:6954–6958.
88. Vasquez-Vivar J, Kalyanaraman B, Martasek P, Hogg N, Masters BS, Karoui H, Tordo P, Pritchard KA Jr. Superoxide generation by endothelial nitric oxide synthase: The influence of cofactors. *Proc Natl Acad Sci USA* 1998;95:9220–9225.
89. Xia Y, Dawson VL, Dawson TM, Snyder SH, Zweier JL. Nitric oxide synthase generates superoxide and nitric oxide in arginine-depleted cells leading to peroxynitrite-mediated cellular injury. *Proc Natl Acad Sci USA* 1996;93:6770–6774.
90. Thoenes M, Forstermann U, Tracey WR, Bleese NM, Nussler AK, Scholz H, Stein B. Expression of inducible nitric oxide synthase in failing and non-failing human heart. *J Mol Cell Cardiol* 1996;28:165–169.
91. Yang X, Chowdhury N, Cai B, Brett J, Marboe C, Sciacca RR, Michler RE, Cannon PJ. Induction of myocardial nitric oxide synthase by cardiac allograft rejection. *J Clin Invest* 1994;94:714–721.
92. Habib FM, Springall DR, Davies GJ, Oakley CM, Yacoub MH, Polak JM. Tumour necrosis factor and inducible nitric oxide synthase in dilated cardiomyopathy. *Lancet* 1996;347:1151–1155.
93. Haywood GA, Tsao PS, von der Leyen HE, Mann MJ, Keeling PJ, Trindade PT, Lewis NP, Byrne CD, Rickenbacher PR, Bishopric NH, et al. Expression of inducible nitric oxide synthase in human heart failure. *Circulation* 1996;93:1087–1094.
94. MacMicking JD, Nathan C, Hom G, Chartrain N, Fletcher DS, Trumbauer M, Stevens K, Xie QW, Sokol K, Hutchinson N, et al. Altered responses to bacterial infection and endotoxic shock in mice lacking inducible nitric oxide synthase. *Cell* 1995;81:641–650.
95. Wei XQ, Charles IG, Smith A, Ure J, Feng GJ, Huang FP, Xu D, Muller W, Moncada S, Liew FY. Altered immune responses in mice lacking inducible nitric oxide synthase. *Nature* 1995;375:408–411.
96. Laubach VE, Shesely EG, Smithies O, Sherman PA. Mice lacking inducible nitric oxide synthase are not resistant to lipopolysaccharide-induced death. *Proc Natl Acad Sci USA* 1995;92:10688–10692.
97. Lee JK, Borhani M, Ennis TL, Upchurch GR Jr, Thompson RW. Experimental abdominal aortic aneurysms in mice lacking expression of inducible nitric oxide synthase. *Arterioscler Thromb Vasc Biol* 2001;21:1393–1401.
98. Cobb JP, Hotchkiss RS, Swanson PE, Chang K, Qiu Y, Laubach VE, Karl IE, Buchman TG. Inducible nitric oxide synthase (iNOS) gene deficiency increases the mortality of sepsis in mice. *Surgery* 1999;126:438–442.
99. Boyle WA III, Parvathaneni LS, Bourlier V, Sauter C, Laubach VE, Cobb JP. iNOS gene expression modulates microvascular responsiveness in endotoxin-challenged mice. *Circ Res* 2000;87:E18–E24.
100. Hollenberg SM, Broussard M, Osman J, Parrillo JE. Increased microvascular reactivity and improved mortality in septic mice lacking inducible nitric oxide synthase. *Circ Res* 2000;86:774–778.
101. Suzuki Y, Deitch EA, Mishima S, Duran WN, Xu DZ. Endotoxin-induced mesenteric microvascular changes involve iNOS-derived nitric oxide: Results from a study using iNOS knock out mice. Shock 2000;13:397–403.
102. Ihrig M, Dangler CA, Fox JG. Mice lacking inducible nitric oxide synthase develop spontaneous hypercholesterolaemia and aortic atheromas. *Atherosclerosis* 2001;156:103–107.
103. Niu XL, Yang X, Hoshiai K, Tanaka K, Sawamura S, Koga Y, Nakazawa H. Inducible nitric oxide synthase deficiency does not affect the susceptibility of mice to atherosclerosis but increases collagen content in lesions. *Circulation* 2001;103:1115–1120.
104. Kuhlencordt PJ, Chen J, Han F, Astern J, Huang PL. Genetic deficiency of inducible nitric oxide synthase reduces atherosclerosis and lowers plasma lipid peroxides in apolipoprotein E-knockout mice. *Circulation* 2001;103:3099–3104.
105. Detmers PA, Hernandez M, Mudgett J, Hassing H, Burton C, Mundt S, Chun S, Fletcher D, Card DJ, Lisnock J, et al. Deficiency in inducible nitric oxide synthase results in reduced atherosclerosis in apolipoprotein E-deficient mice. *J Immunol* 2000;165:3430–3435.
106. Koglin J, Glysing-Jensen T, Mudgett JS, Russell ME. Exacerbated transplant arteriosclerosis in inducible nitric oxide-deficient mice. *Circulation* 1998;97:2059–2065.
107. Gray GA, Sharif I, Webb DJ, Seckl JR. Oestrogen and the cardiovascular system: The good, the bad and the puzzling. *Trends Pharmacol Sci* 2001;22:152–156.
108. Miller VM. Gender, vascular reactivity. *Lupus* 1999;8:409–415.
109. Chyu KY, Dimayuga P, Zhu J, Nilsson J, Kaul S, Shah PK, Cercek B. Decreased neointimal thickening after arterial wall injury in inducible nitric oxide synthase knockout mice. *Circ Res* 1999;85:1192–1198.
110. Tolbert T, Thompson JA, Bouchard P, Oparil S. Estrogen-induced vasoprotection is independent of inducible nitric oxide synthase expression: Evidence from the mouse carotid artery ligation model. *Circulation* 2001;104:2740–2745.
111. Koglin J, Glysing-Jensen T, Mudgett JS, Russell ME. NOS2 mediates opposing effects in models of acute and chronic cardiac rejection: Insights from NOS2-knockout mice. *Am J Pathol* 1998;153:1371–1376.
112. Koglin J, Granville DJ, Glysing-Jensen T, Mudgett JS, Carthy CM, McManus BM, Russell ME. Attenuated acute cardiac rejection in NOS2$^{-/-}$ recipients correlates with reduced apoptosis. *Circulation* 1999;99:836–842.

113. Xi L, Jarrett NC, Hess ML, Kukreja RC. Myocardial ischemia/reperfusion injury in the inducible nitric oxide synthase knockout mice. *Life Sci* 1999;65:935–945.
114. Mungrue IN, Gros R, You X, Pirani A, Azad A, Csont T, Schulz R, Butany J, Stewart DJ, Husain M. Cardiomyocyte overexpression of iNOS in mice results in peroxynitrite generation, heart block, and sudden death. *J Clin Invest* 2002;109:735–743.
115. Heger J, Godecke A, Flogel U, Merx MW, Molojavyi A, Kuhn-Velten WN, Schrader J. Cardiac-specific overexpression of inducible nitric oxide synthase does not result in severe cardiac dysfunction. *Circ Res* 2002;90:93–99.
116. Tsien JZ, Chen DF, Gerber D, Tom C, Mercer EH, Anderson DJ, Mayford M, Kandel ER, Tonegawa S. Subregion- and cell type-restricted gene knockout in mouse brain. *Cell* 1996;87:1317–1326.
117. Lipardi C, Wei Q, Paterson BM. RNAi as random degradative PCR: siRNA primers convert mRNA into dsRNAs that are degraded to generate new siRNAs. *Cell* 2001;107:297–307.

Part II
NO and Cardiovascular Therapeutics

Nitric Oxide and Cardiovascular Protection

Bodh I. Jugdutt
Cardiology Division, Department of Medicine, University of Alberta, Edmonton, Alberta, Canada

Abstract. **Nitric oxide (NO) plays a critical role in ischemic heart disease and ischemia-reperfusion. There is an increasing body of evidence to support the role of NO in myocardial and vascular protection in disease. The finding that NO might act as a trigger of late ischemic preconditioning (IPC) might lead to the development of novel anti-ischemic therapy. The role of NO signaling in the cardioprotective effects of ACE inhibitors and angiotensin II type 1 receptor(AT_1) receptor antagonists is an active area of study.**

Key Words. **nitric oxide, cardiovascular protection, ischemic preconditioning**

Introduction

The nitric oxide (NO) story is fraught with reports of both beneficial and deleterious effects in the same tissue. NO plays important roles in cardiac and vascular physiology and pathology [1,2]. The role of NO in cardioprotection during ischemia-reperfusion [3] as well as its cardiotoxicity via peroxynitrite formation in ischemia-reperfusion injury [4] have been reviewed in the first symposium issue. Evidence indicates that NO is a ubiquitous cellular messenger that acts by several mechanisms, including the activation of soluble guanylate cyclase, nitrosylation of thiols and formation of peroxynitrite. In general, the action of NO depends on the oxidative conditions in the cell. This review is focused on the cardiovascular protective effects of NO in 3 key areas.

NO and Ischemic Preconditioning

Cumulative evidence indicates that NO plays a critical role in ischemic preconditioning (IPC). This phenomenon, whereby brief episodes of myocardial ischemia render the heart resistant to subsequent episodes has been viewed as an adaptive defence mechanism. The two phases of IPC were elegantly reviewed by Bolli [5,6]. An early phase of IPC that lasts between a few minutes to 3 hours after the stimulus seems to be initiated by receptor-ligand interaction and followed by activation of intracellular targets mediated by G-protein [1,7]. In contrast, a late phase of IPC (Fig. 1) that develops after 12 hours and lasts for 3 to 4 days, involves several signaling molecules and downstream synthesis and activation of several enzymes and effector proteins [6,8]. Key signaling molecules in late IPC (Table 1) include NO [6,9–11], the protein kinase C (PKC)-ε isoform or PKCε [12], and several others [13–19]. Importantly, several NO donors were shown to mimic IPC [20,21].

Several studies over nearly 2 decades support the concept of IPC-induced cardioprotection and the role of NO. Early IPC, consisting of four 5-minute episodes of ischemia separated by 10-minute reperfusion periods, was first shown to limit infarct size following 40 minutes of coronary occlusion in dogs by Murry et al. [22]. Late IPC was then shown to protect against myocardial stunning and infarction using several animal models [23–28].

The longer duration of protection afforded by late IPC has triggered clinical interest and extensive research (for review see Dawn and Bolli) [29]. Although several molecules (including adenosine, reactive oxygen species [ROS] and NO) have been implicated in late IPC using various inhibitors, recent studies suggest that NO acts as a trigger for late IPC [5,16,17,30], possibly via generation of peroxynitrite and/or secondary ROS. Recently, nitroglycerin was shown to induce late IPC in humans [31]. This finding provides an explanation of some for the beneficial effects of the drug observed in the setting of more prolonged ischemic injury as in myocardial infarction with or without reperfusion [30,32–37]. Importantly, this new protective effect of nitroglycerin was not affected by the development of nitrate tolerance [31].

NO in Ischemia-Reperfusion

Myocardial reperfusion after prolonged ischemic episodes of 20 minutes or more is associated with reperfusion injury [38,39], myocardial stunning [40–42] and variable amounts of necrosis [43,44]

Address for correspondence: Dr. Bodh I. Jugdutt, 2C2.43 Walter Mackenzie Health Sciences Centre, Division of Cardiology, Department of Medicine, University of Alberta, Edmonton, Alberta, Canada T6G 2R7. Tel.: (780) 407-7729; Fax: (780) 437-3546; E-mail: bjugdutt@ualberta.ca

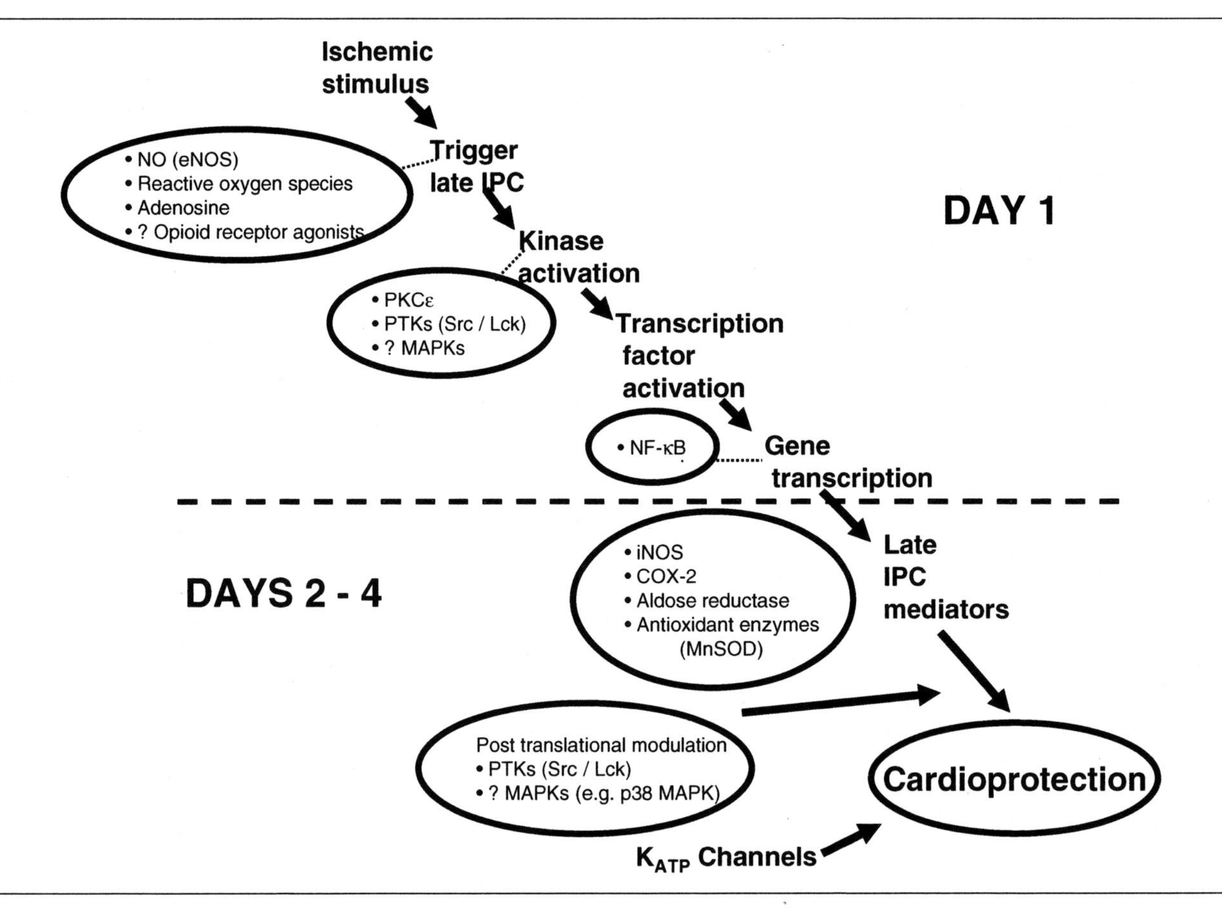

Fig. 1. Postulated role of nitric oxide in the late IPC response. Schematic of cellular mechanisms. Based on reference 29. COX-2, cyclooxygenase-2; IPC, ischemic preconditioning; eNOS, endothelial NO synthase; iNOS, inducible NO synthase, NF-κB, nuclear factor-κB; NO, nitric oxide; PKCε, protein kinase Cε; PTKs, protein tyrosine kinase.

and apoptosis [45,46] depending on the duration of ischemia. Persistent left ventricular dysfunction and myocardial perfusion-function mismatch are significant clinical problems following reperfusion [47]. Although early reperfusion has addressed part of the problem [48], adjunctive therapies are still needed.

The role of NO in ischemia-reperfusion has been reviewed [49]. Cumulative evidence suggests that ischemia-reperfusion results in the production of an excess of oxygen free radicals (OFRs) which contribute to endothelial damage and ventricular dysfunction [49]. The interaction of NO and superoxide results in the formation of peroxynitrite and high concentrations of peroxynitrite are considered cytotoxic [50,51]. Peroxynitrite has also been linked to ischemia-reperfusion injury [52,53], although this has resulted in some controversy [3,49].

The role of NO in apoptosis is controversial. Recently, the inhibition of eNOS (NOS 3) was suggested to potentiate ischemia-reperfusion-induced apoptosis via a caspase-3 dependent pathway [54]. It is pertinent to note that (i) eNOS is expressed in coronary endothelium, endocardium and cardiac myocytes, and (ii) regulates vascular smooth muscle cell tone, permeability and platelet adhesion of endothelial cells, receptor-effector coupling, energetics, contractility as well as apoptosis of cardiomyocytes [55,56]. Recently, Arstall et al. [57] showed cytokine-mediated cardiomyocyte apoptosis associated with iNOS induction and peryoxynitrite generation.

Increased inducible NOS (iNOS or NOS 2) has been implicated in the pathophysiology of myocardial dysfunction in several conditions, including

Table 1. Key signaling molecules in late IPC

Molecule	Reference
Nitric oxide (NO)	[6,9,10]
Protein kinase C (PKCε)	[12]
Protein tyrosine kinase (PTKs), Src, Lck	[13]
Nuclear factor-κB (NF-κB)	[14]
Inducible nitric oxide synthase (iNOS)	[11,15–17]
Cyclooxygenase-2 (COX-2)	[18]
Aldose reductase	[19]

the inflammatory reponse in heart failure [58], myocarditis [59], cardiac rejection [60,61], and other forms of heart failure [62–64]. It is now well recognized that very high levels of NO can be produced by iNOS induction and prove detrimental to cardiomyocytes.

The role of oxidant stress with NO inactivation by superoxide and other reactive oxygen species, and endothelial dysfunction in several cardiovascular diseases has been reviewed [65]. Loss of NO associated with risk factors (such as hypertension, hypercholesterolemia, diabetes mellitus, and cigarette smoking) is thought to predispose to atherosclerosis and its sequelae.

Cardioprotection and Angiotensin II Blockade

The renin-angiotensin system (RAS) and its major effector molecule angiotensin II (AngII) are upregulated during ischemia-reperfusion [66], myocardial infarction [67] and heart failure [68]. The effects of excess AngII are considered deleterious so that AngII blockade is desirable [69]. The RAS present in the endothelium produces vascular AngII [70] which stimulates the membrane-bound NAD/NADPH oxidase system to generate superoxide, which in turn inactivates NO and leads to peroxynitrite formation thereby causing impaired endothelial-dependent relaxation, thrombosis, inflammation and other deleterious effects [65,71,72]. The effects of AngII (physiological and pathological) are mediated mainly through the AngII type 1 (AT_1) and type 2 (AT_2) receptors [73]. The AT_2 receptor, normally abundant in the fetus, is re-expressed in adult hearts after ischemia-reperfusion [74–80], myocardial infarction [81] and heart failure [82]. The AT_2 receptor is thought to antagonize the effects of the AT_1 receptor, thereby exerting vasodilatory, proapoptotic, antigrowth, and antihypertrophic effects [73].

There is now increasing evidence suggesting that the beneficial effects of angiotensin-converting enzyme (ACE) inhibitors are not only due to inhibition of AngII formation by ACE but also to increased bradykinin levels and bradykinin-induced release of NO and prostacyclin [83]. AT_1 receptor antagonists block the effects of AngII mediated by the AT_1 receptor and results in unopposed AT_2 receptor stimulation, which leads to increased bradykinin, NO and cGMP [76,84–86], and PKCε activation [76,86]. This increase in NO is thought to be derived from eNOS. In some studies, AT_1 receptor blockade did not increase flow despite an overall cardioprotective effect after ischemia-reperfusion [87] or failed to improve function [88,89] or decrease infarct size [90]. AT_2 receptor blockade during *in vitro* ischemia-reperfusion also upregulated AT_2 receptor expression [74,89], triggered an increase in PKCε expression and cGMP levels [74] and induced cardioprotection [74,89]. Cardioprotection during IPC was also suggested to involve NO and PKC signaling [91]. Recently, NO was shown to downregulate the AT_1 receptor [92].

We have examined the hypothesis that AT_2 receptor activation and downstream signaling through bradykinin, PKCε, NO and cGMP might contribute to cardioprotection during AT_1 receptor blockade in an *in vivo* dog model of ischemia-reperfusion [79]. We found that AT_1 receptor blockade with candesartan alone improved global systolic and diastolic function, limited acute LV remodeling, decreased infarct size and produced regional increases of the AT_2 receptor and PKCε proteins as well as cGMP in the ischemic zone. Importantly, the AT_2 receptor antagonist and the inhibitors of BK, PKCε and NOS attenuated these beneficial effects of AT_1 receptor blockade as well as the regional increase in AT_2 receptor and PKCε proteins. The overall findings of that study suggest that AT_2 receptor activation and signaling through bradykinin, PKCε, NO and cGMP play a significant role in the cardioprotective effect of AT_1 receptor blockade during ischemia-reperfusion.

In a recent report, the absence of NO in eNOS knock-out mice significantly decreased the cardioprotective effects of ACE inhibitor and AT_1 receptor blockade after MI [92]. In another study [93], cardiac NO production due to ACE inhibition decreased beta-adrenergic myocardial contractility in patients with dilated cardiomyopathy. In addition, impaired NO production in dogs was associated with increased angiostatin and decreased coronary angiogenesis [94].

It therefore appears that PKCε, NO and cGMP signaling may be a common pathway for cardioprotection.

Conclusion

Collective evidence suggests that NO can be cardioprotective or detrimental depending on several factors. In IPC, NO appears to be beneficial. In ischemia-reperfusion, the interactions between NO and the multitude of proteins, peptides, inflammatory cytokines, growth factors, angiotensin II and oxygen free radicals are complex. Therapies that enhance NO availability by increasing endothelial eNOS and new NO donors might be effective for limiting LV dysfunction in reperfused infarction and heart failure.

Acknowledgments

We are grateful for the assistance Catherine Jugdutt for manuscript preparation.

This study was supported in part by a grant from the Canadian Institutes for Health Research, Ottawa, Ontario.

References

1. Gross SS, Wolin MS. Nitric oxide: Pathophysiological mechanisms. *Annu Rev Physiol* 1995;57:737–769.
2. Kelly RA, Balligand JL, Smith TW. Nitric oxide and cardiac function. *Circ Res* 1996;79:363–380.
3. Jugdutt BI. Nitric oxide and cardioprotection during ischemia-reperfusion. *Heart Failure Rev* 2002;7(4):391–406.
4. Lalu MM, Wang W, Schulz R. Peroxynitrite in myocardial ischemia-reperfusion injury. *Heart Failure Rev* 2002;7(4): 359–370.
5. Bolli R. The early and late phases of preconditioning against myocardial stunning and the essential role of oxyradicals in the late phase: An overview. *Basic Res Cardiol* 1996;91:57–63.
6. Bolli R. The late phase of preconditioning. *Circ Res* 2000;87:972–983.
7. Downey JM, Cohen MV, Ytrehus K, Liu Y. Cellular mechanisms in ischemic preconditioning: The role of adenosine and protein kinase C. *Ann NY Acad Sci* 1994;723:82–98.
8. Marber MS, Yellon DM. Myocardial adaptation, stress proteins, and the second window of protection. *Ann N Y Acad Sci* 1996;793:123–141.
9. Bolli R, Bhatti ZA, Tang XL, Qiu Y, Zhang Q, Guo Y, Jadoon AK. Evidence that late preconditioning against myocardial stunning in conscious rabbits is triggered by the generation of nitric oxide. *Circ Res* 1997;81:42–52.
10. Qiu Y, Rizvi A, Tang XL, Manchikalapudi S, Takano H, Jadoon AK, Wu WJ, Bolli R. Nitric oxide triggers late preconditioning against myocardial infarction in conscious rabbits. *Am J Physiol* 1997;273:H2931–H2936.
11. Bolli R, Dawn B, Tang XL, Qiu Y, Ping P, Xuan YT, Jones WK, Takano H, Guo Y, Zhang J. The nitric oxide hypothesis of late preconditioning. *Basic Res Cardiol* 1998;93:325–338.
12. Ping P, Zhang J, Qiu Y, Tang XL, Manchikalapudi S, Cao X, Bolli R. Ischemic preconditioning induces selective translocation of protein kinase C isoforms epsilon and eta in the heart of conscious rabbits without subcellular redistribution of total protein kinase C activity. *Circ Res* 1997;81:404–414.
13. Ping P, Zhang J, Zheng YT, Li RC, Dawn B, Tang XL, Takano H, Balafanova Z, Bolli R. Demonstration of selective protein kinase C-dependent activation of Src and Lck tyrosine kinases during ischemic preconditioning in conscious rabbits. *Circ Res* 1999;85:542–550.
14. Xuan YT, Tang XL, Banerjee S, Takano H, Li RC, Han H, Qiu Y, Li JJ, Bolli R. Nuclear factor-kappaB plays an essential role in the late phase of ischemic preconditioning in conscious rabbits. *Circ Res* 1999;84:1095–1109.
15. Bolli R, Manchikalapudi S, Tang XL, Takano H, Qiu Y, Guo Y, Zhang Q, Jadoon AK. The protective effect of late preconditioning against myocardial stunning in conscious rabbits is mediated by nitric oxide synthase. Evidence that nitric oxide acts both as a trigger and as a mediator of the late phase of ischemic preconditioning. *Circ Res* 1997;81:1094–1107.
16. Takano H, Manchikalapudi S, Tang XL, Qiu Y, Rizvi A, Jadoon AK, Zhang Q, Bolli R. Nitric oxide synthase is the mediator of late preconditioning against myocardial infarction in conscious rabbits. *Circulation* 1998;98:441–449.
17. Guo Y, Jones WK, Xuan YT, Tang XL, Bao W, Wu WJ, Han H, Laubach VE, Ping P, Yang Z, Qiu Y, Bolli R. The late phase of ischemic preconditioning is abrogated by targeted disruption of the inducible NO synthase gene. *Proc Natl Acad Sci USA* 1999;96:11507–11512.
18. Shinmura K, Tang XL, Wang Y, Xuan YT, Liu SQ, Takano H, Bhatnagar A, Bolli R. Cyclooxygenase-2 mediates the cardioprotective effects of the late phase of ischemic preconditioning in conscious rabbits. *Proc Natl Acad Sci USA* 2000;97:10197–10202.
19. Shinmura K, Liu S-Q, Tang XL et al. Aldose reductase is an obligatory mediator of the late phase of ischemic preconditioning. *Circulation* 2000;102(Suppl II):II-120. Abstract
20. Takano H, Tang XL, Qiu Y, Guo Y, French BA, Bolli R. Nitric oxide donors induce late preconditioning against myocardial stunning and infarction in conscious rabbits via an antioxidant-sensitive mechanism. *Circ Res* 1998;83:73–84.
21. Banerjee S, Tang XL, Qiu Y, Takano H, Manchikalapudi S, Dawn B, Shirk G, Bolli R. Nitroglycerin induces late preconditioning against myocardial stunning via a PKC-dependent pathway. *Am J Physiol* 1999;277:H2488–H2494.
22. Murry CE, Jennings RB, Reimer KA. Preconditioning with ischemia: A delay of lethal cell injury in ischemic myocardium. *Circulation* 1986;74:1124–1136.
23. Kuzuya T, Hoshida S, Yamashita N, Fuji H, Oe H, Hori M, Kamada T, Tada M. Delayed effects of sublethal ischemia on the acquisition of tolerance to ischemia. *Circ Res* 1993;72:1293–1299.
24. Sun JZ, Tang XL, Knowlton AA, Park SW, Qiu Y, Bolli R. Late preconditioning against myocardial stunning. An endogenous protective mechanism that confers resistance to postischemic dysfunction 24 h after brief ischemia in conscious pigs. *J Clin Invest* 1995;95:388–403.
25. Tang XL, Qiu Y, Park SW, Sun JZ, Kalya A, Bolli R. Time course of late preconditioning against myocardial stunning in conscious pigs. *Circ Res* 1996;79:424–434.
26. Teschner S, Qiu Y, Tang XL, et al. Late preconditioning against myocardial stunning in conscious rabbits: A dose-related or all-or-none phenomenon? *Circulation* 1996;94(Suppl I):I–423. Abstract
27. Qui Y, Maldonado C, Tang XL et al. Late preconditioning against myocardial stunning in conscious rabbits. *Circulation* 1995;92(Suppl I):I–715. Abstract
28. Tang XL, Qiu Y, Park SW, et al. The early and late phases of ischemic preconditioning: A comparative analysis of their effects on infarct size, myocardial stunning, and arrhythmias in conscious pigs undergoing a 40-minute coronary occlusion. *Circ Res* 1997;80:730–742.
29. Dawn B, Bolli R. Role of nitric oxide in myocardial preconditioning. In: Chiueh CC, Hong J-S, Leong SK (eds.). *Nitric Oxide. Novel Actions, Deleterious Effects, and Clinical Potential.* Annals New York Acad of Sci Vol. 962,2002:18–41.
30. Jugdutt BI, Khan MI, Jugdutt SJ, Blinston GE. Impact of left ventricular unloading after late reperfusion of canine anterior myocardial infarction on remodeling and function

using isosorbide-5-mononitrate. *Circulation* 1995;92:926–934.

31. Hill M, Takano H, Tang XL, Kodani E, Shirk G, Bolli R. Nitroglycerin induces late preconditioning against myocardial infarction in conscious rabbits despite development of nitrate tolerance. *Circulation* 2001;104:694–699.
32. Jugdutt BI, Becker LC, Hutchins GM, Bulkley BH, Reid PR, Kallman CH. Effect of intravenous nitroglycerin on collateral blood flow and infarct size in the conscious dog. *Circulation* 1981;63:17–28.
33. Jugdutt BI, Sussex BA, Warnica JW, Rossall RE. Persistent reduction in left ventricular asynergy in patients with acute myocardial infarction with infusion of nitroglycerin. *Circulation* 1983;68:1264–1273.
34. Jugdutt BI, Warnica JW. Intravenous nitroglycerin therapy to limit myocardial infarct size, expansion and complications: Effect of timing, dosage and infarct location. *Circulation* 1988;78:906–919.
35. Jugdutt BI, Khan MI. Effect of prolonged nitrate therapy on left ventricular remodeling after canine acute myocardial infarction. *Circulation* 1994;89:2297–2307.
36. Jugdutt BI, Schwarz-Michorowski BL, Tymchak WJ, Burton JR. Prompt improvement of left ventricular function and topography with combined reperfusion and intravenous nitroglycerin in acute myocardial infarction. *Cardiology* 1997;88:170–179.
37. Jugdutt BI. Nitroglycerin. In: Bates E (ed). *Thrombolysis and Adjunctive Therapy for Myocardial Infarction*. N. York: Marcel Dekker, 1992:119–144.
38. Hearse DJ. Reperfusion of the ischemic myocardium. *J Mol Cell Cardiol* 1997;9:605–616.
39. Braunwald E, Kloner RA. The stunned myocardium: Prolonged, post-ischemic ventricular dysfunction. *Circulation* 1982;66:1146–1149.
40. Bolli R. Mechanisms of myocardial "stunning." *Circulation* 1990;82:723–738.
41. Kloner RA, Jennings RB. Consequences of brief ischemia: Stunning, preconditioning, and their clinical implications. Part 1. *Circulation* 2001;104:2981–2989.
42. Kloner RA, Jennings RB. Consequences of brief ischemia: Stunning, preconditioning, and their clinical implications. Part 2. *Circulation* 2001;104:3158–3167.
43. Reimer KA, Lowe JE, Rasmussen MM, Jennings RB. The wavefront phenomenon of ischemic cell death. 1. Myocardial infarct size vs duration of coronary occlusion in dogs. *Circulation* 1977;56:786–794.
44. Becker LC, Jeremy RW, Schaper J, Schaper W. Ultrastructural assessment of myocardial necrosis occurring during ischemia and 3-h reperfusion in the dog. *Am J Physiol* 1999;277:H243–H252.
45. Fliss H, Gattinger D. Apoptosis in ischemic and reperfused rat myocardium. *Circ Res* 1996;76:949–956.
46. Gottlieb RA, Gruol DL, Zhu JY, Engler RL. Preconditioning in rabbit cardiomyocytes: Role of pH, vacuolar proton ATPase, and apoptosis. *J Clin Invest* 1996;97:2391–2398.
47. Kim CB, Braunwald E. Potential benefits of late reperfusion of infarcted myocardium. The open artery hypothesis. *Circulation* 1993;88:2426–2436.
48. Topol EJ. Early myocardial reperfusion: An assessment of current strategies in acute myocardial infarction. *Eur Heart J* 1996;17 (Suppl E):42–48.
49. Lefer AM, Hayward R. The role of nitric oxide in ischemia-reperfusion. In: Loscalzo J, Vita JA (eds.). *Contemporary Cardiology*, Vol 4: *Nitric Oxide and the Cardiovascular System*. Humana Press, 2000:357–380.
50. Beckman JS, Beckman TW, Chen J, Marshall PA, Freeman BA. Apparent hydroxyl radical production by peroxynitrite: Implications for endothelial injury from nitric oxide and superoxide. *Proc Natl Acad Sci USA* 1990;87:1620–1624.
51. Radi R, Beckman JS, Bush KM, Freeman BA. Peroxynitrite-induced membrane lipid peroxidation: The cytotoxic potential of superoxide and nitric oxide. Arch Biochem Biophys 1991;288:481–487.
52. Wang P, Zweier JL. Measurement of nitric oxide and peroxynitrite generation in the postischemic heart. Evidence for peroxynitrite-mediated reperfusion injury. *J Biol Chem* 1996;271:29223–29230.
53. Yasmin W, Strynadka KD, Schulz R. Generation of peroxynitrite contributes to ischemia-reperfusion injury in isolated rat hearts. *Cardiovasc Res* 1997;33:422–432.
54. Weiland U, Haendeler J, Ihling C, Albus U, Scholz W, Ruetten H, Zeiher AM, Dimmeler S. Inhibition of endogenous nitric oxide synthase potentiates ischemia-reperfusion-induced myocardial apoptosis via a caspase-3 dependent pathway. *Cardiovasc Res* 2000;45:671—678.
55. Kelly RA, Balligand JL, Smith TW. Nitric oxide and cardiac function. *Circ Res* 1996;79:363–380.
56. Paulus WJ. The role of nitric oxide in the failing heart. *Heart Fail Rev* 2001;6:105–118.
57. Arstall MA, Sawyer DB, Fukazawa R, Kelly RA. Cytokine-mediated apoptosis in cardiac myocytes: The role of inducible nitric oxide synthase induction and peroxynitrite generation. *Circ Res* 1999;85:829–840.
58. Thoenes M, Forstermann U, Tracey WR, Bleese NM, Nussler AK, Scholz H, Stein B. Expression of inducible nitric oxide synthase in failing and non-failing human heart. *J Mol Cell Cardiol*. 1996;28:165–169.
59. Hirono S, Islam MO, Nakazawa M, Yoshida Y, Kodama M, Shibata A, Izumi T, Imai S. Expression of inducible nitric oxide synthase in rat experimental autoimmune myocarditis with special reference to changes in cardiac hemodynamics. *Circ Res* 1997;80:11–20.
60. Lewis NP, Tsao PS, Rickenbacher PR, Xue C, Johns RA, Haywood GA, von der Leyen H, Trindade PT, Cooke JP, Hunt SA, Billingham ME, Valantine HA, Fowler MB. Induction of nitric oxide synthase in the human cardiac allograft is associated with contractile dysfunction of the left ventricle. *Circulation* 1996;93:720–729.
61. Szabolcs M, Michler RE, Yang X, Aji W, Roy D, Athan E, Sciacca RR, Minanov OP, Cannon PJ. Apoptosis of cardiac myocytes during cardiac allograft rejection. Relation to induction of nitric oxide synthase. *Circulation* 1996;94:1665–1673.
62. Haywood GA, Tsao PS, von der Leyen HE, Mann MJ, Keeling PJ, Trindade PT, Lewis NP, Byrne CD, Rickenbacher PR, Bishopric NH, Cooke JP, McKenna WJ, Fowler MB. Expression of inducible nitric oxide synthase in human heart failure. *Circulation* 1996;93:1087–1094.
63. Satoh M, Nakamura M, Tamura G, Makita S, Segawa I, Tashiro A, Satodate R, Hiramori K. Inducible nitric oxide synthase and tumor necrosis factor-alpha in myocardium in human dilated cardiomyopathy. *J Am Coll Cardiol* 1997;29:716–724.
64. Habib FM, Springall DR, Davies GJ, Oakley CM, Yacoub MH, Polak JM. Tumour necrosis factor and inducible

nitric oxide synthase in dilated cardiomyopathy. *Lancet* 1996;347:1151–1155.
65. Cai H, Harrison DG. Endothelial dysfunction in cardiovascular diseases: The role of oxidant stress. *Circ Res* 2000;87:840–844.
66. Youhua Z, Shouchun X. Increased vulnerability of hypertrophied myocardium to ischemia and reperfusion injury. Relation to cardiac renin-angiotensin system. *Chin Med J* 1995;108:28–32.
67. Sun Y, Weber KT. Angiotensin II receptor binding following myocardial infarction in the rat. *Cardiovasc Res* 1994;28:1623–1628.
68. Francis GS, McDonald KM, Cohn JN. Neurohumoral activation in preclinical heart failure. Remodeling and the potential for intervention. *Circulation* 1993;87(5 Suppl):IV90–96.
69. Brunner HR. Experimental and clinical evidence that angiotensin II is an independent risk factor for cardiovascular disease. *Am J Cardiol* 2001;87:3C–9C.
70. Hilgers KF, Veelken R, Muller DN, Kohler H, Hartner A, Botkin SR, Stumpf C, Schmieder RE, Gomez RA. Renin uptake by the endothelium mediates vascular angiotensin formation. *Hypertension* 2001;38:243–248.
71. Griendling KK, Sorescu D, Ushio-Fukai M. NAD(P)H Oxidase: Role in cardiovascular biology and disease. *Circ Res* 2000;86:494–501.
72. Dzau VJ. Tissue angiotensin and pathobiology of vascular disease. A unifying hypothesis. *Hypertension* 2001;37:1047–1052.
73. Matsubara H. Pathophysiological role of angiotensin II type 2 receptor in cardiovascular and renal diseases. *Circ Res* 1998;83:1182–1191.
74. Xu Y, Clanachan AS, Jugdutt BI. Enhanced expression of AT_2R, IP_3R and PKC_ε during cardioprotection induced by AT_2R blockade. *Hypertension* 2000;36:506–510.
75. Xu Y, Menon V, Jugdutt BI. Cardioprotection after angiotensin II type 1 blockade involves angiotensin II type 2 receptor expression and activation of protein kinase C-ε in acutely reperfused myocardial infarction. Effect of UP269-6 and losartan on AT_1 and AT_2 receptor expression, and IP_3 receptor and PKC_ε proteins. *J Renin-Angiotensin Aldosterone System* 2000;1:184–195.
76. Jugdutt BI, Xu Y, Balghith M, Moudgil R, Menon V. Cardioprotection induced by AT_1R blockade after reperfused myocardial infarction: Association with regional increase in AT_2R, IP_3R and PKC_ε proteins and cGMP. *J Cardiovasc Pharmacol & Therapeut* 2000;5:301–311.
77. Moudgil R, Xu Y, Menon V, Jugdutt BI. Effect of chronic pretreatment with AT_1 receptor antagonism on postischemic functional recovery and AT_1/AT_2 receptor proteins in isolated working rat hearts. *J Cardiovasc Pharmacol & Therapeut* 2001;6:183–188.
78. Jugdutt, BI, Xu Y, Balghith M, Menon V. Cardioprotective effects of angiotensin II type 1 receptor blockade with candesartan after reperfused myocardial infarction: Role of angiotensin II type 2 receptor. *J Renin-Angiotensin Aldosterone System* 2001;2:S162–S166.
79. Jugdutt BI, Balghith M. Enhanced regional AT_2 receptor and $PKC\varepsilon$ expression during cardioprotection induced by AT_1 receptor blockade after reperfused myocardial infarction. *J Renin-Angiotensin Aldosterone System* 2001;2:134–140.
80. Moudgil R, Menon V, Xu Y, Musat-Marcu S, Jugdutt BI. Postischemic apoptosis and functional recovery after angiotensin II type 1 receptor blockade in isolated working rat hearts. *J Hypertension* 2001;19:1121–1129.
81. Nio Y, Matsubara H, Murasawa S, Kanasaki M, Inada M. Regulation and gene transcription of angiotensin II receptor subtypes in myocardial infarction. *J Clin Invest* 1995;95:46–54.
82. Haywood GA, Gullestad L, Katsuya T, Hutchinson HG, Pratt RE, Horiuchi M, Fowler MB. AT_1 and AT_2 angiotensin receptor gene expression in human heart failure. *Circulation* 1997;95:1201–1206.
83. Unger T, Gohlke P. Tissue renin-angiotensin systems in the heart and vasculature: possible involvement in the cardiovascular actions of converting enzyme inhibitors. *Am J Cardiol* 1990;65:31–101.
84. Liu YH, Yang XP, Sharov VG, Nass O, Sabbah HN, Peterson E, Carretero OA. Effects of angiotensin-converting enzyme inhibitors and angiotensin II type 1 receptor antagonists in rats with heart failure: Role of kinins and angiotensin type 2 receptors. *J Clin Invest* 1997;99:1926–1935.
85. Jalowy A, Schulz R, Dorge H, Behrends M, Heush G. Infarct size reduction by AT_1-receptor blockade through a signal cascade of AT_2receptor activation, bradykinin and prostaglandins in pigs. *J Am Coll Cardiol* 1998;32:1787–1796.
86. Bartunek J, Weinberg EO, Tajima M, Rohrbach S, Lorell BH. Angiotensin II type 2 receptor blockade amplifies the early signals of cardiac growth response to angiotensin II in hypertrophied hearts. *Circulation* 1999;99:22–25.
87. Dörge H, Behrends M, Schulz R, Jalowy A, Heusch G. Attenuation of myocardial stunning by the AT_1 receptor antagonist candesartan. *Basic Res Cardiol* 1999;94:208–214.
88. Ford WR, Clanachan AS, Jugdutt BI. Opposite effects of angiotensin receptor antagonists on recovery of mechanical function after ischemia-reperfusion in isolated working rat hearts. *Circulation* 1996;94:3087–3089.
89. Xu Y, Dyck J, Ford WR, Clanachan AS, Lopaschuk GD, Jugdutt BI. Angiotensin II type 1 and type 2 receptor protein after acute ischemia-reperfusion in isolated working rat hearts. *Am J Physiol Heart Circ Physiol* 2002;282:H1206–H1215.
90. Liu Y-H, Yang X-P, Sharov VG, Sigmon DH, Sabbah HN, Carretero OA. Paracrine systems in the cardioprotective effect of angiotensin-converting enzyme inhibitors on myocardial ischemia/reperfusion injury in rats. *Hypertension* 1996;27:7–13.
91. Ping P, Takano H, Zhang J, Tang X-L, Qiu Y, Li RCX, Banerjee S, Dawn B, Balafonova Z, Bolli R. Isoform-selective activation of protein kinase C by nitric oxide in the heart of conscious rabbits. A signaling mechanism for both nitric oxide-induced and ischemia-induced preconditioning. *Circ Res* 1999;84:587–604.
92. Liu YH, Xu J, Yang XP, Yang F, Shesely E, Carretero OA. Effect of ACE inhibitors and angiotensin II type 1 receptor antagonists on endothelial NO synthase knockout mice with heart failure. *Hypertension* 2002;39:375–381.
93. Wittstein IS, Kass DA, Pak PH, Maughan WL, Fetics B, Hare JM. Cardiac nitric oxide production due to angiotensin-converting enzyme inhibition decreases beta-adrenergic myocardial contractility in patients with dilated cardiomyopathy. *J Am Coll Cardiol* 2001;38:429–435.
94. Matsunaga T, Weihrauch DW, Moniz MC, Tessmer J, Warltier DC, Chilian WM. Angiostatin inhibits coronary angiogenesis during impaired production of nitric oxide. *Circulation* 2002;105:2185–2191.

Angiotensin II and Nitric Oxide Interaction

Marc de Gasparo, MD
MG Consulting Co, Rossemaison, Switzerland

Abstract. **Nitric oxide degradation linked to endothelial dysfunction plays a central role in cardiovascular diseases. Superoxide producing enzymes such as NADPH oxidase and xanthine oxidase are responsible for NO degradation as they generate a variety of reactive oxygen species (ROS). Moreover, superoxide is rapidly degraded by superoxide dismutase to produce hydrogen peroxide leading to the uncoupling of NO synthase and production of increased amount of superoxide.**

Angiotensin II is an important stimulus of NADPH oxidase. Through its AT_1 receptor, Ang II stimulates the long-term increase of several membrane component of NADPH oxidase such as P_{22} phox or nox-1 and causes an increased activity of NADPH oxidase with inactivation of NO leading to impaired endothelium-dependent vasorelaxation, vascular smooth muscle cell hypertrophy, proliferation and migration, extracellular matrix formation, thrombosis, cellular infiltration and inflammatory reaction. Several preclinical and clinical studies have now confirmed the involvement of the AT_1 receptor in endothelial dysfunction. It is proposed that the AT_2 receptor counterbalances the deleterious effect of the Ang II-induced AT_1 receptor stimulation through bradykinin and NOS stimulation. This mechanism could be especially relevant in pathological cases when the NADPH oxidase activity is blocked with an AT_1 receptor antagonist.

Key Words. **NADPH oxidase, NOS, superoxide, oxidative stress, endothelial dysfunction, angiotensin receptor antagonist**

Introduction

Angiotensin II (Ang II) is a major endocrine hormone and it plays a crucial central role in the pathogenesis of cardiovascular and renal diseases. It increases both vascular resistance and circulating volume. It regulates kidney perfusion pressure producing vasoconstriction in the kidney, mainly of the glomerular efferent vessels, leading to an increased capillary glomerular pressure. It controls tubular reabsorption. In addition to its classical hormonal effects, Ang II also has important and long-term autocrine and paracrine properties. Indeed, Ang II acts well beyond its known hemodynamic targets. It affects directly both vascular smooth muscle and endothelial cells.

The endothelium, although it constitutes only a single cell layer along all blood vessels, can be considered a real organ with a total surface area of 750 m^2 or roughly the size of three tennis courts. It is a central player for cardiovascular homeostasis. Indeed, the endothelium is the source of biologically active mediators with opposite effects, vasoconstrictor and vasodilator substances, growth promoting and inhibiting factors, thrombogenic and thrombolytic agents. The endothelium is also the source of oxidative and antioxidative processes and the cell redox state is now a recognized key determinant of endothelial biology (Fig. 1).

A complete renin-angiotensin system (RAS) is present in the endothelium [1]. Angiotensinogen can be locally formed but is mainly originated from the blood. Intact endothelium can produced Ang I and Ang II from renin taken up from the circulation and locally active angiotensin converting enzyme (ACE). Removal of the endothelium or captopril treatment abolishes indeed the conversion of Ang I to Ang II. Ang II AT_1 and AT_2 receptors have also been detected [1]. Ang II is therefore synthesized by the endothelium where it directly influences its function. If the endothelium is the source of Ang II, it is also responsible for the synthesis of nitric oxide (NO), a potent vasodilator, which can be inactivated by superoxide anions [2].

Both Ang II and NO may be protective of or deleterious for target organs such as the endothelium, vascular SMC and mesangial cells as both affect vascular tone, fluid volume and remodeling. Therefore, an imbalance between Ang II and endothelial NO/reactive oxygen species (ROS) production constitutes a key factor in cardiovascular and renal injury.

Endothelial dysfunction is clearly linked to increased oxidative stress i.e. production of the oxidants peroxynitrite ($ONOO^-$) and hydrogen peroxide (H_2O_2) responsible for alterations in vasorelaxation, SMC proliferation, thrombosis, and an inflammatory reaction with cellular infiltration. These functional alterations lead to structural changes and clinical sequelae such as myocardial infarction, ischemia, congestive heart failure, stroke and death. Thus, excess vascular superoxide reduces NO bioavailability and promotes further vascular oxidative stress.

Address for correspondence: M. de Gasparo, rue es Planches 123, 2842 Rossemaison, Switzerland. Fax: +41 32 422 86 41; E-mail: m.de_gasparo@bluewin.ch

Vasodilation	Vasoconstriction
NO Bradykinin PGI_2	Ang II Endothelin
Growth inhibition	**Growth promotion**
NO Bradykinin PGI_2	Ang II Endothelin
Profibrinolysis	**Antifibrinolysis**
NO Bradykinin PGI_2 tPA	Ang II PAI-1 TF
Antioxidation	**Oxidation**
NO	$O_2^{\cdot} - H_2O_2$ $ONOO^{-}$ - $OH^{\cdot}$

Fig. 1. Endothelium plays a key role in cardiovascular and renal diseases. It is the source of factors having opposite effects.

Interestingly, endothelium dysfunction appears first in life, before development of cardiovascular diseases [3,4]. Clearly, risk factors such as hypertension, dyslipidemia, diabetes or smoking alter both the cell redox system of the vessel and endothelial function. There are a number of recent reviews of endothelium dysfunction [5–10]. This paper will therefore summarize the various constituents involved in oxidative stress and describe the role of Ang II in endothelial dysfunction. A more detailed discussion of the potential function of the two angiotensin receptors, AT_1 and AT_2, in the regulation of NO production also will be presented.

Nitric Oxide Synthase (NOS) and Nitric Oxide

NO is not only responsible for air pollution, acid rain and depletion of the ozone layer, it also plays a major role in cardiovascular and endothelial function. NO is an important messenger in many signal transduction processes. Blood flow across the endothelial cell surface (shear stress) and chemical mediators such as acetyl choline (ACh), bradykinin (BK), Substance P, and β adrenoceptor agonists stimulate the production of the free-radical NO from arginine in a complex reaction that is catalyzed by a constitutive neural (NOS I or nNOS) or endothelial (NOS III or eNOS) nitric oxide synthase (NOS) [11].

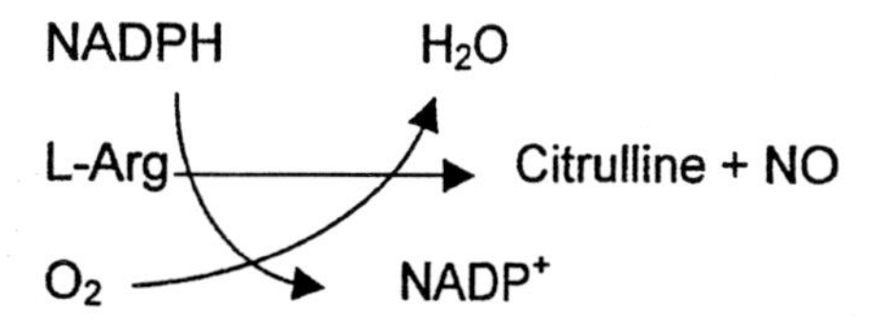

Nitric oxide synthase, a Ca^{++}-dependent enzyme is closely related to cytochrome P_{450} enzymes and requires NADPH (Nicotinamide Adenine Dinucleotide Phosphate) and O_2 as cofactors. In fact, the carboxyl-terminal half of NOS cDNA resembles that of cytochrome P_{450} oxygenase: both contain two binding sites for flavins and one for NADPH [12]. The activity of NOS can be inhibited by endogenous analogues of L-arginine such as asymmetric dimethylarginine.

NO diffuses freely across the abluminal and luminal surfaces of the endothelial cell to vascular SMC and subintimal macrophages. It has a very short half-life, less than a few seconds, because it is highly reactive with reactive oxygen species ROS. There is another NOS, which is inducible by inflammatory cytokines and by Ang II in a Ca^{++}-independent manner (NOS II or iNOS). This iNOS is increased in the failing heart, in perivascular injury and is deleterious as it stimulates neointima growth, myocardial dysfunction and inflammatory reactions [13–15].

At low concentrations, such as after ACh stimulation, NO relaxes arteries by a cGMP independent mechanism, which involves cGMP kinase 1 and a voltage-dependent K^+ channel causing hyperpolarization, increased intracellular K^+ levels and calcium store refilling via the sarco/endoplasmic reticulum calcium ATPase (SERCA) [16,17]. The antihypertrophic effect of NO is also cGMP kinase 1-dependent in cardiomyocytes in culture [18].

In pathophysiological states, NO is present at high concentrations and acts through a cGMP dependent mechanism involving a cAMP kinase or soluble guanylate cyclase [16].

NO regulates blood flow distribution between arterioles and the microvasculature by affecting the diameter of small arteries. It improves arterial elasticity and distensibility with a reduction of pulse wave velocity [19,20]. It also decreases arterial permeability. In the heart, NO produces myocardial relaxation and improves cardiac contractility [21]. In the kidney, NO mainly dilates glomerular afferent vessels and reduces tubular Na^+ reabsorption [22]. It has major antihypertrophic effects and also reduces leukocyte-endothelial cell interactions, platelet adhesion and aggregation and LDL oxidation. It decreases ACE mRNA in the endothelium, down-regulates the AT_1 receptor in vascular smooth muscle cells (SMC) and inhibits endothelin production [23–25]. Chronic inhibition of NOS is clearly linked to activation of the tissue RAS, endothelin production and inflammatory changes associated with increased production of Transforming Growth Factor β (TGFβ), fibronectin, type I collagen, monocyte chemoattractant protein-1 (MCP-1) and monocyte/macrophage

ED-1 [26–29]. In summary, NO can be considered as a physiological antagonist of Ang II.

Redox Signal Mediators

Reactive Oxygen Species (ROS) such as superoxide ($O_2^{\cdot}$), hydrogen peroxide (H_2O_2), peroxinitrite ($ONOO^-$) and hydroxyl radicals ($OH^{\cdot}$) alter many cellular processes. Smooth muscle cells and adventitial fibroblasts account for the most important source of superoxide anions in the vessel wall [30] and various pathways have been described for their production.

One of these, the *mitochondrial respiratory chain*, acts through a NADH dehydrogenase that requires succinate, ADP and O_2 as substrates, and can produce $O_2^{\cdot}$. Physiological concentrations of NO can stimulate this reaction by inhibiting cytochrome oxidase and cytochrome C reductase in rat heart mitochondria [31]. Additional sources of ROS are Cyt P_{450} reductase, lipoxygenase/cyclooxygenase. However, the most important enzymatic pathways responsible for endothelial dysfunction involve xanthine oxidase, the uncoupling of eNOS and NADPH oxidase [7].

Xanthine oxidase, a molybdenum- and iron-containing flavoprotein, oxidizes hypoxanthine to xanthine and then to uric acid. Molecular O_2 is the oxidant in both reactions and is reduced to H_2O_2 , which is further decomposed to H_2O and O_2 by catalase. This mechanism was proposed from studies in spontaneous hypertensive rats (SHR) as the xanthine oxidase inhibitor allopurinol blocked the reaction and decreased blood pressure [32].

Wu and de Champlain [33] have shown that the addition of xanthine oxidase to cultured aortic SMC induced the formation of $O_2^{\cdot}$. A consequent increased tyrosine kinase and phospholipase $C\gamma$ ($PLC\gamma$) activity caused a rise in intracellular Ca^{++} and cell contraction and proliferation. In addition to this mechanism, decreased soluble guanylate cyclase activity leads to a decreased cGMP which then results in stimulation of IP_3 production and increased intracellular Ca^{++}. This effect is significantly greater in cells from SHR than from Wistar Kyoto (WKY) rats. *NOS* itself can also be a paradoxical source of $O_2^{\cdot}$. In either the absence of arginine, with auto-oxidation or a deficiency of tetrahydrobiopterin BH_4, a co-factor essential for NO synthase activity, NOS is uncoupled and generation of $O_2^{\cdot}$ occurs [34].

Therefore, depending on the cell environment, NOS may generate either NO or $O_2^{\cdot}$.

The *membrane-bound NAD/NADPH oxidase* is the most powerful source of endogenous $O_2^{\cdot}$ production [6,35]. It is a key-target for Ang II. Protein kinase C and phospholipase D have been proposed to be involved in the activation of NADPH oxidase [36–38]. NADPH oxidase is a flavocytochrome b heterodimer consisting of two protein subunits, p22-phox and either p91-phox in fibroblasts or Nox1 in smooth muscle cells. These subunits are glycosylated and bind two heme molecules and one flavin adenine dinucleotide (FAD) moiety. Upon activation, which occurs over a period of hours, at least two cytosolic proteins are translated, p47-phox and p67-phox, as well as a small GTP-binding protein rac2. P47-phox is extensively phosphorylated during oxidase activation. These proteins interact with each other and the flavocytochrome. The endothelium, vascular SMC, adventitial fibroblasts as well as inflammatory cells express the element constitutive of the NADPH oxidase in variable amounts [6,39]. After carotid injury, the increased superoxide production derived from medial and neointimal smooth muscle cells and adventitial fibroblasts and the subunits of NADPH oxidase are expressed in a sequential manner [30]. Ang II can also generate growth-promoting ROS through AT_1 receptor mediated epidermal growth factor (EGF) receptor transactivation [40,41] and the involvement of the reactive oxygen species has been demonstrated in this effect [42]. The Tyr kinase cSrc is autophosphorylated by Ang II and constitutes a signaling mechanism that links H_2O_2 and EGF receptor phosphorylation leading to activation of protein kinase B (PKB) and extracellular signal-regulated kinase (ERK) cascades, protein synthesis and cell growth and proliferation (Fig. 2). Other growth-promoting receptors such as plasma-derived growth factor (PDGF) and insulin-like growth factor (IGF) can also be transactivated by Ang II [43,44].

Thus, Ang II activates NADPH oxidase. Various arguments support this interaction. First, this is specific for Ang II as it is not observed after norepinephrine stimulation [45]. Secondly, treatment with liposome encapsulated superoxide dismutase (SOD) restored blood pressure to normal levels in Ang II-infused rats [46]. Thirdly, Ang II-induced stimulation is accompanied by an upregulation of p22-phox, which was abolished by p22-phox antisense, diphenylene iodonium (DPI), an inhibitor of the flavoprotein component of NADPH oxidase, and overexpression of catalase in vascular SMC [35,47,48]. Finally, it is inhibited following RAS blockade [49–51]. In contrast to NO, $O_2^{\cdot}$ is not membrane permeable and the impaired electron passes from one molecule to another in a radical chain reaction. These redox events occur not only in the cytoplasm and the nucleus but also in the interstitial space. There is activation of redox-sensitive kinases and inactivation of specific phosphatases such as Protein-Tyrosine Phosphatase PTB-1B and 2B.

$O_2^{\cdot}$ is rapidly formed in response to Ang II but has limited diffusion capability and is rapidly degraded by superoxide dismutase (SOD) to produce

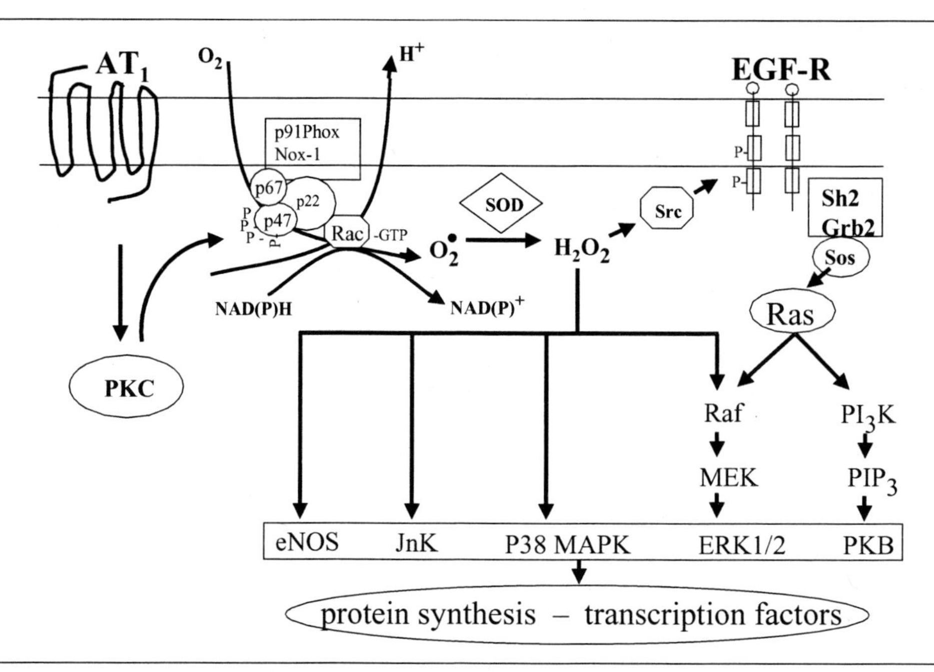

Fig. 2. The angiotensin AT_1 receptor mediated EGF receptor transactivation through stimulation of the reactive oxygen species (ROS).

hydrogen peroxide, which may cause the uncoupling of NOS and further stimulation of $O_2^{\cdot}$ production. Three mammalian SODs have been described [5]. The *cytosolic Cu^{++}/Zn^{++} SOD* is upregulated by laminar shear stress in human endothelial cells in culture [52]. The *mitochondrial* Mn^{++} SOD appears involved in dismutating the $O_2^{\cdot}$ generated by the respiratory chain. Finally, the principal *extracellular Cu^{++}/Zn^{++} SOD* is produced in fibroblasts, in vascular endothelial cells, extracellular matrix and in the lungs and then secreted into the extracellular fluid [53].

The physiological effects of NO and superoxide anions depend on their local concentrations. Under normal conditions, only a small amount of superoxide anions are formed and processed by SOD to generate H_2O_2. Depending on the superoxide:NO stoichiometry, peroxinitrite could also be produced as observed in endothelial cells stimulated with Ang II [54].

Endothelial dysfunction in SHR is associated with an excess of superoxide production compared with normotensive rats rather than a diminished NO generation [5,22,23]. The increased quantity of $O_2^{\cdot}$ overwhelms the production of NO and in such situations, the H_2O_2 production is limited only by the availability of SOD. H_2O_2 can be detected as early as 1 min after addition of Ang II [55]. Ang II induces an almost 3-fold increase in superoxide production and a 5-fold accumulation of H_2O_2 in vascular SMC [55]. Whereas Ang II activates ERK1/2 through $O_2^{\cdot}$, H_2O_2 induces cell proliferation through p38 MAP kinase and PKB stimulation. Studies using NADPH oxidase inhibitors and overexpression or inhibition of catalase activities confirmed these pathways [36,48].

H_2O_2 also increases eNOS mRNA and PLD activity. eNOS causes the formation of NO and a small amount of $O_2^{\cdot}$, which is immediately scavenged by NO. However, in pathological circumstances, large amounts of $O_2^{\cdot}$ are formed, which inactivates NO and leads to the production of peroxynitrite $ONOO^-$, and ultimately to impaired endothelium-dependent vasodilation. Both superoxide anions and H_2O_2 can be transformed into hydroxyl radicals (OH^-), which through cycloxygenase COX1, can catalyze the production of vasoconstrictor prostanoids such as PGH_2. H_2O_2 may also act as an endothelial-derived hyperpolarizing (EDHP) agent, hyperpolarizing the cells and affecting K^+ channels [56]. In addition, massive scavenging of NO stimulates the release of endothelin and accelerates the pathological process [57].

Involvement of the Angiotensin AT_1 Receptor

Ang II and vascular stretch are potent endogenous stimuli of vascular NADPH oxidase [58]. Membrane NADPH oxidase plays a central role in the signaling mechanism in the endothelium. Interestingly, stretch of vascular smooth muscle stimulates Ang II release [52,59]. The majority of the biological effects of Ang II occur following its binding to and activation of the AT_1 receptor.

Production of $O_2^{\cdot}$ and other reactive oxygen species leads to the deleterious effect of Ang II: impaired vasodilation, vascular SMC hypertrophy, proliferation and migration, extracellular matrix formation, thrombosis and cellular infiltration and inflammatory reactions. Through the AT_1 receptor, Ang II induces PDGF, IFG and TGFβ expression and the matrix metalloproteinase activity involved in vascular remodeling [45,60]. Activation of the AT_1 receptor is also responsible for the stimulation of plasminogen activator inhibitor −1 (PAI-1) and tissue factor (TF), both in vitro and in vivo and therefore will affect thrombolytic homeostasis [61–63]. Stimulation of the AT_1 receptor activates the chemoattractant protein MCP-1, adhesion molecules VCAM-1 and ICAM-1, cytokines such as tumor necrosis factor (TNFα) and causes an inflammatory reaction as observed in atherosclerosis or diabetes [63–65]. Increased tissue Ang II and ACE expression occur after vascular injury or inflammatory reactions [66–69] and magnify endothelial dysfunction, further decreasing NO and increasing ROS and result in a deleterious vicious cycle [10]. There is no doubt that blockade of the RAS improves endothelial function in both human and experimental animals and suggests a reduced superoxide production and an increased NO availability [70,71]. Large clinical trials such as TREND have confirmed these observations [72]. Furthermore, the beneficial effect observed in the HOPE trial extends well beyond the 2–3 mm Hg fall in blood pressure and substantiates a beneficial effect of RAS blockade on endothelium function [73]. Several studies have now confirmed that RAS blockade confers additional benefit over that of blood pressure reductions per se. Comparing the effect of losartan and atenolol after 1 year treatment in patients with mild hypertension, Schiffrin et al. [74,75] observed a significant reduction of the medial/luminal ratio of resistance gluteal arteries and improvement of the endothelium-dependent relaxation. These benefits occurred after treatment with the angiotensin receptor blocker only and not after β blockade [74,75].

Following the seminal observation by Griendling et al. [35] showing that Ang II stimulates intracellular superoxide anion formation in vascular SMC in culture through increased NADPH oxidase activity, many reports have confirmed the role of the AT_1 receptor. Losartan decreased p22-phox mRNA expression and NADPH oxidase activity in cultured vascular SMC as well as in aorta from hypertensive rats [46,55,76]. Treatment of one kidney one clip hypertensive rat with losartan decreased superoxide generation by 40% in isolated aortic rings [77]. In SHR, irbesartan administered for 14 weeks reduced p22-phox and superoxide anion to levels similar to those observed in WKY. This normalization accompanied the improvement of ACh-induced vasodilation, aortic media thickness and cross-sectional area [22]. In contrast, amlodipine or hydrochlorothiazide/hydralazine were significanly less effective in reducing superoxide and p21phox than irbesartan, although they all produced a similar fall in blood pressure in stroke prone SHR [78]. Losartan treatment for 16 weeks markedly reduced myocardial oxidative stress in rats with congestive heart failure after myocardial infarction [79]. Similarly, patients with coronary artery disease treated with irbesartan showed a significant decreased superoxide generation and lipid peroxidation and a concomitant reduction in monocyte binding capacity, soluble TNFα and VCM-1 levels [80]. Losartan also improved flow-dependent endothelium mediated relaxation by more than 75% as the SOD activity was increased by more than 200% [51].Thus, there is clear evidence for increased NO bioavailability, decreased oxidative stress and rise of SOD activity after AT_1 receptor antagonism.

Involvement of the AT_2 Receptor

Following blockade of the AT_1 receptor, there is an increased production of renin as the regulatory negative feedback mechanism is no longer operative at the juxta-glomerular cell level. As a result, plasma Ang II increases [81–83] and stimulates the AT_2 receptor, which is unaffected by the selective AT_1 receptor antagonist. Although the AT_2 receptor is usually expressed at low density in adults, it is upregulated in pathological states such as vascular injury, salt depletion, heart failure or cardiac hypertrophy [84–87]. Pharmacological studies indicate that there is crosstalk between AT_1 and AT_2 receptors and stimulation of the AT_2 receptor opposes the effect of the AT_1 in a Yin and Yang manner [88]. Whereas stimulation of the AT_1 receptor leads to cellular growth and hypertrophy, angiogenesis, vasoconstriction, interstitial fibrosis and cardiac remodeling, AT_2 receptor stimulation causes opposite effects, antiproliferation and apoptosis, antiangiogenesis, vasodilation, decreased neointimal formation and inhibition of cardiac remodeling [88–91].

Some of the beneficial effects of AT_2 receptor stimulation appear to be linked to the stimulation of a bradykinin-nitric oxide cascade [92–96]. Guan et al. [97] demonstrated an attenuation of the blood pressure lowering effect of an angiotensin receptor blocker and ACE inibitor following NOS blockade and was probably the first to suggest a contribution of NO to the blood pressure reduction mechanism of the RAS blockers. Liu et al. [98] have confirmed that the bradykinin receptor B2 antagonist icataban, like the AT_2 receptor blocker PD123319, reversed the beneficial effect of losartan in rat with coronary ligature. Similarly, infusion of Ang II produced an increase in cGMP content in aorta of stroke prone SHR treated with losartan [99]. Again, this effect was reduced by icataban or by PD123319 but was not affected by a potassium channel opener minoxidin and is thus, independent of any effect on blood pressure. In AT_2 receptor knockout mice with coronary ligation, valsartan increased ejection fraction and cardiac output, decreased LV dimension and myocyte cross-sectional area and the interstitial collagen fraction was significantly reduced compared with wild type mice following coronary ligature. In contrast, enalapril improved cardiac function and remodeling similarly in both strains [100].

Superoxide formation due to NADPH oxidase stimulation increased by 37% when human endothelial cells were stimulated with Ang II but by 73% in the presence of PD 123319, an AT_2 receptor antagonist [101]. The involvement of the AT_2 receptor was also demonstrated by showing a 30% increase of Tyr phosphatase activity, which was prevented by the selective Tyr phosphatase inhibitor vanadate [101]. Tyr phosphatase is indeed part of the signaling mechanism of the AT_2 receptor [102]. In vascular remodeling, activation of adventitial fibroblasts can lead to conversion into myofibroblasts in the adventia or in the media. These cells then migrate toward the subendothelial space. De-differentiated vascular SMC also migrate to the neointima [103]. Metalloproteinases and their tissue inhibitors (TIMP) control this migration and the mechanism is reminiscent of a fetal gene program. Local application of Ang II to intact carotid arteries of normotensive rat induces adventitial thickening with increased cellularity characterized by DNA synthesis, neovascularization and collagen deposition [104]. This effect was not observed after norepinephrine application. Most importantly, blockade of the AT_1 receptor as well as stimulation of the AT_2 receptor inhibits the effect of Ang II. There is significant upregulation of NADPH oxidase in restenosis after carotid balloon injury [30]. $O_2^{\cdot}$ production was increased in the adventitia and in the innermost medial layer. There was a concomitant rise of the NADPH oxidase subunits, first p22 phox and nox 1 in SMC and later, p91 phox in fibroblasts. Nox 4, a more recently described subunit of NADPH oxidase increased at a later stage and could be responsible for cellular differentiation [30]. The prevention of restenosis with valsartan as seen in the small clinical trial VAL-PREST may be due to the AT_2 receptor acting as a mediator of the beneficial effect of AT_1 receptor blockade [105].

Using a microdialysis technique, renal and interstitial fluid content of bradykinin (BK), the NO end products nitrite and nitrate and cGMP significantly increased following AT_1 receptor blockade with valsartan in sodium depleted conscious rats. AT_2 receptor blockade with PD inhibits this response and thus confirms the role of this receptor. Moreover, blockade of the response after icataban treatment demonstrated the involvement of bradykinin in the increase of renal NO and cGMP [106]. Combination of an angiotensin receptor blocker with an ACE inhibitor to protect bradykinin from degradation leads to potentiation of the BK and cGMP levels in renal interstitial fluid. Again, this effect was blocked by the AT_2 receptor antagonist PD123319 [107].

Thus, interstitial generation of BK modulates NO production. It leads to the activation of cGMP, which is the final link in the biochemical cascade set in motion by AT_2 receptor stimulation following AT_1 receptor blockade. Over expression of the AT_2 receptor in SMC of transgenic rats provided additional evidence of a linkage between the AT_2 receptor and the BK/NO cascade. These mice showed a decreased sensitivity to the Ang II pressure effect, which was restored after AT_2 receptor or BK type2 receptor blockade or NO synthase inhibition with an arginine analogue N^{ω}-nitro-L-arginine methyl ester (L-NAME). The vasodilatory effect of the AT_2 receptor was associated with an increased cGMP production and activation of the kallikrein-kinin system [95].

NO down-regulates the AT_1 receptor expression [68]. It is therefore possible that increased NO production following AT_2 receptor stimulation may decrease the AT_1 receptor responsiveness. Locally produced Ang II produced a flow-induced vasodilation through endothelial AT_2 receptors and NO production in rat mesenteric resistance arteries perfused in situ. Selective blockade of the AT_2 receptor with PD123319 decreased significantly the vessel diameter. This effect was absent after NO synthesis blockade or after endothelial disruption [58].

Barber et al. [108] also reported that the AT_2 agonist CGP 42112 potentiated the blood pressure lowering effect of candesartan in SHR. PD123319 abolished this effect. Similarly, Carey et al. [109] reported that chronic infusion of Ang II in AT_1 receptor blocked conscious Sprague Dawley rats produced a significant AT_2 receptor-mediated fall in blood pressure. This effect was increased by CGP42112, reversed by PD and inhibited by L-NAME, indicating the involvement of NO in the vasodilatory pathway of the AT_2 receptor.

There are however reports conflicting with this general view of a beneficial effect of the AT_2 receptor stimulation.

There was no evidence of an AT_2 receptor-mediated vasoactive response in normotensive rats measured with radioactive microspheres [110]. However, these rats were anaesthetized, PD

123319 was infused at a lower dose and for a shorter period than in Carey's study [109]. Divergent vasoactive responses to AT_2 receptors occur in different vascular beds as proposed by Schuijt et al. [111] who observed that AT_2 receptor stimulation results in vasodilation in the coronary but not in the renal vasculature. Furthermore, this effect was enhanced in rats with myocardial infarction. Such a distinction between vascular beds could be related to differences in Ang II receptor expression after NOS blockade. Both AT_1 and AT_2 receptors were increased in the heart early after L-NAME administration [112] whereas the receptor density was not affected by NOS blockade in the renal cortex but significantly decreased in the medulla [113]. Both receptors in the cortex were however significantly decreased after AT_1 receptor blockade in this model [113]. Moreover, the observation that candesartan and enalapril alone or in combination exerted similar kidney protection after L-NAME treatment suggested that the AT_2 receptor did not appear to be an absolute requirement [114]. The ACE inhibitor enalapril reduced serotonin endothelium-mediated endothelium-dependent vasodilation whereas blockade of the AT_1 receptor with valsartan did not [115]. Also, ceramide, a possible link in the signaling pathway of the AT_2 receptor [116,117], has no effect on NOS activity. It even decreased NO and increased superoxide resulting in inhibition of BK-induced vasorelaxation [118]. Finaly, Ang II stimulated inflammatory reaction in activating NF-κB in vascular SMC through both AT_1 and AT_2 receptors [119] and proximal tubular cell proliferation and apoptosis observed after Ang II infusion in rats was blocked by both AT_1 and AT_2 receptor antagonist [120].

Thus, although cell culture and animal studies have clearly demonstrated a beneficial effect of the AT_2 receptor stimulation, its role in pathophysiology remains puzzling. Further work is needed to clarify the potential benefit of the AT_2 receptor.

Involvement of Other Components of the RAS Cascade

Metabolites of angiotensin are also involved in endothelial function. Ang 1–7 and Ang IV (Ang 3–8), alter endothelial function by inducing NO-mediated vasodilation although specific binding sites for these peptides have not yet been clearly defined [121]. Interestingly, AT_1 receptor blockade inhibited Ang 1–7 induced NO release from primary bovine aortic endothelial cells by 60% whereas PD 123319 decreased it by 90%. Icatabant totally abolished the effect of Ang 1–7 [122]. In contrast, both angiotensin AT_1 (L158809) and AT_2 receptor (PD123319) antagonists had no effect on Ang (1–7) induced dilatation of rabbit afferent arterioles microperfused *in vitro* [123].

Similarly, pretreatment with L-NAME blocked the vasodilatory property of Ang IV [124] but increased NOS activity and cGMP content in endothelial cells [125]. Aldosterone, the last step in the RAS cascade, reduced whereas its blockade increased endothelial NO bioactivity [126].

Conclusion

There is a clear interaction between Ang II and NO and an imbalance between them play a critical role in cardiovascular pathology. Such an imbalance is well established in hypertension, renal insufficiency, dyslipidemia, atherosclerosis, diabetes and insulin resistance as well as in the aging process [44,127,128]. In atherosclerosis, Ang II facilitates the recruitment of monocytes/macrophages in the vessel wall by activation of membrane bound NADPH oxidase and promotion of superoxide radicals. Decreased vascular NO bioactivity promotes abnormal end-organ remodeling as Ang II-induced hypertrophy appears to be mediated by intracellularly produced H_2O_2.

Angiotensin AT_1 and AT_2 receptors are ideal candidates for maintaining a proper balance between the vasodilator agent NO and the scavenger $O_2^{\cdot}$ and other ROS. It is obvious from experimental data that Ang II acting through the AT_1 receptor stimulates a membrane bound NADPH oxidase that causes the accumulation of superoxide, hydrogen peroxide and peroxinitrite. Superoxide scavenges NO and blocks its beneficial properties. As a consequence, there is a stimulation of various MAP kinases, which promote growth and proliferation, as observed in cardiac and vessel hypertrophy and remodeling. Conversely, Ang II stimulates the AT_2 receptor, which results in increased interstitial levels of bradykinin, NO and prostacyclin. There is a regulatory cycle (Fig. 3)

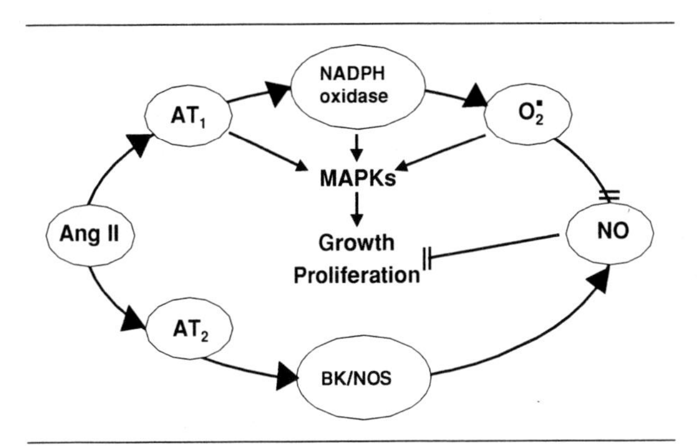

Fig. 3. *Interaction between Ang II and nitric oxide: The AT_1 and AT_2 receptors have opposite effects.*

that involves both the AT_1 and AT_2 receptor. It is therefore proposed that in pathological states, the stimulation of the AT_1 receptor by increased circulating or tissue levels of Ang II, will stimulate growth and proliferation, affect vascular remodeling and produce an inflammatory response. In contrast, blockade of the AT_1 receptor, which is accompanied by increased circulating Ang II levels, will stimulate the AT_2 receptor and oppose the effect of AT_1 receptor activation. This mechanism appears to be involved in the beneficial effects of the angiotensin receptor blockers and deserves further work.

Acknowledgment

The help of Gillian R. Bullock, PhD, and Randy Webb, PhD, in preparing this review was greatly appreciated.

References

1. Hilgers KF, Veelken R, Muller DN, Kohler H, Hartner A, Botkin SR, Stumpf C, Schmieder RE, Gomez RA. Renin uptake by the endothelium mediates vascular angiotensin formation. *Hypertension* 2001;38:243–248.
2. Rubanyi GM, Vanhoutte PM. Superoxide anions and hyperoxia inactivate endothelium-derived relaxing factor. *Am J Physiol* 1986;250:H822–H827.
3. Taddei S, Virdis A, Mattei P, Arzilli F, Salvetti A. Endothelium-dependent forearm vasodilation is reduced in normotensive subjects with familial history of hypertension. *J Card Vasc Pharmacol* 1999;20(Suppl 12):S193–S195.
4. Gaeta G, De Michele M, Cuomo S, Guarini P, Foglia MC, Bond MG, Trevisan M. Arterial abnormalities in the offspring of patients with premature myocardial infarction. *N Engl J Med* 2000;343:840–846.
5. McIntyre M, Bohr DF, Dominiczak AF. Endothelial function in hypertension: The role of superoxide anion. *Hypertension* 1999;34:539–545.
6. Griendling KK, Sorescu D, Ushio-Fukai M. NAD(P)H Oxidase: Role in cardiovascular biology and disease. *Circ Res* 2000;86:494–501.
7. Cai H, Harrison DG. Endothelial dysfunction in cardiovascular diseases: The role of oxidant stress. *Circ Res* 2000;87:840–844.
8. Ruiz-Ortega M, Lorenzo O, Rupérez M, Esteban V, Suzuki Y, Mazzano S, Plaza JJ, Egido J. Role of the renin-angiotensin system in vascular diseases. Expanding the field. *Hypertension* 2001;38:1382–1387.
9. Griendling KK, Ushio-Fukai M. Reactive oxygen species as mediators of angiotensin II signaling. *Regul Pept* 2000;91:21–27.
10. Dzau VJ. Tissue angiotensin and pathobiology of vascular disease. A unifying hypothesis. *Hypertension* 2001;37:1047–1052.
11. John S, Schmieder RE. Impaired endothelial function in arterial hypertension and hypercholesterolemia: Potential mechanisms and differences. *J Hypertens* 2000;18:363–374.
12. Bredt DS, Hwang PM, Glatt CE, Lowenstein C, Redd RR, Snyder SH. Cloned and expressed nitric oxide synthase structurally resembles cytochrome P-450 reductase. *Nature* 1991;351:714–718.
13. Drexler H, Kastner S, Strobel A, Studer R, Brodde OE, Hasenfuss G. Expression, activity and functional significance of inducible nitric oxide synthase in the failing human heart. *J Am Coll Cardiol* 1998;32:955–963.
14. Feng Q, Lu X, Jones DL, Shen J, Arnold JMO. Increased inducible nitric oxide synthase expression contributes to myocardial dysfunction and higher mortality after myocardial infarction in mice. *Circulation* 2001;104:700–704.
15. Chyu KY, Dimayuga P, Zhu J, Nilsson J, Kaul S, Shah PK, Cercek B. Decreased neointimal thickening after arterial wall injury in inducible nitric oxide synthase knockout mice. *Circ Res* 1999;85:1192–1198.
16. Sausbier M, Schubert R, Voigt V, Hirneiss C, Pfeifer A, Korth M, Kleppisch T, Ruth P, Hofmann F. Mechanism of NO/cGMP-dependent vasorelaxation. *Circ Res* 2000;87:825–830.
17. Adachi T, Matsui R, Weisbrod RM, Najibi S, Cohen RA. Reduced sarco/endoplasmic reticulum Ca^{2+} uptake activity can account for the reduced response to NO, but not sodium nitroprusside, in hypercholesterolemic rabbit aorta. *Circulation* 2001;104:1040–1045.
18. Wollert KC, Fiedler B, Gambaryan S, Smolenski A, Heineke J, Butt E, Trautwein C, Lohmann SM, Drexler H. Gene transfer of CGMP-dependent protein kinase I enhances the antihypertrophic effects of nitric oxide in cardiomyocytes. *Hypertension* 2002;39:87–92.
19. Kinlay S, Creager MA, Fukumoto M, Hikita H, Fang JC, Selwyn AP, Ganz P. Endothelium-derived nitric oxide regulates arterial elasticity in human arteries *in vivo*. *Hypertension* 2001;38:1049–1053.
20. Wilkinson IB, Qasem A, McEniery CM, Webb DJ, Avolio AP, Cockcroft JR. Nitric oxide regulates local arterial distensibility *in vivo*. *Circulation* 2002;105:213–217.
21. MacCarthy PA, Grieve DJ, Li JM, Dunster C, Kelly FJ, Shah AM. Impaired endothelial regulation of ventricular relaxation in cardiac hypertrophy: Role of reactive oxygen species and NADPH oxidase. *Circulation* 2001;104:2967–2974.
22. Zalba G, Beaumont FJ, San Jose G, Fortuno A, Fortuno MA, Diez J. Is the balance between nitric oxide and superoxide altered in spontaneously hypertensive rats with endothelial dysfunction? *Nephrol Dial Transplant* 2001;16(Suppl 1):2–5.
23. Wiemer G, Itter G, Malinski T, Linz W. Decreased nitric oxide availability in normotensive and hypertensive rats with failing hearts after myocardial infarction. *Hypertension* 2001;38:1367–1371.
24. Sventek P, Li JS, Grove K, Deschepper CE, Schiffrin EL. Vascular structure and expression of endothelin-1 gene in L-NAME treated spontaneously hypertensive rats. *Hypertension* 1996;27:49–55.
25. Takemoto M, Egashira K, Tomita H, Usui M, Okamoto H, Kitabatake A, Shimokawa H, Sueishi K, Takeshita A. Chronic angiotensin-converting enzyme inhibition and angiotensin II type 1 receptor blockade: Effects on cardiovascular remodeling in rats induced by the long-term blockade of nitric oxide synthesis. *Hypertension* 1997;30:1621–1627.

26. Tomita H, Egashira K, Ohara Y, Takemoto M, Koyanagi M, Katoh M, Yamamoto H, Tamaki K, Shimokawa H, Takeshita A. Early induction of transforming growth factor-β via angiotensin II type 1 receptors contributes to cardiac fibrosis induced by long-term blockade of nitric oxide synthesis in rats. *Hypertension* 1988;32:273–279.
27. Kashiwagi M, Shinozaki M, Hirakata H, Tamaki K, Hirano T, Tokumoto M, Goto H, Okuda S, Fujishima M. Locally activated renin-angiotensin system associated with TGF-beta 1 as a major factor for renal injury induced by chronic inhibition of nitric oxide synthase in rats. *J Am Soc Nephrol* 2000;11:616–624.
28. Koyanagi M, Egashira K, Kitamoto S, Ni W, Shomokawa H, Takeya M, Yoshimura T, Takeshita A. Role of monocyte chemoattractant protein-1 in cardiovascular remodeling induced by chronic blockade of nitric oxide synthesis. *Circulation* 2000;102:2243–2248.
29. Usui M, Egashira K, Tomita H, Koyanagi M, Katoh M, Shimokawa H, Takeya M, Yoshimura T, Matsushima K, Takeshita A. Important role of local angiotensin II activity mediated via type 1 receptor in pathogenesis of cardiovascular inflammatory changes induced by chronic blockade of nitric oxide synthesis in rats. *Circulation* 2000;101:305–310.
30. Szöcs K, Lassègue B, Sorescu D, Hilenski LL, Valppu L, Couse TL, Wilcox JN, Quinn MT, Lambeth JD, Griendling KK. Upregulation of NO_x based NAD(P)H oxidases in restenosis after carotid injury. *Arterioscler Thromb Vasc Biol* 2002;22:21.
31. Poderoso JJ, Carreras MC, Lisdero C, Riobo N, Schopfer F, Boveris. A nitric oxide inhibits electron transfer and increases superoxide radical production in rat heart mitochondria and submitochondrial particles. *Arch Biochem Biophys* 1996;328:85–92.
32. Suzuki H, Swei A, Zweifach BW, Schmid-Schonbein GW. *In vivo* evidence for microvascular oxidative stress in spontaneously hypertensive rats. Hydroethidine microfluorography. *Hypertension* 1995;25:1083–1089.
33. Wu L, deChamplain J. Effects of superoxide on signaling pathways in smooth muscle cells from rats. *Hypertension* 1999;34:1247–1253.
34. Laursen JB, Rajagopalan S, Galis Z, Tarpey M, Freeman BA, Harrison DG. Role of superoxide in angiotensin II induced but not catecholamine-induced hypertension. *Circulation* 1997;95:588–593.
35. Griendling KK, Minieri CA, Ollerenshaw JD, Alexander RW. Angiotensin II stimulates NADH and NADPH oxidase activity in cultured vascular smooth muscle cells. *Circ Res* 1994;74:1141–1148.
36. Baas AS, Berk BC. Differential activation of mitogen-activated protein kinases by H_2O_2 and O_2- in vascular smooth muscle cells. *Circ Res* 1995;77:29–36.
37. Touyz RM, Schiffrin EL. Ang II-stimulated superoxide production is mediated via phospholipase D in human vascular smooth muscle cells. *Hypertension* 1999;34:976–982.
38. Beckman JA, Goldfine AB, Gordon MB, Garrett LA, Creager MA. Inhibition of protein kinase Cβ prevents impaired endothelium-dependent vasodilation caused by hyperglycemia in humans. *Circ Res* 2002;90:107–111.
39. Zalba G, Jose GS, Moreno MU, Fortuno MA, Fortuno A, Beaumont FJ, Diez J. Oxidative stress in arterial hypertension: Role of NAD(P)H oxidase. *Hypertension* 2001;38:1395–1399.
40. Eguchi S, Numaguchi K, Iwasaki H, Matsumoto T, Yamakawa T, Utsunomya H, Motley ED, Kawakatsu H, Owada KM, Hirata Y, Marumo F, Inagami T. Calcium-dependent epidermal growth factor receptor transactivation mediates the angiotensin II-induced mitogen-activated protein kinase activation in vascular smooth muscle cells. *J Biol Chem* 1998;273:8890–8896.
41. Murasawa S, Mori Y, Nozawa Y, Gotoh N, Shibuya M, Masaki H, Maruyama K, Tsutsumi Y, Moriguchi Y, Shibazaki Y, Tanaka Y, Iwasaka T, Inada M, Matsubara H. Angiotensin II type 1 receptor-induced extracellular signal-regulated protein kinase activation is mediated by Ca^{2+}/calmodulin-dependent transactivation of epidermal growth factor receptor. *Circ Res* 1998;82:1338–1348.
42. Ushio-Fukai M, Griendling KK, Becker PL, Hilenski L, Halleran S, Alexander RW. Epidermal growth factor receptor transactivation by angiotensin II requires reactive oxygen species in vascular smooth muscle cells. *Arterioscler Thromb Vasc Biol* 2001;21:489–495.
43. Heeneman S, Haendeler J, Saito Y, Ishida M, Berk BC. Angiotensin II induces transactivation of two different populations of the platelet-derived growth factor beta receptor. Key role for the p66 adaptor protein Shc. *J Biol Chem* 2000;275:15926–15932.
44. Du J, Sperling LS, Marrero MB, Phillips L, Delafontaine P. G-protein and tyrosine kinase receptor cross-talk in rat aortic smooth muscle cells: Thrombin- and angiotensin II-induced tyrosine phosphorylation of insulin receptor substrate-1 and insulin-like growth factor 1 receptor. *Biochem Biophys Res Commun* 1996;218:934–939.
45. Rajagopalan S, Kurz S, Münzel T, Tarpey M, Freeman BA, Griendling KK, Harrison DG. Angiotensin II-mediated hypertension in the rat increases vascular superoxide production via membrane NADH/NADPH oxidase activation. *J Clin Invest* 1996;97:1916–1923.
46. Fukui T, Ishizaka N, Rajagopalan S, Laursen JB, Capers Q, Taylor WR, Harrison DG, De Leon H, Wilcox JN, Griendling KK. p22phox mRNA expression and nadph oxidase activity are increased in aortas from hypertensive rats. *Circ Res* 1997;80:45–51.
47. Ushio-Fukai M, Zafari AM, Fukui T, Ishizaka N, Griendling KK. p22phox is a critical component of the superoxide-generating NADH/NADPH oxidase system and regulates angiotensin iiinduced hypertrophy in vascular smooth muscle cells. *J Biol Chem* 1996;271:23317–23321.
48. Ushio-Fukai M, Alexander RW, Akers M, Griendling KK. p38 Mitogen-activated protein kinase is a critical component of the redox-sensitive signaling pathways activated by angiotensin II. Role in vascular smooth muscle cell hypertrophy. *J Biol Chem* 1998;273:15022–15029.
49. Finta KM, Fischer MJ, Lee L, Gordon D, Pitt B, Webb RC. Ramipril prevents impaired endothelium-dependent relaxation in arteries from rabbits fed an atherogenic diet. *Atherosclerosis* 1993;100:149–156.
50. Munzel T, Keaney JF. Are ACE inhibitors a "magic bullet" against oxidative stress? *Circulation* 2001;104:1571–1574.
51. Hornig B, Landmesser U, Kohler C, Ahlersmann D, Spiekermann S, Christoph A, Tatge H, Drexler H. Comparative effect of ACE inhibition and angiotensin II type 1 receptor antagonism on bioavailability of nitric oxide in

patients with coronary artery disease. Role of superoxide dismutase. *Circulation* 2001;103:799–805.

52. Sadoshima J, Xu Y, Slayter HS, Izumo S. Autocrine release of angiotensin II mediates stretch-induced hypertrophy of cardiac myocytes *in vitro*. *Cell* 1993;75:977–984.
53. Marklund SL. Expression of extracellular superoxide dismutase by human cell lines. *Biochem J* 1990;266:213–219.
54. Pueyo ME, Arnal JF, Rami J, Michel JB. Angiotensin II stimulates the production of NO and peroxynitrite in endothelial cells. *Am J Physiol Cell Physiol* 1998;274:C214–C220.
55. Zafari AM, Ushio-Fukai M, Akers M, Yin Q, Shah A, Harrison DG, Taylor WR, Griendling KK. Role of NADH/NADPH oxidase-derived H_2O_2 in angiotensin II-induced vascular hypertrophy. *Hypertension* 1998;32:488–495.
56. Matoba T, Shimokawa H, Kubota H, Morikawa K, Jujiki T, Kunihiro I, Mukai Y, Hirakawa Y, Takeshita A. Hydrogen peroxide is an endothelium-derived hyperpolarizing factor in human mesenteric arteries. *Biochem Biophys Res Commun* 2002;25:909–913.
57. Vanhoutte P. Endothelium-derived free radicals: For worse and for better. *J Clin Invest* 2001;107:23–25.
58. Matrougui K, Loufrani L, Heymes C, Levy BI, Henrion D. Activation of AT_2 receptors by endogenous angiotensin II is involved in flow-induced dilation in rat resistance arteries. *Hypertension* 1999;34:659–665.
59. Yamazaki T, Komuro I, Kudoh S, Zou Y, Shiojima I, Mizuno T, Takano H, Hiroi Y, Ueki K, Tobe K, Kadowaki T, Nagai R, Yazaki Y. Angiotensin II partly mediates mechanical stress-induced cardiac hypertrophy. *Circ Res* 1995;77:258–265.
60. Takagishi T, Murahashi N, Azagami S, Morimatsu M, Sasaguri Y. Effect of angiotensin II and thromboxane A2 on the production of matrix metalloproteinase Hy human aortic smooth muscle cells. *Biochem Mol Biol Int* 1995;35:265–273.
61. Vaughan DE, Lazos SA, Tong K. Angiotensin II regulates the expression of plasminogen activator inhibitor in cultured endothelial cells. A potential link between the renin-angiotensin and thrombosis. *J Clin Invest* 1995;95:995–1001.
62. Sironi L, Calvio AM, Arnaboldi L, Corsini A, Parolari A, de Gasparo M, Tremoli E, Mussoni L. Effect of valsartan on angiotensin II-induced plasminogen activator inhibitor-1 biosynthesis in arterial smooth muscle cells. *Hypertension* 2001;37:961–966.
63. Oubiña MP, de las Heras N, Vázquez-Pérez S, Cediel E, Sanz-Rosa D, Ruilope LM, Cachofeiro V, Lahera V. Valsartan improves fibrinolytic balance in atheroslerotic rabbits. *J Hypertens* 2000;20:303–310.
64. Tummala PE, Chen XL, Sundell CL, Laursen JB, Hammes CP, Alexander RW, Harrison DG, Medford RIM. Angiotensin II induces vascular cell adhesion molecule-1 expression in rat vasculature: A potential link between the renin-angiotensin system and atherosclerosis. *Circulation* 1999;100:1223–1229.
65. Mervaala EM, Muller DN, Park JK, Schmidt F, Lohn M, Breu V, Dragun D, Ganten D, Haller H, Luft FC. Monocyte infiltration and adhesion molecules in a rat model of high human renin hypertension. *Hypertension* 1999;33:389–395.
66. Rakugi H, Kim DK, Krieger JE, Wang DS, Dzau VJ, Pratt RE. Induction of angiotensin converting enzyme in the neointima after vascular injury. Possible role in restenosis. *J Clin Invest* 1994;93:339–346.
67. Diet F, Pratt RE, Berry GJ, Momose N, Gibbons GH, Dzau VJ. Increased accumulation of tissue ACE in human atherosclerotic coronary disease. *Circulation* 1996;94:2756–2767.
68. Ichiki T, Usui M, Kato M, Funakoshi Y, Ito K, Egashira K, Takeshita A. Downregulation of angiotensin II type 1 receptor gene transcription by nitric oxide. *Hypertension* 1998;31:342–348.
69. Usui M, Ichiki T, Katoh M, Egashira K, Takeshita A. Regulation of angiotensin II receptor expression by nitric oxide in rat adrenal gland. *Hypertension* 1998;32:527–533.
70. Warnholtz A, Nickenig G, Schulz E, Macharzina R, Brasen JH, Skatchkov M, Heitzer T, Stasch JP, Griendling KK, Harrison DG, Bohm M, Meinertz T, Munzel T. Increased NADH-oxidase-mediated superoxide production in the early stages of atherosclerosis: Evidence for involvement of the renin-angiotensin system. *Circulation* 1999;99:2027–2033.
71. Lemay J, Hou Y, deBlois D. Evidence that nitric oxide regulates AT_1 receptor agonist and antagonist efficacy in rat injured carotid artery. *J Cardiovasc Pharmacol* 2000;35:693–699.
72. Mancini GBJ, Henry GC, Macaya C, O'Neill BJ, Pucillo AL, Carere RG, Wargovich TJ, Mudra H, Lüscher TF, Klibaner MI, Haber HE, Uprichard ACG, Pepine CJ, Pitt B. Angiotensin-converting enzyme inhibition with quinapril improves endothelial vasomotor dysfunction in patients with coronary artery disease. The TREND (Trial on Reversing ENdothelial Dysfunction) study. *Circulation* 1996;94:258–265.
73. Svensson P, de Faire U, Sleight P, Yusuf S, Ostergren J. Comparative effects of ramipril on ambulatory and office blood pressures: A HOPE substudy. *Hypertension* 2001;38:E28–E32.
74. Schiffrin EL, Park JB, Intengan HD, Touyz RM. Correction of arterial structure and endothelial dysfunction in human essential hypertension by the angiotensin receptor antagonist losartan. *Circulation* 2000;101:1653–1659.
75. Schiffrin EL, Park JB, Pu Q. Effect of crossing over hypertensive patients from a beta-blocker to an angiotensin receptor antagonist on resistance artery structure and on endothelial function. *J Hypertens* 2002;20:71–78.
76. Boulanger CM, Caputo L, Levy BI. Endothelial AT_1-mediated release of nitric oxide decreases angiotensin II contractions in rat carotid artery. *Hypertension* 1995;26:752–757.
77. Dobrian AD, Schriver SD, Prewitt RL. Role of angiotensin II and free radicals in blood pressure regulation in a rat model of renal hypertension. *Hypertension* 2001;38:361–366.
78. Brosnan MJ, Hamilton CA, Graham D, Lygate CA, Jardine E, Dominiczak AF. Irbesartan lowers superoxide levels and increases nitric oxide bioavailability in blood vessels from spontaneously hypertensive stroke-prone rats. *J Hypertens* 2002;20:281–286.
79. Khaper N, Singal PK. Modulation of oxidative stress by a selective inhibition of angiotensin II type 1 receptors in MI rats. *J Am Coll Cardiol* 2001;37:1461–1466.

80. Khan BV, Navalkar S, Khan QA, Rahman ST, Parthasarathy S. Irbesartan, an angiotensin type 1 receptor inhibitor, regulates the vascular oxidative state in patients with coronary artery disease. *J Am Coll Cardiol* 2001;38:1662–1667.
81. Bunkenburg B, Schnell C, Baum HP, Cumin F, Wood JM. Prolonged angiotensin II antagonism in spontaneously hypertensive rats. Hemodynamic and biochemical consequences. *Hypertension* 1991;18:278–288.
82. Christen Y, Waeber B, Nussberger J, Borland RM, Lee RJ, Maggon K, Shum L, Timmermans PB, Brunner HR. Oral administration of Dup 753, a specific angiotensin II receptor antagonist in normal volunteers. Inhibition of pressore response to exogenous angiotensin I and II. *Circulation* 1991;83:1333–1342.
83. Mazzolai L, Maillard M, Rossat J, Nussberger J, Brunner HR, Burnier M. Angiotensin II receptor blockade in normotensive subjects: A direct comparison of three AT_1 receptor antagonists. *Hypertension* 1999;33:850–855.
84. Rogg H, de Gasparo M, Graedel E, Stulz P, Burkart F, Eberhard M, Erne P. Angiotensin II-receptor subtypes in human atria and evidence for alterations in patients with cardiac dysfunction. *Eur Heart J* 1996;17:1112–1120.
85. Tsutsumi Y, Matsubara H, Ohkubo N, Mori Y, Nozawa Y, Murasawa S, Kijima K Maruyama K, Masaki H, Moriguchi Y, Shibasaki Y, Kamiha H, Inada M, Iwasaka T. Angiotensin II type 2 receptor is upregulated in human heart with interstitial fibrosis and cardiac fibroblasts are the major cell type for its expression. *Circ Res* 1998;83:1035–1046.
86. Hutchinson HG, Hein L, Fujinaga M, Pratt RE. Modulation of vascular development and injury by angiotensin II. *Cardiovasc Res* 1999;41:689–700.
87. Ozono R, Wang ZQ, Moore AF, Inagami T, Siragy HM, Carey RM. Expression of the subtype 2 angiotensin (AT_2) receptor protein in rat kidney. *Hypertension* 1997;30:1238–1246.
88. de Gasparo M, Siragy HM. The AT_2 receptor: Fact, fancy and fantasy. *Regul Pept* 1999;81:11–24.
89. Masaki H, Kurihara T, Yamaki A, Inomata N, Nozawa Y, Mori Y, Murasawa S, Kizima K, Maruyama K, Horiuchi M, Dzau VJ, Takahashi H, Iwasaka T, Inada M, Matsubara H. Cardiac-specific overexpression of angiotensin II AT_2 receptor causes attenuated response to AT_1 receptor-mediated pressor and chronotropic effects. *J Clin Invest* 1998;101:527–535.
90. Carey RM, Wang ZQ, Siragy HM. Update: Role of the angiotensin type-2 (AT(2)) receptor in blood pressure regulation. *Curr Hypertens Rep* 2000;2:198–201.
91. Horiuchi M, Akishita M, Dzau VJ. Recent progress in angiotensin II type 2 receptor research in the cardiovascular system. *Hypertension* 1999;33:613–621.
92. Siragy HM, Carey RM. The subtype-2 (AT_2) angiotensin receptor regulates renal cyclic guanosine 3′, 5′-monophosphate and AT_1 receptor-mediated prostaglandin E2 production in conscious rats. *J Clin Invest* 1996;97:1978–1982.
93. Siragy HM, Jaffa AA, Margolius HS, Carey RM. Renin-angiotensin system modulates renal bradykinin production. *Am J Physiol* 1996;271:R1090–R1095.
94. Siragy HM, Carey RM. The subtype 2 (AT_2) angiotensin receptor mediates renal production of nitric oxide in conscious rats. *J Clin Invest* 1997;100:264–269.
95. Tsutsumi Y, Matsubara H, Masaki H, Kurihara H, Murasawa S, Takai S, Miyazaki M, Nozawa Y, Ozono R, Nakagawa K, Miwa T, Kawada N, Mori Y, Shibasaki Y, Tanaka Y, Fujiama S, Koyama Y, Fujiyama A, Takahashi H, Iwasaka T. Angiotensin II type 2 receptor overexpression activates the vascular kinin system and causes vasodilation. *J Clin Invest* 1999;104:925–935.
96. Yang XP, Liu YH, Mehta D, Cavasin MA, Shesely E, Xu J, Liu F, Carretero OA. Diminished cardioprotective response to inhibition of angiotensin-converting enzyme and angiotensin II type 1 receptor in B(2) kinin receptor gene knockout mice. *Circ Res* 2001;88:1072–1079.
97. Guan H, Cachofeiro V, Pucci ML, Kaminski PM, Wolin MS, Nasjletti A. Nitric oxide and the depressor response to angiotensin blockade in hypertension. *Hypertension* 1996;27:19–24.
98. Liu YH, Yang XP, Sharov VG, Nass O, Sabbah HN, Peterson E, Carretero OA. Effects of angiotensin-converting enzyme inhibitors and angiotensin II type 1 receptor antagonists in rats with heart failure. Role of kinins and angiotensin II type 2 receptors. *J Clin Invest* 1997;99:1926–1935.
99. Gohlke P, Pees C, Unger T. AT_2 receptor stimulation increases aortic cyclic GMP in SHRSP by a kinin-dependent mechanism. *Hypertension* 1998;81:349–355.
100. Wu L, Iwai M, Nakagami H, Chen R, Suzuki J, Akishita M, de Gasparo M, Horiuchi M. Effect of angiotensin II Type 1 receptor blockade on cardiac remodeling in angiotensin ii type 2 receptor null mice. *Arterioscler Thromb Vasc Biol* 2002;22:49–54.
101. Sohn HY, Raff U, Hoffmann A, Gloe T, Heermeier K, Galle J, Pohl U. Differential role of angiotensin II receptor subtypes on endothelial superoxide formation. *Br J Pharmacol* 2000;131:667–672.
102. Nouet S, Nahmias C. Signal transduction from the angiotensin II AT_2 receptor. *Trends Endocrinol Metab* 2000;11:1–6.
103. Sartore S, Chiavegato A, Faggin E, Franch R, Puato M, Ausoni S, Pauletto P. Contribution of adventitial fibroblasts to neointima formation and vascular remodeling: From innocent bystander to active participant. *Circ Res* 2001;89:1111–1121.
104. Scheidegger KJ, Wood JM. Local application of angiotensin II to the rat carotid artery induces adventitial thickening. *J Vasc Res* 1997;34:436–446.
105. Peters S, Gotting B, Trummel M, Rust H, Brattstrom A. Valsartan for prevention of restenosis after stenting of type B2/C lesions: The VAL-PREST trial. *J Invasive Cardiol* 2001;13:93–97.
106. Siragy HM, de Gasparo M, Carey RM. Angiotensin type 2 receptor mediates valsartan-induced hypotension in conscious rats. *Hypertension* 2000;35:1074–1077.
107. Siragy HM, de Gasparo M, El Kersh M, Carey RM. Angiotensin-converting enzyme inhibition potentiates angiotensin II type 1 receptor effects on renal bradykinin and cGMP. *Hypertension* 2001;38:183–186.
108. Barber MN, Sampey DB, Widdop RE. AT_2 receptor stimulation enhances antihypertensive effects of AT1 receptor antagonist in hypertensive rats. *Hypertension* 1999;34:1117–1122.
109. Carey RM, Howell NL, Jin XH, Siragy HM. Angiotensin type 2 receptor-mediated hypotension in angiotensin type-1 receptor-blocked rats. *Hypertension* 2001;38:1272–1277.

110. Schuijt MP, de Vries R, Saxena PR, Danser AH. No vasoactive role of the angiotensin II type 2 receptor in normotensive Wistar rats. *J Hypertens* 1999;17:1879–1884.
111. Schuijt MP, Basdew M, van Veghel R, de Vries R, Saxena PR, Schoemaker RG, Danser AH. AT(2) receptor-mediated vasodilation in the heart: Effect of myocardial infarction. *Am J Physiol Heart Circ Physiol* 2001;281:H2590–H2596.
112. Katoh M, Egashira K, Usui M, Shimokawa H, Rakugi H, Takeshita A. Cardiac angiotensin II receptors are upregulated by long-term inhibition of nitric oxide synthesis in rats. *Circ. Res* 1998;83:743–751.
113. Uhlenius N, Vuolteenabo O, Tikkanen I. Renin-angiotensin blockade improves rebal cGMP production via non-AT2-receptor mediated mechanisms in hypertension-induced by chronic NOS inhibition in rat. *JRAAS* 2001;2:233–239.
114. Nakamura Y, Ono H, Zhou X, Frohlich ED. Angiotensin type 1 receptor antagonism and ACE inhibition produce similar renoprotection in N-nitro-L-arginine methyl ester/spontaneously hypertensive rats. *Hypertension* 2001;37:1262–1267.
115. van Ampting JM, Hijmering ML, Beutler JJ, van Etten RE, Koomans HA, Rabelink TJ, Stroes ES. Vascular effects of ace inhibition independent of the renin-angiotensin system in hypertensive renovascular disease: A randomized, double-blind, crossover trial. *Hypertension* 2001;37:40–45.
116. Gallinat S, Busche S, Schütze S, Krönke M, Unger T. AT_2 receptor stimulation induces generation of ceramides in PC12W cells. *FEBS Lett* 1999;443:75–79.
117. Lehtonen JY, Horiuchi M, Daviet L, Akishita M, Dzau VJ. Activation of the *de novo* biosynthesis of sphyngolipids mediates angiotensin II type 2 receptor-induced apoptosis. *J Biol Chem* 1999;274:16901–16906.
118. Zhang DX, Zou AP, Li PL. Ceramide reduces endothelium-dependentvasodilation by increasing superoxide production in small bovine coronary arteries. *Circ Res* 2001;88:824.
119. Ruiz-Ortega M, Lorenzo O, Rupérez M, König S, Wittig B, Egido J. Angiotensin II activates nuclear transcription factor κB trough AT_1 and AT_2 in vascular smooth muscle cells. Molecular mechanisms. *Circ Res* 2000;86:1266–1272.
120. Cao Z, Kelly DJ, Cox A, Casley D, Forbes JM, Martinello P, Dean R, Gilbert RE, Cooper, ME. Angiotensin type 2 receptor is expressed in the adult rat kidney and promotes cellular proliferation and apoptosis. *Kidney Int* 2000;58:2437–2451.
121. de Gasparo M, Catt KJ, Inagami T, Wright JW, Unger T. International Union of Pharmacology. XXIII. The Angiotensin II receptors. *Pharmacol Rev* 2001;52:415–472.
122. Heitsch H, Brovkovych S, Malinski T, Wiemer G. Angiotensin-(1–7)-stimulated nitric oxide and superoxide release from endothelial cells. *Hypertension* 2001;37:72.
123. Ren YL, Garvin JL, Carretero OA. Vasodilator action of angiotensin-(1–7) on isolated rabbit afferent arterioles. *Hypertension* 2002;39:799–802.
124. Kramar EA, Harding JW, Wright JW. Angiotensin II- and IV-induced changes in cerebral blood flow. Roles of AT_1, AT_2, and AT_4 receptor subtypes. *Regul Pept* 1997;68:131–138.
125. Patel M, Martens JR, Li YD, Gelband CH, Raizada MK, Block ER. Angiotensin IV receptor-mediated activation of lung endothelial NOS is associated with vasorelaxation. *Am J Physiol* 1998;275:L1061–L1068.
126. Farquharson CAJ, Struthers AD. Spironolactone increases nitric oxide bioactivity, improves endothelial vasodilator dysfunction, and suppresses vascular angiotensin I/angiotensin II conversion in patients with chronic heart failure. *Circulation* 2000;101:594–597.
127. Raij L. Hypertension and cardiovascular risk factors. Role of the angiotensin II-nitric oxide interaction. *Hypertension* 2001;37:767–773.
128. Vapaatalo H, Mervaala E. Clinically important factors influencing endothelial function. *Med Sci Monit* 2001;7:1075–1085.

The Nitric Oxide-Endothelin-1 Connection

David Alonso and Marek W. Radomski
Department of Integrative Biology and Pharmacology, University of Texas-Houston, USA

Abstract. **Nitric oxide (NO) and endothelin-1 (ET-1) are endothelium-derived mediators that play important roles in vascular homeostasis. This review is focused on the role and reciprocal interactions between NO and ET-1 in health and diseases associated with endothelium dysfunction. We will also discuss the clinical significance of NO donors and drugs that antagonize ET receptors.**

Key Words. **nitric oxide, endothelin, vascular tone, cardiovascular system, receptor antagonist, heart failure**

Introduction

In addition to being the barrier between the circulating blood and the surrounding tissue, the endothelium plays a crucial role in regulation of vascular tone and cardiovascular homeostasis. The endothelium modulates vascular tone by releasing endothelium-derived vasodilators, including nitric oxide (NO), prostacyclin, bradykinin, and endothelium-derived hyperpolarising factor, and vasoconstrictors, such as endothelin-1 (ET-1) and angiotensin II, in response to a number of biochemical and physical stimuli.

The vasodilator and platelet-regulatory functions of endothelium are impaired during the course of vascular disorders including atherosclerosis, coronary artery disease, essential hypertension, diabetes mellitus and preeclampsia [1]. Nitric oxide and ET-1 are the major vasodilator and vasoconstrictor endothelium-derived substances, respectively, and their interactions in the cardiovascular system have been extensively studied. Recent studies have suggested that an imbalance between NO and ET-1 may contribute to changes in vascular tone observed in these disease states. Indeed, a number of vasculopathies associated with an impaired bioavailability of NO have been found to be linked to enhanced synthesis of ET-1, suggesting a crosstalk between these mediators.

The aim of this article is to review the actions of the NO and ET-1 in cardiovascular physiology and during heart failure. We will also attempt to discuss the current status of therapeutic development of ET-1 receptor antagonists.

Nitric Oxide in the Cardiovascular System

Nitric oxide is a gaseous biological mediator that accounts for the vasodilator activity of endothelium-derived relaxing factor (EDRF) [2,3], a non-prostaglandin vasorelaxant substance first described in the endothelial cells by Furchgott and Zawadzki [4]. Nitric oxide is generated from the guanidino-nitrogen of L-arginine yielding citrulline (for review see [5]), and plays a prominent role in controlling a variety of functions in the cardiovascular, immune, reproductive, and nervous systems [6–8].

Nitric oxide production is catalyzed by three major isoforms of NO synthase (NOS), neuronal (nNOS), inducible (iNOS), and endothelial (eNOS) enzymes [9–11]. The nNOS and eNOS are considered to be constitutively expressed and activated by calcium entry into cells [12,13], whereas iNOS is calcium-independent, and its synthesis is induced in inflammatory and other cell types by stimuli such as endotoxin and proinflammatory cytokines [14,15]. Although cDNAs for the respective proteins are found almost in all mammalian cells, under physiological conditions, eNOS is the major NOS isoform expressed in the endothelial cells [16]. In contrast, inflammation and cell damage are often associated with the expression of iNOS [16]. In the heart, both, eNOS and iNOS, have been involved in signalling pathways that modulate the contractile properties of cardiac myocytes. The eNOS isoform is expressed within the heart in the endothelium both of the endocardium and of the coronary vasculature, in cardiac myocytes, and in specialized cardiac conduction tissue and its activity seems to be regulated by the contractile state of the heart [17,18]. In contrast, iNOS expression is induced by cytokines in cardiac myocytes, endocardial endothelium, infiltrating inflammatory cells, vascular smooth muscle, fibroblast, and microvascular endothelium [18–20].

Address for correspondence: Dr. Marek W. Radomski, University of Texas-Houston, 2121 W. Holcombe Blvd., Houston TX, 77030. E-mail: MAREK.RADOMSKI@UTH.TMC.EDU

Nitric oxide is generated and released from the endothelial cells both under basal and agonist-stimulated conditions. Shear stress and pulsatile flow are major stimuli that cause release of NO under basal conditions [21]. In the cardiovascular system, NO not only causes vessel relaxation, but also inhibits platelet adhesion and aggregation, smooth muscle cell proliferation, monocyte adhesion, expression of different adhesion molecules and ET-1 production [22–25,26]. The effects of NO on myocardial functions are still a matter of extensive investigation. There is now evidence showing that the basal endogenous NO production supports myocardial contractility and heart rate, whereas the expression of iNOS has been reported to have cardiodepressive actions because of the negative inotropic effects of NO at high concentrations (for review [27]).

Endothelins

In the 1980s, a novel endothelial-derived vasoconstrictor was characterized from bovine aortic endothelial cells in culture by its ability to cause sustained, concentration-dependent isometric and endothelium-independent constriction of isolated vascular rings from various species [28,29]. The subsequent isolation, cloning and sequencing of this vasoconstrictor resulted in characterization of a 21-amino-acid vasoconstrictor peptide, that was named endothelin (ET) [30]. Current evidence shows that ET interacts with a number of biological systems including the L-arginine/nitric oxide, renin/angiotensin and sympathetic nervous systems. Therefore, the endothelin system has been shown to play roles, in diverse biological functions such as regulation of vascular tone, sodium balance, neural crest cell development, and neurotransmission (for review see [31]).

Endothelin System

Endothelins are a family of three potent vasoactive peptides that are produced from biologically inactive intermediates, termed big endothelins (big ETs), via a proteolytic processing at Trp-Val/Ile by endothelin-converting enzymes (ECEs) [30]. Three separate genes encode for three structurally and pharmacologically distinct isopeptides, preproendothelin-1 (ppET-1), preproendothelin-2 (ppET-2) and preproendothelin-3 (ppET-3) [32]. The production of the mature active forms, ETs, is the result of a two-step process consisting of an initial proteolytic cleavage at dibasic sites of the proendothelins by furin-like endopeptidases to form the biologically inactive 37- to 41-amino-acid peptides big ETs [33]. Next, big ETs intermediates are further processed to the mature 21-amino-acid active ETs by ECEs, a family of membrane-bound zinc metalloproteases from the thermolysin subfamily of proteases. The conversion of big ET-1 to ET-1 is essential for biological activity [34] and therefore, the administration of phosphoramidon, a metalloproteinase inhibitor, suppresses the hypertensive effect of big endothelin-1 [35]. For these reasons, ECEs represent an important regulatory system in the biosynthesis of these potent vasoconstrictors.

To date, two family members involved in big-ETs processing, ECE-1 and ECE-2, have been identified [36–38]. ECE-1 is found in a variety of cells; however, the vascular endothelium constitutes the major site of constitutive EC-1 expression [39]. EC-1 is fully active at neutral pH, which makes it suitable for intracellular and cell surface processing of big ETs [37]. Recently, four different isoforms of ECE-1 (termed ECE-1a, ECE-1b, ECE-1c and ECE-1d), with identical efficiency but different subcellular localization, have been cloned from a single gene through the use of alternative promoters [40,41]. On the other hand, ECE-2 is an intracellular-processing enzyme, with an optimal activity at pH 5.8, found in diverse cell types including neurons. Both ECE-1 and ECE-2 process big endothelin-1 more efficiently than either big endothelin-2 or big endothelin-3. In addition to these two ECEs, other ECE-independent pathways have been described to process big-ETs producing different vasoactive peptides. In this respect, we have recently reported that vascular matrix metalloproteinase-2 and matrix metalloproteinase-9 can cleave the Gly32-Leu33 bond of big ET-1 yielding the novel vasoconstrictor peptide, ET-1 [1–32,42,43].

The regulation of the endothelin system takes place at the transcription levels of both the endothelins and the receptors. The ppET gene promoter contains binding sites for nuclear factor-1, activator protein-1 (AP-1), and GATA-2 and is regulated by hormones and physical factors [44–46].

Once ETs are produced, they exert their actions in a paracrine-autocrine fashion by interacting with two different G-protein-coupled receptors, i.e. ET_A and ET_B, to induce different physiological responses [47,48]. Therefore, ET-1 acts through smooth muscle ET_A and ET_B receptors producing vasoconstriction, cell growth and cell adhesion. Conversely, the binding of ET-1 to endothelial ET_B receptors stimulates the release of NO and prostacyclin, prevents apoptosis, inhibits ECE-1 expression in endothelial cells and has an important role in ET-1 clearance. ET_A receptors are found, among others, in vessel and airway smooth muscle cells, cardiomyocytes, fibroblasts, hepatocyes and neurons, and present higher affinities for ET-1 and

ET-2 than for ET-3 [47]. On the other hand, ET_B receptors are mainly localized on endothelial cells and smooth muscle cells, but also in cardiomyocytes, fibroblasts hepatocyes, different epithelial cells, neurons, etc., and have equal subnanomolar affinities for all endothelin peptides [48]. Both ET_A and ET_B receptors activate different G proteins, leading to activation of phospholipase C which results in an accumulation of inositol triphosphate and diacylglycerol, and an increase of intracellular calcium [49]. First, there is a rapid increase of intracellular Ca^{2+}concentration from intracellular stores followed by a more sustained increase due to extracellular Ca^{2+} entry through different Ca^{2+} channels. Moreover, there is also activation of phospholipase A2 and D and protein kinase C [50,51].

Endothelins in the Cardiovascular System

The endothelin system participates in both physiology and pathology of the cardiovascular system [52], although it is important to note that the extent of this contribution is both species- and vascular bed-dependent. The endothelin system has a role in the regulation of basal vascular tone [53,54] and altered expression/activity of ET-1 could contribute to the development of diseases such as hypertension, atherosclerosis, and vasospasm after subarachnoid hemorrhage [52]. Of three endothelin isoforms only ET-1 is produced constitutively by endothelial cells and, therefore, it plays the most important role in regulating vascular function. Interestingly, under inflammatory conditions vascular smooth muscle cells have the capacity to generate ET-1 [55]. A number of stimuli, including thrombin, insulin, cyclosporine, epinephrine, angiotensin II, cortisol, inflammatory mediators, hypoxia and vascular shear stress have been shown to incresase ET-1 levels [44]. Endothelin-1 is secreted mainly on the basal side of the endothelial cells [56] to act on ET_A receptors on the underlying smooth muscles cells, as well as on ET_B receptors on endothelial and on some smooth muscle cells. Wright and Fozard showed that endothelin produced vasodilatation when administered to anaesthetized, spontaneously hypertensive rats [57]. This transient depression of blood pressure was explained via the production of NO and prostacyclin by the stimulation of ET_B receptors located in endothelial cells [58]. The subsequent vasoconstritive response is mediated by the action of ET-1 on ET_A and ET_B receptors on vascular smooth muscle cells [59].

ET-1 is also the main cardiac endothelin produced by cardiomyocytes, endothelial cells, and cardiac fibroblasts. It is interesting to note that ET_A receptors represent most of the endothelin receptors present on cardiomyocytes. The endothelin system in the heart seems to affect inotropy and chronotropy, but also mediate cardiac hypertrophy and remodeling in congestive heart failure through its mitogenic properties [60]. Figure 1 shows the actions of NO and ET-1 on the vascular wall.

Crosstalk Between NO and ET-1 in Heart Failure

Reduced cardiac function and endothelial dysfunction are either consequences or causal factors of heart failure, depending on the etiology of the disease, but in any case, there is no doubt about the strong correlation among them. The right balance between NO and ET-1 production seems to be crucial in maintaining cardiovascular homeostasis and preventing endothelial dysfunction [61].

Accordingly, an attenuated endothelium-dependent vasodilation in the peripheral [62], pulmonary, and coronary circulation have been reported in patients with chronic heart failure. However, the mechanisms involved in the effects of NO in heart failure are complex and not completely understood. In the myocardium, this impaired endothelium-dependent vasodilator response was found in the microcirculation [63], but also in large conduit vessels [64] and might result in ischemia and further myocardial damage and dysfunction. The eNOS downregulation in the vascular endothelium has been proposed to be involved in the alteration of endothelial function. Recent evidence indicates that in heart failure endothelial-derived NO could be inactivated by oxygen free radicals [65]. In addition, heart failure is associated with an increase in circulating cytokines, particularly tumor necrosis factor alpha (TNFα), and expression of inducible NO synthase in the myocardium. Evidence from both animal models and patients suggests that the induction of iNOS, resulting in a high NO production, exerts a negative inotropic effect in heart failure [27]. Furthermore, TNFα has been shown to decrease eNOS mRNA levels by increasing the rate of mRNA degradation [66]. Moreover, de Belder and colleagues reported a significant activity of inducible enzyme, accompanied by a low activity of the constitutive NO synthase in right ventricular tissue from patients with dilated cardiomyopathy [67].

The endothelium-derived NO impairment is usually accompanied by an increase in plasma ET-1 concentration in several cardiovascular diseases. Thus, increased plasma concentrations of ET-1 and its precursor, big ET-1, have been reported in heart failure, suggesting a possible

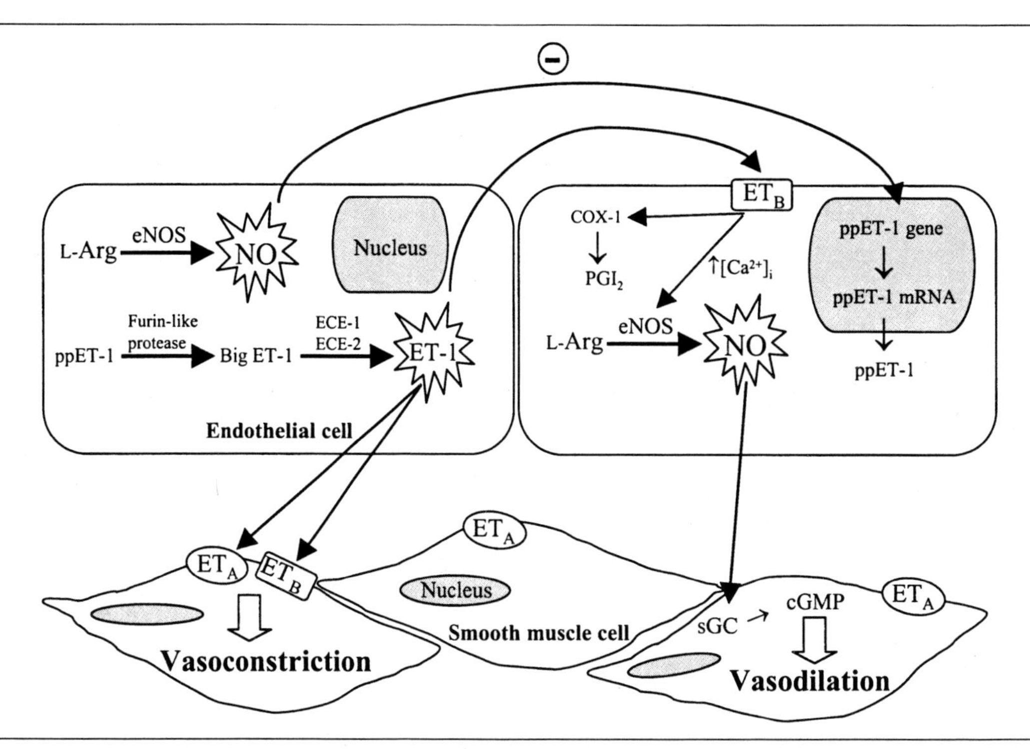

Fig. 1. Overview of the roles of NO and endothelin in regulation of vascular tone. Nitric oxide generated from L-arginine (L-arg) by the endothelial cells activates the soluble gunaylate cyclase (sGC) to produce vasodilation. Nitric oxide can also inhibit ET-1 synthesis at the level of transcription. Conversely, ET-1 is produced by endothelial cells and activates two types of receptors ET_A and ET_B. In smooth muscle cells, activation of ET_A and ET_B receptors leads to vascular vasoconstriction. The activation of ET_B receptors in endothelial cells leads to NO and prostacyclin production.

pathophysiological significance [68–71]. In this respect, Pacher and colleagues reported an inverse correlation between increased big ET-1 plasma levels and survival in advanced heart failure, providing a more powerful prognostic marker for mortality compared to other hemodynamic variables [72]. The expression of ET-1 and its receptors in cardiomyocytes has been shown to be increased in heart failure, suggesting the involvement of the cardiac ET-1 system in cardiovascular diseases. Accordingly, an increase in the density of ET_A and ET_B receptors on cardiomyocytes and ventricular membranes and an overexpression of mRNA for ppET-1 and ECE-1 on cardiomyocytes have been reported in patients with ischemic cardiomyopathy [73]. Moreover, oxidative stress has been shown to increase the synthesis of big ET-1 in endothelial cells by affecting ET-1 promoter activity [74].

Several lines of evidence suggest the crosstalk between the NO and the endothelin systems in the cardiovascular system. Thus, it is well established that ET-1 enhances NO production via ET_B receptors located in endothelial cells [75,76]. The signal transduction mechanism of NO production triggered by ET_B receptors activation in endothelial cells involves the increase in intracellular calcium by the activation of PLC, leading to up-regulation of eNOS, although a PKC-dependent pathway has also been proposed.

Conversely, enhanced vasoconstriction response to ET and an increase in circulating ET-1 levels result from inhibition of endogenous NO formation [77,78], indicating that NO decreases ET production or its effects. Furthermore, L-NAME-induced vasoconstriction was reversed by selective ET_A or dual ET_A/ET_B receptor blockade [79] indicating a predominant role for the endothelin system in hypertension associated with reduced NO production. Thus, the inhibition of ET-1 production by endothelium-derived NO, NO donor drugs and dilator prostanoids appears to be at the level of transcription, via a cGMP-dependent pathway [80–82]. On the other hand, prolonged exposure to NO has been reported to upregulate ET_A receptors and to increase the affinity of ET-1 for this type of receptor acting via cGMP-dependent activation of the cAMP-dependent protein kinase [83]. Figure 2 depicts the effects of the endothelin and NO system in vascular pathology.

Clinical Trials with Endothelin Blockers

Endothelin receptor antagonists represent a novel class of agents that are used for the treatment of

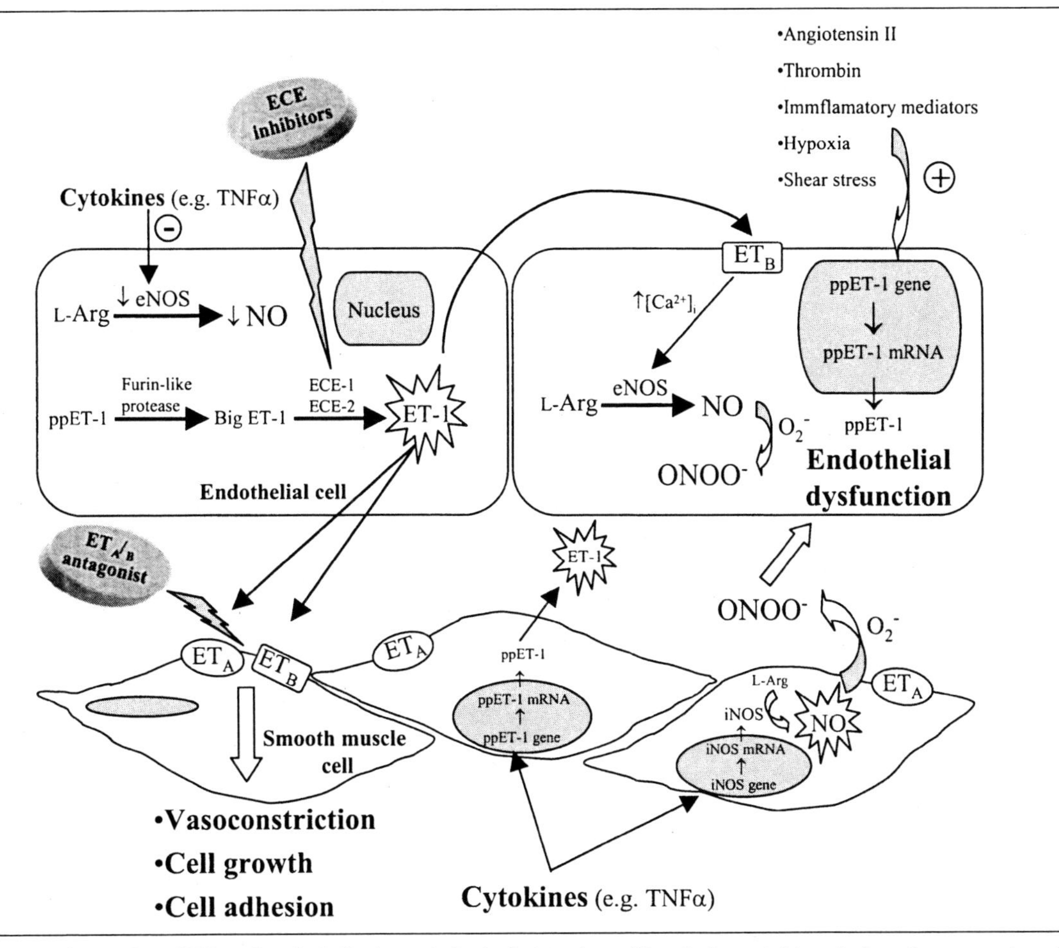

Fig. 2. *Overview of the roles of NO and endothelin in pathological situations. Circulating cytokines induce the expression of iNOS and ET-1 in smooth muscle cells. At the same time, they also downregulate eNOS expression in endothelial cells. Nitric oxide generated in high amounts by iNOS can react with superoxide anion (O_2^-) to form peroxynitrite ($ONOO^-$), which is associated with further endothelial dysfunction. Some pharmacological tools are available to inhibit the production or actions of ET-1.*

heart failure. The levels of ET, that are elevated in heart failure, can be normalize using inhibitors of ECE [84–86] and ET-receptor antagonists. However, the therapeutic usefulness of ECE inhibitors is limited by the existence of ECE-independent ET synthesis. Therefore, the main efforts in drug development have been focused on ET-receptor antagonists. Both selective ET_A receptor antagonists as well as nonselective ET_A-ET_B receptor antagonists are now under investigation. Animal models and early-phase clinical trials have provided evidence that these agents could be effective in the treatment of heart failure, essential hypertension, pulmonary hypertension, and atherosclerosis. Indeed, preliminary clinical trials in heart failure patients have demonstrated beneficial systemic and pulmonary hemodynamic effects of short-term oral administration of a nonpeptide competitive dual ET_A/ET_B-receptor antagonist, bosentan (Ro 47-0203) [87]. In long-term studies, the preliminary data from the REACH-1 trial (Research on Endothelin Antagonism in Chronic heart failure) showed symptomatic improvement, reduced hospitalization rates, and reduced hospital stays in patients with severe chronic heart failure [88]. However, the trial had to be prematurely discontinued due to the occurrence of asymptomatic albeit reversible increases in liver enzyme levels. In the light of these results, and the fact that lower dosages of bosentan were also hemodynamically efficacious, bosentan has recently moved into Phase III clinical trial [89]. Therefore, ENABLE-1 and 2 (ENdothelin Antagonist Bosentan for Lowering cardiac Events in heart failure), two large, multicenter, mortality/morbidity studies are already in progress. In addition, another ET-receptor antagonist is also being tested, tezosentan, an intravenous dual ET-receptor antagonist. Tezosentan has significantly increased cardiac index and decreased pulmonary and systemic vascular resistances without changes in heart rate [90,91]. Moreover, the phenomenon of hemodynamic rebound, as a result of the increase in circulating ET-1 due to receptor-blockage, was not observed. Currently, the RITZ clinical trials program (Randomized Intravenous TeZosentan) is evaluating tezosentan in patients with a variety of acute heart failure syndromes.

On the other hand, the differences in biological effects mediated by ET_A and ET_B receptors, suggested the therapeutic use of ET_A selective receptor antagonists. The short-term administration of BQ-123, a selective endothelin ET_A-receptor antagonist, in chronic heart failure led to systemic vasodilatation, that was associated with decreased pulmonary artery preassure [92]. Gonon and colleagues attributed the cardioprotective effects of selective ET_A blockade in ischemia-reperfusion injury to increased generation of NO production [93]. More recently, a multicenter, double-blind, placebo-controlled trial with sitaxsentan, a selective ET_A receptor antagonist, in patients with chronic stable heart failure have showed a selective pulmonary vasodilation [94].

Despite the fact that organic nitrates have been used for over a century in patients with heart failure, controlled clinical trials with NO or NO donors in heart failure have not been performed [95]. However, the nitric oxide donors 3-morpholinylsydnonimine (SIN-1), S-nitroso-N-acetylpenicillamine (SNAP), diethylamine NONOate (DEA/NO), and diethylenetriamine NONOate (DETA/NO) have been reported to antagonize the effects of ET-1 in human arteries suggesting that NO donors might be useful therapeutic agents in the treatment of heart failure [96]. We have recently reviewed the use of NO, NO donors, organic nitrates, and novel nitroderivatives as therapeutic tools in the treatment of myocardial infarction and reperfusion [97].

In conclusion, the American College of Cardiology (ACC) and the American Heart Association (AHA) guidelines for the evaluation and management of chronic heart failure in the adult recommend the use of ACE inhibitors and beta-blockers as drugs for routine use in patients at different stages of heart failure, such as those with symptomatic left ventricular dysfunction [98]. However, the use of endothelin antagonists is still not recommended for use in patients with heart failure until the final results of clinical trials currently under evaluation are known.

Acknowledgments

This work is supported by the Canadian Institutes of Health Research (CIHR) and the Secretaria de Estado de Educacion y Universidades, cofunded by the European Social Fund. DA is a post-doctoral fellow of Spanish Ministry of Education, MWR is a CIHR scientist. We thank Mr. David Wong for his help in preparation of this manuscript.

References

1. Cooke M, John P, Dzau M, Victor J. Nitric oxide synthase: Role in the genesis of vascular disease. *Annu Rev Medicine* 1997;48:489–509.
2. Palmer RM, Ferrige AG, Moncada S. Nitric oxide release accounts for the biological activity of endothelium-derived relaxing factor. *Nature* 1987;327:524–526.
3. Moncada S, Radomski MW, Palmer RM. Endothelium-derived relaxing factor. Identification as nitric oxide and role in the control of vascular tone and platelet function. *Biochem Pharmacol* 1988;37:2495–2501.
4. Furchgott RF, Zawadzki JV. The obligatory role of endothelial cells in the relaxation of arterial smooth muscle by acetylcholine. *Nature* 1980;288:373–376.
5. Marletta MA. Nitric oxide synthase: Aspects concerning structure and catalysis. *Cell* 1994;78:927–930.
6. Knowles RG, Palacios M, Palmer RM, Moncada S. Formation of nitric oxide from L-arginine in the central nervous system: A transduction mechanism for stimulation of the soluble guanylate cyclase. *Proceedings of the National Academy of Sciences of the United States of America* 1989;86:5159–5162.
7. Bredt DS, Snyder SH. Isolation of nitric oxide synthetase, a calmodulin-requiring enzyme. *Proceedings of the National Academy of Sciences of the United States of America* 1990;87:682–685.
8. Moncada S, Higgs EA. Endogenous nitric oxide: Physiology, pathology and clinical relevance. *Eur J Clin Invest* 1991;21:361–374.
9. Forstermann U, Schmidt HH, Pollock JS, Sheng H, Mitchell JA, Warner TD, Nakane M, Murad F. Isoforms of nitric oxide synthase. Characterization and purification from different cell types. *Biochem Pharmacol* 1991;42:1849–1857.
10. Moncada S, Palmer RM, Higgs EA. Nitric oxide: Physiology, pathophysiology, and pharmacology. *Pharmacol Rev* 1991;43:109–142.
11. Pollock JS, Forstermann U, Mitchell JA, Warner TD, Schmidt HH, Nakane M, Murad F. Purification and characterization of particulate endothelium-derived relaxing factor synthase from cultured and native bovine aortic endothelial cells. *Proc Natl Acad Sci USA* 1991;88:10480–10484.
12. Ignarro LJ, Buga GM, Wood KS, Byrns RE, Chaudhuri G. Endothelium-derived relaxing factor produced and released from artery and vein is nitric oxide. *Proc Natl Acad Sci USA* 1987;84:9265–9269.
13. Palmer RM, Ashton DS, Moncada S. Vascular endothelial cells synthesize nitric oxide from L-arginine. *Nature* 1988;333:664–666.
14. Marletta MA, Yoon PS, Iyengar R, Leaf CD, Wishnok JS. Macrophage oxidation of L-arginine to nitrite and nitrate: Nitric oxide is an intermediate. *Biochem* 1988;27:8706–8711.
15. Szabo C, Thiemermann C. Regulation of the expression of the inducible isoform of nitric oxide synthase. *Adv Pharmacol* 1995;34:113–153.
16. Radomski MW, Salas E. Nitric oxide-biological mediator, modulator and factor of injury: Its role in the pathogenesis of atherosclerosis. *Atherosclerosis* 1995;118 (suppl):S69–S80.

17. Schulz R, Smith JA, Lewis MJ, Moncada S. Nitric oxide synthase in cultured endocardial cells of the pig. *Br J Pharmacol* 1991;104:21–24.
18. Balligand JL, Kobzik L, Han X, Kaye DM, Belhassen L, O Hara DS, Kelly RA, Smith TW, Michel T. Nitric oxide-dependent parasympathetic signaling is due to activation of constitutive endothelial (type III) nitric oxide synthase in cardiac myocytes. *J Biol Chem* 1995;270:14582–14586.
19. Schulz R, Nava E, Moncada S. Induction and potential biological relevance of a Ca(2+)-independent nitric oxide synthase in the myocardium. *Br J Pharmacol* 1992;105:575–580.
20. Balligand JL, Ungureanu_Longrois D, Simmons WW, Pimental D, Malinski TA, Kapturczak M, Taha Z, Lowenstein CJ, Davidoff AJ, Kelly RA. Cytokine-inducible nitric oxide synthase (iNOS) expression in cardiac myocytes. Characterization and regulation of iNOS expression and detection of iNOS activity in single cardiac myocytes *in vitro*. *J Biol Chem* 1994;269:27580–27588.
21. Cooke JP, Rossitch E, Andon NA, Loscalzo J, Dzau VJ. Flow activates an endothelial potassium channel to release an endogenous nitrovasodilator. *J Clin Invest* 1991;88:1663–1671.
22. Radomski MW, Palmer RM, Moncada S. Comparative pharmacology of endothelium-derived relaxing factor, nitric oxide and prostacyclin in platelets. *Br J Pharmacol* 1987;92:181–187.
23. Radomski MW, Palmer RM, Moncada S. Endogenous nitric oxide inhibits human platelet adhesion to vascular endothelium. *Lancet* 1987;2:1057–1058.
24. Radomski MW, Palmer RM, Moncada S. The role of nitric oxide and cGMP in platelet adhesion to vascular endothelium. *Biochem Biophys Res Commun* 1987;148:1482–1489.
25. Radomski MW, Palmer RM, Moncada S. The anti-aggregating properties of vascular endothelium: Interactions between prostacyclin and nitric oxide. *Br J Pharmacol* 1987;92:639–646.
26. Taddei S, Virdis A, Ghiadoni L, Salvetti A. Vascular effects of endothelin-1 in essential hypertension: Relationship with cyclooxygenase-derived endothelium-dependent contracting factors and nitric oxide. *J Cardiovasc Pharmacol* 2000;35(4):S37–S40.
27. Kojda G, Kottenberg K. Regulation of basal myocardial function by NO. *Cardiovasc Res* 1999;41:514–523.
28. Hickey KA, Rubanyi G, Paul RJ, Highsmith RF. Characterization of a coronary vasoconstrictor produced by cultured endothelial cells. *Am J Physiol* 1985;248:C550–C556.
29. O'Brien RF, Robbins RJ, McMurtry IF. Endothelial cells in culture produce a vasoconstrictor substance. *J Cell Physiol* 1987;132:263–270.
30. Yanagisawa M, Kurihara H, Kimura S, Tomobe Y, Kobayashi M, Mitsui Y, Yazaki Y, Goto K, Masaki T. A novel potent vasoconstrictor peptide produced by vascular endothelial cells. *Nature* 1988;332:411–415.
31. Kedzierski RM, Yanagisawa M. Endothelin system: The double-edged sword in health and disease. *Annu Rev Pharmacol Toxicol* 2001;41:851–876.
32. Inoue A, Yanagisawa M, Kimura S, Kasuya Y, Miyauchi T, Goto K, Masaki T. The human endothelin family: Three structurally and pharmacologically distinct isopeptides predicted by three separate genes. *Proc Natl Acad Sci USA* 1989;86:2863–2867.
33. Denault JB, Claing A, D Orleans-Juste P, Sawamura T, Kido T, Masaki T, Leduc R. Processing of proendothelin-1 by human furin convertase. *FEBS Lett* 1995;362:276–280.
34. Kimura S, Kasuya Y, Sawamura T, Shinimi O, Sugita Y, Yanagisawa M, Goto K, Masaki T. Conversion of big endothelin-1 to 21-residue endothelin-1 is essential for expression of full vasoconstrictor activity: Structure-activity relationships of big endothelin-1. *J Cardiovasc Pharmacol* 1989;13 (suppl. 5):S5–S7; discussion S18.
35. Matsumura Y, Hisaki K, Takaoka M, Morimoto S. Phosphoramidon, a metalloproteinase inhibitor, suppresses the hypertensive effect of big endothelin-1. *Eur J Pharmacol* 1990;185:103–106.
36. Shimada K, Takahashi M, Tanzawa K. Cloning and functional expression of endothelin-converting enzyme from rat endothelial cells. *J Biol Chem* 1994;269:18275–18278.
37. Xu D, Emoto N, Giaid A, Slaughter C, Kaw S, deWit D, Yanagisawa M. ECE-1: A membrane-bound metalloprotease that catalyzes the proteolytic activation of big endothelin-1. *Cell* 1994;78:473–485.
38. Emoto N, Yanagisawa M. Endothelin-converting enzyme-2 is a membrane-bound, phosphoramidon-sensitive metalloprotease with acidic pH optimum. *J Biol Chem* 1995;270:15262–15268.
39. Korth P, Bohle RM, Corvol P, Pinet F. Cellular distribution of endothelin-converting enzyme-1 in human tissues. *J Histochem Cytochem* 1999;47:447–462.
40. Schweizer A, Valdenaire O, Nelbock P, Deuschle U, Dumas-Milne-Edwards JB, Stumpf JG, Loffler BM. Human endothelin-converting enzyme (ECE-1): Three isoforms with distinct subcellular localizations. *Biochem J* 1997;328 (pt. 3):871–877.
41. Valdenaire O, Lepailleur-Enouf D, Egidy G, Thouard A, Barret A, Vranckx R, Tougard C, Michel J-B. A fourth isoform of endothelin-converting enzyme (ECE-1) is generated from an additional promoter: Molecular cloning and characterization. *Eur J Biochem* 1999;264:341–349.
42. Fernandez-Patron C, Radomski MW, Davidge ST. Vascular matrix metalloproteinase-2 cleaves big endothelin-1 yielding a novel vasoconstrictor. *Circ Res* 1999;85:906–911.
43. Fernandez-Patron C, Zouki C, Whittal R, Chan JS, Davidge ST, Filep JG. Matrix metalloproteinases regulate neutrophil-endothelial cell adhesion through generation of endothelin-1 [1–32]. *FASEB J* 2001;15:2230–2240.
44. Inoue A, Yanagisawa M, Takuwa Y, Mitsui Y, Kobayashi M, Masaki T. The human preproendothelin-1 gene. Complete nucleotide sequence and regulation of expression. *J Biol Chem* 1989;264:14954–14959.
45. Lee ME, Bloch KD, Clifford JA, Quertermous T. Functional analysis of the endothelin-1 gene promoter. Evidence for an endothelial cell-specific cis-acting sequence. *The Journal of Biological Chemistry* 1990;265:10446–10450.
46. Kawana M, Lee ME, Quertermous EE, Quertermous T. Cooperative interaction of GATA-2 and AP1 regulates transcription of the endothelin-1 gene. *Mol Cell Biol* 1995;15:4225–4231.
47. Arai H, Hori S, Aramori I, Ohkubo H, Nakanishi S. Cloning and expression of a cDNA encoding an endothelin receptor. *Nature* 1990;348:730–732.
48. Sakurai T, Yanagisawa M, Takuwa Y, Miyazaki H, Kimura S, Goto K, Masaki T. Cloning of a cDNA encoding a non-isopeptide-selective subtype of the endothelin receptor. *Nature* 1990;348:732–735.
49. Takuwa Y, Kasuya Y, Takuwa N, Kudo M, Yanagisawa M, Goto K, Masaki T, Yamashita K. Endothelin receptor is

coupled to phospholipase C via a pertussis toxin-insensitive guanine nucleotide-binding regulatory protein in vascular smooth muscle cells. *J Clin Invest* 1990;85:653–658.
50. Griendling KK, Tsuda T, Alexander RW. Endothelin stimulates diacylglycerol accumulation and activates protein kinase C in cultured vascular smooth muscle cells. *J Biol Chem* 1989;264:8237–8240.
51. Simonson MS, Dunn MJ. Cellular signaling by peptides of the endothelin gene family. *FASEB J* 1990;4:2989–3000.
52. Haynes WG, Webb DJ. Endothelin as a regulator of cardiovascular function in health and disease. *J Hypertens* 1998;16:1081–1098.
53. Haynes WG, Webb DJ. Contribution of endogenous generation of endothelin-1 to basal vascular tone. *Lancet* 1994;344:852–854.
54. Haynes WG, Ferro CJ, O Kane KP, Somerville D, Lomax CC, Webb DJ. Systemic endothelin receptor blockade decreases peripheral vascular resistance and blood pressure in humans. *Circulation* 1996;93:1860–1870.
55. Woods M, Mitchell JA, Wood EG, Barker S, Walcot NR, Rees GM, Warner TD. Endothelin-1 is induced by cytokines in human vascular smooth muscle cells: Evidence for intracellular endothelin-converting enzyme. *Molecular Pharmacology* 1999;55:902–909.
56. Wagner OF, Christ G, Wojta J, Vierhapper H, Parzer S, Nowotny PJ, Schneider B, Waldhausl W, Binder BR. Polar secretion of endothelin-1 by cultured endothelial cells. *J Biol Chem* 1992;267:16066–16068.
57. Wright CE, Fozard JR. Regional vasodilation is a prominent feature of the haemodynamic response to endothelin in anaesthetized, spontaneously hypertensive rats. *Eur J Pharmacol* 1988;155:201–203.
58. de Nucci G, Thomas R, D Orleans-Juste P, Antunes E, Walder C, Warner TD, Vane JR. Pressor effects of circulating endothelin are limited by its removal in the pulmonary circulation and by the release of prostacyclin and endothelium-derived relaxing factor. *Proc Natl Acad Sci USA* 1988;85:9797–9800.
59. Haynes WG, Strachan FE, Webb DJ. Endothelin ETA and ETB receptors cause vasoconstriction of human resistance and capacitance vessels *in vivo*. *Circulation* 1995;92:357–363.
60. Miyauchi T, Masaki T. Pathophysiology of endothelin in the cardiovascular system. *Annu Rev Physiol* 1999;61:391–415.
61. Cardillo C, Kilcoyne CM, Cannon RO, Panza JA. Interactions between nitric oxide and endothelin in the regulation of vascular tone of human resistance vessels *in vivo*. *Hypertension* 2000;35:1237–1241.
62. Carville C, Adnot S, Sediame S, Benacerraf S, Castaigne A, Calvo F, de Cremou P, Dubois-Rande JL. Relation between impairment in nitric oxide pathway and clinical status in patients with congestive heart failure. *J Cardiovasc Pharmacol* 1998;32:562–570.
63. Treasure CB, Vita JA, Cox DA, Fish RD, Gordon JB, Mudge GH, Colucci WS, Sutton MG, Selwyn AP, Alexander RW. Endothelium-dependent dilation of the coronary microvasculature is impaired in dilated cardiomyopathy. *Circulation* 1990;81:772–779.
64. Drexler H, Hayoz D, Munzel T, Just H, Zelis R, Brunner HR. Endothelial function in congestive heart failure. *Am Heart J* 1993;126:761–764.
65. Arimura K, Egashira K, Nakamura R, Ide T, Tsutsui H, Shimokawa H, Takeshita A. Increased inactivation of nitric oxide is involved in coronary endothelial dysfunction in heart failure. *Am J Physiol Heart Circ Physiol* 2001;280:H68–H75.
66. Yoshizumi M, Perrella MA, Burnett JC, Lee ME. Tumor necrosis factor downregulates an endothelial nitric oxide synthase mRNA by shortening its half-life. *Circ Res* 1993;73:205–209.
67. de Belder AJ, Radomski MW, Why HJ, Richardson PJ, Bucknall CA, Salas E, Martin JF, Moncada S. Nitric oxide synthase activities in human myocardium. *Lancet* 1993;341:84–85.
68. Miyauchi T, Yanagisawa M, Tomizawa T, Sugishita Y, Suzuki N, Fujino M, Ajisaka R, Goto K, Masaki T. Increased plasma concentrations of endothelin-1 and big endothelin-1 in acute myocardial infarction. *Lancet* 1989;2:53–54.
69. McMurray JJ, Ray SG, Abdullah I, Dargie HJ, Morton JJ. Plasma endothelin in chronic heart failure. *Circulation* 1992;85:1374–1379.
70. Rodeheffer RJ, Lerman A, Heublein DM, Burnett JC. Increased plasma concentrations of endothelin in congestive heart failure in humans. *Mayo Clin Proc* 1992;67:719–724.
71. Wei C, Lerman A, Rodeheffer R, McGregor C, Brandt R, Wright S, Heublein D, Kao P, Edwards W, Burnett J, Jr. Endothelin in human congestive heart failure. *Circulation* 1994;89:1580–1586.
72. Pacher R, Stanek B, Hulsmann M, Koller_Strametz J, Berger R, Schuller M, Hartter E, Ogris E, Frey B, Heinz G, Maurer G. Prognostic impact of big endothelin-1 plasma concentrations compared with invasive hemodynamic evaluation in severe heart failure. *J Am Coll Cardiol* 1996;27:633–641.
73. Serneri GGN, Cecioni I, Vanni S, Paniccia R, Bandinelli B, Vetere A, Janming X, Bertolozzi I, Boddi M, Lisi GF, Sani G, Modesti PA. Selective upregulation of cardiac endothelin system in patients with ischemic but not idiopathic dilated cardiomyopathy: Endothelin-1 system in the human failing heart. *Circ Res* 2000;86:377–385.
74. Kahler J, Mendel S, Weckmuller J, Orzechowski HD, Mittmann C, Koster R, Paul M, Meinertz T, Munzel T. Oxidative stress increases synthesis of big endothelin-1 by activation of the endothelin-1 promoter. *J Mol Cell Cardiol* 2000;32:1429–1437.
75. Suzuki S, Kajikuri J, Suzuki A, Itoh T. Effects of endothelin-1 on endothelial cells in the porcine coronary artery. *Circ Res* 1991;69:1361–1368.
76. Verhaar MC, Strachan FE, Newby DE, Cruden NL, Koomans HA, Rabelink TJ, Webb DJ. Endothelin-A receptor antagonist-mediated vasodilatation is attenuated by inhibition of nitric oxide synthesis and by endothelin-B receptor blockade. *Circulation* 1998;97:752–756.
77. Lerman A, Sandok EK, Hildebrand FL, Burnett JC. Inhibition of endothelium-derived relaxing factor enhances endothelin-mediated vasoconstriction. *Circulation* 1992;85:1894–1898.
78. Ahlborg G, Lundberg JM. Nitric oxide-endothelin-1 interaction in humans. *J Appl Physiol* 1997;82:1593–1600.
79. Banting JD, Friberg P, Adams MA. Acute hypertension after nitric oxide synthase inhibition is mediated primarily by increased endothelin vasoconstriction. *J Hypertens* 1996;14:975–981.

80. Boulanger C, Luscher TF. Release of endothelin from the porcine aorta. Inhibition by endothelium-derived nitric oxide. *J Clin Invest* 1990;85:587–590.
81. Warner TD, Schmidt HH, Murad F. Interactions of endothelins and EDRF in bovine native endothelial cells: Selective effects of endothelin-3. *Am J Physiol* 1992;262:H1600–H1605.
82. Gray GA, Webb DJ. The endothelin system and its potential as a therapeutic target in cardiovascular disease. *Pharmacol Ther* 1996;72:109–148.
83. Redmond EM, Cahill PA, Hodges R, Zhang S, Sitzmann JV. Regulation of endothelin receptors by nitric oxide in cultured rat vascular smooth muscle cells. *J Cell Physiol* 1996;166:469–479.
84. Ahn K, Sisneros AM, Herman SB, Pan SM, Hupe D, Lee C, Nikam S, Cheng XM, Doherty AM, Schroeder RL, Haleen SJ, Kaw S, Emoto N, Yanagisawa M. Novel selective quinazoline inhibitors of endothelin converting enzyme-1. *Biochem Biophys Res Commun* 1998;243:184–190.
85. Martin P, Tzanidis A, Stein Oakley A, Krum H. Effect of a highly selective endothelin-converting enzyme inhibitor on cardiac remodeling in rats after myocardial infarction. *J Cardiovasc Pharmacol* 2000;36:S367–S370.
86. Trapani AJ, Beil ME, Bruseo CW, De Lombaert S, Jeng AY. Pharmacological properties of CGS 35066, a potent and selective endothelin-converting enzyme inhibitor, in conscious rats. *J Cardiovasc Pharmacol* 2000;36:S40–S43.
87. Sutsch G, Kiowski W, Yan XW, Hunziker P, Christen S, Strobel W, Kim JH, Rickenbacher P, Bertel O. Short-term oral endothelin-receptor antagonist therapy in conventionally treated patients with symptomatic severe chronic heart failure. *Circulation* 1998;98:2262–2268.
88. Packer M, Caspi A, Charlon V, La Roche H, Cohen-Solal A, Kiowski W. Multicenter, double-blind, placebo-controlled study of long-term endothelin blockade with bosentan in chronic heart failure- results of the REACH-1 trial. *Circulation* 1998;98 (suppl.):1–3.
89. Ono K, Matsumori A. Endothelin antagonism with bosentan: Current status and future perspectives. *Cardiovasc Drug Rev* 2002;20:1–18.
90. Schalcher C, Cotter G, Reisin L, Bertel O, Kobrin I, Guyene TT, Kiowski W. The dual endothelin receptor antagonist tezosentan acutely improves hemodynamic parameters in patients with advanced heart failure. *Am Heart J* 2001;142:340–349.
91. Torre Amione G, Durand JB, Nagueh S, Vooletich MT, Kobrin I, Pratt C. A pilot safety trial of prolonged (48 h) infusion of the dual endothelin-receptor antagonist tezosentan in patients with advanced heart failure. *Chest* 2001;120:460–466.
92. Cowburn PJ, Cleland JG, McArthur JD, MacLean MR, McMurray JJ, Dargie HJ. Short-term haemodynamic effects of BQ-123, a selective endothelin ET(A)-receptor antagonist, in chronic heart failure. *Lancet* 1998;352:201–202.
93. Gonon AT, Gourine AV, Pernow J. Cardioprotection from ischemia and reperfusion injury by an endothelin A-receptor antagonist in relation to nitric oxide production. *J Cardiovasc Pharmacol* 2000;36:405–412.
94. Givertz MM, Colucci WS, LeJemtel TH, Gottlieb SS, Hare JM, Slawsky MT, Leier CV, Loh E, Nicklas JM, Lewis BE. Acute endothelin A receptor blockade causes selective pulmonary vasodilation in patients with chronic heart failure. *Circulation* 2000;101:2922–2927.
95. Elkayam U, Karaalp IS, Wani OR, Tummala P, Akhter MW. The role of organic nitrates in the treatment of heart failure. *Prog Cardiovasc Dis* 1999;41:255–264.
96. Wiley KE, Davenport AP. Novel nitric oxide donors reverse endothelin-1-mediated constriction in human blood vessels. *J Cardiovasc Pharmacol* 2000;36:S151–S152.
97. Alonso D, Radomski MW. Nitric oxide, platelet function, myocardial infarction and reperfusion therapies. *Heart Fail Rev* 2002.
98. Hunt SA, Baker DW, Chin MH, Cinquegrani MP, Feldmanmd AM, Francis GS, Ganiats TG, Goldstein S, Gregoratos G, Jessup ML, Noble RJ, Packer M, Silver MA, Stevenson LW, Gibbons RJ, Antman EM, Alpert JS, Faxon DP, Fuster V, Jacobs AK, Hiratzka LF, Russell RO, Smith SC, and ACC/AHA Guidelines for the Evaluation and Management of Chronic Heart Failure in the Adult: Executive summary. A report of the American College of Cardiology/American Heart Association Task Force on Practice Guidelines (Committee to Revise the 1995 Guidelines for the Evaluation and Management of Heart Failure): Developed in Collaboration with the International Society for Heart and Lung Transplantation; endorsed by the Heart Failure Society of America. *Circulation* 2001;104:2996–3007.

The Therapeutic Effect of Natriuretic Peptides in Heart Failure; Differential Regulation of Endothelial and Inducible Nitric Oxide Synthases

Angelino Calderone
Centre de Recherche de l'Institut de Cardiologie de Montréal, et Département de Physiologie, Université de Montréal, Montréal, Québec, Canada

Abstract. **The abnormal regulation of nitric oxide synthase activity represents an underlying feature of heart failure. Increased peripheral vascular resistance, and decreased renal function may be in part related to impaired endothelium-dependent nitric oxide (NO) synthesis. Paradoxically, the chronic production of NO by inducible nitric oxide synthase (iNOS) in heart failure exerts deleterious effects on ventricular contractility, and circulatory function. Consequently, pharmacologically improving endothelium-dependent NO synthesis and the concomitant inhibition of iNOS activity would be therapeutically advantageous. Interestingly, natriuretic peptides have been shown to differentially regulate endothelial NOS (eNOS) and iNOS activity. Moreover, in both patients and animal models of heart failure, pharmacologically increasing plasma natriuretic peptide levels ameliorated vascular tone, renal function, and ventricular contractility. Based on these observations, the following review will explore whether the therapeutic benefit of the natriuretic peptide system in heart failure may occur in part via the amelioration of endothelium-dependent NO synthesis, and the concomitant inhibition of cytokine-mediated iNOS expression.**

Key Words. **atrial natriuretic peptide, endothelial nitric oxide synthase, inducible nitric oxide synthase, tumour necrosis factor-α, heart failure**

Abbreviations. **ACE: angiotensin converting enzyme, ANP: atrial natriuretic peptide, BNP: brain natriuretic peptide, eNOS: endothelial nitric oxide synthase, iNOS: inducible nitric oxide synthase, LPS: lipopolysaccharide, NPR: natriuretic peptide receptor, NEP: neutral endopeptidase, RAAS: renin-angiotensin II-aldosterone system, TNF-α: tumour necrosis factor-α.**

Introduction

Recruitment of the renin-angiotensin II-aldosterone system (RAAS), and the sympathetic system represent an adaptive compensatory response following an acute decrease in cardiac function, regardless of the hemodynamic stress. However, the long-term and marked activation of these systems lead to excessive vasoconstriction, impaired renal function, marked ventricular dysfunction, and maladaptive cardiac remodelling leading to subsequent heart failure. Based on these observations, RAAS and the sympathetic system have been regarded as integral events implicated in the progression of heart failure, and as such have become therapeutic targets. Indeed, targeting either the synthesis (e.g. angiotensin converting enzyme (ACE) inhibitors), and/or antagonizing agonist-receptor complex formation (e.g. AT1, or β-adrenergic receptor antagonists) improved vascular tone, renal function, and cardiac contractility in heart failure patients. An alternative pharmacological approach receiving considerable attention involves exploiting the biological actions of natriuretic peptides. The potential therapeutic benefit of natriuretic peptides was initially considered by their ability to counteract the physiological actions of vasoconstrictive systems via direct and indirect effects on the vasculature, and kidney. Indeed, the infusion of natriuretic peptides, and/or their inhibition of degradation in patients and animal models of heart failure markedly improved circulatory hemodynamics, and ventricular function. An underlying mechanism which may in part explain the therapeutic effect of natriuretic peptides involves the reciprocal regulation of constitutive (e.g. eNOS) and iNOS activity. Akin to natriuretic peptides, the inorganic molecule NO, synthesized by constitutive NOS can exert similar physiological actions. The biological commonality resides at the cellular level, as natriuretic peptide binding to its cognate receptor, and nitric oxide activation of soluble guanylate cyclase increased the synthesis of the second messenger, cyclic GMP. Moreover, several studies support the premise that the physiological actions of natriuretic peptides may also occur in

Address for correspondence: Angelino Calderone, Institut de Cardiologie de Montréal, Centre de Recherche, 5000 rue Belanger est, Montréal, Québec, Canada H1T 1C8. Tel.: 514-376-3330, ext. 3710; Fax: 514-376-1355; E-mail: calderon@icm.umontreal.ca part via a nitric oxide pathway following receptor-mediated stimulation of eNOS. By contrast, acting via a cyclic GMP-independent pathway, the chronic synthesis of NO by iNOS has been characterized as nefarious in the setting of heart failure. Consequently, the diminished levels of physiological NO by eNOS, and the concomitant cytotoxic effects associated with iNOS represent integral events implicated in the progression of heart failure. In this regard, the following review will explore whether the therapeutic action of natriuretic peptides in heart failure may occur in part via the reciprocal regulation of eNOS and iNOS activity.

Classification of Natriuretic Peptides, Their Cognate Receptors, and Nitric Oxide Synthases

The Family of Natriuretic Peptides

The discovery that the *in vivo* administration of atrial extracts elicited natriuresis, diuresis, and hypotension led to the rapid isolation and characterization of atrial natriuretic peptide (ANP) [1,2]. In atrial cells, ANP is stored in dense granules as a 126 amino acid prohormone, and upon secretion into the circulation, the prohormone is cleaved into the active peptide (99–126 amino acid fragment), and inactive N-terminal fragment (1–98 amino acid fragment) [3]. In the normal heart, the atria represents the primary site of ANP synthesis, and to a much lesser extent, ANP-containing granules have been identified in the ventricle [1,3,4]. In porcine brain extracts, Sudoh and colleagues identified a closely related peptide to ANP, designated brain natriuretic peptide (BNP) [5]. BNP was also identified in atrial cells, but its abundance is much lower than ANP [6,7]. Analogous to ANP, atrial BNP is stored as a prohormone of 108 amino acids, and upon release into the circulation is cleaved to an active peptide of 32 amino acids, and an inactive N-terminal fragment (1–76 amino acid fragment) [6–8]. However, especially under pathological conditions, BNP synthesis occurred primarily in the ventricles [9,10]. Analogous to BNP, a third related member of the natriuretic peptide family was isolated in the brain, and designated as C-type natriuretic peptide (CNP) [11,12]. In contrast to ANP, and BNP, the brain and vascular endothelium are the principal sites of synthesis, and circulating plasma levels of CNP are low [12,13].

Natriuretic Peptide Receptor Subtypes, and Tissue Distribution

The biological actions of natriuretic peptides are mediated by at least three receptor subtypes designated natriuretic peptide receptor-A (NPR-A), -B (NPR-B), and -C (NPR-C) [14–16]. The NPR-A, and NPR-B subtypes possess intrinsic guanylate cyclase activity, and following agonist binding, stimulate the synthesis of the second messenger cyclic GMP [14,15]. By contrast, the NPR-C subtype does not possess intrinsic guanylate cyclase activity [16]. Ligand affinity studies have demonstrated that the NPR-A receptor subtype mediated the physiological actions of ANP, and BNP (ANP $\geq$ BNP > CNP), whereas the NPR-B receptor subtype represented the preferred target of CNP (CNP > ANP $\geq$ BNP) [14,17]. Lastly, the relative affinity of natriuretic peptides for the NPR-C receptor subtype are similar [18]. Several studies suggested that the apparent physiological role of the NPR-C receptor subtype was to bind and remove natriuretic peptides from the circulation, thereby designating this subtype as a clearance receptor. However, biological activity, including the suppression of cyclic AMP synthesis, has been ascribed to the NPR-C receptor following the identification of a specific ligand, denoted C-ANP_{4-23} [18].

Consistent with their physiological actions, natriuretic peptide receptor subtypes have been localized in the kidney, adrenal gland, juxtaglomerular cells, presynaptic terminals, vascular smooth muscle cells, and endothelial cells [19]. Moreover, the NPR-A, NPR-B, and NPR-C transcripts were identified in isolated cardiac myocytes by RT-PCR, and cyclic GMP production was observed either with ANP or BNP administration, but not with CNP [20]. These data suggest that the NPR-A receptor represented the predominant guanylate cyclase coupled receptor subtype expressed in rat cardiac myocytes [20]. Analogous to cardiac myocytes, the transcripts of NPR-A, NPR-B, and NPR-C were also expressed in rat cardiac fibroblasts [20]. However, in contrast to myocytes, both NPR-A, and NPR-B subtypes were physiologically functional, as assessed via the production of cyclic GMP by ANP, and CNP [20]. Moreover, radioligand binding assay revealed the NPR-C was the predominant subtype expressed in cardiac fibroblasts and biologically relevant [21]. In the human heart, transcripts of NPR-A, NPR-B, and NPR-C were detected in the atria, and left ventricle [22]. However, in the ventricle, the myocyte, and non-myocyte distribution of these receptor subtypes remain unknown.

The Family of Nitric Oxide Synthases

Nitric oxide is a free radical gas synthesized from L-arginine via the action of the enzyme nitric oxide synthase (NOS) [23]. The family of NOS consists of three isoforms, of which I (also known as neuronal or brain NOS) and III (also known as endothelial NOS; eNOS) are constitutively expressed in a wide variety of tissue (e.g. heart, endothelium, kidney, brain) and regulated by calcium/calmodulin [23]. By contrast isoform II (inducible NOS; iNOS) is not constitutively expressed but induced by various factors including cytokines, and bacterial lipopolysaccharide (LPS), and its activation occurred via a calcium/calmodulin-independent process [23]. The immediate second messenger is cyclic GMP, synthesized following NO binding and subsequent activation of the enzyme soluble guanylate cyclase [24]. Interestingly, nitric oxide produced by constitutive nitric oxide synthase is quantitatively lower than iNOS [25]. The intermittent low production of NO synthesis by constitutive nitric oxide synthase results in local physiological effects. By contrast, the large amount of NO produced by iNOS can influence a variety of cellular functions, and in some instances can be cytotoxic in nature [26]. This latter premise has been documented in heart failure, as NO, in the presence superoxide anion O_2^-, will lead to the formation of peroxynitrite, a potent oxidant free radical with nefarious actions on circulatory hemodynamics, and ventricular contractility [27,28]. The cellular consequences of increased peroxynitrite formation include S-nitrosylation of cysteine residues, and tyrosine nitration of proteins implicated in cellular proliferation, apoptosis, vascular smooth muscle contraction, and calcium homeostasis [26].

Biological Actions of Natriuretic Peptides and Nitric Oxide

Natriuretic Peptide Regulation of Vascular Tone, and Renal Function

Atrial distension represents the primary stimulus coupled to increased natriuretic peptide synthesis, and release [29]. Based on these observations, natriuretic peptides may represent an endogenous compensatory mechanism limiting blood pressure increase. Indeed, natriuretic peptides, acting via a particulate guanylate cyclase receptor (e.g. NPR-A, and/or NPR-B) stimulate vascular smooth muscle cell relaxation [30,31]. Consistent with this premise, an increase in total peripheral resistance was observed in mice lacking ANP expression [32]. Secondly, natriuretic peptides elicit natriuresis, and diuresis by increasing the glomerular filtration rate, inhibiting sodium retention in the collecting ducts, and vasopressin-mediated water reabsorption [33]. Moreover, natriuretic peptides can antagonize the physiological actions of RAAS by inhibiting renin and aldosterone release from juxtaglomerular cells, and the adrenal glomerulosa, respectively [34,35]. At the pre-synaptic terminal, natriuretic peptides inhibited catecholamine synthesis and release, and directly influenced sympathetic modulation of vascular tone [36–38]. Lastly, natriuretic peptides exert an antiproliferative action on mitogen-stimulated vascular smooth muscle growth, thereby negatively influencing maladaptive vascular remodelling [39,40].

The Role of Nitric Oxide in Circulatory Homeostasis

Akin to natriuretic peptides, endothelium release of NO exerted a potent vasodilator action via a cyclic GMP-dependent mechanism following soluble guanylate cyclase activation of the underlying vasculature. However, a recent study by Weisbrod and colleagues demonstrated that NO-mediated vascular smooth muscle relaxation occurred in part via a cyclic GMP-independent pathway [41]. Regardless of the mechanism, the integral role of NO in the regulation of vascular tone and blood pressure has been firmly established *in vivo*, as the inhibition of endothelium-derived nitric oxide synthesis with NOS inhibitors resulted in systemic hypertension [42]. Secondly, nitric oxide can influence renal function, as NO inhibited sodium reabsorption, and antagonized vasopressin-mediated water permeability in the collecting ducts [43]. Analogous to ANP, NO attenuated the basal release of aldosterone from the adrenal glomerulosa [44]. However, endothelium-derived NO, and/or NO-generating compounds either inhibited or stimulated renin release from juxtaglomerular cells [45,46]. The underlying mechanism implicated in this disparate effect of NO remains undefined. Regardless of the effect on renin, NO suppressed the conversion of angiotensin I to angiotensin II via the inhibition of ACE activity, and decreasing ACE mRNA expression [47,48]. In addition, NO-elevating agents reduced AT1 receptor expression in vascular smooth muscle cells, and a similar paradigm was documented in the myocardium following the *in vivo* infusion of the L-arginine analog inhibitor N^G-Nitro-L-arginine methyl ester (L-NAME) [49,50]. Akin to natriuretic peptides, NO can negatively regulate the sympathetic system via the inhibition of catecholamine release, the suppression of carrier mediated uptake (uptake-1) transport of norepinephrine in presynaptic terminals, and antagonizing vascular smooth muscle contraction [38,51–54]. Lastly, NO is a potent antiproliferative factor of vascular smooth muscle cells [39].

Cardiac Natriuretic Peptide Expression and the Regulation of Endothelial NOS in Hypertrophy, and Heart Failure

The Transition from Compensated Cardiac Hypertrophy to Heart Failure, and the Regulation of Ventricular Natriuretic Peptide Expression

When ventricular contractile function remains compromised because of a sustained hemodynamic overload, the chronic deformation of the heart, and the accompanying increase in wall stress will trigger hypertrophy [55,56]. Cardiac hypertrophy represents a physiological growth response to counteract the increase in wall stress. In response to a chronic volume-overload, an eccentric pattern of cardiac hypertrophy will ensue [55–57]. By contrast, a chronic pressure-overload will lead to a concentric pattern of cardiac hypertrophy [57,58]. At the cellular level, both forms of cardiac hypertrophy are associated with the synthesis of new sarcomeres and increased production of ATP as a consequence of increased number of mitochondria per cardiac myocyte [55,56]. In concentric cardiac hypertrophy, sarcomere replication occurs in a parallel fashion leading to an increase in cell width, whereas eccentric cardiac hypertrophy is characterized by a series pattern of sarcomere replication resulting in an increase in cell length [59,60]. A hallmark feature of concentric and eccentric cardiac hypertrophy is the re-induction of ventricular ANP [3]. During the maturation of the heart from neonatal to adult, the temporal expression of ventricular prepro-ANP mRNA gradually diminishes [3]. However, the mechanical deformation (e.g. stretch) of the myocardium because of either volume- or pressure-overload represents a powerful stimulus for the re-expression of ventricular prepro-ANP in cardiac myocytes [61]. Consistent with this concept, a sustained and/or intermittent mechanical stretch of isolated cardiac myocytes was shown to induce prepro-ANP mRNA expression [62]. Moreover, AII, and the sympathetic system stimulated prepro-ANP mRNA expression, and cardiac myocyte hypertrophy [63–65]. Moreover, mechanical stretch of cardiac myocytes upregulated angiotensinogen, ACE, and renin expression, thereby providing an autocrine pathway coupling AII to ANP expression [66]. Lastly, concomitant with ventricular prepro-ANP mRNA expression, a marked induction of ventricular prepro-BNP mRNA has been identified, and represents the predominant source of circulating BNP [10,67]. Thus, the early progression of cardiac hypertrophy is associated with a robust and sustained increase of ventricular natriuretic peptide expression.

Although cardiac myocyte hypertrophy represents an adaptive response to diminished contractile function, this process is finite, and thus cannot maintain normal cardiac output in the presence of a sustained hemodynamic overload. Consequently, worsening of cardiac function leads to progressive ventricular dilatation, decompensated cardiac hypertrophy, and subsequent heart failure [55,56]. The transition of compensated hypertrophy to heart failure is characterized by the overt activation of RAAS, and the sympathetic system [55,56,68]. In the failing heart, ventricular prepro-ANP and -BNP mRNA expression remained elevated, resulting in chronically elevated plasma levels [55,56,69,70]. In a study by Takahashi and colleagues, a significant correlation was observed between the decrease in sarcoplasmic reticulum Ca^{2+}-ATPase (SERCA) mRNA levels, and the expression of prepro-ANP, and prepro-BNP mRNAs in the left ventricle of end-stage heart failure patients [69]. SERCA plays an integral role in calcium uptake following systole, and decreased activity has been implicated in diastolic heart failure. Indeed, in the failing heart of various species, including human, the reduction in SERCA protein expression and accompanied decrease of Ca^{2+}-ATPase stimulated calcium uptake appears to represent a conserved event [69,71,72]. These initial observations suggested that depressed cardiac function, and thus the accompanying increase in wall stress may represent an important stimulus of natriuretic peptide expression. Indeed, plasma BNP levels were found to be a powerful marker of left ventricular systolic and diastolic dysfunction, and left ventricular mass in heart failure patients [73,74]. Consistent with this latter premise, prognostic assessment in patients with heart failure recently identified plasma BNP as a potent mortality predictor [75]. Based on these observations, the success of a pharmacological approach to treat heart failure patients may be indirectly assessed by the measure of plasma BNP levels [74]. Indeed, at least one study has demonstrated that pharmacotherapy for heart failure guided by plasma BNP levels was encouraging [76].

The Regulation of Endothelial Nitric Oxide Synthase Activity in the Kidney, and Vasculature in Heart Failure

In the rapid pacing overdrive dog model of heart failure, and the myocardial infarct rat, eNOS-dependent renal oxygen consumption, and renal blood flow were markedly depressed [77,78]. Likewise, reduced vascular eNOS activity in patients and animal models of heart failure has been documented [26,79,80]. A similar pathophysiological mechanism underscored the decreased vasodilatory response of the coronary arterial bed. In heart failure patients, and the rapid pacing overdrive dog model of heart failure, endothelium-dependent vasodilation, and nitrite production were attenuated in isolated coronary microvessels [81–83]. Likewise, bradykinin, and adenosine mediated eNOS-dependent vasodilation of the coronary arterial bed was diminished in guinea pigs with decompensated left ventricular hypertrophy following abdominal aortic banding [84]. Thus, impaired endothelium-dependent NO biological activity in the kidney and vasculature represent hallmark features of heart failure. The mechanisms attributed to the decreased biological actions of eNOS are multifactorial, and include AII-mediated suppression of eNOS protein expression, production of reactive oxygen species

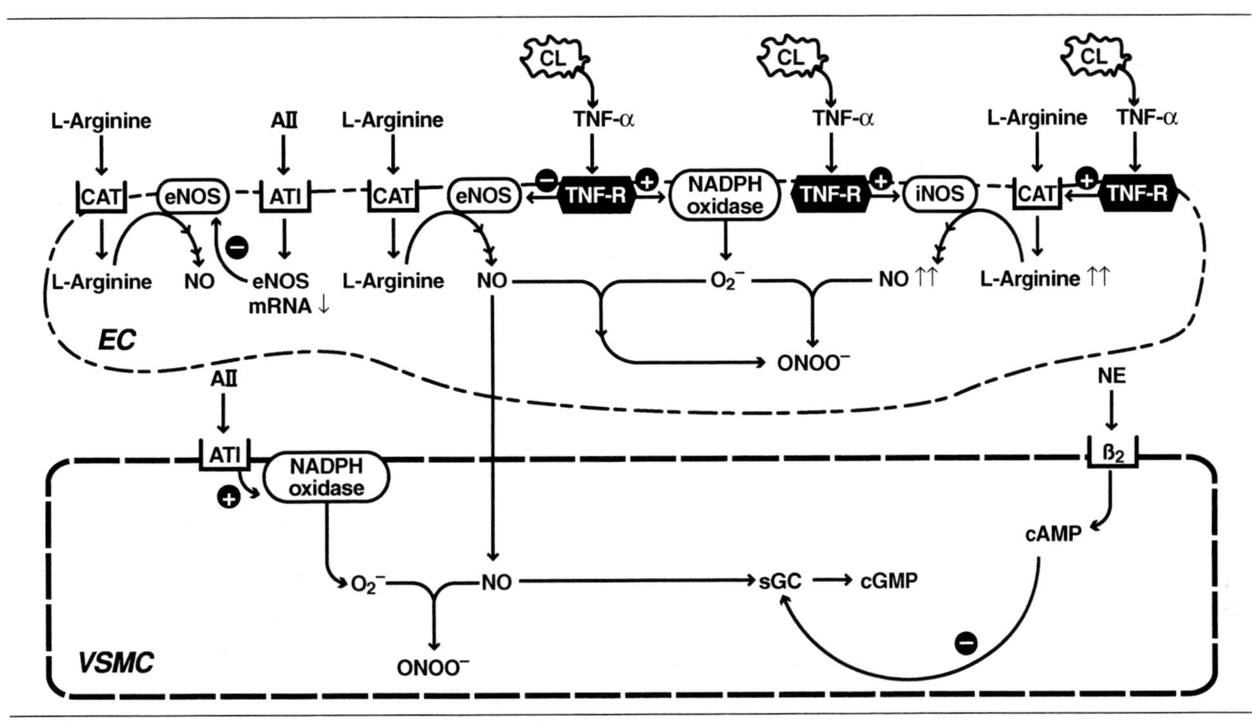

Fig. 1. Cellular mechanisms implicated in the diminished biological activity of eNOS in heart failure. *The following figure summarizes the mechanisms involved in the impaired biological activity of endothelium-dependent NO. Angiotensin II (AII), acting via the AT1 receptor, can stimulate the downregulation of eNOS mRNA levels in endothelial cells (EC). Likewise, the increase release of TNF-α from circulating lymphocytes (CL), acting via a receptor (TNF-R)-dependent process can lead to eNOS mRNA degradation. Moreover, TNF-α stimulates the synthesis of reactive oxygen species (e.g. O_2^-) via the activation of NADPH oxidase which will react with NO derived either from eNOS or iNOS to form peroxynitrite ($ONOO^-$). The sustained production of NO by iNOS is facilitated by the induction of the cationic amino acid transporter (CAT), thereby increasing L-arginine uptake. In vascular smooth muscle cells (VSMC), AII, acting via the AT1 receptor, can stimulate the synthesis of O_2^- via the activation of NADPH oxidase, which will in turn bind to NO to form peroxynitrite. Lastly, the sympathetic system, acting via the β_2-adrenergic receptor, can promote the downregulation of soluble guanylate cyclase (sGC) via a cyclic AMP (cAMP)-dependent pathway.*

which react with nitric oxide to form peroxynitrite, and sympathetic-mediated reduction of soluble guanylate cyclase expression via a cyclic AMP-dependent pathway in vascular smooth muscle cells [85–90] (Fig. 1).

Therapeutic Benefit of Natriuretic Peptides and the Potential Role of Nitric Oxide in Heart Failure

Natriuretic Peptide Infusion and Neutral Endopeptidase Inhibition Improve Circulatory Hemodynamics

Despite elevated plasma levels of natriuretic peptides during the progression of heart failure, the marked increase of vascular tone and diminished renal function persists. The apparent ineffectiveness of natriuretic peptides may be related to the overtly discernible activation of RAAS, and the sympathetic system [55,56,68]. Indeed, excessive vasoconstriction, fluid retention, maladaptive cardiac remodelling, and contractile dysfunction elicited by the chronic activation of RAAS, and the sympathetic system may be implicated in the transition of compensated cardiac hypertrophy to heart failure [55,56]. Based on these clinical observations, it was considered counterintuitive that a further elevation of endogenous plasma natriuretic peptides would be therapeutically advantageous in end-stage heart failure patients. However, in severe acute heart failure patients, decreased vascular resistance, increased diuresis, and cardiac output was observed 48 hrs following ANP infusion, and maintained 7-days after initiation [91]. Colucci and colleagues demonstrated that a 6-hour intravenous infusion of nesiritide, a human recombinant BNP, improved pulmonary-capillary wedge pressure, and global clinical status in patients with decompensated congestive heart failure [92]. In this latter study, the beneficial effects of nesiritide were maintained in a 7-day therapy protocol and the most common side-effect was a dose-related asymptomatic hypotension. The authors concluded that nesiritide could prove to be a valuable tool in the initial treatment of patients with decompensated heart failure [92]. Likewise, in the pacing-overdrive dog model of heart failure, subcutaneous BNP administration for a period of 10-days caused diuresis, natriuresis, a decrease in blood pressure, and a concomitant increase in cardiac output [93]. The physiological action of BNP most likely occurred at least in part via a direct action on the kidney, and vasculature, but may also be related to a sympathoinhibitory effect. A study by Brunner-La Rocca and colleagues demonstrated that the infusion of a low dose of BNP reduced cardiac sympathetic activity in both normal and stable heart failure patients [94]. The administration of a high dose of BNP in heart failure patients reduced renal sympathetic nerve activity, whereas no effect was observed in healthy patients [94]. This finding is of particular interest, since increased renal sympathetic activity plays an integral role in renin release from the juxtaglomerular cells, leading to subsequent AII synthesis. However, because the high dose of BNP reduced cardiac filling pressures in heart failure patients, a direct inhibitory action of renal sympathetic activity could not be ascertained. Lastly, a novel interaction was observed between β-adrenergic receptor antagonist therapy and natriuretic peptides in left ventricular hypertrophy and heart failure. In the study by Luchner and colleagues, a prominent increase in plasma ANP, BNP, and cyclic GMP was observed in patients with left ventricular hypertrophy following the chronic treatment with β-adrenergic receptor blockers [95]. The authors speculated that this latter therapeutic effect on left ventricular remodelling, and possible left ventricular dysfunction may have occurred in part via the increase of natriuretic peptides. Consistent with this latter premise, the β-adrenergic receptor blocker carvedilol increased natriuretic peptide mRNA expression in the myocardium, which correlated with elevated plasma levels in the rat [96]. Collectively, these data underscore the observation that despite elevated plasma natriuretic peptide levels in heart failure, a further increase can be clinically beneficial.

Although continuous natriuretic peptide infusion was therapeutically beneficial, this mode of treatment as a long-term approach is not practical. Secondly, the oral administration of natriuretic peptides is limited because of inactivation. Recently, a membrane-bound metalloenzyme identified as neutral endopeptidase (NEP;24.11) was detected in numerous tissues including the kidney, lung, heart, and shown to degrade and inactivate natriuretic peptides [97]. Moreover, in patients with aortic valve stenosis, and heart failure due to dilated cardiomyopathy, NEP mRNA levels and enzymatic activity were increased in left ventricular biopsies [98]. An immunofluorescence approach identified NEP in myocytes and non-myocytes of the human heart [98]. Consequently, increased NEP activity may represent an important intrinsic mechanism diminishing the physiological actions of natiuretic peptides in heart failure. Collectively, these observations provided the impetus for the design of new pharmacological agents to specifically target NEP in heart failure. Indeed, the administration of the NEP inhibitors in patients with mild heart failure, increased plasma ANP, urinary cyclic GMP levels, and was associated with natriuresis, diuresis, and reductions in atrial and pulmonary wedge pressures [99–103]. However, it is important to note that several studies demonstrated that NEP can degrade the peptide bradykinin, an important vasodilator, and activator of nitric oxide and prostacyclin [104,105]. In this regard, bradykinin may in part contribute to the beneficial hemodynamic effects associated with NEP inhibition. Interestingly, mean arterial pressure was only modestly influenced following NEP inhibition [106,107]. Likewise, a modest effect of NEP inhibitors on blood pressure was observed in hypertensive patients, despite elevated plasma ANP levels [108]. Paradoxically, in a study by Ferro et al., the infusion of structurally distinct NEP inhibitors in the forearm of healthy and hypertensive subjects elicited a slow progressive vasoconstriction [109]. The co-administration of the ET_A receptor antagonist BQ-123 prevented the NEP inhibitor-mediated forearm vasoconstriction, supporting endothelin-1 as a substrate of NEP degradation [109,110]. Moreover, the administration of the NEP inhibitor ecadotril in either normal rats, or after coronary artery ligation increased plasma angiotensin I, and II levels, and tissue angiotensin I content in the heart [111,112]. Likewise, the NEP inhibitor candoxatril increased plasma angiotensin II levels, and the pressor response to exogenous angiotensin II was exaggerated in normal human subjects [113]. Consequently, NEP can degrade both angiotensin I, and endothelin-1, which may in part explain the lack of effect on blood pressure, and furthermore counteract the beneficial biological actions of natriuretic peptides in heart failure.

Dual Inhibition of NEP and ACE, and Therapeutic Benefit in Heart Failure

The secondary increase of vasoconstrictors with NEP inhibitor therapy clearly represented an unexpected consequence. In this regard, it was postulated that the dual inhibition of ACE and NEP may be more efficacious. Indeed, the diuretic, and natriuretic action of the NEP inhibitor ecadotril was potentiated by the co-administration of the ACE inhibitor perindopril in the rat [111]. In two separate studies, the combination of ACE and NEP inhibitor therapy were additive with regard to hemodynamic and renal function in pacing overdrive dogs [114,115]. Based on the efficacy of combination ACE and NEP inhibitor therapy, and the topographical proximity of these enzymes, drugs simultaneously inhibiting both ACE and NEP were generated [116]. The modification of the NEP inhibitor thiorphan resulted in the synthesis of the dual NEP/ACE inhibitors glycoprilat, and alatrioprilat, and when infused in rats, promoted natriuresis, diuresis, and blocked the vasoconstrictive action of AII [116]. The most extensively studied dual NEP/ACE inhibitor is omapatrilat; a vasopepetidase inhibitor containing a mercaptoacyl derivative of a bicyclic thazepinone dipeptide displaying an equipotent inhibitory action on both enzymes [97,105]. In the pacing-overdrive model of heart failure, omapatrilat exerted profound and sustained beneficial hemodynamic and renal effects [117]. Moreover, a superiority of omapatrilat to conventional ACE inhibitor was demonstrated, as a greater natriuretic effect was observed following the acute administration of omapatrilat versus ACE inhibition with fosinoprilat in pacing-induced dogs [118]. The effects of omapatrilat were associated with elevated plasma cGMP, and natriuretic peptides (ANP, & BNP), increased urine cyclic GMP, and ANP excretion, and augmented glomerular filtration rate. To confirm the beneficial effects were attributed to natriuretic peptides, the intrarenal administration of the natriuretic receptor-A antagonist HS-142-1 attenuated the renal actions of omapratilat [118]. The beneficial effects of omapatrilat in heart failure have been reaffirmed in the clinical setting. In the study by McClean and colleagues, omapatrilat was given to heart failure patients (NYHA class II-III) receiving ACE inhibitor therapy, and following 3 months of treatment, an improvement of clinical status, and a reduction of acute heart failure episodes were observed [119]. The latter improvements were associated with increased left ventricular ejection fraction, reduction in peripheral vascular resistance, augmented renal function, reduction in blood volume, and an apparent attenuation of the sympathetic system. These changes observed with omapatrilat therapy were dose-dependent, and associated with increased plasma ANP, and cyclic GMP levels. Interestingly, plasma BNP levels were found modestly decreased. As previously discussed, plasma BNP levels are a prognostic marker of left ventricular dysfunction. In this regard, the improvement of left ventricular ejection fraction may be the primary mechanism reducing BNP expression despite NEP inhibition. In the IMPRESS clinical trial, a direct comparison of the efficacy of an ACE inhibitor versus a dual ACE/NEP inhibitor was examined [120]. In NYHA class II-IV patients, either omapatrilat or the ACE inhibitor lisinopril was given for a period of 24 weeks and the primary endpoint was an improvement in the maximum exercise treadmill test at 12 weeks, with secondary endpoints examining events indicative of worsening heart failure. A similar level of improvement of exercise tolerance was observed in both groups, albeit there was a significant benefit with omapatrilat regarding composite of death, admission, and discontinuation of study treatment for worsening of heart failure. Lastly, based on the observations that both BNP infusion and omapatrilat are clinically beneficial, Chen and colleagues designed a study examining the therapeutic benefit of their co-administration in pacing-overdrive heart failure dogs [121]. The combined effect of subcutaneous administration of BNP, and systemic infusion of omapatrilat on urinary sodium excretion, glomerular filtration rate, and cardiac output was greater as compared to either drug alone [121]. Collectively, these data support the premise that maximizing the action of natriuretic peptides with the concomitant suppression of AII synthesis with dual NEP/ACE inhibitors represents a novel therapeutic strategy in the treatment of heart failure.

The Potential Contribution of Endothelium-Dependent Nitric Oxide in the Therapeutic Action of Dual NEP/ACE Inhibitors

An important consequence of ACE inhibitor therapy is the suppression of bradykinin degradation, resulting in elevated plasma levels in heart failure. Bradykinin may contribute to the therapeutic effect of ACE inhibitor therapy via the stimulation of NO synthesis by eNOS. As previously discussed, constitutive NO production can counteract the deleterious action of vasoconstrictive mechanisms, and directly improve vascular tone, vascular remodelling, and kidney function. However, impaired eNOS activity in the vasculature and kidney has been documented in heart failure. In this regard, either ACE or dual NEP/ACE therapy can improve endothelium-dependent activity by increasing bradykinin activation of eNOS. Indeed, evidence supporting this latter premise was documented in the myocardial infarcted heart of the eNOS-deficient mouse, as the therapeutic effect of ACE inhibition was lost [122]. In the rapid-pacing overdrive model of heart failure, the combined administration of an NEP and ACE inhibitor improved coronary artery endothelial-derived NO production [81]. This latter effect was antagonized by L-NAME, and the bradykinin B_2 receptor antagonist HOE 140 [81]. Likewise, in coronary microvessels isolated from end-stage heart failure patients, the administration of various ACE inhibitors increased nitrite production, and was blocked by L-NAME, and HOE 140 [82]. By contrast, increased NO production in the myocardium with ACE inhibition may be deleterious. In patients with idiopathic dilated cardiomyopathy, ACE inhibition decreased β-adrenergic stimulated myocardial contractility, and this negative inotropic effect was abolished with the NOS inhibitor L-NAME [123].

The contribution of NO in the therapeutic action of ACE inhibitor therapy may be further amplified with dual NEP/ACE inhibition. Analogous to bradykinin, natriuretic peptides represent a physiological activator of NO synthesis via an endothelium-dependent process. In the rat, the hypotensive effect of an *in vivo* infusion of ANP was partially reversed with the prior treatment with L-NAME [124]. To confirm that this effect of ANP proceeded via an NO-dependent pathway, elevated staining of NADPH-diaphorase activity (an indicator of NOS activity) was detected in the aorta, and increased urinary excretion of NO end products was observed [124]. A similar paradigm was described for natriuretic peptide-mediated dilation of the coronary bed in the rat heart [125]. Likewise, in human proximal tubular cells, ANP treatment dose-dependently increased nitric oxide production, an effect mimicked by the specific NPR-C ligand C-ANP_{4-23} [126]. In gastrointestinal smooth muscle cells, the NPR-C subtype was coupled to eNOS via a G_i-dependent pathway [127]. Analogous to ANP, BNP-mediated arterial vasodilation in the forearm of healthy male subjects was partially dependent on NO [128,129]. In ET-1 pre-constricted porcine coronary resistance arteries, the vasodilatory action of BNP was accentuated, and partially mediated via a nitric oxide-dependent pathway [130]. By contrast, BNP-mediated vasodilation of the porcine epicardial coronary artery was insensitive to NO inhibition, but was inhibited by the ATP-sensitive potassium channel blocker glibenclamide [130]. Moreover, the suppression of AII synthesis and/or antagonism of its action by natriuretic peptides would alleviate the downregulation of eNOS protein expression, and significantly reduce NADPH oxidase mediated production of reactive oxygen species (Fig. 2). Lastly, the inhibition of catecholamine release, and the suppression of cyclic

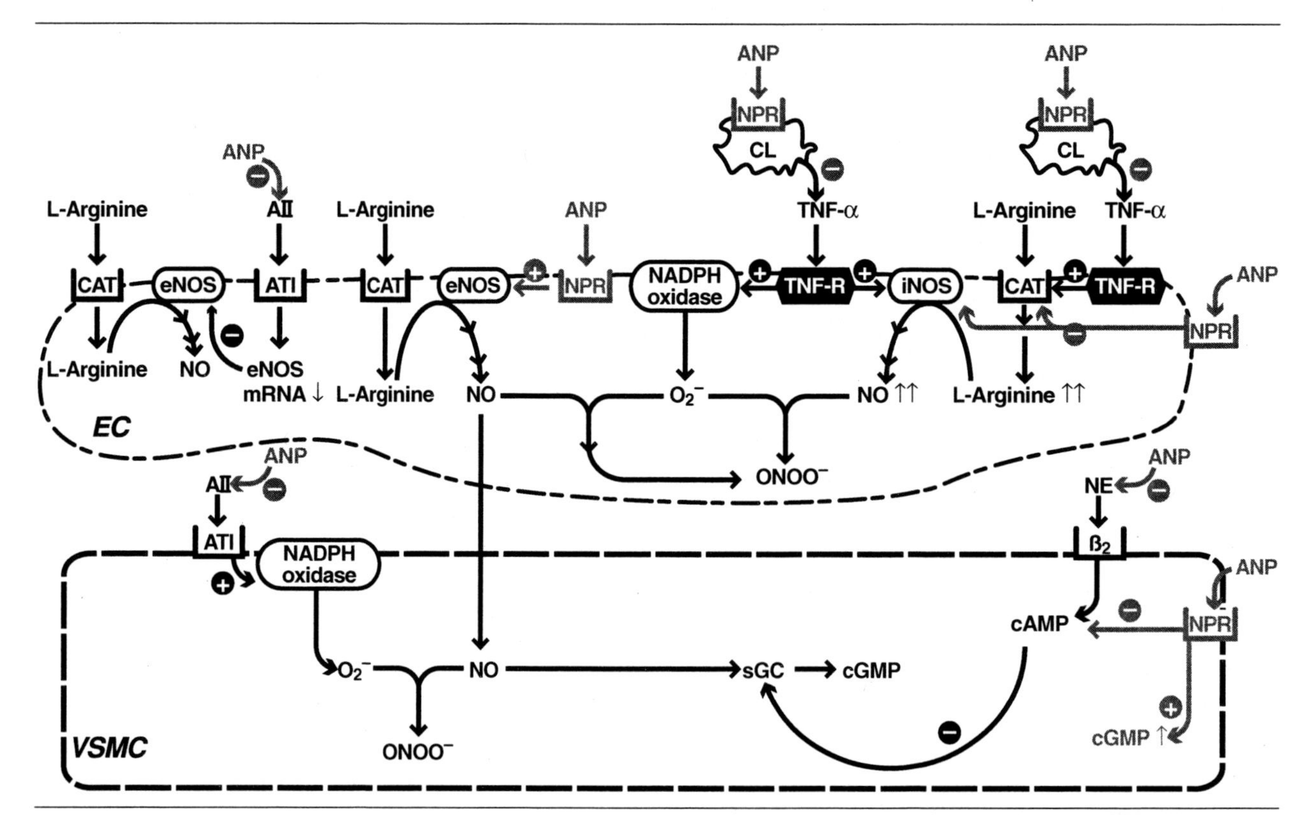

Fig. 2. The differential regulation of eNOS, and iNOS activity by natriuretic peptides. *Atrial natriuretic peptide (ANP), and BNP, acting via a receptor (NPR)-dependent process stimulate eNOS activity. Moreover, the suppression of AII synthesis by natriuretic peptides will attenuate the downregulation of eNOS mRNA in endothelial cells (EC), and suppress the production of reactive oxygen species in vascular smooth muscle cells (VSMC). Natriuretic peptide receptors are present on circulating lymphocytes (CL), and ANP can inhibit the synthesis and release of TNF-α. This latter effect will lead to a decrease of iNOS expression, the inhibition of cationic amino acid transporter (CAT) induction, and the suppression of O_2^- synthesis. Moreover, ANP can directly promote iNOS mRNA degradation, and inhibit L-arginine uptake by CAT. Lastly, ANP can attenuate sympathetic-mediated downregulation of soluble guanylate cyclase (sGC) via the inhibition of cyclic AMP (cAMP) production.*

AMP synthesis by natriuretic peptides would attenuate sympathetic-mediated reduction of soluble guanylate cyclase protein levels [18,36,37] (Fig. 2). Thus, in the setting of heart failure, the therapeutic benefit attributed to natriuretic peptides may occur at least in part via the amelioration of endothelium-dependent NO synthesis, and subsequent biological activity.

A Potential Therapeutic Action of ANP May Involve the Negative Modulation of Tumour Necrosis Factor-α (TNF-α) Stimulation of iNOS Expression

TNF-α Expression in Heart Failure

It has been suggested that advanced heart failure reflects a state of chronic inflammation, as evidenced by the increased secretion of pro-inflammatory cytokines from circulating lymphocytes, in particular tumour necrosis factor-α (TNF-α) [131–134]. As previously discussed, impaired endothelium-derived NO synthesis represents a hallmark feature of heart failure, and TNF-α may play a critical role [135]. In isolated endothelial cells, TNF-α promoted apoptosis, and consistent in part with this premise, TNF-α serum levels correlated with endothelium apoptosis in patients with advanced heart failure [136]. Second, TNF-α has been directly implicated in impaired endothelium-mediated vasodilation. The administration of etanercept, a recombinant TNF-α receptor which binds and inactivates TNF-α significantly ameliorated endothelium-dependent relaxation in the forearm of patients with advanced heart failure [135]. The mechanism(s) attributed to impaired endothelium-dependent vasodilation include TNF-α mediated downregulation of eNOS protein levels, and the formation of peroxynitrite following the interaction between NO and reactive oxygen species generated by NADPH oxidase [134,137–140] (Fig. 1).

Although increased synthesis and secretion of TNF-α has been documented in circulating lymphocytes of heart failure patients, the production of this cytokine has also been demonstrated in the myocardium. In patients with dilated cardiomyopathy, TNF-α mRNA and protein expression were detected in cardiac myocytes, and the endocardium [141]. Likewise, TNF-α mRNA expression was detected in cardiac myocytes, as well as non-myocytes (e.g. fibroblasts, vascular smooth muscle cells, endothelial cells) in the rat heart following myocardial infarction [142,143]. Moreover, resident cardiac macrophages, and infiltrating neutrophils, and monocytes in the necrotic region following myocardial infarction represent an additional source of TNF-α [142,144–146]. The pathophysiological consequences of increased myocardial TNF-α expression include cardiac myocyte apoptosis, disruption of excitation-contraction coupling, desensitization of β-adrenergic receptor responsiveness to sympathetic stimulation, and a negative inotropic effect [131,144]. Consequently, TNF-α has been characterized as a critical factor implicated in the progression of heart failure. Consistent with this latter premise, the phenotype of congestive heart failure was recapitulated in transgenic mice overexpressing cardiac TNF-α [147].

ANP Regulation of TNF-α Synthesis

Despite the beneficial biological actions of either natriuretic peptide infusion, or dual NEP/ACE inhibitory therapy in heart failure patients, the potential regulation of TNF-α synthesis in circulating lymphocytes, or the myocardium has not been evaluated. Interestingly, ANP treatment of macrophages significantly reduced the synthesis, and secretion of TNF-α following an LPS challenge [148]. Moreover, in whole human blood, ANP attenuated TNF-α secretion [148] (Fig. 2). An analogous paradigm has been identified in kupffer cells, the resident macrophage of the liver. The pre-treatment of kupffer cells with ANP inhibited LPS-mediation synthesis of TNF-α via a cyclic GMP-dependent pathway [149]. The cellular mechanism implicated in ANP-mediated inhibition of TNF-α may occur in part at the transcriptional level. The nuclear factor-kappaB (NF-kappaB) is an integral transcriptional factor implicated in the expression of pro-inflammatory genes, including TNF-α [144]. In a quiescent cell, NF-kappaB exists in a latent state in the cytosol bound to an inhibitory protein denoted I-kappaB. However, in response to the appropriate stimuli, I-kappaB is phosphorylated and degraded, and NF-kappaB translocates to the nucleus, and stimulates transcription [144]. In the isolated rat liver, ischemia-reperfusion mediated expression of TNF-α was inhibited by ANP, and the DNA binding capacity of NF-kappaB was significantly reduced [150]. A similar paradigm was identified with regard to ANP inhibition of TNF-α synthesis in macrophages [151]. Moreover, natriuretic peptide inhibition of TNF-α synthesis would alleviate the downregulation of eNOS, and the production of reactive oxygen species, thereby indirectly improving endothelium-dependent NO synthesis, and biological activity. Lastly, natriuretic peptides may also inhibit TNF-α synthesis via the negative modulation of RAAS, as AII induced the expression of this cytokine in cardiac myocytes [152]. This latter observation is consistent with the reported anti-cytokine action of ACE inhibitors, and AT1 receptor antagonists in heart failure [153,154].

The Expression of iNOS in Heart Failure, and Regulation by ANP

Inducible NOS has been implicated in the pathophysiological actions of TNF-α in heart failure. Nitric oxide acting via a cyclic GMP-independent mechanism has been linked to cellular apoptosis, diminished ventricular contractility, and impaired endothelium-dependent biological activity [26–28,144,155,156]. Consistent with this latter premise, iNOS mRNA and protein have been detected in the myocardium, and vasculature of patients and animal models of heart failure [156–160]. Consequently, if natriuretic peptides can suppress and/or least attenuate TNF-α synthesis, the secondary decrease of iNOS expression would be therapeutically advantageous. Indeed, in LPS-challenged macrophages, ANP-mediated inhibition of TNF-α synthesis was associated with the concomitant decrease of iNOS expression [148,161,162]. This latter action of ANP required the activation of both the NPR-A and NPR-C subtypes [148,161,162]. Likewise, in human proximal tubular cells, ANP inhibited cytokine-induced nitric oxide synthesis via a cyclic GMP-independent pathway involving the NPR-C subtype [163]. Interestingly, ANP-stimulated eNOS activity remained intact, despite the NPR-C selective inhibition of cytokine-mediated iNOS expression [163]. Moreover, natriuretic peptides can directly reduce iNOS mRNA expression via a post-transcriptional process of degradation [148,162,163]. An alternative mechanism implicated in ANP-mediated inhibition of iNOS activity entails the suppression of substrate availability. As previously discussed, iNOS generation of nitric oxide is quantitatively greater than constitutive NOS. This disparate effect requires a greater amount of substrate, and in this regard both LPS and cytokines have been shown to stimulate the expression of the cationic amino acid transporter (CAT-2B), thereby increasing L-arginine uptake [164,165] (Fig. 1). In macrophages, ANP, acting via the NPR-A subtype, dose-dependently inhibited LPS-mediated L-arginine uptake [165] (Fig. 2).

Summary

The direct physiological effects of natriuretic peptides on the kidney and vasculature, and the antagonism of RAAS, and the sympathetic system in part explain the therapeutic effect in heart failure. However, the amelioration of endothelium-dependent NO synthesis and biological activity may represent an underlying feature of natriuretic peptides in heart failure. Moreover, the ability to suppress TNF-α induced expression of iNOS highlights a unique biological role of natriuretic peptides, and may in part contribute to its beneficial effect in heart failure. Accordingly, future studies examining the status of eNOS-dependent function, and the concomitant regulation of TNF-α and iNOS in heart failure following either natriuretic peptide infusion or dual NEP/ACE inhibition warrants further investigation.

Acknowledgments

This work was supported by the Heart & Stroke Foundation of Canada & Quebec, and Les Fonds de Recherche de l'Institut de Cardiologie de Montréal. A. Calderone is a Chercheur-Boursier Junior II de la Fonds de Recherche de la Sante de Quebec. The author would like to thank Drs. Celine Fiset, and Anique DuCharme for the critical reading of the manuscript, France Theriault for secretarial assistance and Sylvie Bolduc for the conceptual design of the figures.

References

1. de Bold AJ, Borenstein B, Veress AT, Sonnenberg H. A rapid and potent natriuretic response to intravenous injection of atrial myocardial extract in rat. *Life Sci* 1981;28:89–94.
2. Brenner BM, Ballerman BJ, Gunning ME, Zeidel ML. Diverse biological actions of atrial natriuretic peptide. *Physiol Rev* 1990;70:665–690.
3. Knowlton KU. Atrial natriuretic factor: A molecular marker for cardiac specific, embryonic, and inducible gene expression. *Heart Failure* 1992;121–128.
4. de Bold AJ. Tissue fractionation studies on the relationship between an atrial natriuretic factor and specific atrial granules. *Can J Physiol Pharmacol* 1982;60:324.
5. Sudoh T, Kangawa K, Minamino N, Matsuo H. A new natriuretic peptide in porcine brain. *Nature* 1988;332:78–81.
6. Kambayashi Y, Nakano K, Mukoyama M, Saito Y, Ogawa Y, Shiono S, Inouye K, Yoshida N, Imura H. Isolation and sequence determination of human brain natriuretic peptide in human atrium. *FEBS Lett* 1990;259:341–345.
7. Ogawa Y, Nakao K, Mukoyama M, Shirakami G, et al. Rat brain natriuretic peptide-tissue distribution and molecular form. *Endocrinology* 1990;126:2225–2227.
8. Sudoh T, Maekawa K, Kojima M, Minamino N, Kangawa K, Matsuo H. Cloning and sequence analysis of cDNA encoding a precursor for human brain natriuretic peptide. *Biochem Biophys Res Commun* 1989;159:1427–1434.
9. Hama N, Itoh H, Shirakami G, Nakagawa O et al. Rapid ventricular induction of brain natriuretic peptide gene expression in experimental acute myocardial infarction. *Circulation* 1995;92:1558–1564.
10. Luchner A, Stevens TL, Borgeson DD, Redfield M, Wei CM, Porter JG, Burnett JC, Jr. Differential atrial and ventricular expression of myocardial BNP during evolution of heart failure. *Am J Physiol Heart Circ Physiol* 1998;274:H1684–H1689.
11. Sodoh T, Minamino N, Kangawa K, Matsuo H. C-type natriuretic peptide (CNP): A new member of natriuretic

peptide family identified in porcine brain. *Biochem Biophys Res Commun* 1990;168:863–870.
12. Minamino N, Makino Y, Tateyama K, Kangawa K, Matsuo H. Characterization of immunoreactive human C-type natriuretic peptide in brain and heart. *Biochem Biophys Res Commun* 1991;179:535–542.
13. Suga S, Nakao K, Itoh H, Komatsu Y, Ogawa Y, Hama N, Imura H. Endothelial production of C-type natriuretic peptide and its augmentation by transforming growth factor-β. *J Clin Invest* 1992;90:1145–1149.
14. Chang M, Lowe DG, Lewis M, Hellmiss R, Chen E, Goeddel DV. Differential activation by atrial and brain natriuretic peptides of two different receptor guanylyl cyclases. *Nature* 1989;341:68–71.
15. Schulz S, Singh S, Bellet RA, Singh G, Tubb J, Chin H, Garbers DL. The primary structure of a plasma membrane guanylate cyclase demonstrates diversity within this new receptor family. *Cell* 1989;58:1155–1162.
16. Fuller F, Porter JG, Arfsten AE, Miller J, Schilling JW, Scarborough RM, Lewicki JA, Shenk DB. Atrial natriuretic peptide clearance receptor. *J Biol Chem* 1988;263:9395–9401.
17. Koller KJ, Lowe DG, Bennett GL, Minamino N, Kangawa K, Matsuo H, Goeddel DV. Selective activation of the B natriuretic peptide receptor by the C-type natriuretic peptide (CNP). *Science* 1991;252:120–123.
18. Anand-Srivastava MB. Atrial natriuretic peptide-C receptor and membrane signalling in hypertension. *J Hypertens* 1997;15:815–826.
19. Silberbach M, Roberts CT, Jr. Natriuretic peptide signaling: Molecular and cellular pathways to growth regulation. *Cell Signal* 2001;4:221–231.
20. Lin X, Hanze J, Heese F, Sodmann R, Lang RE. Gene expression of natriuretic peptide receptors in myocardial cells. *Circ Res* 1995;77:750–758.
21. Cao L, Gardner DG. Natriuretic peptides inhibit DNA synthesis in cardiac fibroblasts. *Hypertension* 1995;25:227–234.
22. Nunez DJR, Dickson MC, Brown MJ. Natriuretic peptide receptor mRNAs in the rat and human heart. *J Clin Invest* 1992;90:1966–1971.
23. Kone BC. Molecular biology of natriuretic peptides and nitric oxide synthases. *Cardiovasc Res* 2001;51:429–441.
24. Katsuki S, Arnold W, Mittal C, Murad F. Stimulation of guanylate cyclase by sodium nitroprusside, nitroglycerin, and nitric oxide in various tissue preparations and comparison to the effects of sodium azide and hydroxylamine. *J Cyclic Nucleotide Res* 1977;3:23–35.
25. Lyons CR. The role of nitric oxide in inflammation. *Adv Immunol* 1995;60:323–371.
26. Davis KL, Martin E, Turko IV, Murad F. Novel effects of nitric oxide. *Ann Rev Pharmacol Toxicol* 2001;41:203–236.
27. Bauersachs J, Bouloumie A, Fraccarollo D, Hu K, Busse R, Ertl G. Endothelial dysfunction in chronic myocardial infarction despite increased vascular endothelial nitric oxide synthase, and soluble guanylate cyclase expression: Role of enhanced vascular superoxide production. *Circulation* 1999;100:292–298.
28. Ferdinandy P, Danial Hm Ambrus I, Rothery RA, Schulz R. Peroxynitrite is a major contributor to cytokine-induced myocardial contractile failure. *Circ Res* 2000;87:241–247.
29. Ruskoaho H. Atrial natriuretic peptide: Synthesis, release and metabolism. *Pharmacol Rev* 1992;44:479–602.
30. Drewett JG, Fendly BM, Garbers DL, Lowe DG. Natriuretic peptide receptor-B (guanylyl cyclase-B) mediates C-type natriuretic peptide relaxation of precontracted rat aorta. *J Biol Chem* 1995;270:4668–4674.
31. Imura R, Sano T, Goto J, Yamada K, Matsuda Y. Inhibition by HS-142-1, a novel nonpeptide atrial natriuretic peptide antagonist of microbial origin, of atrial natriuretic peptide-induced relaxation of isolated rabbit aorta through the blockade of guanyl cyclase–linked receptors. *Mol Pharmacol* 1992;42:982–990.
32. Melo LG, Veress AT, Ackerman U, Pang SC, Flynn TG, Sonnenberg H. Chronic hypertension in ANP knockout mice: Contribution of peripheral resistance. *Regul Pept* 1999;79:109–115.
33. Kalra PR, Anker SD, Coats AJS. Water and sodium regulation in chronic heart failure: The role of natriuretic peptides and vasopressin. *Cardiovasc Res* 2001;51:495–509.
34. Kurtz A, Bruna RD, Pfeilschifter J, Taugner R, Bauer C. Atrial natriuretic peptide inhibits renin release from juxtaglomerular cells by a cGMP-dependent process. *Proc Natl Acad Sci USA* 1986;83:4769–4773.
35. Goodfriend TL, Elliott M, Atlas SA. Actions of synthetic atrial natriuretic factor on bovine adrenal glomerulosa. *Life Sci* 1984;35:1675–1682.
36. Floras JS. Sympathoinhibitory effects of atrial natriuretic factor in normal humans. *Circulation* 1990;81:1860–1873.
37. Anand-Srivastava MB, Trachte GJ. Atrial natriuretic factor receptors and signal transduction mechanisms. *Pharmacol Rev* 1993;45:455–497.
38. Rapoport RM. Cyclic guanosine monophosphate inhibition of contraction may be mediated through inhibition of phosphatidylinositol hydrolysis in rat aorta. *Circ Res* 1986;58:407–410.
39. Garg UC, Hassid A. Nitric oxide-generating vasodilators and 8-bromo-cyclic guanosine monophosphate inhibit mitogenesis and proliferation of cultured rat vascular smooth muscle cells. *J Clin Invest* 1989;83:1774–1777.
40. Morishita R, Gibbons GH, Pratt RE, Tomita N, Kaneda Y, Ogihara T, Dzau VJ. Autocrine and paracrine effects of atrial natriuretic peptide gene transfer on vascular smooth muscle and endothelial cellular growth. *J Clin Invest* 1994;94:824–829.
41. Weisbrod RM, Griswold MC, Yaghoubi M, Komalavilas P, Lincoln TM, Cohen RA. Evidence that additional mechanisms to cyclic GMP mediate the decrease in intracellular calcium and relaxation of rabbit aortic smooth muscle to nitric oxide. *Br J Pharmacol* 1998;125:1695–1707.
42. Rees DD, Palmer RM, Moncada S. The role of endothelium-derived nitric oxide in the regulation of blood pressure. *Proc Natl Acad Sci USA* 1989;86:3375–3378.
43. Ortiz PA, Garvin JL. Role of nitric oxide in the regulation of nephron transport. *Am J Physiol Renal Physiol* 2002;282:F777–F784.
44. Hanke CJ, O'Brien T, Pritchard KA, Jr, Campbell WB. Inhibition of adrenal cell aldosterone synthesis by endogenous nitric oxide release. *Hypertension* 2000;35:324–328.
45. Sayago CM, Beierwaltes WH. Nitric oxide synthase and cGMP-mediated stimulation of renin secretion.

Am J Physiol Regulatory Integrative Com Physiol 2001;281:R1146–R1151.
46. Vidal MJ, Romero JC, Vanhoutte PM. Endothelium-derived relaxing factor inhibits renin release. *Eur J Pharmacol* 1988;149:401–402.
47. Ackermann A, Fernandez-Alfonso MS, Snachez de Rojas R, Ortega T, Paul M, Gonzalez C. Modulation of angiotensin-converting enzyme by nitric oxide. *Br J Pharmacol* 1998;124:291–298.
48. Youn TJ, Kim HS, Kang HJ, Kim DW, Cho MC, Oh BH, Lee MM, Park YB. Inhibition of nitric oxide synthesis increases apoptotic cardiomyocyte death and myocardial angiotensin-converting enzyme gene expression in ischemia/reperfusion-injured myocardium of rats. *Heart Vessels* 2001;16:12–19.
49. Ichiki T, Usui M, Kato M, Funakoshi Y, Ito K, Egashira K, Takeshita A. Downregulation of angiotensin II type 1 receptor gene transcription by nitric oxide. *Hypertension* 1998;31:342–348.
50. Katoh M, Egashira K, Usui M, Ichiki T, Tomita H, Shimokawa H, Rakugi H, Takeshita A. Cardiac angiotensin II receptors are upregulated by long-term inhibition of nitric oxide synthesis in rats. *Circ Res* 1998;83:743–751.
51. Schwarz P, Diem R, Dun NJ, Forstermann U. Endogenous and exogenous nitric oxide inhibits norepinephrine release from rat heart sympathetic nerves. *Circ Res* 1995;77:841–848.
52. Kaye DM, Wiviott SD, Kobzik L, Kelly RA, Smith TW. S-nitrosothiols inhibit neuronal norepinephrine transport. *Am J Physiol* 1997;272:H875–H883.
53. Nishida Y, Ding J, Zhou MS, Chen QH, Murakami H, Wu XZ, Kosaka H. Role of nitric oxide in vascular hyperresponsiveness to norepinephrine in hypertensive Dahl rats. *J Hypertens* 1998;16:1611–1618.
54. Zonta F, Barbieri A, Reguzzoni M, Calligara A. Quantitative changes in pharmacodynamic parameters of noradrenaline in different rat aorta preparations: Influence of endogenous EDRF. *J Auton Pharmacol* 1998;18:129–138.
55. St. John Sutton MG, Sharpe N. Left ventricular remodelling after myocardial infarction: Pathophysiology and therapy. *Circulation* 2000;101:2981–2988.
56. Piano MR, Bondmass M, Schwertz DW. The molecular and cellular pathophysiology of heart failure. *Heart Lung* 1998;27:3–19.
57. Grossman W, Jones D, McLaurin LP. Wall stress and patterns of hypertrophy in the human left ventricle. *J Clin Invest* 1975;56:56–64.
58. Anversa P, Ricci R, Olivetti R. Quantitative structural analysis of the myocardium during physiologic growth and induced cardiac hypertrophy: A review. *J Am Coll Cardiol* 1987;7:1140–1149.
59. Morkin E. Postnatal muscle fiber assembly: Localization of newly synthesized myofibrillar proteins. *Science* 1970;167:1499–1501.
60. Gerdes AM, Campbell SE, Hilbelink DR. Structural remodeling of cardiac myocytes in rats with arteriovenous fistulas. *Lab Invest* 1988;59:857–861.
61. Calderone A, Takahashi N, Izzo NJ, Jr, Thaik CM, Colucci WS. Pressure- and volume-induced left ventricular hypertrophies are associated with distinct molecular phenotypes and differential induction of peptide growth factor mRNAs. *Circulation* 1995;92:2385–2390.
62. Sadoshima J, Jahn L, Takahashi T, Kulik TJ, Izumo S. Molecular characterization of the stretch-induced adaptation of cultured cardiac cells, an *in vitro* model of load-induced cardiac hypertrophy. *J Biol Chem* 1992;267:10551–10560.
63. Sadoshima J, Izumo S. Molecular characterization of angiotensin II-induced hypertrophy of cardiac myocytes and hyperplasia of cardiac fibroblasts. *Circ Res* 1993;73:413–423.
64. Knowlton KU, Michel MC, Itani M, Shubeita HE, Ishihara K, Brown JH, Chien KR. The $alpha_{1A}$-adrenergic receptor subtype mediates biochemical, molecular, and, morphological features of cultured myocardial cell hypertrophy. *J Biol Chem* 1993;268:15374–15380.
65. Morisco C, Zebrowski DC, Vatner DE, Vatner SF, Sadoshima J. Beta-adrenergic cardiac hypertrophy is mediated primarily by the beta(1)-subtype in the rat heart. *J Mol Cell Cardiol* 2001;33:561–573.
66. Malhotra R, Sadoshima J, Brosius FC III, Izumo S. Mechanical stretch and angiotensin II differentially upregulate the renin-angiotensin system in cardiac myocytes *in vitro*. *Circ Res* 1999;85:137–146.
67. Calderone A, Abdelaziz N, Colombo F, Schreiber KL, Rindt H. A farnesyltransferase inhibitor attenuates cardiac myocyte hypertrophy and gene expression. *J Mol Cell Cardiol* 2000;32:1127–1140.
68. Francis GS, Goldsmith SR, Olivari MT, Levine TB, Cohn JN. The neurohormonal axis in congestive heart failure. *Ann Intern Med* 1984;101:370–377.
69. Takahashi T, Allen PD, Izumo S. Expression of A-, B-, C-type natriuretic peptide genes in failing and developing human ventricles: Correlation with expression of the Ca^{2+}-ATPase gene. *Circ Res* 1992;71:9–17.
70. Kling R, Hystad M, Kjekshus J, Karlberg BE, Djoseland O, Aakvaag A, Hall C. An experimental study of cardiac natriuretic peptides as markers of development of congestive heart failure. *Scand J Clin Lab Invest* 1998;58:683–692.
71. Mercadier JJ, Lompre AM, Duc P, Boheler KR, Fraysse JB, Wisnewsky C, Allen PD, Komajda M, Schwartz K. Altered sarcoplasmic reticulum Ca^{2+}-ATPase gene expression in the human ventricle during end-stage heart failure. *J Clin Invest* 1990;85:305–309.
72. de la Bastie D, Levitsky D, Rappaport L, Mercadier JJ, Marotte F, Winewsky C, Brovkovich V, Schwartz K, Lompre AM. Function of the sarcoplasmic reticulum and expression of its Ca^{2+}-ATPase gene in pressure-overload-induced cardiac hypertrophy in the rat. *Circ Res* 1990;66:554–564.
73. Yamamoto K, Burnett JC, Jr, Jougasaki M, Nishimura RA, Bailey KR, Saito Y, Nakao K, Redfield MM. Superiority of brain natriuretic peptide as a hormonal marker of ventricular systolic and diastolic dysfunction and ventricular hypertrophy. *Hypertension* 1996;28:988–994.
74. Maisel A. B-type natriuretic peptide levels: A potential novel "white count" for congestive heart failure. *J Card Fail* 2001;7:183–193.
75. Tsutamoto T, Wada A, Maeda K, Hisanaga T et al. Attenuation of compensation of endogenous cardiac natriuretic peptide system in chronic heart failure. Prognostic role of plasma brain natriuretic peptide concentration in patients with chronic symptomatic left ventricular dysfunction. *Circulation* 1997;96:509–516.

76. Troughton RW, Frampton CM, Yandle TG, Espiner EA, Nicholls GA, Richards AM. Treatment of heart failure guided by plasma aminoterminal brain natriuretic peptide (N-BNP) concentrations. *Lancet* 2000;355:1126–1132.
77. Adler S, Huang H, Loke K, Xu X, Laumas A, Hintze TH. Modulation of renal oxygen consumption by nitric oxide is impaired after development of congestive heart failure in dogs. *J Cardiovasc Pharmacol* 2001;37:301–309.
78. Ikenaga H, Ishii N, Didion SP, Zhang K, Cornish KG, Patel KP, Mayhan WG, Carmines PK. Suppressed impact of nitric oxide on renal arteriolar function in rats with chronic heart failure. *Am J Physiol* 1999;276:F79–F87.
79. Hornig B, Arakawa N, Drexler H. Effect of ACE inhibition on endothelial dysfunction in patients with chronic heart failure. *Eur Heart J* 1998;19(suppl. G):G48–G53.
80. Smith CJ, Sun D, Hoegler C, Roth BS, Zhang X, Zhao G, Xu XB, Kobari Y, Pritchard K, Jr, Sessa WC, Hintze TH. Reduced gene expression of vascular endothelial NO synthase and cycloxygenase-1 in heart failure. *Circ Res* 1996;78:58–64.
81. Kichuk MR, Zhang X, Oz M, Michler R, Kaley G, Nasjletti A, Hintze TH. Angiotensin-converting enzyme inhibitors promote nitric oxide production in coronary microvessels from failing explanted human hearts. *Am J Cardiol* 1997;80(3A):137A–142A.
82. Zhang X, Recchia FA, Bernstein R, Xu X, Nasjletti A, Hintze TH. Kinin-mediated coronary nitric oxide production contributes to the therapeutic action of angiotensin-converting enzyme and neutral endopeptidase inhibitors and amlodipine in the treatment of heart failure. *J Pharmacol Exp Ther* 1999;288:742–751.
83. Arimura K, Egashira K, Nakamura R, Ide T, Tsutsui H, Shimokawa H, Takeshita A. Increased inactivation of nitric oxide is involved in coronary endothelial dysfunction in heart failure. *Am J Physiol Heart Circ Physiol* 2001;280:H68–H75.
84. Grieve DJ, MacCarthy PA, Gall NP, Cave AC, Shah AM. Divergent biological actions of coronary endothelial nitric oxide during progression of cardiac hypertrophy. *Hypertension* 2001;38:267–273.
85. Kobayashi N, Mori Y, Nakano S, Tsubokou Y, Kobayashi T, Shirataki H, Matsuoka H. TCV-116 stimulates eNOS and caveolin-1 expression and improves coronary microvascular remodelling in normotensive and angiotensin II-induced hypertensive rats. *Atherosclerosis* 2001;158:359–368.
86. Li D, Tomson K, Yang B, Mehta P, Croker BP, Mehta JL. Modulation of constitutive nitric oxide synthase, bcl-2, and Fas expression in cultured human coronary endothelial cells exposed to anoxia-reoxygenation and angiotensin II: Role of AT1 receptor. *Cardiovasc Res* 1999;41:109–115.
87. Nakane H, Miller FJ, Jr, Faraci FM, Toyoda K, Heistad DD. Gene transfer of endothelial nitric oxide synthase reduces angiotensin II-induced endothelial dysfunction. *Hypertension* 2000;35:595–601.
88. Griendling KK, Minieri CA, Ollerenshaw JD, Alexander RW. Angiotensin II stimulates NADH and NADPH oxidase activity in cultured vascular smooth muscle cells. *Circ Res* 1994;74:1141–1148.
89. Shimouchi A, Janssens SP, Bloch DB, Zapol WM, Bloch KD. cAMP regulates soluble guanylate cyclase β1-subunit gene expression in RFL-6 rat fetal lung fibroblasts. *Am J Physiol* 1993;265:L456–L461.
90. Papapetropoulos A, Marczin N, Mora G, Milici A, Murad F, Catravas JD. Regulation of vascular smooth muscle soluble guanylate cyclase activity, mRNA, and protein levels by cAMP-elevating agents. *Hypertension* 1995;26:696–704.
91. Kitashiro S, Sugiura T, Takayama Y, Tsuka Y, Izuoka T, Tokunaga S, Iwasaka T. Long-term administration of atrial natriuretic peptide in patients with acute heart failure. *J Cardiovasc Pharmacol* 1999;33:948–952.
92. Colucci WS, Elkayam U, Horton DP, Abraham WT et al. Intravenous nesiritide, a natriuretic peptide, in the treatment of decompensated congestive heart failure. Nesiritide study group. *N Engl J Med* 2000;343:246–253.
93. Chen HH, Grantham JA, Schirger JA, Jougasaki M, Redfield MM, Burnett JC, Jr. Subcutaneous administration of brain natriuretic peptide in experimental heart failure. *J Am Coll Cardiol* 2000;36:1706–1712.
94. Brunner-La Rocca HP, Kaye DM, Woods RL, Hastings J, Esler MD. Effects of intravenous brain natriuretic peptide on regional sympathetic activity in patients with chronic heart failure as compared with healthy control subjects. *J Am Coll Cardiol* 2001;37:1221–1227.
95. Luchner A, Burnett JC, Jr, Jougasaki M, Hense HW, Reigger GA, Schunkert H. Augmentation of the cardiac natriuretic peptides by beta-receptor antagonism: Evidence from a population-based study. *J Am Coll Cardiol* 1998;32:1839–1844.
96. Ohta Y, Watanabe K, Nakazawa M, Yamamota T et al. Carvedilol enhances atrial and brain natriuretic peptide mRNA expression and release in rat heart. *J Cardiovasc Pharmacol* 2000;36(suppl. 2):S19–S23.
97. Burnett JC, Jr. Vasopeptidase inhibition: A new concept in blood pressure management. *J Hypertens* 1999;17 (suppl. 1):S37–S43.
98. Fielitz J, Dendorfer A, Pregla R, Ehler E, Zurbrugg HR, Bartunek J, Hetzer R, Regitz-Zagrosek V. Neutral endopeptidase is activated in cardiac myocytes in human aortic valve stenosis and heart failure. *Circulation* 2002;105:286–289.
99. Northbridge DB, Jardine AG, Findlay IN, Archibald M, Dilly SG, Dargie HJ. Inhibition of the metabolism of atrial natriuretic factor causes diuresis and natriuresis in chronic heart failure. *Am J Hypertens* 1990;3(9):682–687.
100. Northbridge DB, Jardine A, Henderson E, Dilly SG, Dargie HJ. Increasing circulating atrial natriuretic factor concentrations in patients with chronic heart failure after inhibition of neutral endopeptidase: Effects on diastolic function. *Br Heart J* 1992;68:387–391.
101. Northbridge DB, Newby DE, Rooney E, Norrie J, Dargie HJ. Comparison of the short-term effects of candoxatril, an orally active neutral endopeptidase inhibitor, and frusemide in the treatment of patients with chronic heart failure. *Am Heart J* 1999;138:1149–1157.
102. Westheim AS, Bostrom P, Christensen CC, Parikka H, Rykke EO, Toivonen L. Hemodynamic and neuroendocrine effects for candoxatril and frusemide in mild stable chronic heart failure. *J Am Coll Cardiol* 1999;34:1794–1801.
103. Kimmelsteil CD, Perrone R, Kilcoyne L, Souhrada J, Udelson J, Smith J, de Bold A, Griffith J, Konstam MA. Effects of renal neutral endopeptidase inhibition on

sodium excretion, renal hemodynamics and neurohormonal activation in patients with congestive heart failure. *Cardiology* 1996;87:46–53.
104. Kokkonen JO, Kuoppala A, Saarinen J, Lindstedt KA, Kovanen PT. Kallidin- and bradykinin-degrading pathways in human heart: Degradation of kallidin by aminopeptidase M-like activity and bradykinin by neutral endopeptidase. *Circulation* 1999;99:1984–1990.
105. Lapointe N, Rouleau JL. Cardioprotective effects of vasopeptidase inhibitors. *Can J Cardiol* 2001;18:415–420.
106. Maki T, Nasa Y, Yamaguchi F, Yoshida H et al. Long-term treatment with neutral endopeptidase inhibitor improves cardiac function and reduces natriuretic peptides in rats with chronic heart failure. *Cardiovasc Res* 2001;51:608–617.
107. Ando S, Rahman MA, Butler GC, Senn BL, Floras JS. Comparison of candoxatril and atrial natriuretic factor in healthy men: Effects on hemodynamics, symapthetic activity, heart rate variability and endothelin. *Hypertension* 1995;26:1160–1166.
108. Bevan EG, Connell JM, Doyle Carmichael HA, Davies DL, Lorimer AR, McInnes GT. Candoxatril, a neutral endopeptidase inhibitor: Efficacy and tolerability in essential hypertension. *J Hypertens* 1992;10:607–613.
109. Ferro CJ, Spratt JC, Haynes WG, Webb DJ. Inhibition of neutral endopeptidase causes vasoconstriction of human resistance vessels *in vivo*. *Circulation* 1998;97:2323–2330.
110. McDowell G, Coutie W, Shaw C, Buchanan KD, Struthers AD, Nicholls DP. The effect of the neutral endopeptidase inhibitor drug, candoxatril, on circulating levels of the two most potent vasoactive peptides. *Br J Clin Pharmacol* 1997;43:329–332.
111. Campbell DJ, Anastasopoulos F, Duncan AM, James GM, Kladis A, Briscoe TA. Effects of neutral endopeptidase inhibition and combined angiotensin converting enzyme and neutral endopeptidase inhibition on angiotensin and bradykinin peptides in rats. *J Pharmacol Exp Ther* 1998;287:567–577.
112. Duncan AM, James GM, Anastasopoulos F, Kladis A, Briscoe TA, Campbell DJ. Interaction between neutral endopeptidase and angiotensin converting enzyme inhibition in rats with myocardial infarction: Effects on cardiac hypertrophy and angiotensin and bradykinin peptide levels. *J Pharmacol Exp Ther* 1999;289:295–303.
113. Richards AM, Wittert GA, Espiner EA, Yandle TG, Ikram H, Frampton C. Effect of inhibition of endopeptidase 24.11 on responses to angiotensin II in human volunteers. *Circ Res* 1992;71:1501–1507.
114. Seymour AA, Asaad MM, Lanoce VM, Langenbacher KM, Fennel SA, Rogers WL. Systemic hemodynamics, renal function and hormonal levels during inhibition of neutral endopeptidase 3.4.24.11 and angiotensin-converting enzyme in conscious dogs with pacing-induced heart failure. *J Pharmacol Exp Ther* 1993;266:872–883.
115. Margulies KB, Perrella MA, McKinley LJ, Burnett JC, Jr. Angiotensin inhibition potentiates the renal responses to neutral endopeptidase inhibition in dogs with congestive heart failure. *J Clin Invest* 1991;88:1636–1642.
116. Gros C, Noel N, Souque A, Schwartz JC, Danvy D, Plaquevent JC, Duhamel L, Duhamel P, Lecomte JM, Bralet J. Mixed inhibitors of angiotensin-converting enzyme (EC 3.4.15.1) and enkephalinase (EC 3.4.24.11): Rational design, properties, and potential cardiovascular applications of glycopril and alatriopril. *Proc Natl Acad Sci USA* 1991;88:4210–4214.
117. Troughton RW, Rademaker MT, Powell JD, Yandle TG, Espiner EA, Frampton CM, Nicholls MG, Richards AM. Beneficial renal and hemodynamic effects of omapatrilat in mild and severe heart failure. *Hypertension* 2000;36:523–530.
118. Chen HH, Lainchbury JG, Matsuda Y, Harty G, Burnett JC, Jr. Endogenous natriuretic peptides participate in renal and humoral actions of acute vasopeptidase inhibition in experimental mild heart failure. *Hypertension* 2001;38:187–191.
119. McClean DR, Ikram H, Garlick AH, Richards AM, Nicholls MG, Crozier IG. The clinical, cardiac, renal, arterial, and neurohormonal effects of omapatrilat, a vasopeptidase inhibitor, in patients with chronic heart failure. *J Am Coll Cardiol* 2000;36:479–486.
120. Rouleau JL, Pfeffer MA, Stewart DJ, Isaac D, Sestier F, Kerut EK, Porter CB, Proulx G, Qian C, Block AJ. Comparison of vasopeptidase inhibitor, ompatrilat, and lisinopril on exrcise tolerance and morbidity in patients with heart failure. IMPRESS randomised trial. *Lancet* 2000;356:615–620.
121. Chen HH, Lainchbury JG, Harty GJ, Burnett JC, Jr. Maximizing the natriuretic peptide system in experimental heart failure: Subcutaneous brain natriuretic peptide and acute vasopeptidase inhibition. *Circulation* 2002;105:999–1003.
122. Liu YH, Xu J, Yang XP, Yang F, Shesely E, Carretero OA. Effect of ACE inhibitors and angiotensin II type 1 receptor antagonists on endothelial NO synthase knockout mice with heart failure. *Hypertension* 2002;39(2 Pt 2):375–381.
123. Wittstein IS, Kass DA, Pak PH, Maughan WL, Fetics B, Hare JM. Cardiac nitric oxide production due to angiotensin-converting enzyme inhibition decreases beta-adrenergic myocardial contractility in patients with dilated cardiomyopathy. *J Am Coll Cardiol* 2001;38:429–435.
124. Costa MD, Bosc LV, Majowicz MP, Vidal NA, Balaszczuk AM, Arranz CT. Atrial natriuretic peptide modifies arterial blood pressure through nitric oxide pathway in rats. *Hypertension* 2000;35:1119–1123.
125. Brunner F, Wolkart G. Endothelial NO/cGMP system contributes to natriuretic peptide-mediated coronary artery and peripheral vasodilation. *Microvasc Res* 2000;61:102–110.
126. McLay JS, Chatterjee PK, Jardine AG, Hawksworth GM. Atrial natriuretic factor modulates nitric oxide production: An ANF-C receptor-mediated effect. *J Hypertens* 1995;13:625–630.
127. Murthy KS, Teng B, Jin J, Makhlouf GM. G-protein-dependent activation of smooth muscle eNOS via natriuretic peptide clearance receptor. *Am J Physiol* 1998;275:C1409–C1416.
128. Sugamori T, Ishibashi Y, Shimada T, Sakane T et al. Nitric oxide-mediated vasodilatory effect of atrial natriuretic peptide in forearm vessels of healthy humans. *Clin Exp Pharmacol Physiol* 2002;29:92–97.
129. van Der Zander K, Houben AJ, Kroon AA, De Mey JG, Smits PA, de Leeuw PW. Nitric oxide and potassium channels are involved in brain natriuretic peptide induced vasodilatation in man. *J Hypertens* 2002;20:493–499.
130. Zellner C, Protter AA, Ko E, Pothireddy MR, DeMarco T, Hutchinson SJ, Chou TM, Chatterjee K, Sudhir K.

Coronary vasodilator effects of BNP: Mechanisms of action in coronary conductance and resistance arteries. *Am J Physiol* 1999;276:H1049–H1057.
131. Das UN. Free radicals, cytokines and nitric oxide in cardiac failure and myocardial infarction. *Mol Cell Biochem* 2000;215:145–152.
132. Lissoni P, Pelizzoni F, Mauri O, Perego M, Pittalis S, Barni S. Enhanced secretion of tumour necrosis factor in patients with myocardial infarction. *Eur J Med* 1992;1:277–280.
133. Vaddi K, Nicolini FA, Mehta P, Mehta JL. Increased secretion of tumour necrosis factor-α by mononuclear leukocytes in patients with ischemic heart disease. *Circulation* 1994;90:694–699.
134. Agnoletti L, Curello S, Bachetti T, Malacarne F et al. Serum from patients with severe heart failure downregulates eNOS and is proapoptotic: Role of tumour necrosis factor-alpha. *Circulation* 1999;100:1983–1991.
135. Fichtlscherer S, Rossig L, Breuer S, Vasa M, Dimmeler S, Zeiher AM. Tumour necrosis factor antagonism with etanercept improves systemic endothelial vasoreactivity in patients with advanced heart failure. *Circulation* 2001;104:3023–3025.
136. Rossig L, Haendeler J, Mallat Z, Hugel B, Freyssinet JM, Tedgui A, Dimmeler S, Zeiher AM. Congestive heart failure induces endothelial cell apoptosis: Protective role of carvedilol. *J Am Coll Cardiol* 2000;36:2081–2090.
137. Yoshizumi M, Perrella MA, Burnet JC, Jr, Lee ME. Tumour necrosis factor downregulates and endothelial nitric oxide synthase mRNA by shortening its half-life. *Circ Res* 1993;73:205–209.
138. De Frutos T, Sanchez De Miguel L, Garcia-Duran F, Gonzalez-Fernandez F et al. NO from smooth muscles cells decrease NOS expression in endothelial cells: Role of TNF-α. *Am J Physiol* 1999;277:H1317–H1325.
139. Frey RS, Rahman A, Kefer JC, Minshall RD, Malik AB. PKCξ regulates TNF-α-induced activation of NADPH oxidase in endothelial cells. *Circ Res* 2002;90:1012–1019.
140. Li J-M, Mullen AM, Yun S, Wientjes F, Brouns GY, Thrasher AJ, Shah AM. Essential role of the NADPH oxidase subunit $p47^{phox}$ in endothelial superoxide production in response to phorbol ester and tumour necrosis factor-α. *Circ Res* 2002;90:143–150.
141. Satoh M, Nakamura M, Tamura G, Makita S, Segawa I, Tashiro A, Satodate R, Hiramori K. Inducible nitric oxide synthase and tumour necrosis factor-alpha in myocardium in human dilated cardiomyopathy. *J Am Coll Cardiol* 1997;29:716–724.
142. Irwin MW, Mak S, Mann DL, Qu R, Penninger JM, Yan A, Dawood F, Wen WH, Shou Z, Liu P. Tissue expression and immunolocalization of tumour necrosis factor-alpha in postinfarction dysfunctional myocardium. *Circulation* 1999;99:1492–1498.
143. Yue P, Massie BM, Simpson PC, Long CS. Cytokine expression increases in nonmyocytes from rats with postinfarction heart failure. *Am J Physiol* 1998;275:H250–H258.
144. Meldrum DR. Tumour necrosis factor in the heart. *Am J Physiol* 1998;274:R577–595.
145. Frangogiannis NG, Smith CW, Entman ML. The inflammatory response in myocardial infraction. *Cardiovasc Res* 2002;53:31–47.
146. Takimoto Y, Aoyama T, Keyamura R, Shinoda E, Hattori R, Yui Y, Sasayama S. Differential expression of three types of nitric oxide synthase in both infracted and non-infarcted left ventricles after myocardial infarction. *Int J Cardiol* 2000;76:135–145.
147. Kubota T, McTiernan CF, Frye CS, Slawson SE, Lemster BH, Koretsky AP, Demetris AJ, Feldman AM. Dilated cardiomyopathy in transgenic mice with cardiac-specific overexpression of tumour necrosis factor-alpha. *Circ Res* 1997;81:627–635.
148. Kiemer AK, Vollmar AM. The atrial natriuretic peptide regulates the production of inflammatory mediators in macrophages. *Ann Rheum Dis* 2001;60(suppl. 3):iii68–70.
149. Kiemer AK, Baron A, Gerbes AL, Bilzer M, Vollmar AM. The atrial natriuretic peptide as a regulator of Kupffer cell functions. *Shock* 2002;16:365–371.
150. Kiemer AK, Vollmar AM, Bilzer M, Gerwig T, Gerbes AL. Atrial natriuretic peptide reduces expression of TNF-alpha mRNA during reperfusion of the rat liver upon decreased activation of NF-kappaB, and AP-1. *J Hepatol* 2000;33:236–246.
151. Tsukagoshi H, Shimizu Y, Kawata T, Hisada T et al. Atrial natriuretic peptide inhibits tumour necrosis factor-alpha production by interferon-gamma activated macrophages via suppression of p38 mitogen-activated protein kinase and nuclear factor-kappa B activation. *Regul Pept* 2001;99:21–29.
152. Kalra D, Sivasubramanian N, Mann DL. Angiotensin II induces tumour necrosis factor biosynthesis in the adult mammalian heart through a protein kinase C-dependent pathway. *Circulation* 2002;105:2198–2205.
153. Zhao SP, Xie XM. Captopril inhibits the production of tumour necrosis factor-alpha by human mononuclear cells in patients with congestive heart failure. *Clin Chim Acta* 2001;304:85–90.
154. Tsutamoto T, Wada A, Maeda K, Mabuchi N et al. Angiotensin II type 1 receptor antagonist decreases plasma levels of tumour necrosis factor alpha, interleukin-6 and soluble adhesion molecules in patients with chronic heart failure. *J Am Coll Cardiol* 2000;35:714–721.
155. Arstall MA, Sawyer DB, Fukazawa R, Kelly RA. Cytokine-mediated apoptosis in cardiac myocytes: The role of inducible nitric oxide synthase induction and peroxynitrite generation. *Circ Res* 1999;85:829–840.
156. Gealekman O, Abassi Z, Rubinstein I, Winaver J, Binak O. Role of myocardial indicible nitric oxide synthase in contractile dysfunction and β-adrenergic hyporesponsiveness in rats with experimental volume-overload heart failure. *Circulation* 2002;105:236–243.
157. Habib FM, Springall DR, Davies GJ, Oakley CM, Yacoub MH, Polak JM. Tumour necrosis factor and inducible nitric oxide synthase in dilated cardiomyopathy. *Lancet* 1996;347:1151–1155.
158. Haywood GA, Tsao PS, vo der Leyen HE, Mann MJ et al. Expression of inducible nitric oxide synthase in human heart failure. *Circulation* 1996;93:1087–1094.
159. Heba G, Krzeminski T, Porc M, Grzyb J, Ratajska A, Dembinska-Kiec A. The time course of tumour necrosis factor-alpha, inducible nitric oxide synthase and vascular endothelial growth factor expression in an experimental model of chronic myocardial infarction in rats. *J Vasc Res* 2001;38:288–300.
160. Fukuchi M, Hussain SNA, Giaid A. Heterogeneous expression and activity of endothelial and inducible

nitric oxide synthases in end-stage human heart failure. *Circulation* 1998;98:132–139.
161. Kiemer AK, Hartung T, Vollmar AM. cGMP-mediated inhibition of TNF-alpha production by the atrial natriuretic peptide in murine macrophages. *J Immunol* 2000;165:175–181.
162. Kiemer AK, Vollmar AM. Autocrine regulation of inducible nitric oxide synthase in macrophages by atrial natriuretic peptide. *J Biol Chem* 1998;273:13444–13451.
163. Chatterjee PK, Hawksworth, McLay JS. Cytokine-stimulated nitric oxide production in the human renal proximal tubule and its modulation by natriuretic peptides: A novel immunomodulatory mechanism? *Exp Nephrol* 1999;7:438–448.
164. Irie K, Tsukahara F, Fujii E, Uchida Y, Yoshioka T, He WR, Shitashige M, Murota S, Muraki T. Cationic amino acid transporter-2 mRNA induction by tumour necrosis factor-alpha in vascular endothelial cells. *Eur J Pharmacol* 1997;339:289–293.
165. Kiemer AK, Vollmar AM. Induction of L-arginine transport is inhibited by atrial natriuretic peptide; a peptide hormone as a novel regulator of inducible nitric oxide synthase substrate availability. *Mol Pharmacol* 2001;60:421–426.

Mechanisms Underlying Nitrate-Induced Endothelial Dysfunction: Insight from Experimental and Clinical Studies

Ascan Warnholtz,[1] Nikolaus Tsilimingas,[2] Maria Wendt,[1] and Thomas Münzel[1]
[1]Division of Cardiology and [2]Cardiovascular Surgery, University Hospital Hamburg-Eppendorf, Hamburg, Germany

Abstract. **The hemodynamic and anti-ischemic effects of nitroglycerin (NTG) are rapidly blunted due to the development of nitrate tolerance. With initiation of nitroglycerin therapy one can detect neurohormonal activation and signs for intravascular volume expansion. These so called pseudotolerance mechanisms may compromise nitroglycerin's vasodilatory effects. Long-term treatment with nitroglycerin is also associated with a decreased responsiveness of the vasculature to nitroglycerin's vasorelaxant potency suggesting changes in intrinsic mechanisms of the tolerant vasculature itself may also contribute to tolerance. More recent experimental work defined new mechanisms of tolerance such as increased vascular superoxide production and increased sensitivity to vasoconstrictors secondary to an activation of the intracellular second messenger protein kinase C. As potential superoxide producing enzymes, the NADPH oxidase and the nitric oxide synthase have been identified. Nitroglycerin-induced stimulation of oxygen-derived free radicals together with NO derived from nitroglycerin may lead to the formation of peroxynitrite, which may be responsible for the development of tolerance as well as for the development of cross tolerance to endothelium-dependent vasodilators. The oxidative stress concept of tolerance and cross tolerance may explain why radical scavengers such as vitamin C or substances which reduce oxidative stress, such as ACE-inhibitors, AT1 receptor blockers or folic acid, are able to beneficially influence both tolerance and nitroglycerin-induced endothelial dysfunction. New aspects concerning the role of oxidative stress in nitrate tolerance and nitrate induced endothelial dysfunction and the consequences for the NO/cyclicGMP downstream target, the cGMP-dependent protein kinase will be discussed.**

Key Words. **nitroglycerin, nitrate tolerance, endothelial dysfunction, atherosclerosis**

Vasodilator Action of Nitroglycerin

Nitrates are still widely used in the management of coronary artery disease including in patients with stable and unstable angina, acute myocardial infarction as well as congestive heart failure. The therapeutic efficacy is due to peripheral venous and arterial dilation that results in decreased myocardial oxygen consumption. Nitrates also dilate large coronary arteries and collaterals while having minimal or no effect on arteriolar tone. It is assumed that nitroglycerin induces vasorelaxation by releasing the vasoactive principle nitric oxide (NO) via an enzymatic biotransformation step. NO, an endothelium-derived relaxing factor (EDRF), activates the target enzyme soluble guanylyl cyclase (sGC) and increases tissue levels of the second messenger cGMP. Cyclic GMP in turn activates a cGMP-dependent protein kinase (cGK-I) which has been shown to mediate vasorelaxation via phosphorylation of proteins that regulate intracellular Ca^{2+} levels [1]. NO released from nitroglycerin may also beneficially influence the process of atherosclerosis by reducing neutrophil adhesion to the endothelium and by inhibiting the expression of adhesion molecules and platelet aggregation. Despite the nitroglycerin's beneficial hemodynamic profile and potential anti-atherogenic effects, the results concerning the efficacy of this kind of treatment in patients with coronary artery disease remain disappointing [1–3]. Recent meta-analysis even indicates that the long-term use of nitrates may be deleterious to patients with ischemic heart disease [4].

Enhanced Sensitivity to Vasoconstriction Contributes to Nitrate Tolerance

Though acute application of nitroglycerin exhibits high vasodilator and anti-ischemic efficacy, this activity is rapidly lost upon chronic nitroglycerin treatment due to the development of nitrate tolerance [5,6]. The mechanisms underlying

Address for correspondence: Thomas Münzel M.D., Universitätsklinikum Hamburg-Eppendorf, Medizinische Klinik III, Schwerpunkt Kardiologie und Angiologie, Martinistr. 52, 20246 Hamburg, Germany. Tel.: 49-40-42803-3988; Fax: 49-40-42803-5862; E-mail: muenzel@uke.uni-hamburg.de

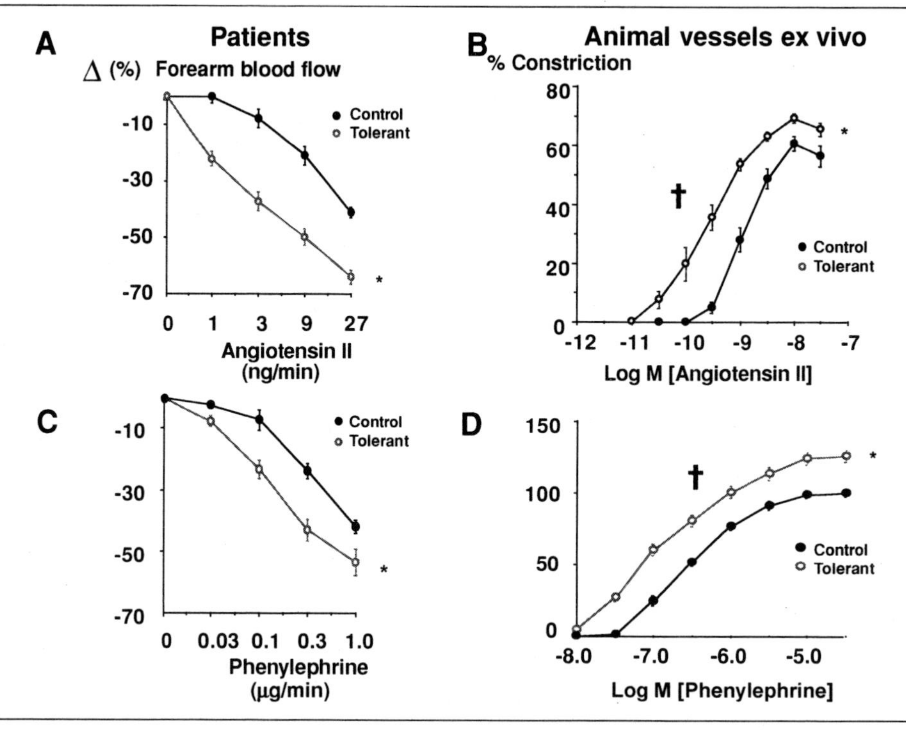

Fig. 1. Effects of chronic nitroglycerin treatment on vasoconstrictor sensitivity of forearm resistance vessels in patients with coronary artery disease and rabbit aorta in response to angiotensin II and phenylephrine. Nitroglycerin treatment for 3d led to a marked hypersensitivity to vasoconstricting agents such as angiotensin II and phenylephrine in patients with coronary artery disease (A,C) [9] as well as in experimental animals (B,D) [7].

this phenomenon are likely to be multifactorial and may involve neurohormonal counter-regulatory mechanisms, impaired nitroglycerin biotransformation or changes intrinsic to the vasculature.

An interesting new aspect of chronic therapy with organic nitrates was the recent demonstration of increased sensitivity to vasoconstrictors such as serotonin, phenylephrine, angiotensin II and thromboxane. This has been shown to occur in animals treated with nitroglycerin for a 3 day period in a clinically relevant concentration of 1.5 μg/kg/min [7] and in rats with chronic nitroglycerin infusion [8] (Fig. 1B,D). Reports from patients with coronary artery disease also indicate that these observations may have clinical relevance. Heitzer et al. observed that reductions in forearm blood flow in response to intra-arterial (brachial artery) angiotensin II and phenylephrine (Fig. 1A,C) infusion were markedly enhanced in patients pretreated with nitroglycerin for a 48 h period (0.5 μg/kg/min) [9]. Interestingly, these hypercontractile responses could be blocked by concomitant treatment with the ACE-inhibitor captopril, suggesting that the renin angiotensin system is, at least in part, responsible for this phenomenon [9]. The demonstration of such hypercontractile responses of the forearm microcirculation within 48 h of continuous nitroglycerin treatment may indicate that increases in sensitivity to constrictors represents a major mechanism responsible for the attenuation of nitroglycerin's vasodilator effects [10]. Using an animal model of nitrate tolerance we could normalize vasoconstrictor responses using *in vitro* inhibitors of protein kinase C (PKC), an important second messenger for smooth muscle contraction [7]. This concept is supported by more recent data demonstrating that *in vivo* treatment with protein kinase C inhibitors prevented the increase in sensitivity to vasoconstrictors and in parallel nitrate tolerance [8].

Nitroglycerin Treatment Induces Endothelial Dysfunction

A phenomenon related to nitrate tolerance is cross-tolerance to other endothelium-dependent and -independent nitrovasodilators. This has been observed most commonly in situations where nitroglycerin was administered chronically *in vivo* in experimental animal models [11–13] (Fig. 2A) and is not encountered in situations where nitrate tolerance is produced by short-term exposure of vascular segments *in vitro* [14].

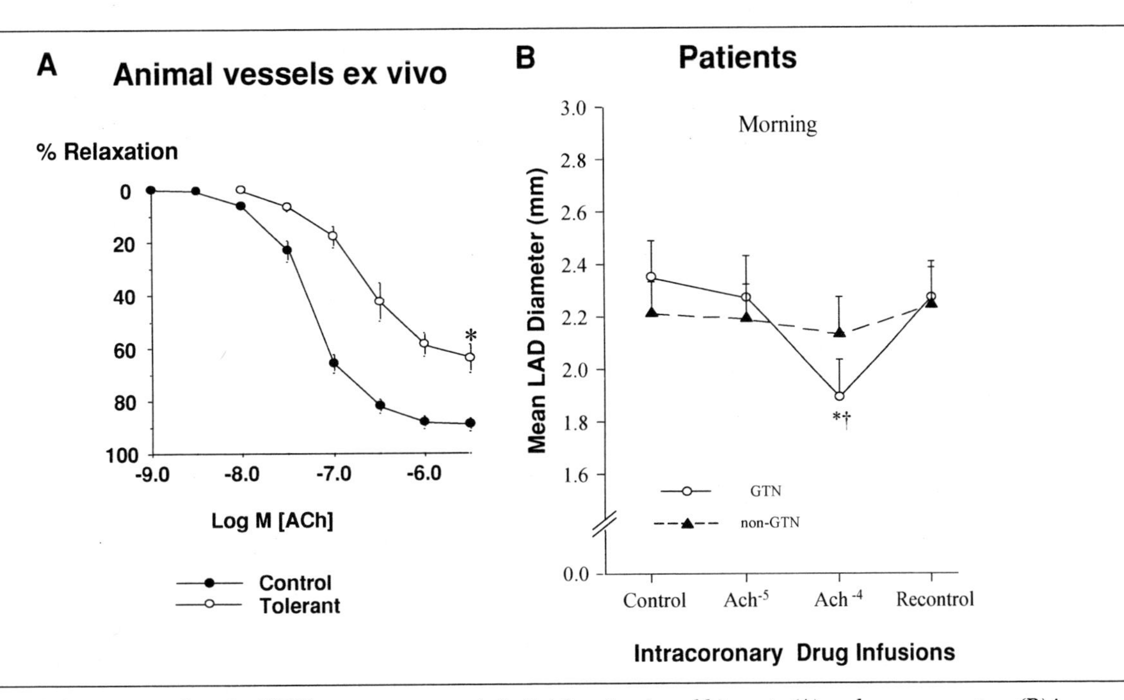

Fig. 2. *Effects of chronic nitroglycerin (NTG) treatment on endothelial function in rabbit aorta (A) and coronary artery (B) in patients with coronary artery disease. Nitroglycerin treatment for 3d resulted in a significant shift of the acetylcholine (Ach) dose response relationship in rabbit aorta compatible with endothelial dysfunction [11]. Likewise, nitroglycerin treatment for 6d with patches led to stronger acetylcholine-induced coronary constriction compared to patients treated with placebo, compatible with nitroglycerin-induced endothelial dysfunction [15].*

The question remains whether similar negative effects on endothelial function can be observed in humans during prolonged nitroglycerin therapy. Using acetylcholine-induced vasoconstriction as a surrogate parameter for endothelial function in large coronary arteries, Caramori et al. recently found that continuous treatment with nitroglycerin (5d, nitroglycerin patches; Fig. 2B) leads to enhanced acetylcholine-induced constriction [15]. The interpretation of these results may be confounded by the fact that chronic nitroglycerin treatment also causes a hypersensitivity to vasoconstricting agonists [9] e.g. by activating the second messenger protein kinase C and by stimulating the expression of endothelin-1 within the smooth muscle layer [7]. Thus by studying acetylcholine-induced constriction in the coronary circulation, it may be difficult to differentiate whether this phenomenon is due to endothelial dysfunction or secondary to nitroglycerin-induced hyperreactivity of the smooth muscle cells to constricting agonists, a question which was addressed by a subsequent study by John Parkers group.

In this study, Gori et al. examined whether chronic nitroglycerin treatment may affect endothelial function of the forearm circulation of healthy volunteers [16]. Endothelial function was assessed using strain gauge plethysmography. The nitroglycerin dose was 0.6 mg/h/d, which averages a nitroglycerin concentration of about 0.1 μg/kg/min. The treatment period was 6d and the study was designed in an investigator-blinded parallel fashion. Flow responses of the brachial artery were studied in response to intra-arterial infusion of the endothelium-dependent vasodilator acetylcholine and the inhibitor of the nitric oxide synthase (NOS), N^G-monomethyl-L-arginine (L-NMMA) (Fig. 3). Continuous treatment with nitroglycerin patches for 6d resulted in a marked inhibition of acetylcholine-induced increases in forearm blood flow as compared to the control group without nitroglycerin pretreatment [16]. Likewise, the N^G-monomethyl-L-arginine induced vasoconstriction was significantly blunted in volunteers treated with nitroglycerin. Using the lowest concentration, N^G-monomethyl-L-arginine was even able to cause a paradoxical dilation. Based on these findings, the authors concluded that nitroglycerin treatment has an inhibitory effect on basal as well as stimulated vascular NO-bioavailability and that this is, at least in part, due to abnormalities in nitric oxide synthase III function [16]. Taken together, we believe that there is no doubt that chronic nitroglycerin treatment is causing endothelial dysfunction, which may have important clinical implications since endothelial dysfunction [17] as well as oxidative stress leading to endothelial dysfunction [18] have been shown

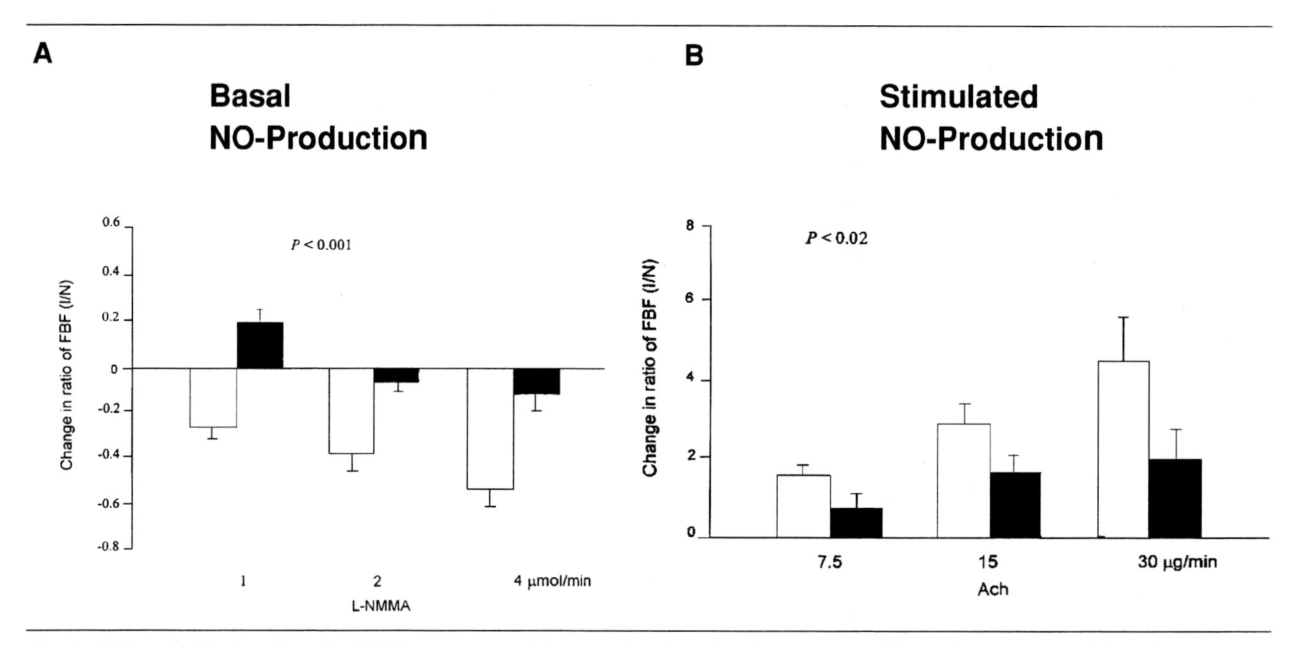

Fig. 3. Effects of 6d nitroglycerin (NTG) treatment with patches on basal and stimulated NO-production in healthy volunteers. N^G-monomethyl-L-arginine (L-NMMA)-induced reduction in forearm blood flow were drastically reduced in nitroglycerin-treated subjects, reflecting a marked decrease in basal NO-production. Likewise, acetylcholine (Ach)-induced increases in forearm blood flow were significantly reduced by in vivo nitroglycerin-treatment, compatible with reductions in agonist-induced endothelial function (modified from [37]).

to be predictors of adverse long-term outcome in patients with coronary artery disease.

Oxidative Stress Accounts for Endothelial Dysfunction

The demonstration of attenuated acetylcholine forearm blood flow responses as well as of attenuated N^G-monomethyl-L-arginine responses in healthy volunteers as a result of chronic nitroglycerin treatment indicates diminished vascular NO-bioavailability.

Several phenomena may account for this, namely a decrease in the expression of nitric oxide synthase, a dysfunctional nitric oxide synthase due to intracellular L-arginine or tetrahydrobiopterin (BH_4) deficiency, increased vascular superoxide production or a desensitization of the smooth muscle cell target enzyme, soluble guanylyl cyclase as suggested by Ferid Murad's group [12]. At first glance, it seems difficult to see how these quite different mechanisms may fit into one solid concept. A closer look, however, indicates that all of these findings may be explained by a nitroglycerin-induced increase in vascular superoxide production.

Evidence for a role of oxidative stress leading to cross-tolerance was first provided by experimental studies showing that superoxide dismutase (SOD) was able to improve cross tolerance to acetylcholine [11]. Subsequently, tolerance has been shown to be associated with increased superoxide levels in endothelial as well as smooth muscle cells [19]. As a potential superoxide producing enzyme, the NADPH oxidase has been identified [20]. The stimulation of vascular superoxide production by nitroglycerin therapy may have several consequences. Increased superoxide levels in endothelial and smooth muscle cells may inhibit the nitroglycerin's vasodilator potency simply by degrading nitric oxide released from nitroglycerin during the biotransformation process.

Superoxide stimulated by nitroglycerin-treatment may also react with NO in a diffusion limited reaction ($k = 1.9 \times 10^{10}\ M^{-1} \cdot s^{-1}$) that is about 10 times faster than the dismutation of superoxide by superoxide dismutase. This reaction produces peroxynitrite, a compound with limited NO-like bioactivity thereby "shunting" NO away from its typical target actions such as vasodilation and inhibition of platelets. *In vitro* as *in vivo* data indicate, that nitroglycerin treatment increases vascular [21] and urinary nitrotyrosine levels [22], which can be considered as a marker of peroxynitrite-dependent oxidative damage. There is a growing body of evidence showing that increased vascular peroxynitrite formation may also have deleterious consequences for the function of nitric oxide synthase III. Peroxynitrite is a strong stimulus for the oxidation of the nitric oxide synthase III cofactor tetrahydrobiopterin to dihydrobiopterin [23]. The resulting intracellular tetrahydrobiopterin deficiency may lead to an uncoupling of nitric oxide synthase III [23,24]. Thus, nitroglycerin therapy may switch nitric oxide synthase III from a NO to a superoxide producing enzyme, which may further increase oxidative stress in vascular tissue in a positive feedback fashion. Indeed, we were recently able to demonstrate an uncoupled nitric oxide synthase in an animal model of nitrate tolerance since an inhibitor of nitric oxide synthase, N^G-nitro-L-arginine was able to significantly reduce vascular superoxide production in tolerant vessels [25]. In addition, supplementation of nitroglycerin-treated rats with tetrahydrobiopterin was able to reverse nitroglycerin-induced endothelial dysfunction [26] further indicating that chronic nitroglycerin treatment causes endothelial dysfunction, which is at least in part secondary to intracellular depletion of the nitric oxide synthase III cofactor tetrahydrobiopterin (Fig. 4). The clinical relevance of these experimental findings has recently been highlighted by Gori et al. [16]. In this study, the authors could not

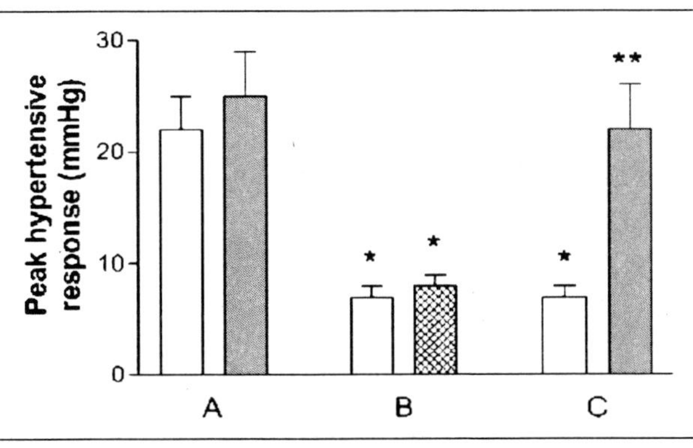

Fig. 4. Effects of tetrahydrobiopterin on N^G-monomethyl-L-arginine (L-NMMA)-induced increases in control and nitroglycerin (NTG)-treated rats. Peak hypertensive response (mm Hg) to N^G-monomethyl-L-arginine in normal, non-tolerant rats and nitroglycerin tolerant rats before and after a 2 h infusion of tetrahydrobiopterin (BH4, filled columns) or placebo (crosshatched column). Tetrahydrobiopterin treatment did not modify the N^G-monomethyl-L-arginine induced hypertension in controls (A). In the setting of tolerance, the N^G-monomethyl-L-arginine induced increases in blood pressure were markedly blunted compatible with a decrease in basal NO production (B). Treatment with tetrahydrobiopterin (C) but not with placebo (B) restored the N^G-monomethyl-L-arginine responses in NTG-treated animals. These findings clearly indicate that nitroglycerin treatment decreases baseline NO production due to the induction of intracellular tetrahydrobiopterin deficiency and that reduction in baseline NO production may be in part due to nitric oxide synthase III uncoupling (modified from [26]).

only demonstrate that nitroglycerin-induced endothelial dysfunction greatly responds to treatment with folic acid but they also found a great improvement of nitrate tolerance in forearm vessels of healthy volunteers. Interestingly, recent *in vitro* studies with the isolated nitric oxide synthase III enzyme indicate, that folic acid is a compound that restores nitric oxide synthase III function by increasing depleted intracellular tetrahydrobiopterin levels [27] (Fig. 5).

Another mechanism by which nitric oxide synthase III uncoupling may occur is intracellular depletion of L-arginine [28]. Interestingly, incubation of cultured endothelial cells with nitroglycerin has been shown to reduce nitroglycerin's vasodilator potency, to deplete intracellular L-arginine levels [29], and to stimulate endothelial cells to produce superoxide [30]. Since endothelial superoxide production was blocked by nitric oxide synthase III inhibitors and tolerance improved with L-arginine, the authors concluded that nitroglycerin-induced increases in superoxide production may be, at least in part, secondary to nitric oxide synthase III uncoupling [30]. It is not very likely that endothelial dysfunction is secondary to decreased expression of nitric oxide synthase III, since recent experimental studies have shown that nitric oxide synthase, in the setting of tolerance, is not at all modified in eNOS knockout mice [31] and rather up- than downregulated in tolerant rabbit aorta [25]. Superoxide has also recently been shown to be a potent stimulus for the activation of protein kinase C in endothelial cells [32]. Protein kinase C in turn may phosphorylate nitric oxide synthase leading to an inhibition of nitric oxide synthase activity and therefore to an inhibition of NO production by the enzyme [33], all of which may contribute to nitroglycerin-induced endothelial dysfunction.

Nitroglycerin-induced increases in oxidative stress may also lead to increased production of endothelin-1 within endothelial and smooth muscle cells leading to protein kinase C activation which in turn may trigger enhanced constrictor responses to almost every agonist [34,35].

That nitroglycerin therapy indeed stimulates superoxide production in human tissue was recently shown by Sage et al. [36] for the internal mammary artery in patients undergoing bypass surgery. The authors failed, however, to demonstrate any cross-tolerance to endothelium dependent and independent vasodilators and *in vitro* modulation of vascular superoxide production did not modify the nitroglycerin-dose response relationship. In addition, they established a decreased tissue content of 1,2 glyceryl dinitrate in tolerant tissue, which was used to argue that impaired nitroglycerin-biotransformation specifically accounts for tolerance and that the endothelial

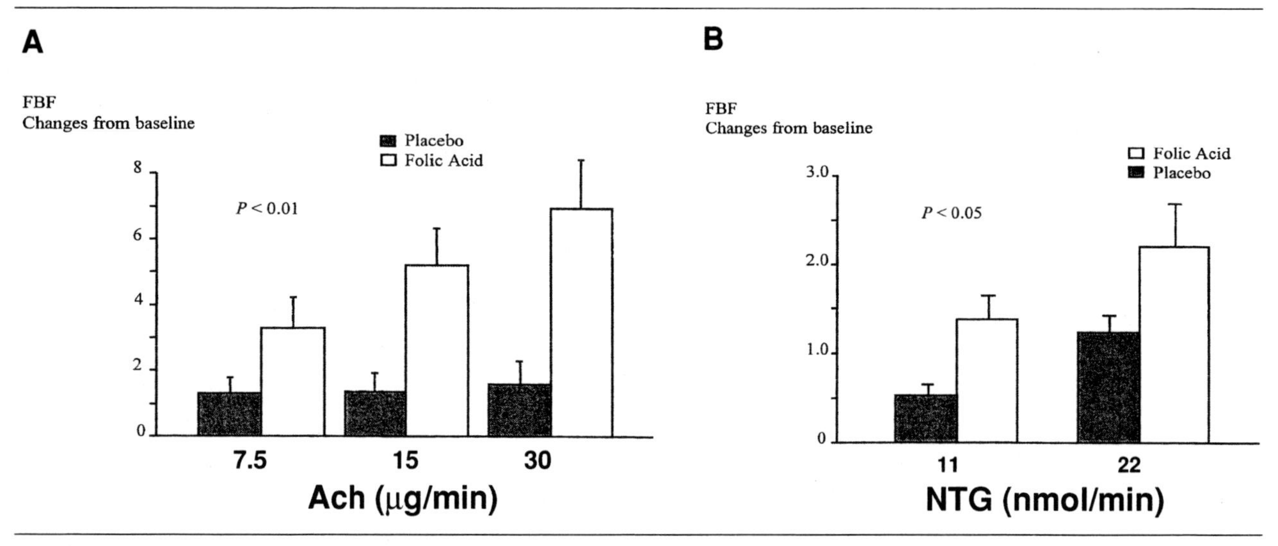

Fig. 5. Responses of forearm blood flow to intra-arterial acetylcholine (Ach) (A) and nitroglycerin (NTG) (B) in healthy controls treated or not treated with folic acid (10 mg/day for 10 days). Folic acid treatment markedly improved nitroglycerin-induced endothelial dysfunction and improved tolerance. Since folic acid restores depleted tetrahydrobiopterin (BH4) levels, the authors concluded that nitroglycerin-induced tolerance as well as cross tolerance is reversed, at least in part, by a restoration of intracellular BH4 bioavailability (modified from [16]).

function is preserved [36]. Differences in the vessel region studied (conductance versus resistance vessels) as well as differences in the duration of nitroglycerin treatment (1 versus 6 days) make it difficult, however, to compare the studies by Sage [36] and Gori [37].

NTG-Induced Superoxide/Peroxynitrite Production Inhibits the Activity of Cyclic GMP-Dependent Protein Kinase I (cGK-I)

How does nitroglycerin-induced stimulation of superoxide production influence intracellular signaling downstream from NO/cyclicGMP? Recent studies in cGK-I deficient mice highlighted the crucial role of this enzyme in mediating NO/cGMP stimulated vasodilation [38]. In cGK-I knockout animals, endothelium-dependent as well as endothelium-independent nitrovasodilators failed to induce significant relaxations of vascular tissue. This observation clearly indicates that the activity and/or expression of cGK-I has to be monitored in the setting of tolerance and endothelial dysfunction. The phosphorylation of the 46/50-kDa vasodilator stimulated phosphoprotein (VASP) and in particular the phosphorylation of the vasodilator stimulated phosphoprotein at serine 239 (P-VASP) has been shown to be a useful monitor for cGK-I activity in intact cells [39]. The vasodilator stimulated phosphoprotein, a protein highly expressed in vascular cells including platelets, endothelial cells and vascular smooth muscle cells, is phosphorylated at three distinct sites (serine 157, serine 239 and threonine 278) by both cGK-I and cyclicAMP-dependent protein kinases with overlapping specificity and efficiency [39]. Experiments with cGK-I-deficient human endothelial cells and platelets established that cGMP-mediated VASP phosphorylation is mediated by cGK-I [40,41]. In addition, recent studies by our group indicate that changes in P-VASP closely follow changes in endothelial function and oxidative stress, suggesting that P-VASP can be used as a novel, biochemical surrogate parameter for vascular NO-bioavailability [42].

To assess whether cGK-I, the downstream-target of soluble guanylyl cyclase, was affected by chronic nitroglycerin treatment, we analyzed the expression of this protein by western blotting. We could not detect any changes in cGK-I expression in aortas from nitroglycerin-tolerant rats and rabbits compared to controls.

To analyze the activity of the cGK-I we studied the expression of P-VASP [39]. Nitroglycerin-treatment of rabbits and rats for three days caused a striking reduction of P-VASP in aortic tissue of these animals when compared to untreated controls (Fig. 6). These findings clearly indicate that the NO/cGMP pathway is functionally inhibited during nitroglycerin-induced tolerance while cGK-I and VASP expression was not modified at all. To address a role for oxidative stress in inhibiting the activity of cGK-I, the P-VASP was quantified in response to *in vitro* and *in vivo* treatment with

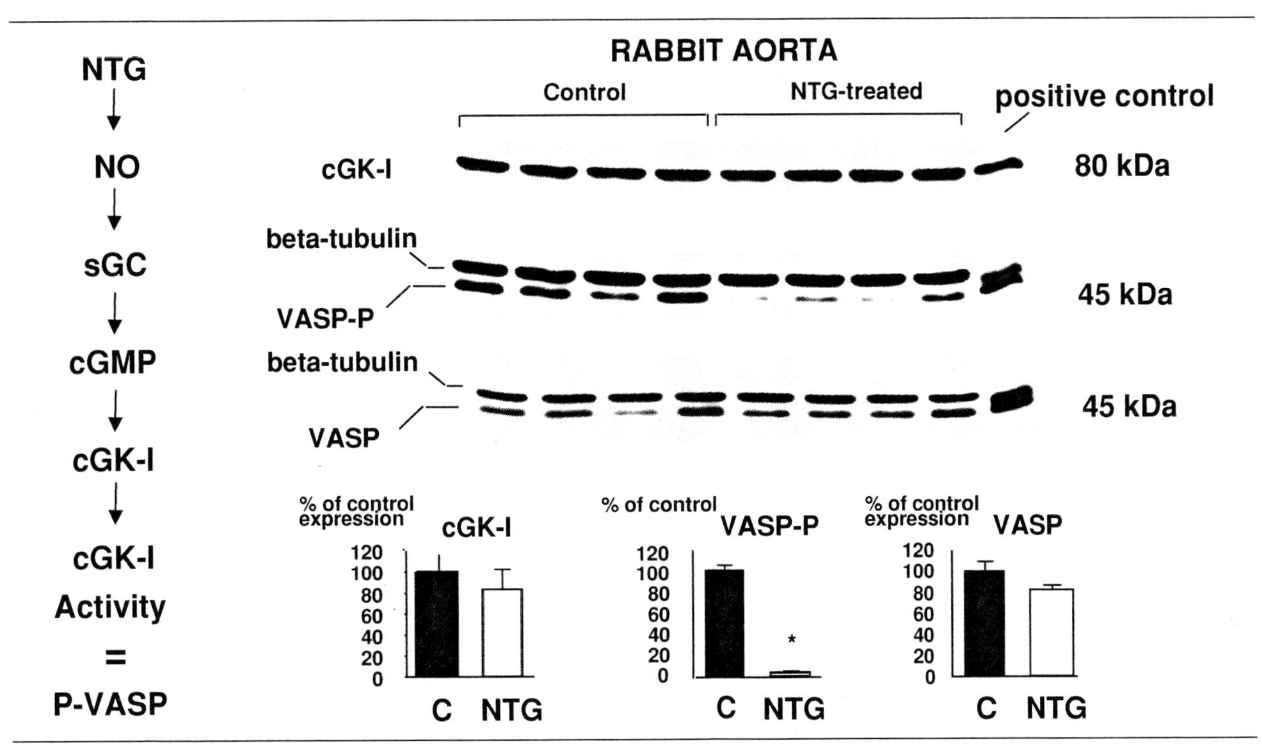

Fig. 6. Effects of chronic nitroglycerin (NTG) treatment on the expression and activity of the cGMP-dependent protein kinase (cGK-I). cGK-I activity was assessed using monoclonal antibodies against vasodilator stimulated phosphoprotein-ser239 (P-VASP) as described recently [42]. Nitroglycerin treatment for 3d markedly decreased P-VASP, while having no significant effect on the expression of the cGK-I (modified from [19]).

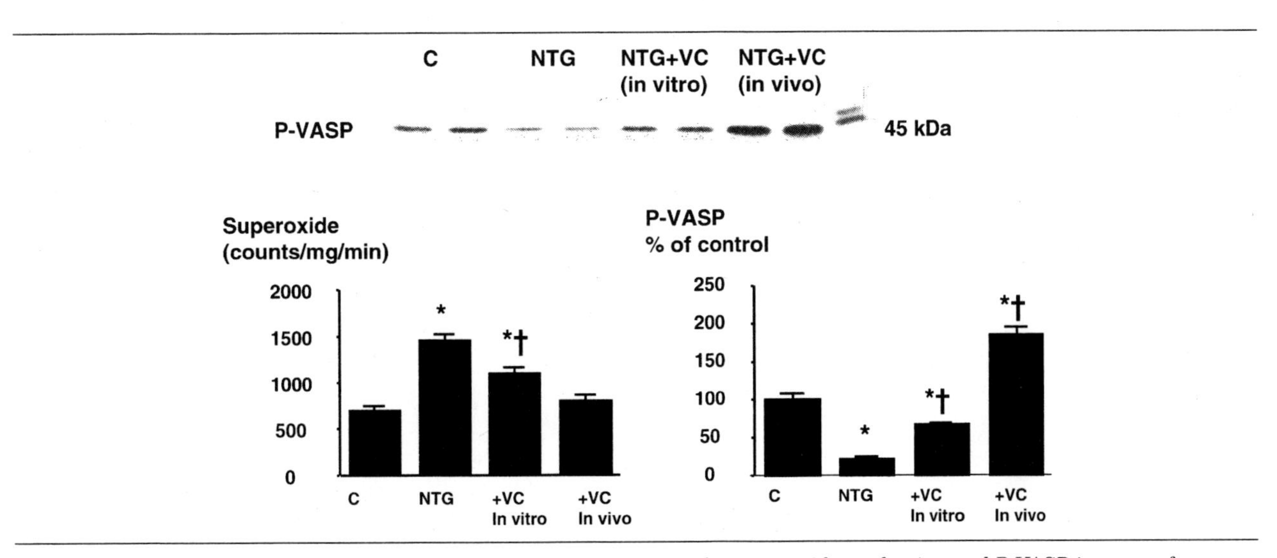

Fig. 7. Effects of in vivo and in vitro vitamin C (VC) treatment on vascular superoxide production and P-VASP in aortas from nitroglycerin (NTG)-treated rabbits. Treatment with the antioxidant restored P-VASP indicating that nitroglycerin (NTG)-induced oxidative stress contributes at least in part to the inhibition of the activity of the cGK-I (modified from [19]). C; control.

the superoxide/peroxynitrite scavenger vitamin C. With these studies we could demonstrate that vitamin C markedly restored P-VASP, decreased superoxide levels and accordingly restored nitroglycerin sensitivity in vessels from nitroglycerin-treated animals (Fig. 7).

We therefore propose the following sequence of events leading to tolerance as well as to endothelial dysfunction in vascular tissue in response to chronic nitroglycerin treatment: Nitroglycerin therapy stimulates vascular superoxide production during the biotransformation process

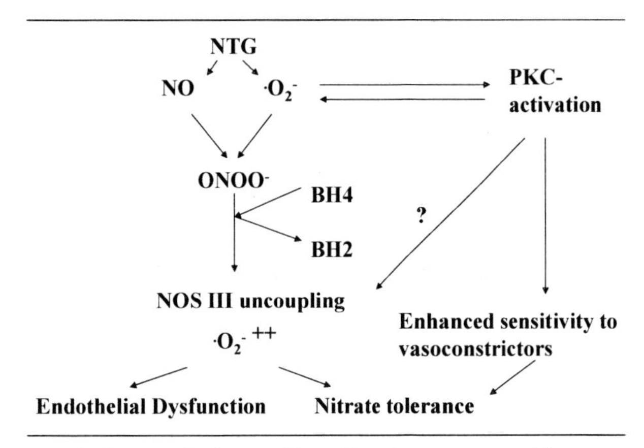

Fig. 8. *Proposed scheme of how chronic treatment with nitroglycerin (NTG) may cause tolerance and endothelial dysfunction. Nitroglycerin therapy stimulates vascular superoxide production during the biotransformation process (cytochrome P450-mediated?) or via activation of the NADPH oxidase in endothelial as well as in smooth muscle cells. Superoxide in turn is a strong stimulus for the activation of protein kinase C (PKC) in the endothelium and in the smooth muscle cell layer, leading to nitric oxide synthase (NOS III) inhibition and to enhanced sensitivity to vasoconstricting agonists. The nitric oxide/superoxide reaction product, peroxynitrite ($ONOO^-$), may cause NOS III uncoupling due to tetrahydrobiopterin (BH4) oxidation and may potently inhibit the activity of soluble guanylyl cyclase (sGC), all of which may contribute, at least in part, to tolerance and cross tolerance in response to chronic treatment with nitroglycerin.*

(cytochrome P450-mediated?) or via activation of the NADPH oxidase in endothelial as well as in smooth muscle cells. Superoxide in turn is a strong stimulus for the activation of protein kinase C in the endothelium and in the smooth muscle cell layer leading to nitric oxide synthase III inhibition and to enhanced sensitivity to vasoconstricting agonists. The nitric oxide/superoxide reaction product peroxynitrite may cause nitric oxide synthase III uncoupling due to tetrahydrobiopterin oxidation and may potently inhibit the activity of the soluble guanylyl cyclase, all of which may contribute, at least in part, to tolerance and cross tolerance in response to chronic treatment with nitroglycerin (Fig. 8).

Strategies to Prevent the Development of Tolerance and Cross-Tolerance

Which strategy is the best to prevent the development of tolerance and cross-tolerance? The most widely accepted approach to prevent the tolerance phenomenon has been suggested to be a nitrate free interval. Intermittent administration of nitroglycerin patches allowing a nitrate free interval of 8–12 h has been shown to retain nitrate sensitivity, with the disadvantage of a lack of protection during this period. Another potential problem of a nitrate free interval may also be the development of rebound ischemia. Treatment of experimental animals with a nitrate free interval was not able to normalize endothelial dysfunction and hypersensitivity to vasoconstrictors [43] (Fig. 9A). During the nitrate free interval the frequency of angina symptoms as well as of silent angina is significantly increased [44]. By treating patients intermittently with nitroglycerin patches for a 5d period, Azevedo et al. have recently shown that this kind of regimen may prevent the development of tolerance but acute nitrate withdrawal increases the coronary vasomotor responses to acetylcholine suggesting that the rebound phenomena may be secondary to the development of endothelial dysfunction [45] (Fig. 9B). These data clearly indicate that the phenomenon of nitroglycerin-induced endothelial dysfunction cannot be prevented by a nitrate free interval.

If oxidative stress is important for tolerance and cross-tolerance, antioxidants or drugs, which are able to reduce oxidative stress within vascular tissue should be able to positively influence both phenomena. Recent studies in patients with coronary artery disease and heart failure indeed demonstrated, that the development of tolerance is beneficially influenced by vitamin C [46,47], vitamin E [48], ACE-inhibitors [49,50] and by hydralazine [51].

Summary and Clinical Implications

In summary, there is mounting evidence that systemic therapy with NO via organic nitrates induces endothelial dysfunction in patients with coronary artery disease [15,45] and even in healthy controls [37]. One mechanism contributing to this phenomenon may be a nitrate-induced stimulation of vascular superoxide and/or peroxynitrite production. This may represent a kind of biochemical baroreflex, in which the nitroglycerin-induced increases in vascular NO are diminished due to local degradation by superoxide. Treatment of coronary artery disease patients with ACE-inhibitors and HMG-CoA inhibitors has been shown to improve endothelial dysfunction and simultaneously to improve prognosis in these patients. Several experimental and clinical studies indicate that chronic nitroglycerin treatment worsens endothelial function and a recent meta-analysis indicates that nitrates may worsen prognosis in patients with ischemic heart disease [4]. Further studies are required, however, to understand the precise mechanisms underlying nitroglycerin-induced endothelial dysfunction in

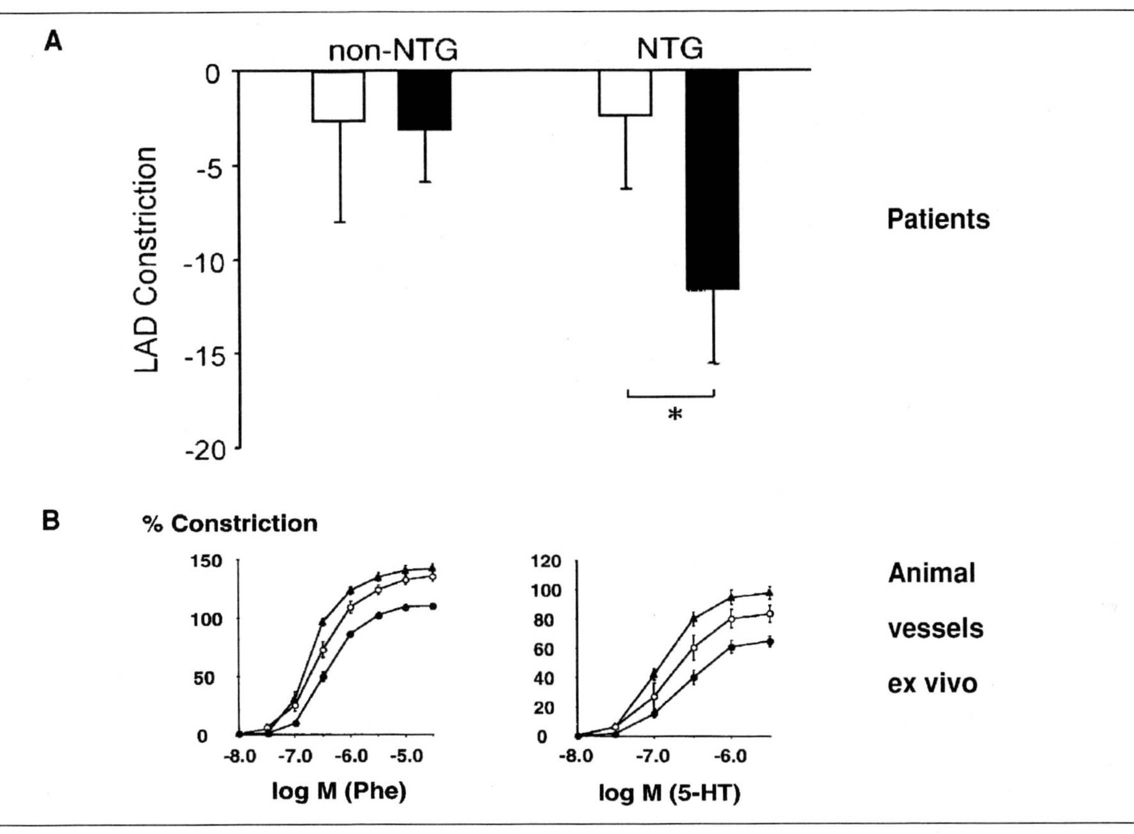

Fig. 9. Effects of a nitrate free interval on coronary endothelial function in patients with coronary artery disease (A) and on vasoconstrictor sensitivity of rabbit aorta to phenylephrine (PE) and serotonin (5-HT), (B). Patients received active nitroglycerin (NTG) (A, right hand side) treatment or no treatment (A, left hand side) followed by a 12 h nitrate free interval. Intracoronary acetylcholine was applied in the presence of the patch (open columns) and 3 h after the removal of the patch (filled columns) [45]. In patients on active nitroglycerin-therapy, patch removal lead to an acetylcholine-induced coronary constriction, which was markedly stronger than in patients not on active nitroglycerin treatment. Likewise, in experimental animals, a 12 h nitrate free interval did not prevent the development of hypersensitivity to phenylephrine and serotonin. The agonist-induced constriction was even stronger in animals treated with a nitrate free interval (triangles) as compared to animals treated continuously with nitroglycerin (open circles) (modified from [43]). Thus, in both conditions, a nitrate free interval was not able to prevent the development of endothelial dysfunction in patients with coronary artery disease and also failed to prevent the development of a hypersensitivity to vasoconstrictors.

order to develop strategies to prevent the resultant side effects.

Acknowledgment

T.M. is supported by Deutsche Forschungsgemeinschaft Mu 1079 3-1 and 4-1.

References

1. Lincoln TM, Dey N, Sellak H. Invited Review: cGMP-dependent protein kinase signaling mechanisms in smooth muscle: From the regulation of tone to gene expression. *J Appl Physiol* 2001;91:1421–1430.
2. Gruppo Italiano per lo Studio della Sopravvivenza nell'infarto Miocardico. GISSI-3: Effects of lisinopril and transdermal glyceryl trinitrate singly and together on 6-week mortality and ventricular function after acute myocardial infarction. *Lancet* 1994;343:1115–1122.
3. ISIS-4 (Fourth International Study of Infarct Survival) Collaborative Group. ISIS-4: A randomised factorial trial assessing early oral captopril, oral mononitrate, and intravenous magnesium sulphate in 58,050 patients with suspected acute myocardial infarction. *Lancet* 1995;345:669–685.
4. Nakamura Y, Moss AJ, Brown MW, Kinoshita M, Kawai C. Long-term nitrate use may be deleterious in ischemic heart disease: A study using the databases from two large-scale postinfarction studies. Multicenter Myocardial Ischemia Research Group. *Am Heart J* 1999;138:577–585.
5. Elkayam U, Kulick D, McIntosh N, Roth A, Hsueh W, Rahimtoola SH. Incidence of early tolerance to hemodynamic effects of continuous infusion of nitroglycerin in patients with coronary artery disease and heart failure. *Circulation* 1987;76:577–584.
6. Munzel T, Heitzer T, Kurz S, et al. Dissociation of coronary vascular tolerance and neurohormonal adjustments

during long-term nitroglycerin therapy in patients with stable coronary artery disease. *J Am Coll Cardiol* 1996;27:297–303.

7. Munzel T, Giaid A, Kurz S, Stewart DJ, Harrison DG. Evidence for a role of endothelin 1 and protein kinase C in nitroglycerin tolerance. *Proc Natl Acad Sci* 1995;92:5244–5248.
8. Zierhut W, Ball HA. Prevention of vascular nitroglycerin tolerance by inhibition of protein kinase C. *Br J Pharmacol* 1996;119:3–5.
9. Heitzer T, Just H, Meinertz T. Brockhoff C, Munzel T. Chronic angiotensin converting enzyme inhibition with captopril prevents nitroglycerin induced hypersensitivity to vasoconstrictors in patients with stable coronary artery disease. *J Am Coll Cardiol* 1998.
10. Heitzer T, Just H, Brockhoff C, Meinertz T, Olschewski M, Munzel T. Long-term nitroglycerin treatment is associated with supersensitivity to vasoconstrictors in men with stable coronary artery disease: Prevention by concomitant treatment with captopril. *J Am Coll Cardiol* 1998;31:83–88.
11. Munzel T, Sayegh H, Freeman BA, Tarpey MM, Harrison DG. Evidence for enhanced vascular superoxide anion production in nitrate tolerance. A novel mechanism underlying tolerance and cross-tolerance. *J Clin Invest* 1995;95:187–194.
12. Molina CR, Andresen JW, Rapoport RM, Waldman S, Murad F. Effect of *in vivo* nitroglycerin therapy on endothelium-dependent and independent vascular relaxation and cyclic GMP accumulation in rat aorta. *J Cardiovasc Pharmacol* 1987;10:371–378.
13. Laursen JB, Boesgaard S, Poulsen HE, Aldershvile J. Nitrate tolerance impairs nitric oxide-mediated vasodilation *in vivo*. *Cardiovascular Research* 1996;31:814–819.
14. Mulsch A, Busse R, Bassenge E. Desensitization of guanylate cyclase in nitrate tolerance does not impair endothelium-dependent responses. *Eur J Pharmacol* 1988;158:191–198.
15. Caramori PR, Adelman AG, Azevedo ER, Newton GE, Parker AB, Parker JD. Therapy with nitroglycerin increases coronary vasoconstriction in response to acetylcholine. *J Am Coll Cardiol* 1998;32:1969–1974.
16. Gori T, Burstein JM, Ahmed S, et al. Folic acid prevents nitroglycerin-induced nitric oxide synthase dysfunction and nitrate tolerance: A human *in vivo* study. *Circulation* 2001;104:1119–1123.
17. Schachinger V, Britten MB, Zeiher AM. Prognostic impact of coronary vasodilator dysfunction on adverse long-term outcome of coronary heart disease. *Circulation* 2000;101:1899–1906.
18. Heitzer T, Schlinzig T, Krohn K, Meinertz T, Munzel T. Endothelial dysfunction, oxidative stress, and risk of cardiovascular events in patients with coronary artery disease. *Circulation* 2001;104:2673–2678.
19. Mulsch A, Oelze M, Kloss S, et al. Effects of *in vivo* nitroglycerin treatment on activity and expression of the guanylyl cyclase and cGMP-dependent protein kinase and their downstream target vasodilator-stimulated phosphoprotein in aorta. *Circulation* 2001;103:2188–2194.
20. Munzel T, Kurz S, Rajagopalan S, et al. Hydralazine prevents nitroglycerin tolerance by inhibiting activation of a membrane-bound NADH oxidase. A new action for an old drug. *J Clin Invest* 1996;98:1465–1470.
21. Mihm MJ, Coyle CM, Jing L, Bauer JA. Vascular peroxynitrite formation during organic nitrate tolerance. *J Pharmacol Exp Ther* 1999;291:194–198.
22. Skatchkov M, Larina LL, Larin AA, Fink N, Bassenge E. Urinary nitrotyrosine content as a marker of peroxynitrite-induced tolerance to organic nitrates. *J Cardiovasc Pharmacol Thera* 1997;2:85–96.
23. Laursen JB, Somers M, Kurz S, et al. Endothelial regulation of vasomotion in apoE-deficient mice: Implications for interactions between peroxynitrite and tetrahydrobiopterin. *Circulation* 2001;103:1282–1288.
24. Xia Y, Tsai AL, Berka V, Zweier JL. Superoxide generation from endothelial nitric-oxide synthase. A Ca^{2+}/calmodulin-dependent and tetrahydrobiopterin regulatory process. *J Biol Chem* 1998;273:25804–25808.
25. Munzel T, Li H, Mollnau H, et al. Effects of long-term nitroglycerin treatment on endothelial nitric oxide synthase (NOS III) gene expression, NOS III-mediated superoxide production, and vascular NO bioavailability. *Circ Res* 2000;86:E7–E12.
26. Gruhn N, Aldershvile J, Boesgaard S. Tetrahydrobiopterin improves endothelium-dependent vasodilation in nitroglycerin-tolerant rats. *Eur J Pharmacol* 2001;416:245–249.
27. Stroes ES, van Faassen EE, Yo M, et al. Folic acid reverts dysfunction of endothelial nitric oxide synthase. *Circ Res* 2000;86:1129–1134.
28. Leber A, Hemmens B, Klosch B, et al. Characterization of recombinant human endothelial nitric oxide synthase from the yeast pichia pastoris. *J Biol Chem* 1999;274:37658–37664.
29. Abou-Mohamed G, Kaesemeyer WH, Caldwell RB, Caldwell RW. Role of L-arginine in the vascular actions and development of tolerance to nitroglycerin. *Br J Pharmacol* 2000;130:211–218.
30. Kaesemeyer WH, Ogonowski AA, Jin L, Caldwell RB, Caldwell RW. Endothelial nitric oxide synthase is a site of superoxide synthesis in endothelial cells treated with glyceryl trinitrate. *Br J Pharmacol* 2000;131:1019–1023.
31. Wang EQ, Lee WI, Fung HL. Lack of critical involvement of endothelial nitric oxide synthase in vascular nitrate tolerance in mice. *Br J Pharmacol* 2002;135:299–302.
32. Nishikawa T, Edelstein D, Du XL, et al. Normalizing mitochondrial superoxide production blocks three pathways of hyperglycaemic damage. *Nature* 2000;404:787–790.
33. Fleming I, Fisslthaler B, Dimmeler S, Kemp BE, Busse R. Phosphorylation of Thr(495) regulates Ca(2+)/calmodulin-dependent endothelial nitric oxide synthase activity. *Circ Res* 2001;88:E68–E75.
34. Kahler J, Mendel S, Weckmuller J, et al. Oxidative stress increases synthesis of big endothelin-1 by activation of the endothelin-1 promoter. *J Mol Cell Cardiol* 2000;32:1429–1437.
35. Kahler J, Ewert A, Weckmuller J, et al. Oxidative stress increases endothelin-1 synthesis in human coronary artery smooth muscle cells. *J Cardiovasc Pharmacol* 2001;38:49–57.
36. Sage PR, de la Lande IS, Stafford I, et al. Nitroglycerin tolerance in human vessels: Evidence for impaired nitroglycerin bioconversion. *Circulation* 2000;102:2810–2815.
37. Gori T, Mak SS, Kelly S, Parker JD. Evidence supporting abnormalities in nitric oxide synthase function induced by nitroglycerin in humans. *J Am Coll Cardiol* 2001;38:1096–1101.

38. Pfeifer A, Klatt P, Massberg S, et al. Defective smooth muscle regulation in cGMP kinase I-deficient mice. *Embo J* 1998;17:3045–3051.
39. Smolenski A, Bachmann C, Reinhard K, et al. Analysis and regulation of vasodilator-stimulated phosphoprotein serine 239 phosphorylation *in vitro* and in intact cells using a phosphospecific monoclonal antibody. *J Biol Chem* 1998;273:20029–20035.
40. Massberg S, Sausbier M, Klatt P, et al. Increased adhesion and aggregation of platelets lacking cyclic guanosine 3′,5′-monophosphate kinase I. *J Exp Med* 1999;189:1255–1264.
41. Draijer R, Vaandrager AB, Nolte C, de Jonge HR, Walter U, van Hinsbergh VW. Expression of cGMP-dependent protein kinase I and phosphorylation of its substrate, vasodilator-stimulated phosphoprotein, in human endothelial cells of different origin. *Circ Res* 1995;77:897–905.
42. Oelze M, Mollnau H, Hoffmann N, et al. Vasodilator-stimulated phosphoprotein serine 239 phosphorylation as a sensitive monitor of defective nitric oxide/cGMP signaling and endothelial dysfunction. *Circ Res* 2000;87:999–1005.
43. Munzel T, Mollnau H, Hartmann M, et al. Effects of a nitrate-free interval on tolerance, vasoconstrictor sensitivity and vascular superoxide production. *J Am Coll Cardiol* 2000;36:628–634.
44. Pepine CJ, Lopez LM, Bell DM, Handberg-Thurmond EM, Marks RG, McGorray S. Effects of intermittent transdermal nitroglycerin on occurrence of ischemia after patch removal: Results of the second transdermal intermittent dosing evaluation study (TIDES-II). *J Am Coll Cardiol* 1997;30:955–961.
45. Azevedo ER, Schofield AM, Kelly S, Parker JD. Nitroglycerin withdrawal increases endothelium-dependent vasomotor response to acetylcholine. *J Am Coll Cardiol* 2001;37:505–509.
46. Watanabe H, Kakihana M, Ohtsuka S, Sugishita Y. Randomized, double-blind, placebo-controlled study of the preventive effect of supplemental oral vitamin C on attenuation of development of nitrate tolerance. *J Am Coll Cardiol* 1998;31:1323–1329.
47. Bassenge E, Fink N, Skatchkov M, Fink B. Dietary supplement with vitamin C prevents nitrate tolerance. *J Clin Invest* 1998;102:67–71.
48. Watanabe H, Kakihana M, Ohtsuka S, Sugishita Y. Randomized, double-blind, placebo-controlled study of supplemental vitamin E on attenuation of the development of nitrate tolerance. *Circulation* 1997;96:2545–2550.
49. Mehra A, Ostrzega E, Shotan A, Johnson JV, Elkayam U. Persistent hemodynamic improvement with short-term nitrate therapy in patients with chronic congestive heart failure already treated with captopril. *Am J Cardiol* 1992;70:1310–1314.
50. Elkayam U, Johnson JV, Shotan A, et al. Double-blind, placebo-controlled study to evaluate the effect of organic nitrates in patients with chronic heart failure treated with angiotensin-converting enzyme inhibition. *Circulation* 1999;99:2652–2657.
51. Gogia H, Mehra A, Parikh S, et al. Prevention of tolerance to hemodynamic effects of nitrates with concomitant use of hydralazine in patients with chronic heart failure. *J Am Coll Cardiol* 1995;26:1575–1580.

Statins and the Role of Nitric Oxide in Chronic Heart Failure

Stephan von Haehling,[1,2] *Stefan D. Anker,*[1,2] *and Eberhard Bassenge*[3]

[1] *Department of Clinical Cardiology, National Heart & Lung Institute, Imperial College, School of Medicine, London, UK;* [2] *Franz Volhard Klinik (Charité, Campus Berlin-Buch) at Max Delbrück Centrum for Molecular Medicine, Berlin, Germany;* [3] *Institute of Applied Physiology, Albert-Ludwigs University, Freiburg, Germany*

Abstract. **Endothelial dysfunction plays an important role in a number of cardiovascular diseases. An important pathogenetic factor for the development of endothelial dysfunction is lack of nitric oxide (NO), which is a potent endothelium-derived vasodilating substance. 3-Hydroxy-3-methylglutaryl-coenzyme A reductase inhibitors (statins), originally designed to lower plasma cholesterol levels, seem to ameliorate endothelial dysfunction by a mechanism so far only partly understood. However, statins increase nitric oxide synthase activity. It has been speculated that this and other "side effects" of statin treatment are due to inhibition of Rho, an intracellular signalling protein that initiates Rho kinase transcription. Moreover, statins possess anti-inflammatory characteristics. Some statins have proven to lower plasma levels of C-reactive protein, which is induced by pro-inflammatory cytokines. Other statins have been demonstrated to directly inhibit pro-inflammatory cytokine induction. Finally, some data suggest that statins might be able to counterbalance an increased production of oxygen free radicals. Since chronic heart failure is accompanied not only by endothelial dysfunction, but also by pro-inflammatory cytokine activation and enhanced formation of oxygen free radicals, it is tempting to speculate that statins might be an ideal candidate to treat certain features of this disease. The doses needed to achieve the desired effects might be much lower than those needed to treat hypercholesterolemia.**

Key Words. **chronic heart failure, statin, nitric oxide, endothelial dysfunction**

Introduction

The endothelium is the largest autocrine, paracrine, and endocrine organ [1]. It covers approximately 700 m^2 and weighs 1.5 kg. Endothelial dysfunction has therefore vast consequences. This condition, mainly characterised by a lack of nitric oxide (NO), is a common feature of the aging process as well as of a number of chronic illnesses. It is seen in chronic heart failure (CHF), hypercholesterolemia, atherosclerosis, hypertension, and certain inflammatory diseases. Recently, endothelial dysfunction has been proposed to be a useful marker of early cardiovascular disease [2]. Elevated levels of serum cholesterol have been implicated in the progression of endothelial dysfunction. 3-Hydroxy-3-methylglutaryl-coenzyme A (HMG-CoA) reductase inhibitors (statins), originally designed to lower serum cholesterol levels (Fig. 1), seem to ameliorate endothelial dysfunction by a mechanism not yet fully understood. These drugs were found to reduce recurrent events as early as 16 weeks after acute coronary events [3]. This time frame is probably too short to attribute the risk reduction to cholesterol lowering alone. It has therefore been suggested that statins may have beneficial effects beyond the reduction of plasma low-density lipoproteins (LDL).

The purpose of this review is to shed further light on the pleiotropic effects of statin treatment. It is tempting to embark on the role of these drugs in atherosclerosis, because a lot of data has been published about statins in this disease. However, this review will focus on the potential role of statin treatment in CHF. This multi-system disorder not only affects the cardiovascular system but also the musculoskeletal, renal, neuroendocrine and immune systems. Some evidence points to the fact that statins might be able to improve certain symptoms in patients with CHF.

Nitric Oxide and the Endothelium

Far from being inert, the vascular endothelium is a monolayer of cells, which is deeply involved

Address for correspondence: Stephan von Haehling, Department of Clinical Cardiology, National Heart & Lung Institute, Dovehouse Street, London, SW3 6LY, UK. Tel.: +44 20 7351 8127; Fax: +44 20 7351 8733; E-mail: stephan.von.haehling@web.de

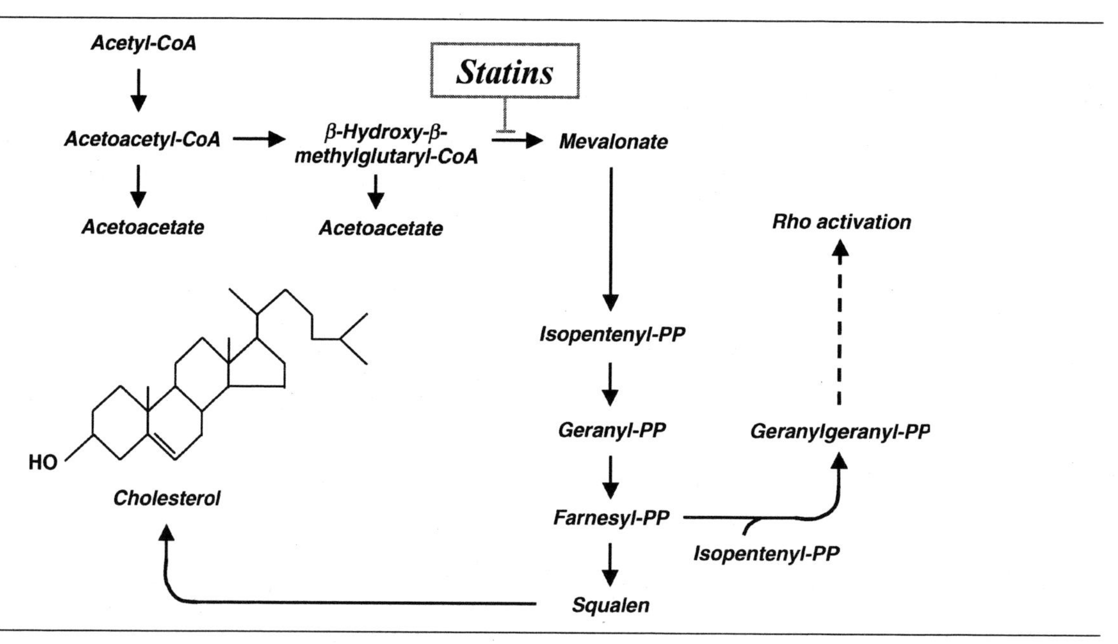

Fig. 1. *Pathway of cholesterol biosynthesis. Initially, acetyl-CoA is transported from the mitochondrium to the cytosol. The rate limiting step is 3-hydroxy-3-methylglutaryl-CoA (HMG-CoA) reductase activity, which is competitively inhibited by statins. Intermediates in the pathway are used for the synthesis of different proteins. For example, geranylgeranyl-pyrophosphate is needed for the activation of Rho. PP—Pyrophosphate.*

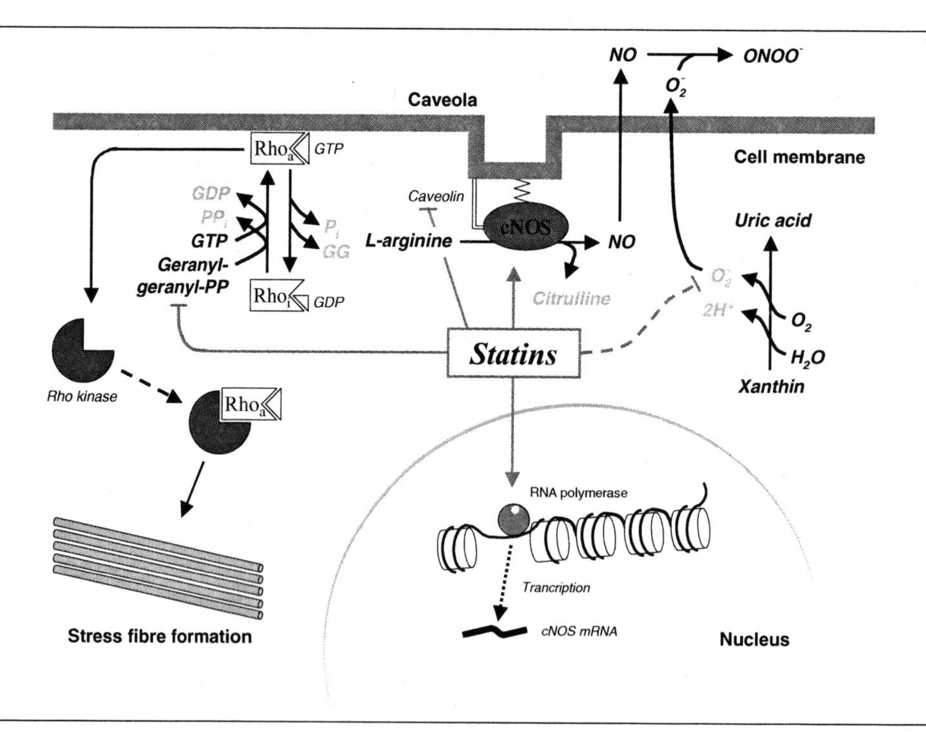

Fig. 2. *Statin-mediated effects in endothelial cells. Statins interfere with geranylgeranyl-pyrophosphate biosynthesis, which is a byproduct from cholesterol synthesis. This effect leads to the accumulation of inactive Rho in the cytosol thus inhibiting the activation of Rho kinase. Statins also augment cNOS activity and induce cNOS gene transcription. Caveolin, which binds to cNOS in caveolae, acts as a negative regulator of the enzyme. Statins prevent LDL-induced caveolin-expression in endothelial cells. NO generation is eventually increased. The statin-mediated effect on superoxide generation seems to be indirect. Therefore, the rapid deactivation of NO appears to be attenuated. NOS—Nitric Oxide Synthase, $ONOO^-$—Peroxynitrite, PP—Pyrophosphate.*

in cardiovascular physiology and various chronic diseases (Fig. 2). The endothelium has a central role in vascular homeostasis, and both NO and endothelin-1 (ET-1) play a pivotal role in mediating its function. NO is a potent vasodilator in the vasculature. It also inhibits platelet adherence [4] and regulates endothelial permeability to lipoproteins and other plasma constituents [5].

NO was first described by Furchgott and Zawadski in 1980 and termed *endothelium-derived relaxing factor* (EDRF) [6]. This substance is a lipophilic, freely diffusible, soluable gas. It has a short half life of less than four seconds in biological solutions, because it readily reacts with oxygen-derived free radicals, namely superoxide anion, to form peroxynitrite ($ONOO^-$) (Fig. 2) [7]. NO is produced from the amino acid L-arginine by nitric oxide synthase (NOS). Three isoforms of this enzyme have been identified with the endothelial (constitutive) isoform (cNOS, also known as eNOS) producing a continuous amount of NO. This basal production is increased by different chemical and physical stimuli. cNOS was found to be primarily associated with the plasma membrane [8]. Specific microdomains of the plasma membrane, termed caveolae, are known to display a NOS activity 9- to 10-fold greater than other areas [9]. These caveolae are characterised by a specific marker protein, caveolin, which is exclusively found in caveolae membranes (Fig. 2). Caveolin domains seem to interact directly with cNOS domains.

The neuronal isoform (nNOS) is expressed by nerves in the adventitia of certain vessels. Whilst cNOS and nNOS are found in healthy vascular cells, the inducible isoform (iNOS) is mainly produced upon inflammatory stimuli. In contrast to cNOS, iNOS is mainly found in the cytosol [9]. For example, interferon-γ (IFN-γ), tumor necrosis factor-α (TNFα), and bacterial lipopolysaccharide (LPS) are potent stimulators of iNOS gene transcription not only in endothelial cells [10]. Since the intracellular NO content of macrophages rises during bacterial infection, NO is thought to play an additional role in microbial killing in innate immune responses [11]. Moreover, small quantities of iNOS-derived NO are thought to be involved in signal transduction during infection [11].

After synthesis in the endothelium, NO diffuses across cell membranes and enters vascular smooth muscle cells [10]. In these cells, NO leads to increased concentrations of cyclic guanosine-3',5'-monophosphate (cGMP) via activation of guanylate cyclase. cGMP serves as a second messenger, which is crucially involved in control of vascular tone, platelet function, and other NO-mediated functions. The inhibition of the contractile machinery is mainly controlled by a cGMP-dependent protein kinase, which increases calcium extrusion from the cytosol of smooth muscle cells [12]. Whilst NO, being produced by nitric oxide synthase within endothelial cells, is a potent vasodilator, ET-1 is an endothelium-derived vasoconstrictor and pressor peptide. Some evidence has accumulated in the past years that endothelial dysfunction and in particular an imbalance between the two mediators play a crucial role in the development of some cardiovascular diseases.

Mechanisms Underlying Chronic Heart Failure

CHF is accompanied by a variety of pathophysiological changes which trigger disease progression [13,14]. The development of cardiac remodelling is maintained by neurohormonal activation and synthesis of pro-inflammatory cytokines. In this context, TNFα is thought to play a predominant role, and increased levels of TNFα relate to reduced peripheral blood flow [15], apoptosis [16], and lower skeletal muscle mass [17]. In addition, pro-inflammatory cytokine levels relate to prognosis in CHF [18]. Nevertheless, induction of iNOS and free radical formation yield additional effects. In the course of the disease, impairment in skeletal muscle metabolism and function become obvious. Overexpression of both ET-1 and its receptor worsens this condition [19,20]. Indeed, the endothelium-dependent dilation of coronary and peripheral resistance vessels is blunted in CHF [21,22]. This ultimately leads to an impaired reactive hyperemia in various vascular beds [23], an impairment in tissue perfusion and reduced muscular endurance [24]. These abnormalities can ultimately lead to weight loss and cardiac cachexia, a serious complication of CHF with increased mortality [25], which is associated with the highest plasma levels of TNFα seen in CHF patients [26]. Therefore, it seems likely that cachectic patients would benefit from immunosuppressive therapy [27].

It has been observed that iNOS is expressed in the myocardium of all patients with CHF [28,29]. It has therefore been speculated that low NO concentrations may account for protective effects, which rescue myocytes from deleterious stimuli such as mechanical stress or noradrenaline. Higher concentrations of NO, on the other hand, may contribute to a loss of myocytes. Nevertheless, increased iNOS activity leads to depression of myocardial activity [30,31]. Since dyspnea is one of the major factors limiting exercise capacity in CHF, it has been proposed that reduced perfusion of the ventilated lung might contribute to this symptom [14]. Moreover, NO might be a key regulator of lung function. Thus, dyspnea, which is frequently observed during exercise, might be attributed to reduced pulmonary vasodilation due to a smaller increase in NO production compared to healthy control subjects.

Pleiotropic Effects of Statins

The advent of statins has revolutionised the treatment of hypercholesterolemia. In 1987 lovastatin was the first HMG-CoA reductase inhibitor to be approved by the Food and Drug Administration (FDA). It is extracted from the fungus *Aspergillus terreus*. Since then, five other statins have been synthesised, and the development of new substances is well underway. So far, lovastatin, simvastatin, pravastatin, fluvastatin, and atorvastatin are available. Like lovastatin, simvastatin and pravastatin are derived from fungal fermentation [32]. The other statins are entirely synthetic. All statins are highly effective in reducing LDL. Moreover, these drugs are well tolerable and generally safe. As a rare side effect, rhabdomyolysis can occur, but the statin with the highest side effect rate of this kind (cerivastatin) has been withdrawn from the market. The primary mechanism of action is competitive inhibition of HMG-CoA reductase, the rate-limiting enzyme in cholesterol biosynthesis (Fig. 1) [33]. This yields a reduction in hepatocyte cholesterol concentration. Thus, the expression of hepatic LDL receptors is increased, and LDL and its precursors are cleared more effectively from the circulation [34]. An additional increase in HDL levels has been observed [32]. Triglyceride lowering correlates to baseline triglyceride levels and to the LDL-lowering potency of the respective drug [35]. In general, LDL is reduced by another 7% with each doubling of the dose [36]. However, lipoprotein a levels, another marker of increased cardiovascular risk, are either not altered during statin treatment [37] or raised by as much as 34% [38]. It is unclear whether this yields an additional effect on future cardiovascular events.

Statin-Mediated Effects on Endothelial Cells

Statins have proven beneficial in the treatment of coronary artery disease, and these substances can ameliorate morbidity and mortality irrespective of serum cholesterol levels [39]. These data suggest that the beneficial actions of statins are independent of low density lipoprotein (LDL) cholesterol reduction. It seems therefore likely that certain (pleiotropic) "side effects" yield an improvement in endothelial dysfunction, although the precise mechanism is only partly understood (Fig. 2). Nevertheless, it appears that both direct and indirect mechanisms are involved. However, not all statins share the same characteristics, although all substances from this class improve endothelial dysfunction [2]. For example, *in vitro* pravastatin appears to have an effect superior to simvastatin [40]. The different actions on endothelial dysfunction may be attributable to different effects on NO production: it has been demonstrated that lovastatin and simvastatin induce cNOS gene transcription in human endothelial cells (Fig. 2) [41]. Both drugs also play an important role in the protection from experimental stroke, an effect which seems to be mediated by increased production of NO [42]. Moreover, endothelial function seems to improve in monkeys receiving a dose of pravastatin that does not reduce LDL levels [43]. Whilst the coronary arteries of the pravastatin-fed monkeys dilated significantly as quantified using angiography compared to the status before treatment, those of the control group constricted.

A reduced vasodilation is observed during exercise in CHF. Moreover, increased vasoconstriction occurs, and this reaction seems to be mediated by enhanced synthesis of ET-1 [10]. It is most likely that an increased vascular generation of superoxide anion inactivates NO to a great extent. Recently, it has been proposed that statins might be useful to reduce oxidative stress. The pathways involved seem to cover inhibition of vascular NADH oxidase and cNOS-mediated superoxide production [44]. Pravastatin, for example, increases the bioavailability of NO in atheroslerotic arterial walls [43], and it activates cNOS independently of its cholesterol-lowering features [40]. In addition to sole NO effects, an imbalance between ET-1 and NO is thought to play a role in the pathophysiology of CHF. Circulating ET-1 levels are normally low. Some studies point to the fact that statins might be able to decrease ET-1 synthesis in endothelial cells [45]. Atorvastatin and simvastatin inhibit the expression of pre-pro-ET-1 mRNA and reduce ET-1 levels [46].

Statin-Mediated Effects on Caveolin Abundance

A recent study by Feron and coworkers suggests that atorvastatin is able to reduce caveolin expression in endothelial cells (Fig. 2) [47]. Treatment with normocholesterolemic or hypercholesterolemic levels of LDL-cholesterol induces caveolin expression after serum starvation in endothelial cells. This leads to an impairment in NO release due to stabilisation of the inhibitory caveolin/cNOS complex. Atorvastatin, despite decreasing net caveolin levels, increases the amount of unbound cNOS for activation and NO production. However, some questions remain to be addressed [48]: If atorvastatin indeed decreases caveolin expression—does this reduce the number of caveolae? May this be detrimental for cNOS activation? Future studies will hopefully provide information on this new field of statin action.

Anti-Inflammatory Effects of Statins

Inflammation seems to play a predominant role in the progression of CHF, and some evidence points to the fact that pro-inflammatory cytokine activation is triggered by LPS [49,50]. In CHF patients, LPS could enter the circulation through the edematous gut wall [51]. However, it is only when the inflammatory response becomes overactive that it begins to have harmful effects. Thus, TNFα can induce left-ventricular impairment and septic cardiomyopathy [52,53]. Interestingly, statins seem to possess anti-inflammatory features. Pravastatin, for example, lowers the levels of C-reactive protein (CRP) after myocardial infarction and in hypercholesterolemic patients [54,55]. CRP is an acute phase reactant that is produced by the liver in response to pro-inflammatory cytokines such as interleukin-6 and TNFα. This has to be kept in mind, because lovastatin has been found to inhibit the induction of pro-inflammatory cytokines in macrophages [56]. In atherosclerotic plaques, statins seem to reduce the number of inflammatory cells [39]. This finding is in keeping with a recent study reporting on statin-binding to a counterreceptor for intercellular adhesion molecule-1 (ICAM-1) on leukocytes thus reducing adhesion to endothelial cell [57,58].

Statins and Rho Family Proteins

Rho proteins (Ras homologous) form a family of proteins that activate certain kinases (Fig. 2). These proteins are GTP-binding molecules that have been identified to be involved in the regulation of the actin cytoskeleton (Fig. 2). Extracellular stimuli convert the inactive form of Rho (GDP-Rho) to the active form (GTP-Rho) [59]. Rho thus appears to coordinate specific cellular responses by interacting with a number of downstream targets [60]. Binding of the active form of Rho to Rho kinase activates this serine/threonine kinase [61]. Thus, Rho family proteins regulate gene expression by activating different kinases, such as Rho kinases, which are involved in smooth muscle contraction and stress fibre formation [61].

Statins inhibit the synthesis of L-mevalonic acid from HMG-CoA. This also prevents the synthesis of other important intermediates, such as farnesyl pyrophosphate and geranylgeranyl pyrophosphate (Fig. 1) [62]. The latter is an important lipid attachment for the posttranslational modification of Rho. Statin treatment therefore interferes with Rho activation and leads to the accumulation of inactive Rho in the cytoplasm [59]. This affects intracellular transport, membrane trafficking, mRNA stability, and gene transcription. It is therefore tempting to speculate that some of the pleiotropic effects of statins are attributed to Rho inhibition, although only limited data are available on this topic.

Oxidative Stress

Oxidative stress appears to be an important feature in the development of endothelial dysfunction. However, the value of antioxidants, such as vitamins, is controversial and not sufficiently documented. The most important oxygen free radical derived from cell metabolism is superoxide anion. It is generated by oxidases during cellular metabolism. The formation of oxygen free radicals is known to be increased in patients with CHF [22,63,64]. Physical training, however, reverses endothelial dysfunction [65]. Increased expression of antioxidative enzyme genes seems to be one of the factors mediating this effect [66]. Dietary supplementation of L-arginine yields additional effects [67]. Evidence suggests that oxygen free radicals can upregulate sensitivity to LPS [68], thus underscoring the importance of pro-inflammatory cytokine activation. TNFα itself causes a rapid rise in intracellular reactive oxygen intermediates [69].

Xanthine oxidase is one of the oxidases generating superoxide anion (Fig. 2). Xanthine oxidase triggers the last step in purine breakdown which yields uric acid. Superoxide anion is a byproduct. Interestingly, a significant elevation of uric acid is frequently observed in patients with CHF [70]. This is seen as a marker of elevated xanthine oxidase activity. Allopurinol, a xanthine oxidase inhibitor, might be able to improve endothelial dysfunction in CHF by reducing free radical generation [22,71].

A diversity of antioxidant systems, such as superoxide dismutase and catalase, counteract the continuous generation of reactive oxygen species. A decreased expression of both systems has been observed simultaneously in CHF [72]. Some evidence has accumulated that statins might be able to reduce vascular production of reactive oxygen species [73]. The authors, however, believe that this effect might be mediated via statin-induced down-regulation of vascular angiotensin receptors, because stimulation of this receptor leads to an increase of free radical production. In another model, vascular superoxide anion generation was unaffected by statin treatment, but it was increased during withdrawal [74].

Conclusion

Statins have proven beneficial in the treatment of a number of cardiovascular diseases. Although it was reasonable to attribute these benefits to lipid lowering, the evidence suggested otherwise. Statins are potent in improving endothelial dysfunction due to enhancing cNOS gene transcription, and the decrease in caveolin expression seems to contribute to increased NO production. Moreover, statins seem to possess anti-inflammatory features, and some evidence suggests that a direct suppression of proinflammatory cytokines may occur. A reduction in oxygen free radicals, which has been reported for some statins, seems to contribute to the beneficial effects of statins. However, some statin actions require further elucidation. We do not completely understand the impact of Rho inactivation by statins, although it is commonly believed that Rho inhibition contributes to some of the beneficial effects of statins. Since CHF is accompanied by endothelial dysfunction, pro-inflammatory cytokine activation, and excessive production of oxygen free radicals, it would appear to use statins in the treatment of this chronic illness, independently of the underlying etiology.

After simvastatin treatment fewer instances of new-onset CHF were observed in the Scandinavian Simvastatin Survival Study [75]. Again this effect seems to be independent of the lipid-lowering features of simvastatin, because changes in the lipoprotein profiles and baseline characteristics were similar in patients with or without future events. The anti-inflammatory, anti-proliferative, and anti-oxidative effects of statins might play an important role in the prevention of CHF by statins [32,76].

Interestingly, lower lipoprotein concentrations are independent predictors of impaired survival in patients with CHF [77,78]. It has been speculated that lipoproteins may in fact confer some benefits in CHF patients, by forming micells around LPS and preventing its biologic action [79]. This may explain why in CHF patients higher levels of cholesterol levels are related to better not worse mortality. Hence lower doses of statins that still confer pleiotropic effects (on NO, free radicals and the endothelium) but do not or to a lesser extent lower lipoprotein plasma levels may be of particular benefit in CHF patients.

References

1. Vogel RA. Cholesterol lowering and endothelial dysfunction. *Am J Med* 1999;107:479–487.
2. Aengevaeren WR. Beyond lipids—the role of the endothelium in coronary artery disease. *Atherosclerosis* 1999;147(Suppl 1):S11–S16.
3. Schwartz GG, Olsson AG, Ezekowitz MD, Ganz P, Oliver MF, Waters D, Zeiher A, Chaitman BR, Leslie S, Stern T. Effects of atorvastatin on early recurrent ischemic events in acute coronary syndromes: The MIRACL study: A randomised controlled trial. *JAMA* 2001;285:1711–1718.
4. Cooke JP, Tsao PS. Is NO an endogenous antiatherogenic molecule? *Arterioscler Thromb* 1994;30:325–333.
5. Ross R. Atherosclerosis—an inflammatory disease. *N Engl J Med* 1999;340:115–126.
6. Furchgott RF, Zawadski JV. The obligatory role of endothelial cells in the relaxation of arterial smooth muscle by acetylcholine. *Nature* 1980;288:373–376.
7. Moncada S, Higgs EA. The L-arginine-nitric oxide pathway. *N Engl J Med* 1993;329:2002–2012.
8. Hecker M, Mulsch A, Bassenge E, Forstermann U, Busse R. Subcellular localization and characterization of nitric oxide synthase(s) in endothelial cells: Physiological implications. *Biochem J* 1994;299:247–252.
9. Shaul PW. Regulation of endothelial nitric oxide synthase: Location, location, location. *Annu Rev Physiol* 2002;64:749–774.
10. Vallance P, Chan N. Endothelial function and nitric oxide: Clinical relevance. *Heart* 2001;85:342–350.
11. Bogdan C, Rollinghoff M, Diefenbach A. The role of nitric oxide in innate immunity. *Immunol Rev* 2000;173:17–26.
12. Mombouli JV, Vanhoutte PM. Endothelial dysfunction: From physiology to therapy. *J Mol Cell Cardiol* 1999;31:61–74.
13. Bassenge E, Fink B, Schwemmer M. Oxidative stress, vascular dysfunction and heart failure. *Heart Fail Rev* 1999;4:133–145.
14. Sharma R, Coats AJS, Anker SD. The role of inflammatory mediators in chronic heart failure: Cytokines, nitric oxide, and endothelin-1. *Int J Cardiol* 2000;72:175–186.
15. Anker SD, Volterrani M, Egerer KR, Felton CV, Kox WJ, Poole-Wilson PA, Coats AJ. Tumour necrosis factor alpha as a predictor of impaired peak leg blood flow in patients with chronic heart failure. *QJM* 1998;91:199–203.
16. Ceconi C, Curello S, Bachetti T, Corti A, Ferrari R. Tumor necrosis factor in congestive heart failure: A mechanism of disease for the new millennium? *Prog Cardiovasc Dis* 1998;41(1 Suppl 1):25–30.
17. Anker SD, Ponikowski PP, Clark AL, Leyva F, Rauchhaus M, Kemp M, Teixeira MM, Hellewell PG, Hooper J, Poole-Wilson PA, Coats AJ. Cytokines and neurohormones relating to body composition alterations in the wasting syndrome of chronic heart failure. *Eur Heart J* 1999;20:683–693.
18. Rauchhaus M, Doehner W, Francis DP, Davos C, Kemp M, Liebenthal C, Niebauer J, Hooper J, Volk HD, Coats AJ, Anker SD. Plasma cytokine parameters and mortality in patients with chronic heart failure. *Circulation* 2000;102:3060–3067.
19. Sakai S, Miyauchi T, Kobayashi M, Yamaguchi I, Goto K, Sugishita Y. Inhibition of myocardial endothelin pathway improves long-term survival in heart failure. *Nature* 1996;384:353–355.
20. Stewart D. Update on endothelin. *Can J Cardiol* 1998;14(Suppl D):11D–13D.
21. Drexler H, Hayoz D, Munzel T, Just HJ, Zelis R, Brunner HR. Endothelial function in congestive heart failure. *Am Heart J* 1993;126:761–764.
22. Doehner W, Schoene N, Rauchhaus M, Leyva-Leon F, Pavitt DV, Reaveley DA, Schuler G, Coats AJ, Anker SD, Hambrecht R. Effects of xanthine oxidase inhibition with allopurinol on endothelial function and peripheral blood flow in hyperuricemic patients with chronic heart failure: Results from 2 placebo-controlled studies. *Circulation* 2002;105:2619–2624.

23. Hayoz D, Drexler H, Munzel T et al. Flow mediated arterial dilation is abnormal in congestive heart failure. *Circulation* 1993;87:VII-92–VII-96.
24. Drexler H, Hayoz D, Munzel T et al. Endothelial function in chronic congestive heart failure. *Am J Cardiol* 1992;69:1596–1601.
25. Anker SD, Ponikowski P, Varney S, Chua TP, Clark AL, Webb-Peploe KM, Harrington D, Kox WJ, Poole-Wilson PA, Coats AJ. Wasting as independent risk factor for mortality in chronic heart failure. *Lancet* 1997;349:1050–1053.
26. Anker SD, Chua TP, Ponikowski P, Harrington D, Swan JW, Kox WJ, Poole-Wilson PA, Coats AJ. Hormonal changes and catabolic/anabolic imbalance in chronic heart failure and their importance for cardiac cachexia. *Circulation* 1997;96:526–534.
27. von Haehling S, Genth-Zotz S, Anker SD, Volk HD. Cachexia: A therapeutic approach beyond cytokine antagonism. *Int J Cardiol* 2002;85:173–183.
28. Haywood GA, Tsao PS, von der Leyen HE, Mann MJ, Keeling PJ, Trindade PT, Lewis NP, Byrne CD, Rickenbacher PR, Bishopric NH, Cooke JP, McKenna WJ, Fowler MB. Expression of inducible nitric oxide synthase in human heart failure. *Circulation* 1996;93:1087–1094.
29. Satoh M, Nakamura M, Tamura G, Makita S, Segawa I, Tashiro A, Satodate R, Hiramori K. Indurcible nitric oxide synthase and tumor necrosis factor-alpha in myocardium in human dilated cardiomyopathy. *J Am Coll Cardiol* 1997;29:716–724.
30. Francis SE, Holden H, Holt CM, Duff GW. Interleukin-1 in myocardium and coronary arteries of patients with dilated cardiomyopathy. *J Mol Cell Cardiol* 1998;30:215–223.
31. Vejlstrup NG, Bouloumie A, Boesgarrd S, Andersen CB, Nielsen-Kudsk JE, Mortensen SA, Kent JD, Harrison DG, Busse R, Aldershvile J. Inducible nitric oxide synthase (iNOS) in the human heart: Expression and localization in congestive heart failure. *J Mol Cell Cardiol* 1998;30:1215–1223.
32. Maron DJ, Fazio S, Linton MF. Current perspectives on statins. *Circulation* 2000;101:207–213.
33. Endo A, Tsujita Y, Kuroda M, Tanzawa K. Inhibition of cholesterol synthesis *in vitro* and *in vivo* by ML-236° and ML-236B, competitive inhibitors of 3-hydroxy-3-methylglutaryl-coenzyme A reductase. *Eur J Biochem* 1977;77:31–36.
34. Brown MS, Goldstein JL. A receptor-mediated pathway for cholesterol homeostasis. *Science* 1986;232:34–47.
35. Stein EA, Lane M, Laskarzewski P. Comparison of statins in hypertriglyceridemia. *Am J Cardiol* 1998;81:66B–69B.
36. Roberts WC. The rule of 5 and the rule of 7 in lipid-lowering by statin drugs. *Am J Cardiol* 1997;82:106–107.
37. Kostner GM, Gavish D, Leopold B, Bolzano K, Weintraub MS, Breslow JL. HMG-CoA reductase inhibitors lower LDL cholesterol without reducing Lp(a) levels. *Circulation* 1989;80:1313–1319.
38. Hunninghake CB, Stein EA, Mellies MJ. Effects of one year of treatment with pravastatin, an HMG-CoA reductase inhibitor, on lipoprotein(a). *J Clin Pharmacol* 1993;33:574–580.
39. Vaughan CJ, Gotto AM, Basson CT. The evolving role of statins in the management of atherosclerosis. *J Am Coll Cardiol* 2000;35:1–10.
40. Kaesemeyer WH, Caldwell RB, Huang J, Caldwell RW. Pravastatin sodium activates endothelial nitric oxide synthase independent of its cholesterol-lowering actions. *J Am Coll Cardiol* 1999;33:234–241.
41. Laufs U, Fata VL, Plutzky J, Liao JK. Upregulation of endothelial nitric oxide synthase by HMG CoA reductase inhibitors. *Circulation* 1998;97:1129–1135.
42. Endres M, Laufs U, Huang Z, Nakamura T, Huang P, Moskowitz MA, Liao JK. Stroke protection by 3-hydroxy-3-methylglutaryl (HMG)-CoA reductase inhibitors mediated by endothelial nitric oxide synthase. *Proc Natl Acad Sci USA* 1998;95:8880–8885.
43. Williams JK, Sukhova GK, Herrington DM, Libby P. Pravastatin has cholesterol-lowering independent effects on the artery wall of atherosclerotic monkeys. *J Am Coll Cardiol* 1998;31:684–691.
44. Molinau H, Meinertz T, Hink U, Muenzel T. HMG-CoA reductase inhibition inhibits vascular NADH oxidase activity, prevents uncoupling of nitric oxide synthase and improves endothelial dysfunction in cholesterol fed rabbits (abstract). *Eur Heart J* 2000;21:16.
45. Puddu P, Puddu GM, Muscari A. HMG-CoA reductase inhibitors: Is the endothelium the main target? *Cardiology* 2001;95:9–13.
46. Hernandez-Perera O, Perez-Sala D, Navarro-Antolin J, Sanchez-Pascuala R, Hernandez G, Diaz C, Lamas S. Effects of the 3-hydroxy-3-methylglutaryl-CoA reductase inhibitors, atorvastatin and simvastatin, on the expression of endothelin-1 and endothelial nitric oxide synthase in vascular endothelial cells. *J Clin Invest* 1998;101:2711–2719.
47. Feron O, Dessy C, Desager JP, Balligand JL. Hydroxymethylglutaryl-coenzyme A reductase inhibition promotes endothelial nitric oxide synthase activation through a decrease in caveolin abundance. *Circulation* 2001;103:113–118.
48. Sessa WC. Can modulation of endothelial nitric oxide synthase explain the vasculoprotective actions of statins? *Trends Mol Med* 2001;7:189–191.
49. Niebauer J, Volk HD, Kemp M, Dominguez M, Schumann RR, Rauchhaus M, Poole-Wilson PA, Coats AJ, Anker SD. Endotoxin and immune activation in chronic heart failure: A prospective cohort study. *Lancet* 1999;353:1838–1842.
50. Genth-Zotz S, von Haehling S, Bolger AP, Kalra PR, Coats AJS, Anker SD. Pathophysiological quantities of endotoxin induce tumor necrosis factor release in whole blood from patients with chronic heart failure. *Am J Cardiol*, in press.
51. Anker SD, Egerer KR, Volk HD, Kox WJ, Poole-Wilson PA, Coats AJS. Elevated soluble CD14 receptors and altered cytokines in chronic heart failure. *Am J Cardiol* 1997;79:1426–1430.
52. Suffredini AF, Fromm RE, Parker MM, Brenner M, Kovacs JA, Wesley RA, Parrillo JE. The cardiovascular response of normal humans to the administration of endotoxin. *N Engl J Med* 1989;3:280–287.
53. Hegewish S, Weh HJ, Hossfeld DK. TNF-induced cardiomyopathy. *Lancet* 1990;2:294–295.
54. Ridker PM, Rifai N, Pfeffer MA, Sacks F, Braunwald E. Long-term effects of pravastatin on plasma-concentrations of C-reactive protein. *Circulation* 1999;100:230–235.
55. Musial J, Undas A, Gajewski P, Jankowski M, Sydor W, Szczeklik A. Anti-inflammatory effects of simvastatin in subjects with hypercholesterolemia. *Int J Cardiol* 2001;77:247–253.
56. Pahan K, Sheikh FG, Namboodiri AM, Singh I. Lovastatin and phenyl-acetate inhibit the induction of nitric oxide synthase and cytokines in rat primary astrocytes,

microglia, and macrophages. *J Clin Invest* 1997;100:2671–2679.

57. Weitz-Schmidt G, Welzenbach K, Brinkmann V, Kamata T, Kallen J, Bruns C, Cottens S, Takada Y, Hommel U. Statins selectively inhibit leukocyte function antigen-1 by binding to a novel regulatory integrin site. *Nat Med* 2001;7:687–692.
58. Niwa S, Totsuka T, Hayashi S. Inhibitory effects of fluvastatin, an HMG-CoA reductase inhibitor, on the expression of adhesion molecules on human monocytes cell line. *Int J Immunopharmacol* 1996;18:669–675.
59. Takemoto M, Liao JK. Pleiotropic effects of 3-hydroxy-3-methylglutaryl coenzyme A reductase inhibitors. *Arterioscler Thromb Vasc Biol* 2001;21:1712–1719.
60. Ridley AJ. Rho family proteins: Coordinating cell responses. *Trends Cell Biol* 2001;11:471–477.
61. Amano M, Fukata Y, Kaibuchi K. Regulation and functions of Rho-associated kinase. *Exp Cell Res* 2000;261:44–51.
62. Goldstein JL, Brown MS. Regulation of the mevalonate pathway. *Nature* 1990;343:425–430.
63. Belch JJ, Bridges AB, Scott N, Chopra M. Oxygen free radicals and congestive heart failure. *Br Heart J* 1991;65:245–248.
64. Webb DJ, McMurray JJ. Enhanced basal nitric oxide production in heart failure. *Lancet* 1994;344:887–888.
65. Hambrecht R, Fiehn E, Weigl C, Gielen S, Hamann C, Kaiser R, Yu J, Adams V, Niebauer J, Schuler G. Regular physical exercise corrects endothelial dysfunction and improves exercise capacity in patients with chronic heart failure. *Circulation* 1998;98:2709–2715.
66. Ennezat PV, Malendowicz SL, Testa M, Colombo PC, Cohen-Solal A, Evans T, LeJemtel TH. Physical training in patients with chronic heart failure enhances the expression of genes encoding antioxidative enzymes. *J Am Coll Cardiol* 2001;38:194–198.
67. Hambrecht R, Hilbrich L, Erbs S, Gielen S, Fiehn E, Schoene N, Schuler G. Correction of endothelial dysfunction in chronic heart failure: Additional effects of exercise training and oral L-arginine supplementation. *J Am Coll Cardiol* 2000;35:706–713.
68. Mendez C, Garcia I, Maier R. Oxidants augment endotoxin-induced activation of alveolar macrophages. *Shock* 1996;6:157–163.
69. Ferrari R, Agnoletti L, Comini L, Gaia G, Bachetti T, Cargnoni A, Ceconi C, Curello S, Visioli O. Oxidative stress during myocardial ischemia and heart failure. *Eur Heart J* 1998;19:B2–B11.
70. Leyva F, Anker SD, Godsland IF, Teixeira M, Hellewell PG, Kox WJ, Poole-Wilson PA, Coats AJ. Uric acid in chronic heart failure: A marker of chronic inflammation. *Eur Heart J* 1998;19:1814–1822.
71. Farquharson CA, Butler R, Hill A, Belch JJ, Struthers AD. Allopurinol improves endothelial dysfunction in chronic heart failure. *Circulation* 2002;106:221–226.
72. Dhalla AK, Hill MF, Singal PK. Role of oxidative stress in transition of hypertrophy to heart failure. *J Am Coll Cardiol* 1996;28:506–514.
73. Wassmann S, Laufs U, Baumer AT, Muller K, Ahlbory K, Linz W, Itter G, Rosen R, Bohm M, Nickenig G. HMG-CoA reductase inhibitors improve endothelial dysfunction in normocholesterolemic hypertension via reduced production of reactive oxygen species. *Hypertension* 2001;37:1450–1457.
74. Vecchione C, Brandes RP. Withdrawal of 3-hydroxy-3-methylglutaryl coenzyme a reductase inhibitors elicits oxidative stress and induces endothelial dysfunction in mice. *Circ Res* 2002;91:173–179.
75. Kjekshus J, Pedersen TR, Olsson AG, Faergeman O, Pyorala K. The effects of simvastatin on the incidence of heart failure with coronary heart disease. *J Card Fail* 1997;3:207–213.
76. Vaughan CJ, Murphy MB, Buckley BM. Statins do more than just lower cholesterol. *Lancet* 1996;348:1079–1082.
77. Richartz BM, Radovancevic B, Frazier OH, Vaughn WK, Taegtmeyer H. Low serum cholesterol levels predict high perioperative mortality in patients supported by a left-ventricular assist system. *Cardiology* 1998;89:184–188.
78. Rauchhaus M, Koloczek V, Volk H, Kemp M, Niebauer J, Francis DP, Coats AJ, Anker SD. Inflammatory cytokines and the possible immunological role for lipoproteins in chronic heart failure. *Int J Cardiol* 2000;76:125–133.
79. Rauchhaus M, Coats AJ, Anker SD. The endotoxin-lipoprotein hypothesis. *Lancet* 2000;356:930–933.

Role of Nitric Oxide in Matrix Remodeling in Diabetes and Heart Failure

Suresh C. Tyagi[1] *and Melvin R. Hayden*[2]
[1] *Department of Physiology and Biophysics, University of Mississippi Medical Center, Jackson, MS, USA;*
[2] *Camdenton Cardiovascular Research Center, Camdenton, MO, USA*

Abstract. **Accumulation of oxidized-matrix between the endothelium and myocytes is associated with endocardial endothelial (EE) dysfunction in diabetes and heart failure. High levels of circulating homocysteine (Hcy) have been demonstrated in diabetes mellitus (DM). These high levels of Hcy (hyperhomocysteinemia, HHcy) have a negative correlation with peroxisome proliferator activated receptor (PPAR) expression. Studies have demonstrated that Hcy decreases bioavailability of endothelial nitric oxide (eNO), generates nitrotyrosine, and activates latent matrix metalloproteinase (MMP), instigating EE dysfunction. PPAR ligands ameliorate endothelial dysfunction and DM. In addition Hcy competes with PPAR ligands. The understanding of molecular, cellular, and extracellular mechanisms by which Hcy amplifies DM will have therapeutic ramifications for diabetic cardiomyopathy.**

Key Words. **MMP, TIMP, PPAR, collagen, redox stress, eNOS, ADMA, DDAH**

Abbreviations. **AA: arachidonic acid, BH_4: tetrahydrobiopterin, CBS: cystathionine β synthase, CIMP: cardiac inhibitor of metalloproteinase, CZ: ciglitazone, DM: diabetic mellitus, ECM: extracellular matrix, EE: endocardial endothelium, eNOS: endothelial nitric oxide synthase, Hcy: homocysteine, HHcy: hyperhomocysteinemia, LB_4: leukotriene B_4: L-NAME: N-nitro-l-arginine methyl ester, LV: left ventricle, MMP: matrix metalloproteinase, NADH: nicotinamide adenine dinucleotide (reduced), NFkB: nuclear factor kappa B, NO: nitric oxide, $ONOO^-$: peroxynitrite, PCR: polymerase chain reaction, PGJ2: 15-deoxy 12, 14-prostaglandin J2, PPAR: peroxisome proliferator activated receptor, Redox: reduction-oxidation, ROS: reduced oxygen species, RXR: retinoid X receptor, SOD: superoxide dismutase, SH-homo: thiol-homocysteine, Homo-s-s-Homo: homocystine, TIMP: tissue inhibitor of metalloproteinase.**

Matrix Remodeling and EE Dysfunction

Sixteen percent of the myocardium is composed of capillaries (including lumen and endothelium) [1]. Capillary endothelium, strategically located between the superfusing luminal blood and the underlying cardiac muscle, plays an important role in controlling the myocardial performance [2–7]. Several lines of evidence suggest that the frequency of apoptosis of capillary microvessel cells predict the development of the histologic lesions in diabetes [8]. The capillary surface area is reduced and tissue thickness from capillaries to myocytes is increased in the LV of diabetic rats [9]. There is increased tissue thickness between the endothelium and cardiomyocytes in CHF (Fig. 1). In DM the alteration of ultrastructure of the myocardium has been a hallmark of cardiovascular dysfunction [10–14]. Several lines of evidence indicate structural pathological manifestations in diabetic cardiomyopathy and its reversal by insulin treatment in rats [15].

Oxidative Stress and Matrix Remodeling

Remodeling implies synthesis and degradation of ECM. Latent MMP are activated by oxidant and instigate remodeling (Fig. 2). The composition of ECM (collagen and elastin) in the basement membrane of capillary endothelium contributes to the accumulation of oxidized-matrix between endothelium and myocytes. An abnormal collagen glycation, and chamber tissue stiffness, affecting diastolic function are appeared to be a major factor in impaired glucose tolerance in DM [16,17]. Alteration in LV diastolic filling is also associated with reciprocal changes in LV collagen gene [18], and accumulation of myocardial collagen in insulin-resistant syndrome [19]. There is enhanced ECM production in rat heart endothelial cells [20], and the levels of MMP activity are excessive in DM [21], especially MMP-9 [22]. The elastinolytic proteinase is upregulated in the basement membrane of microvessels of diabetes [23]. Both MMP-2 and -9 degrade elastin [24], and MMP-2 also degrades ultrastructure

Address for correspondence: Suresh C. Tyagi, Department of Physiology & Biophysics, The University of Mississippi Medical Center, 2500 North State Street, Jackson, MS 39216-4505, USA. Tel.: (601) 984-1899; Fax (601) 984-1817; E-mail: styagi@physiology.umsmed.edu

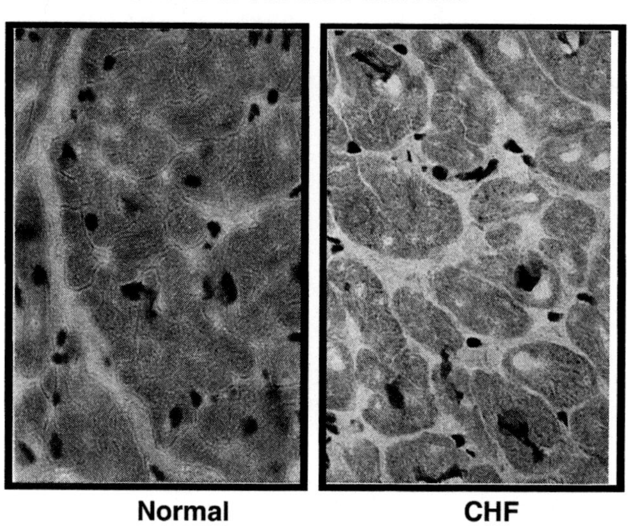

Fig. 1. Decreased capillary endothelial cell density and cardiac muscle hypertrophy in congestive heart failure (CHF): H & E staining. ECM deposition between endothelium and myocyte is a hallmark of CHF. IC (interstitial cells) and the numbers on the left refer to % by volume in the normal heart [10].

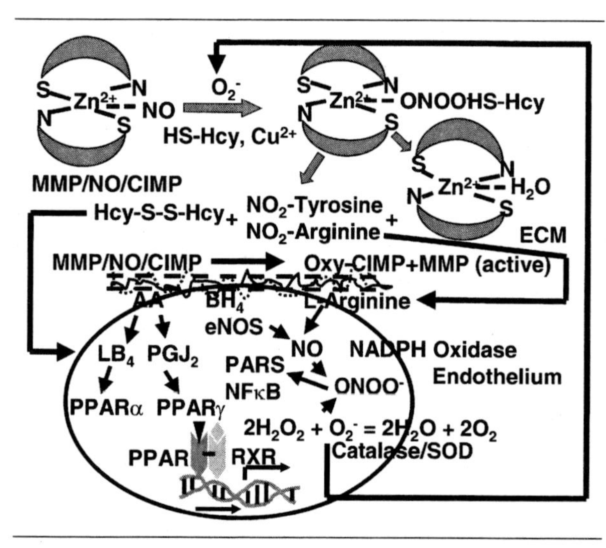

Fig. 2. Hypothesis: EE dysfunction in DM is due to increased levels of reduced oxygen species (ROS), Hcy, nitrotyrosine, MMP activity, collagenolysis, and decreased levels of eeNO in response to antagonizing PPARγ. The O_2^-, Cu^{2+}, and Hcy activate latent MP in MP/NO/CIMP complex, and generates NO_2-tyrosine and NO_2-arginine. Peroxisome proliferator, AA generates LB_4 (ligand of PPARα) and PGJ_2 (ligand of PPARγ). Hcy represses PPAR by competing with its ligands. BH_4, eNOS and L-arginine synthesize NO. Hcy induces NFkB and NADH/NAD oxidase. O_2^- and NO generate $ONOO^-$, NO_2-tyrosine, NO_2-arginine, and inhibit eNOS. $ONOO^-$ activates PARS, and PARS induces apoptosis. PPAR in cordinance with RXR induces antioxidant enzymes (SOD & catalase) and represses NADH/NAD oxidase. The overlap and garbled area in the middle represent ECM and its degradation.

collagen [25]. Because elastin turnover is remarkably lower than collagen [26], elastin and ultrastructural collagen are replaced by oxidatively modified stiffer collagen, leading to increased distance between endothelium and myocyte, thus impairing eeNO diffusion from endothelium to myocyte. This increased distance acts as a diffusion barrier to eeNO and may be considered a space occupying lesion within the myocardium.

Genetic Variations and Hetero- and Homozygosity in eNOS and EE Dysfunction

The eNOS knockout (–/–) in mice causes hypertension, hyperlipidemia, and insulin resistance [27]. Asymmetrical dimethylarginine (ADMA) is an endogenous inhibitor of eNOS [28]. Elevated levels of ADMA have been observed in hypertension, hyperlipidemia, hyperglycemia, renal failure, and hyperhomocysteinemia. Dimethylarginine dimethylaminohydrolase (DDAH) hydrolyses ADMA to L-citrulline and dimethylamine. Hcy inhibits DDAH with resultant ADMA elevation and endothelial cell dysfunction. ADMA levels were positively correlated with triglyceride levels and redox stress. The PPAR gamma agonist rosiglitizone was able to improve insulin resistance and decreases ADMA levels [29,30]. Accumulation of Hcy decreases eNO and instigates cardiovascular dysfunction [31]. There are four ways by which Hcy is accumulated (Fig. 3).

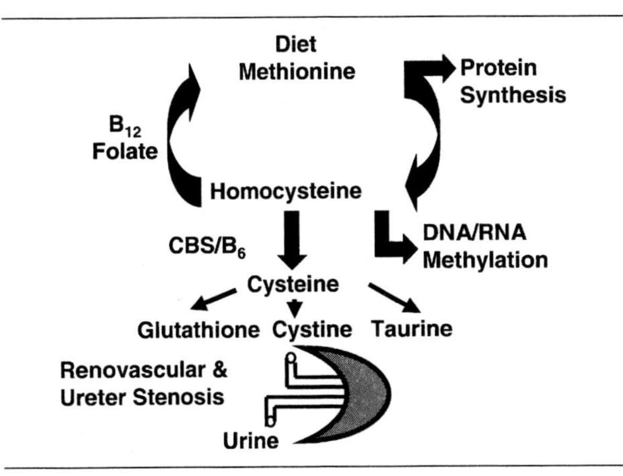

Fig. 3. *Excretion pathways of homocysteine: dietary methionine produces protein and homocysteine. Approximate 10% of homocysteine is required for DNA / RNA methylation. The remainder is converted to methionine by B12 and folate-dependent pathways. In addition, deficiency of cystathionine beta synthase (CBS) leads to accumulation of homocysteine. The volume retention by renovascular stenosis adds to homocysteine accumulation.*

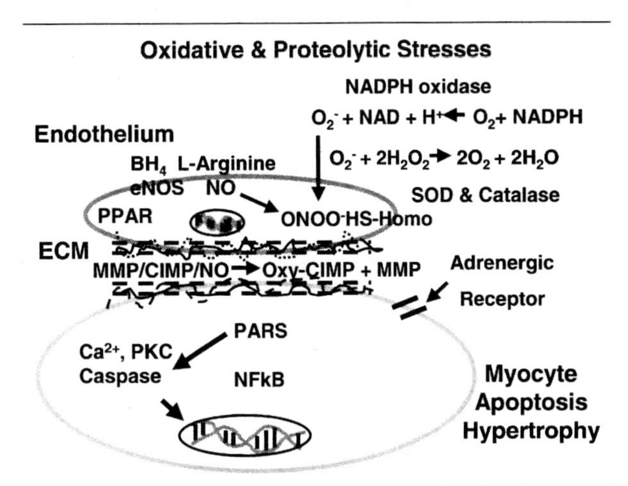

Fig. 4. *Role of oxidative and proteolytic stresses on EE and cardiac muscle: Latent MMP / NO / CIMP complex is oxidized to generate nitrotyrosine in CIMP and activates MMP. This leads to decrease eeNO concentration. Peroxynitrite activates poly(ribose-ADP)synthase, PARS, and induces apoptosis in some myocytes, and hypertrophy in others. The overlap and garbled area in the middle represent ECM and its degradation.*

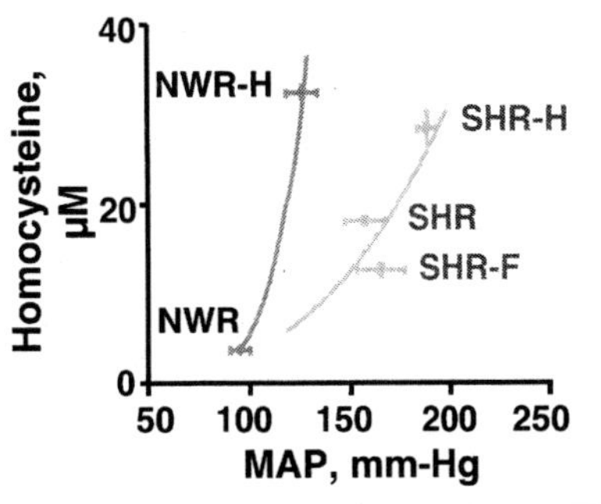

Fig. 5. *There is a relationship between blood pressure and redox stress: In NWR, homocysteine increases blood pressure. In contrast, in the SHR, similar increase in the level of homocysteine are associated with higher blood pressure. NWR-H and SHR-H are NWR and SHR treated with Hcy, respectively. SHR-F, SHR treated with folic acid.*

HHcy is associated with an increased risk of cardiovascular disease and DM when the associated nephropathy develops and GFR begins to decrease [32–34]. In DM and in HHcy-hypertrophic myocardium, the capillary endothelial cell density is reduced [35,36]. PPAR mediates the metabolism of various fatty acids and induces genes related to fatty acid metabolism and also has anti-inflammatory effects [37]. DM and Hcy facilitate ROS and inflammation [38,39], and cause elastinolysis/collagenolysis [40,41], including endothelial cell desquamation [42]. Hcy activates MMP [43], modulation of insulin resistance is accomplished by PPARγ [44,45], and PPAR inhibits MMP activation [46]. Although endothelial dysfunction is ameliorated, the levels of Hcy are increased by PPAR therapy in HHcy [47–49]. Recent results from our laboratory have suggested displacement of PPAR agonists by Hcy [50]. A gradient of eNO concentration, i.e. high in EE and low in midmyocardium has been demonstrated, suggesting a role of eeNO in beating heart [51]. Increased ROS decreases eNO availability, and generates nitrotyrosine in diabetes [52,53]. The decrease in eNO increases MMP activity [54]. Therefore, to reduce the wall stress, in the absence of eeNO, the latent MMP is activated to dilate the heart. Although increased chamber size increases wall stress, the unabated MMP activation leads to cardiomyopathy [55,56]. Although the endothelial cell density is decreased, ECM is increased, Hcy is elevated, and PPAR is not activated in DM, the mechanism of EE dysfunction in DM is unclear. It is possible that increased oxidized-matrix accumulation (diffusion barrier) between endothelium and myocyte may decrease bioavailability of eeNO (Fig. 4). Increased Hcy is associated with hypertension and proteinuresis (Figs. 5 and 6), suggesting a role of Hcy in cardiovascular dysfunction.

Gene and Embryonic Stem Cell Factor Therapy for EE Dysfunction

PPARγ agonists ameliorate diabetic mellitus [44,45], and recent clinical trials demonstrate a reduction in restenotic events post homocysteine lowering [57,58]. A positive correlation between

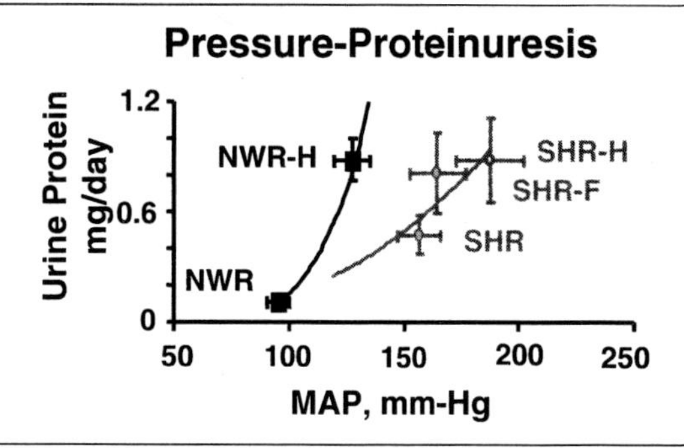

Fig. 6. *Increase in urinary protein excretion is associated with hypertension. In NWR proteinuresis is associated with hypertension. In contrast, in the SHR, similar increase in the level of proteinuresis is associated with higher blood pressure. NWR-H and SHR-H are NWR and SHR treated with Hcy, respectively. SHR-F, SHR treated with folic acid.*

Hcy and plasma creatinine might be related with the increase in muscle mass in DM patients [59]. Hcy induces cardiac hypertrophy in rats [35,36]. Ventricular pressure overload studies in mice have demonstrated reduced expression of PPAR during cardiac hypertrophy [60], and peroxisome proliferators reduce cardiac hypertrophy [61]. Hcy induces endocardial remodeling, structure and function changes, by defining their links to nitric oxide metabolism in DM. Hcy antagonizes PPARγ. This in turn increases oxidative stress, decreases eeNO, and increases nitrotyrosine content and MMP activity in endomyocardial wall. The heterozygosity in the activity of CBS is one cause of increased oxidative stress in humans [62].

Replacement of injured tissue by implanting skeletal muscle cell has been shown to be compatible with the conversion of skeletal muscle to cardiac muscle fibers and improvement of cardiac function post injury [63]. Cardiac dysfunction post MI was attenuated in MMP-9 knockout (–/–) mice [64]. It is of great interest to generate CIMP transgenic mice and measure the cardiac function during chronic heart failure. The mechanism of reverse remodeling by MMP inhibition specifically targeting the heart is a relatively important area. Stem cell therapy has been proposed for heart failure. However, it would be of great interest to identify the embryonic stem cell factor that causes robust growth and proliferation of injured cardiac muscle. MMP inhibition is undergoing clinical trials, and is therefore of great importance.

Acknowledgment

This work was supported, in part, by NIH grants GM-48595 and HL-71010.

References

1. Hoppeler H, Kayar SR. Capillary and oxidative capacity of muscles. *News Physiol Sci* 1988;3:113–116.
2. Roberts JT, Wearn JT. Quantitative changes in the capillary-muscle relationship in human hearts during normal growth and hypertrophy. *Am Heart J* 1941;617–633.
3. Hayden MR, Tyagi SC. Remodeling of the endocrine pancreas: The central role of amylin and insulin resistance. *Southern Med J* 2000;93:24–28.
4. Hayden MR, Tyagi SC. "A" is for amylin and Amyloid in type 2 diabetes mellitus. *J Pancreas* 2001;2(4):124–139.
5. Henderson AH, Lewis MJ, Shah AM, Smith JA. Endothelium, endocardium, and cardiac contraction. *Cardiovasc Res* 1992;26(4):305–308.
6. Smith JA, Shah AM, Fort S, Lewis MJ. The influence of endocardial endothelium on myocardial contraction. *Trends Pharmacological Sci* 1992;13:113–116.
7. Mebazaa A, Wetzel R, Cherian M, Abraham M. Comparison between endothelial and great vessel endothelial cells: Morphology, growth, and prostaglandin release. *Am J Physiol* 1995;268:H250–H259.
8. Kern TS, Tang J, Mizutani M, Kowluru RA et al. Response of capillary cell death to aminoguanidine predicts the development of retinopathy: Comparison of diabetes and galactosemia. *Invest Ophthalmol* 2000;41(12):3972–3978.
9. Warley A, Powell JM, Skepper JN. Capillary surface area is reduced and tissue thickness from capillaries to myocytes is increased in LV of STZ-diabetic rats. *Diabetologia* 1995;38:413–421.
10. Zak R. Cell proliferation during cardiac growth. *Am J Cardiol* 1973;31:211–219.
11. Di Bello V, Giampietro O, Matteucci E, Talarico L, Giorgi D et al. Ultrasonic video-densitometric analysis in type I diabetic myocardium. *Coron Artey Dis* 1996;7(12):895–901.
12. Sokolov EI, Zaichikova OS, Tsyplenkova VG. Ultrastructure of the myocardium in patients with cardiac pathology complicated by diabetes mellitus. *Arkh Patol* 1998;60(1):49–54.
13. Dhalla NS, Prierce GN, Innes IR, Beamish RE. Pathogenesis of cardiac dysfunction in diabetes mellitus. *Canad J Cardiol* 1985;1(4):263–281.
14. Abe T, Ohga Y, Tabayashi N et al. LV diastolic dysfunction in type 2 DM model rats. *Am J Physiol* 2002;282:H138–H148.
15. Thompson EW. Structural manifestations of diabetic cardiomyopathy in the rat and its reversal by insulin treatment. *Am J Anat* 1998;182:270–282.
16. Avendano GF, Agarwal RK, Bashey RI, Effects of glucose intolerance on myocardial function and collagen-lonked glycation. *Diabetes* 1999;48:1443–1447.
17. Turk Z, Misur I, Turk N, Benko B. Rat tissue collagen modified by advanced glycation: Co-relation with duration of diabetes and glycemic control. *Clin Chem Lab Med* 1999;37:813–820.
18. Katayama S, Abe M, Negishi K et al. Reciprocal changes in LV collagen alpha 1 chain gene expression between type I and IV in spontaneously diabetic rats. *Diabetes Res Clin Pract* 1994;26:163–169.
19. Mizushige K, Yao L, Noma T et al. Alteration in LV diastolic filling and accumulation of mycardial collagen at insulin-restant prediabetic stage of a type 2 diabetic rat model. *Circulation* 2000;101:899–907.

20. He Q, Spiro MJ. Isolation of rat heart endothelial cells and pericytes: Evaluation of their role in the formation of ECM components. *J Mol Cell Cardiol* 1995;27(5):1173–1183.
21. Ryan ME, Usman A et al. Excessive matrix metalloproteinase activity in diabetes: Inhibition by tetracycline analogues with zinc reactivity. *Curr Med Chem* 2001;8(3):305–316.
22. Uemura S, Matsushita H, Li W, Glassford AJ, Asagami T, Lee KH, Harrison DG, Tsao PS. Diabetes mellitus enhances vascular matrix metalloproteinase activity: Role of oxidative stress. *Circ Res* 2001;88:1291–1298.
23. Kwan CY, Wang RR, Beazley JS, Lee RM. Alterations of elastin and elastase-like activities in aortas of diabetic rats. *Biochim Biophys Acta* 1988;967:322–325.
24. Senior RM, Griffin GL, Eliszar CJ, Shapiro SD, Goldberg GI, Welgus HG. Human 92- and 72- kilodalton type IV collagenases are elastases. *J Biol Chem* 1991;266:7870–7875.
25. Aimes RT, Quigley JP. MMP-2 is an interstitial collagenase. *J Biol Chem* 1995;270:5872–5876.
26. Rucklidge GJ, Milne G, McGaw BA, Milne E, Robins SP. Turnover rates of different collagen types measured by isotope ratio mass spectrometry. *Biochim Biophys Acta* 1992;11:1156–1157.
27. Duplain H, Burcelin R, Sartori C, Cook S, Egli M, Lepori M, Vollenweider P et al. Insulin resistance, hyperlipidemia, and hypertension in mice lacking eNOS. *Circulation* 2001;104:342–345.
28. Stuhlinger MC, Abbasi F, Chu JW et al. Relationship between insulin resistance and an endogenous nitric oxide synthase inhibitor. *J Am Med Assoc* 2002;287:1420–1426.
29. Stuhlinger MC, Tsao PS, Her JH et al. Homocysteine impairs the nitric oxide pathway: Role of asymmetric dimethylarginine. *Circulatuon* 2001;104:2568–2575.
30. Nash DT. Insulin resistance, ADMA levels, and cardiovascular disease. *J Am Med Assoc* 2002;287:1451–1452.
31. Tyagi SC. Homocyst(e)ine and heart disease: Pathophysiology of extracellular matrix. *Clin and Exper Hypertension* 1999;21:181–198.
32. Gallistl S, Sudi K, Mangge H, Erwa W, Borkenstein M. Insulin is an independent correlate of plasma Hcy levels in obese children and adolescents. *Diabetes Care* 2000;23:1348–1352.
33. Hoogeveen EK, Kostense PJ, Beks PJ et al. HHcy is associated with an increased risk of cardiovascular disease, especially in NIDDM. *Arterios Throm Vasc Biol* 1998;18:133–138.
34. Drzewoski J, Czupryniak L, Chwatko G, Bald E. Hyperhomocysteinemia in poorly controlled type 2 diabetes patients. *Diabet Nutr Metab* 2000;13(6):319–324.
35. Miller A, Mujumdar V, Shek E, Guillot J, Angelo M, Pakmer L, Tyagi SC. Hyperhomocysteinemia induces multiorgan damage. *Heart and Vessels* 2000;15(3):135–143.
36. Miller A, Mujumdar V, Palmer L, Bower JD, Tyagi SC. Reversal of endocardial endothelial dysfunction by folic acid in homocysteinemic hypertensive rats. *Am J Hyperten* 2002;15:157–163.
37. Nolte RT, Wisely GB, Westin S, Cobb JE et al. Ligand binding and co-activator assembly of the PPARγ. *Nature* 1998;395:137–143.
38. Zhang X, Li H, Jin H, Ebin Z, Brodsky S, Goligorsky MS. Effects of homocysteine on endothelial nitric oxide production. *Am J Physiol* 2000; 279:F671–F678.
39. Salas A, Panes J, Elizalde JL et al. Mechanisms responsible for enhanced inflammatory response to ischemia-reperfusion in diabetes. *Am J Physiol* 1998;275:H1773–H1781.
40. Konecky N, Malinow MR, Tunick PA, Freedberg RS, Rosenzweig BP et al. Correlation between plasma homocysteine and aortic atherosclerosis. *Am Heart J* 1997;133:534–540.
41. Rolland PH, Friggi A, Barlatier A, Piquet P, Latrille V, Faye MM, Guillou J, Charpioy P, Bodard H, Ghininghelli O, Calaf R, Luccioni R, Garcon D. Hyperhomocysteinemia-induced vascular damage in the minipigs. *Circulation* 1995;91:1161–1174.
42. Starkebaum G, Harlan JM. Endothelial cell injury due to Cu-catalyzed H_2O_2 generation from homocysteine. *J Clin Invest* 1986;77:1370–1376.
43. Tyagi SC, Smiley LM, Mujumdar VS, Clonts B, Parker JL. Reduction-oxidation (redox) and vascular tissue level of homocyst(e)ine in human coronary atherosclerotic lesions and role in vascular ECM remodeling and vascular tone. *Mol Cell Biochem* 1998;181:107–116.
44. Lebovitz HE, Manerji MA, Insulin resistance and its treatment by thiazolidinediones. *Recent Prog Horm Res* 2001;56:265–294.
45. Itoh H, Doi K, Tanaka T, Fukunaga Y, Hosoda K et al. Hypertension and insulin resistance: Role of PPARγ. *Clin Exp Pharmacol Physiol* 1999;26:558–560.
46. Marx N, Sukhova G, Murphy C, Libby P, Plutzky J. Macrophages in human atheroma conain PPAR: Differential dependent peroxisomal proliferator activated receptorγ expression and reduction of MMP-9 activity through PPAR activation in mononuclear phaocytes *in vitro*. *Am J Pathol* 1998;153:17–23.
47. Blane GF. Comparative toxicity and safety profile of fenofibrate and other fibric acid derivatives. *Am J Med* 1987;83:26–36.
48. Dierkes J, Westphal S, Luley S. Serum homocysteine increases after therapy with fenofibrate or bezafibrate. *Lancet* 1999;354:219–220.
49. Bissonnette R, Treacy E, Rozen R, Boucher B, Cohn JS, Genest J. Fenofibrate raises plasma homocysteine levels in the fased and fed states. *Atherosclerosis* 2001;155:455–462.
50. Hunt MJ, Tyagi SC. Peroxisome proliferator compete and ameliorate Hcy-mediated EE cells activation. *Am J Physiol* 2002;283:C1073–C1079.
51. Pinsky DJ, Patton S, Mesaros S, Brovkovych V, Kubaszewski E, Grunfeld S, Malinski T. Mechanical transduction of NO synthesis in the beating heart. *Circ Res* 1997;81:372–379.
52. Kajstura J, Fiordaliso F et al. IGF-1 overexpression inhibits the development of diabetic cardiomyopathy and angII-mediated oxidative stress. *Diabetes* 2001;50:1414–1424.
53. Rosen R, Du X, Tschope D. Role of ROS for vascular dysfunction in the diabetic hearts. *Mol Cell Biochm* 1998;188:103–111.
54. Radomski A, Sawicki G, Olson DM, Radomski MW. The role of nitric oxide and metalloproteinases in the pathogenesis of hyperoxia-induced lung injury in newborn rats. *Brit J Pharmacol* 1998;125:1455–1462.
55. Tyagi SC, Campbell SE, Reddy HK, Tjahja E, Voelker DJ. Matrix metalloproteinase activity expression in infarcted,

noninfarcted and dilated cardiomyopathic human hearts. *Mol Cell Biochem* 1996;155:13–21.
56. Tyagi SC, Haas SJ, Kumar SG, Reddy HK, Voelker DJ, Hayden MR, Demmy TL, Schmaltz RA, Curtis JJ. Post-transcriptional regulation of extracellular matrix metalloproteinase in human heart end-stage failure secondary to ischemic cardiomyopathy. *J Mol Cell Cardiol* 1996;28:1415–1428.
57. Hackam DG, Peterson JC, Spence JD. What level of plasma homocysteine should be treated? *Am J Hyperten* 2000;13:105–110.
58. Schnyder G et al. Decreased rate of coronary restenosis after lowering of plasma homocysteine levels. *N Eng J Med* 2001;345:1593–1600.
59. Pavia C, Ferrer I, Valls C et al. Total homocysteine in patients with type I diabetes. *Diabetes Care* 2000;23:84–87.
60. Barger PM, Brandt JM, Leone TC, Weinheimer CJ, Kelly DP. Deactivation of PPARα during cardiac hypertrophic growth. *J Clin Invest* 2000;105:1723–1730.
61. Yamamoto K, Ohki R, Lee RT. Ikeda U, Shimada K. PPARγ activators inhibit cardiac hypertrophy in cardiomyocytes. *Circulation* 2001;104:1670–1675.
62. Davi G, Di Minno G, Coppola A, Andria G, Cerbone AM et al. Oxidative stress and platelet activation in homozygous homocysteinuria. *Circulation* 2001;104:1124–1128.
63. Taylor DA, Atkins BZ, Hungspreugs P et al. Regenerating functional myocardium: Improved performance after skeletal myoblast transplantation. *Nature Med* 1998;4(10):1200.
64. Ducharme A, Frantz S, Aikawa M et al. Targeted deletion of MMP-9 attenuates LV enlargement and collagen accumulation after experimental myocardial infarction. *I Clin Invest* 2000;106(1):55–62.

Peroxynitrite in Myocardial Ischemia-Reperfusion Injury

Manoj M. Lalu,[1] Wenjie Wang,[1] and Richard Schulz[1,2]
Departments of [1]Pharmacology and [2]Pediatrics, Cardiovascular Research Group, University of Alberta, Edmonton, Alberta, Canada

Abstract. **Peroxynitrite is a highly reactive oxidant which is produced during reperfusion of the ischemic heart. The role that this molecule plays in reperfusion injury has been controversial. Many investigations have demonstrated toxic effects of peroxynitrite, whereas others have found it to be protective during reperfusion. This review surveys evidence supporting both sides and proposes that peroxynitrite is a dichotomous molecule with beneficial and detrimental effects on the reperfused heart. Its toxic effects are mediated by modification and activation of a variety of targets (including poly (ADP) ribose synthetase and matrix metalloproteinases) while its beneficial effects are primarily mediated through its reaction with thiols, resulting in the formation of NO donor compounds (S-nitrosothiols).**

Key Words. **peroxynitrite, ischemia-reperfusion, myocardium, nitric oxide**

Reperfusion of the Ischemic Heart

Reperfusion of the ischemic myocardium must be performed in order to prevent cellular damage and necrosis. Unfortunately, associated with this restoration of blood flow which provides oxygen and nutrients to the ischemic tissue is a spectrum of deleterious events, including arrhythmias, mechanical dysfunction, and microvascular damage [1]. A great deal of attention has been given to the mechanisms underlying reperfusion injury, as it is a clinically relevant event which can occur following angina, myocardial infarction, and cardioplegic arrest [2]. Of the many theories regarding the development of reperfusion injury, the enhanced generation of highly reactive oxygen species by the heart during the acute reperfusion phase is an appealing one which is supported by a large foundation of experimental evidence. These reactive species react with an array of biochemical targets inside and outside the cell, including lipids, proteins, and DNA, resulting in altered cellular function. However, questions remain as to whether the generation of reactive oxygen species is entirely detrimental, or whether it may in fact have some beneficial effects. This debate has focused on the enhanced production of nitric oxide and other reactive oxygen species during early reperfusion, in particular the reaction product of nitric oxide (NO) and superoxide anion ($O_2^{\cdot -}$), peroxynitrite.

NO Biology and Pathophysiology

NO is a labile gas with a half-life of a few seconds at 37°C in the biological milieu. NO is generated by three isoforms of nitric oxide synthase (NOS), endothelial NOS (eNOS), inducible NOS (iNOS), and neuronal NOS (nNOS). There is some evidence of NO being produced by nNOS in the heart [3], however, under normal physiological conditions it is primarily generated by eNOS in cardiac myocytes, endocardial cells [4], and vascular endothelial cells [5]. eNOS catalyzes a five-electron oxidation of L-arginine to produce NO and the by-product L-citrulline. This catalytic activity is tightly regulated by intracellular calcium levels and, as a result, only small quantities of NO are produced for brief periods when intracellular Ca^{2+} levels are elevated. *In vivo*, this low level of NO production exerts a number of regulatory and cytoprotective effects: (1) decreasing intracellular calcium levels by increasing cyclic GMP production [6]; (2) promoting coronary vasodilation [7]; (3) decreasing adhesion of platelets [8] and neutrophils [9] to the endothelium; and (4) regulating cellular metabolism by reversibly inhibiting mitochondrial respiration [10] and enzymes involved in glycolysis [11]. During reperfusion of the ischemic myocardium, intracellular calcium overload and shear stress along the endothelium stimulate eNOS in a variety of cells in the heart to produce a large amount of NO over a relatively brief period of time (seconds to minutes).

NO can also be produced in high concentrations under conditions of inflammatory stress (for example, during exposure to proinflammatory

Address for correspondence: Dr. Richard Schulz, Cardiovascular Research Group, 4-62 Heritage Medical Research Centre, University of Alberta, Edmonton, Alberta, Canada, T6G 2S2. Tel.: +1 (780) 492-6581; Fax: +1 (780) 492-9753; E-mail: richard.schulz@ualberta.ca cytokines), following the expression of inducible NOS (iNOS) in endocardial endothelium [4], vascular endothelial cells [12], cardiac myocytes [13,14], vascular smooth muscle [15] and neutrophils [16]. iNOS, unlike eNOS, is independent of intracellular calcium levels and thus produces higher rates of NO formation which is sustained over several hours. In experimental models of myocardial ischemia-reperfusion there has been some debate surrounding the contribution of iNOS to reperfusion injury, as it may, or may not, be expressed during the short time frame of acute ischemia and reperfusion [17]. However, coronary artery disease patients who would be susceptible to ischemia-reperfusion injury have been shown to have iNOS protein expressed in their coronary vasculature [18] and iNOS mRNA in their myocardium [19]. Thus, the overproduction of NO by iNOS is a potentially significant event during myocardial reperfusion. As well, neutrophils which infiltrate the myocardium later in reperfusion are likely another potential source of NO generated from iNOS.

The Formation of Peroxynitrite

During the first seconds to minutes of reperfusion, the myocardium produces a large concentrated burst of NO (Fig. 1) which can be measured by electron spin trap techniques [20]. At the same time that NO is being produced, large amounts of superoxide anion ($O_2^{\cdot-}$) are also generated [20,21]. Possible sources of $O_2^{\cdot-}$ are NAD(P)H oxidase, xanthine oxidase [22–24], uncoupled mitochondrial respiration [25], and NOS itself if there is a lack of either its cofactor tetrahydrobiopterin or its substrate L-arginine [26–28]. NO and $O_2^{\cdot-}$ rapidly react during early reperfusion to form peroxynitrite [20,29], a potent oxidant. The quenching of available NO by $O_2^{\cdot-}$ likely explains why some early investigations observed a decrease in myocardial NO production upon reperfusion [30].

The kinetics of the reaction between NO and $O_2^{\cdot-}$ must be examined in order to understand why peroxynitrite is preferentially formed during reperfusion. The body's natural defence against $O_2^{\cdot-}$ is superoxide dismutase (SOD). There are three forms of SOD which catalyze the dismutation of $O_2^{\cdot-}$ to H_2O_2 and water: extracellular copper-zinc SOD, nuclear or cytosolic copper-zinc SOD, and mitochondrial manganese SOD. NO is the only biological molecule known to outcompete SOD for $O_2^{\cdot-}$ [31]. Under normal physiological conditions the lower amounts of NO being produced in the heart are not of a sufficient concentration to compete effectively with SOD for the $O_2^{\cdot-}$ which may be available. Moreover, there is likely only a low concentration of $O_2^{\cdot-}$ available [20,32,33]. Consequently,

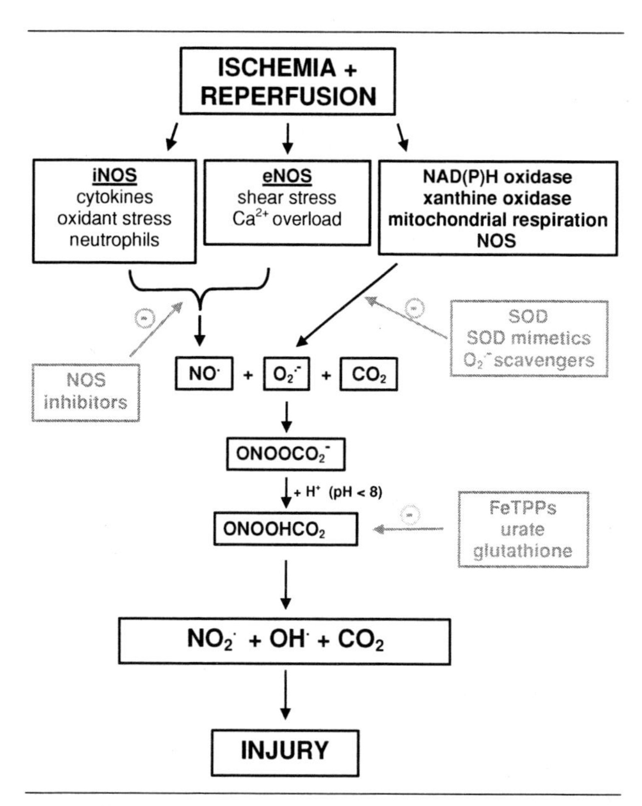

Fig. 1. *Peroxynitrite formation: Reperfusion of a previously ischemic area of the myocardium causes the stimulation of NO production (primarily from eNOS and partially from iNOS) and $O_2^{\cdot-}$ production (from NAD(P)H oxidase, xanthine oxidase, uncoupled mitochondrial respiration, and NOS under conditions of L-arginine or tetrahydrobiopterin depletion). These two molecules react to form peroxynitrite, which then adds carbon dioxide to form nitrosoperoxycarbonate anion ($ONOOCO_2^-$). This moiety is protonated at physiological pH to form nitrosoperoxycarbonic acid ($ONOOHCO_2$) which is highly unstable and rapidly decomposes into highly toxic free radical species. A number of pharmacological interventions can inhibit the formation of peroxynitrite, including NOS inhibitors and SOD or its mimetics. Peroxynitrite can also be directly scavenged by 5,10,15,20-tetrakis-[4-sulfonatophenyl]-porphyrinato-iron[III] (FeTPPs), urate, or thiols such as glutathione.*

under normal physiological conditions in the heart, the dismutation of $O_2^{\cdot-}$ by SOD will predominate over the formation of peroxynitrite. However, when NO (and $O_2^{\cdot-}$) production increases during reperfusion it can outcompete endogenous SOD for the reaction with $O_2^{\cdot-}$. Thus, during the first minutes of reperfusion, the rate of formation of peroxynitrite will predominate over the dismutation of $O_2^{\cdot-}$. There may even be a sustained production of peroxynitrite, as total myocardial SOD activity decreases during reperfusion [34] while available concentrations of $O_2^{\cdot-}$ increase. It should also be noted that in a physiological CO_2/bicarbonate environment peroxynitrite reacts with CO_2 to form nitrosoperoxycarbonate anion ($ONOOCO_2^-$)

[35–37]. This molecule is then protonated at physiological pH to form nitrosoperoxycarbonic acid. This highly unstable intermediate spontaneously decomposes by homolytic cleavage to give rise to nitrogen dioxide, hydroxyl radicals, and CO_2. Both nitrogen dioxide and hydroxyl radicals are far more chemically reactive than peroxynitrite and thus represent the actual species responsible for the detrimental effects of peroxynitrite.

The actual concentration of peroxynitrite that is produced in the reperfused heart has been a point of contention. Some investigators have speculated that perhaps only low micromolar concentrations of peroxynitrite may be formed, and that concentrations above these levels would probably not be formed *in vivo* [38]. Other investigators have alternatively suggested that peroxynitrite may be formed in much higher micromolar concentrations, at least over brief periods of time in certain compartments of the cell due to the localization of the production of both NO and $O_2^{\cdot-}$ [39,40]. Unfortunately, there is no accurate method to directly measure the actual concentration of peroxynitrite in biological milieu, owing in part to the extremely short half-life of this molecule [31]. Thus, the true *in vivo* concentration of peroxynitrite remains a speculation. However, based on the available techniques to estimate its formation, peroxynitrite production during reperfusion is undoubtedly increased relative to preischemic levels [20,29]. Moreover, it should be noted that these methods likely underestimate endogenous peroxynitrite formation by at least one or more orders of magnitude [29].

The Contentious Role of Peroxynitrite in Myocardial Reperfusion Injury

The increased production of peroxynitrite has an array of effects in the reperfused myocardium which have been the subject of a number of reviews [41–44]. However, there has been some controversy in regards to whether peroxynitrite is harmful or protective in the setting of ischemia-reperfusion injury. This review will seek to synthesize these seemingly disparate views into a coherent explanation of peroxynitrite's role in myocardial reperfusion injury. To summarize, peroxynitrite which is produced endogenously (that is, within the myocardium) is likely cytotoxic, and acts through a number of different mechanisms (discussed below) to contribute to reperfusion injury. Peroxynitrite produced outside of the myocardium (from neutrophils and other circulating cells or administered in experimental settings) reacts with endogenous thiols containing compounds, in particular glutathione, to produce NO donors which are cytoprotective and ameliorate cardiac dysfunction. Thus, the objective of this review will be to demonstrate that peroxynitrite is a dichotomous molecule with both detrimental and beneficial effects in myocardial reperfusion injury.

Peroxynitrite is Toxic and Contributes to Reperfusion Injury

The targets of peroxynitrite in the cell include proteins [45], lipids [46], carbohydrates [47], and nucleic acids [48]. The reaction of peroxynitrite with these biomolecules results in lipid peroxidation [46,49], protein modification by oxidization of sulfhydryl groups [50] and nitration of tyrosine residues [51]. The detrimental effects of peroxynitrite include structural damage, enzyme dysfunction, ion channel and transporter malfunction and eventually cell death.

Wang and Zweier [20] demonstrated an increased production of peroxynitrite upon reperfusion of the ischemic rat heart by using an electron paramagnetic resonance spin trap technique. By detecting dityrosine, a fluorescent reaction product between L-tyrosine and peroxynitrite, we showed a rapid release of peroxynitrite into the coronary perfusate during reperfusion following ischemia which peaks within the first minute [29]. This production of peroxynitrite was blocked by the addition of N^G-monomethyl-L-arginine (L-NMMA, a NOS inhibitor) [29] or the addition of superoxide dismutase [20,29]. Nitrotyrosine residues within proteins, a footprint of peroxynitrite production, was also detected in both rat and canine myocardium using *in vivo* models of ischemia-reperfusion [17,52]. Increased myocardium derived nitrotyrosine was also detected in human patients undergoing cardiopulmonary bypass, a clinical situation of ischemia-reperfusion injury [53]. Blocking the production of peroxynitrite conferred functional protection on the heart [20,29]. We observed that low concentrations of NOS inhibitors significantly improved mechanical function during reperfusion following global ischemia in isolated working rabbit hearts perfused in the presence of physiologically relevant concentrations of fatty acid [32] as well as in isolated rat hearts perfused according to the method of Langendorff with fatty acid free crystalloid buffer [29].

The cytotoxic effects of peroxynitrite can also be seen by its infusion directly into isolated hearts. Continuous infusion of 40 μM peroxynitrite into isolated working rat hearts caused a significant decrease in cardiac efficiency (as measured by a decrease in mechanical function with no corresponding change in oxygen consumption) [40]. There was a lag time of 20–25 minutes before the decline in mechanical function began, and this loss was already irreversible by 45 minutes of infusion.

This suggests that there is likely a depletion of anti-oxidant reserves before irreversible changes occur in the heart. Ma et al. [54] found that the infusion of 3-morpholinosydnonimine (SIN-1, a spontaneous donor of both NO and $O_2^{\cdot-}$ and thus considered to be a peroxynitrite donor) during reperfusion significantly aggravated mechanical dysfunction and the loss of tissue creatine kinase. Interestingly, the simultaneous infusion of superoxide dismutase with SIN-1 (which would effectively remove any $O_2^{\cdot-}$ produced by SIN-1 and render it an NO donor) significantly improved both measures of ischemia-reperfusion injury [54].

The controversial effect of NOS inhibitors in ischemia-reperfusion resides in the dual roles of NO. The basal generation of NO in the coronary vasculature acts both as an anti-oxidant and as a cytoprotective molecule in myocardial ischemia-reperfusion injury [29]. The enhanced simultaneous production of NO and superoxide, leading to enhanced peroxynitrite formation, is detrimental to the cardiac myocyte. This partially explains the reported findings of a lack of protection from reperfusion injury with NOS inhibitors, particularly when used at high concentrations [55–60]. We showed that the relationship between the concentration of NOS inhibitor versus protection from ischemia-reperfusion is in the form of a steep bell-shaped curve [29]. By using high concentrations of NOS inhibitors, all NO production (including the basal, physiologically necessary production of NO) will be eliminated. A similar response was also noted by Depré et al. [61] using a low-flow ischemia model in isolated rabbit hearts. Thus, the basal production NO is protective and obligatory for proper cardiac function.

There is a complete absence of literature showing any cardioprotective effects of peroxynitrite formed endogenously in the heart. Such endogenous peroxynitrite formation is involved not only in reperfusion injury, but also pro-inflammatory cytokine-induced heart failure [62], doxorubicin-associated cardiotoxicity [63], allograft rejection [64], and myocardial inflammation [65]. These investigations provide a consensus that myocardial peroxynitrite formation contributes to a variety of myocardial injuries, including ischemia-reperfusion induced injury.

Emerging Effectors of Peroxynitrite Induced Reperfusion Injury

As mentioned previously, peroxynitrite oxidizes and nitrates a large number of moieties, including lipids, proteins, and DNA (Fig. 2). Peroxynitrite can directly initiate lipid peroxidation in polyunsaturated fatty acids and form nitrated lipids [46]. This leads to instability of the cellular membrane

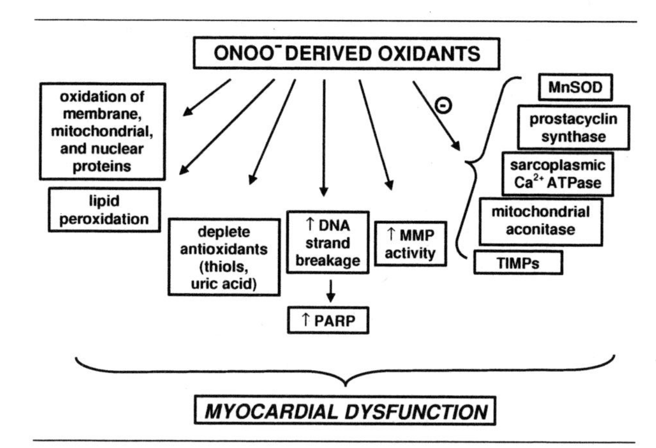

Fig. 2. Mechanisms of peroxynitrite induced cardiac dysfunction: Peroxynitrite derived oxidants have been shown to interact with a variety of intracellular targets in vivo and in vitro. Peroxynitrite oxidizes a number of membrane, mitochondrial, and nuclear proteins. The inhibition of manganese SOD (MnSOD), prostacyclin synthase, sarcoplasmic Ca^{2+}-ATPase, and mitochondrial aconitase contribute to cellular dysregulation. Peroxynitrite can also contribute to myocardial dysfunction by directly activating matrix metaloproteinases (MMPs), inactivating tissue inhibitors of metalloproteinase (TIMPs), and by activating poly (ADP-ribose) polymerase (PARP) indirectly by causing DNA strand breaks.

and may contribute to increased membrane permeability following reperfusion. Peroxynitrite can also disrupt and inhibit the function of proteins with iron-sulfur centres. For instance, mitochondrial aconitase, an enzyme involved in the Krebs cycle, is irreversibly inhibited by peroxynitrite [66]. Inhibition of this enzyme might contribute to the disrupted metabolic state of the myocardium following reperfusion [67]. Other reported actions of peroxynitrite that may contribute to myocardial reperfusion injury include: (1) the inactivation of manganese SOD, and thus a decrease in $O_2^{\cdot-}$ scavenging [68]; (2) the inactivation of the sarcoplasmic Ca^{2+}-ATPase and subsequent disregulation of calcium homeostasis [69]; (3) the inactivation of prostacyclin synthase, reducing the production of cytoprotective prostacyclin [70]; (4) oxidation of mitochondrial and nuclear proteins; and (5) nitration of tyrosine residues, which disrupts protein structure by changing a normally hydrophobic residue into a hydrophilic one [31]. Many more potential targets exist, however, the limited scope of this review prevents a full examination of these interactions. Instead, two exciting and emerging downstream effectors of peroxynitrite are discussed below.

Matrix Metalloproteinases

Matrix metalloproteinases (MMPs) are a family of zinc-containing endopeptidases which, when activated, can cleave a variety of extracellular matrix proteins. Based on their *in vitro* substrate preference and structural homology, MMPs are categorized into five groups: collagenases, gelatinases, stromelysins, membrane-type MMPs (MT-MMPs) and miscellaneous MMPs [71]. Each group has distinct features but also share common properties with other group members. MMP-2 (gelatinase A, type IV collagenase) is of particular interest due to its unique characteristics. MMP-2 is a constitutive enzyme which is ubiquitously expressed at a higher level than any other MMP [72]. Moreover, MMP-2 has been recognized to play an important role in a number of physiological and pathological conditions such as embryogenesis, wound healing, angiogenesis, atherosclerosis, and heart failure [72].

We recently uncovered the role of MMP-2 in acute myocardial ischemia-reperfusion injury [73]. We found that there is a basal release of MMP-2 into the coronary effluent of aerobically perfused rat hearts. During reperfusion following ischemia, there is a rapidly enhanced release of MMP-2 within the first minute which was accompanied by a loss in its tissue level. This peak in MMP-2 release follows the peak in peroxynitrite release by about 30–120 seconds, as observed in nearly identical perfusion protocols [29,74]. The activity of the released MMP-2 correlates directly with the ischemic time, and there is a negative correlation between the release of MMP-2 activity in reperfusion and the recovery of cardiac mechanical function. Inhibitors of MMPs (either o-phenanthroline, doxycycline or a neutralizing anti-MMP-2 antibody) protected the heart from ischemia-reperfusion injury [74].

How MMP-2 activity is increased during ischemia-reperfusion is not known. The critical step of MMP activation is the disruption of bond between a cysteine residue in the propeptide domain and its catalytic zinc centre which can be achieved by two distinct pathways *in vitro*. Proteases such a plasminogen, trypsin, and other MMPs can activate MMPs by proteolysis of the propeptide domain [75]. The active MMP then has a lower molecular weight. Many chemical reagents are also able to activate MMPs by disrupting the cysteinyl-zinc bond without a change in molecular weight. With its strong oxidation potential, peroxynitrite activates MMP-2, MMP-9, and MMP-8 in a concentration-dependent fashion [76–78]. More recently, S-glutathiolation of MMPs has been suggested to be involved in this peroxynitrite-dependent activation process [79]. Interestingly, peroxynitrite can also inactivate the tissue inhibitors of MMPs (TIMPs), the endogenous inhibitors of MMPs [80]. The striking similarity between the profiles of peroxynitrite production and MMP-2 activation during ischemia-reperfusion injury leads us to hypothesize that peroxynitrite is indeed involved in the activation of MMP-2 [29,74].

We explored the role of peroxynitrite mediated activation of MMPs in crystalloid-buffer perfused, isolated rat hearts [74]. Peroxynitrite was continuously infused into these aerobically perfused Langendorff hearts at concentrations of 30 or 80 μM for 15 minutes. Infusion of 80 μM peroxynitrite caused a progressive decline of cardiac mechanical function after a lag time of 5–10 minutes. Enhanced release of MMP-2 activity was also seen to precede the decline of cardiac mechanical function in a concentration-dependent manner. Scavenging peroxynitrite with glutathione prevented the increased release of MMP-2 activity. The mechanical dysfunction caused by peroxynitrite infusion was ameliorated by the co-infusion of an MMP inhibitor [74]. Similar myocardial activation of MMPs has been reported *in vivo* using porcine and canine models of regional myocardial ischemia [81,82]. In human atrial biopsy samples taken from patients undergoing cardiac bypass surgery, MMP-2 activity has also been shown to be upregulated in an acute time frame which parallels peroxynitrite production [83]. These findings suggest that peroxynitrite induced myocardial injury is mediated, at least in part, through MMPs. How does the oxidative activation and subsequent release of MMPs result in ischemia-reperfusion injury?

The proteolysis of the regulatory element of contractile proteins such as troponin I is involved in the pathophysiology of myocardial ischemia-reperfusion injury [84]. The enzyme responsible for the proteolytic degradation of troponin I remains unknown, although some evidence suggests a possible role for calpain, a calcium-activated protease [85]. We found that troponin I was highly susceptible to proteolysis by MMP-2 *in vitro* [86]. Reduced levels of troponin I was found in hearts subjected to ischemia-reperfusion compared to aerobically perfused hearts. Inhibition of MMPs using either o-phenanthroline or doxycycline not only improved the recovery of cardiac mechanical function but also prevented the degradation and loss of troponin I. We also found a co-localization of MMP-2 within sarcomeres using immunogold and confocal microscopy, as well as by immunoprecipitation studies. This is the first evidence for both the intracellular localization and biological action of MMPs, a family of enzymes previously considered to have actions outside of cells [86]. Taken together, these data clearly suggest a critical role of MMP-2 in acute myocardial ischemia-reperfusion injury.

In addition to contractile proteins, MMP-2 may also adversely affect myocardial microvascular tone by processing big-endothelin into a novel peptide termed medium endothelin, which is a more potent vasoconstrictor than endothelin-1 itself [87]. MMP-2 can also inactivate the vasodilator calcitonin gene related peptide [88]. These effects of MMP-2 released from the heart would lead to an overall vasoconstrictive effect and could contribute to the decrease in coronary flow that is seen following reperfusion.

Poly (ADP-ribose) Polymerase

Poly (ADP-ribose) polymerase (PARP) is a chromatin bound enzyme that is found in the nuclei of a variety of cells, including cardiac myocytes [89]. Under normal physiological conditions it is involved in a number of processes, such as gene expression and cellular differentiation, which rely on DNA replication and repair [90]. However, under conditions of oxidant stress, peroxynitrite can serve as a powerful trigger of DNA single strand breaks with the consequent dimerization and activation of PARP [91]. When activated under these conditions, PARP catalyzes the transfer of ADP ribose from NAD to a variety of nuclear proteins in an attempt to repair the single strand breaks. Since NAD is essential for proper mitochondrial respiration, the depleted NAD must be restored using an ATP dependent process.

It is believed that when the myocardium is reperfused, peroxynitrite causes significant DNA damage that activates PARP. The depletion of both NAD and ATP caused by this activation leads to cellular dysfunction and death. This sequence of events was mimicked by the addition of peroxynitrite to H9C2 rat cardiac myocytes. In this system, 500 μM of peroxynitrite caused a significant activation of PARP and a subsequent depression of mitochondrial respiration [92]. However, the deleterious effects of peroxynitrite could be blocked by the addition of 3-aminobenzamide, a PARP inhibitor which reacts with thiols in the zinc finger DNA binding domain of the enzyme [90].

Further whole organ and *in vivo* studies have confirmed the importance of PARP activation in mediating myocardial ischemia and reperfusion injury. The addition of 3-aminobenzamide to isolated rabbit hearts decreased infarct size and cardiac mechanical dysfunction following 30 minutes of ischemia [93]. Similar protection was noted by Halmosi et al. [94] when they used PARP inhibitors (either 3-aminobenzamide, nicotinamide, BGP-15, or 4-hydroxyquinazoline) in isolated perfused rat hearts subjected to ischemia-reperfusion. Using an *in vivo* rat model of ischemia-reperfusion, Zingarelli et al. [92] demonstrated that pretreatment with 3-aminobenzamide significantly decreased myocardial creatine kinase loss and increased myocardial ATP levels. Finally, isolated working hearts from PARP (−/−) knockout mice were resistant to functional depression following hypoxia-reoxygenation injury [95].

Interestingly, the inhibition of PARP also protects the myocardium from neutrophil infiltration and the oxidative stress that these cells elicit [90]. The inhibition of PARP decreased the expression of the adhesion glycoproteins (P-selectin, ICAM-1) responsible for neutrophil adherence and infiltration following myocardial ischemia-reperfusion [96]. Furthermore, the administration of 3-aminobenzamide to rats subjected to myocardial ischemia-reperfusion *in vivo* decreased neutrophil infiltration and nitrotyrosine staining [92,96]. This suggests that PARP is not only a downstream effector of peroxynitrite induced damage, but also an upstream trigger for further oxidative damage. Thus, pharmacologically inhibiting the interaction between peroxynitrite and PARP, and/or the selective inhibition of PARP, may be a promising new avenue to prevent myocardial ischemia-reperfusion injury.

Peroxynitrite Ameliorates Reperfusion Injury Through the Formation of NO Donors. Peroxynitrite applied directly through the coronary vasculature may also have some potentially beneficial effects in myocardial ischemia-reperfusion injury (Fig. 3). In order to demonstrate these effects the heart must be perfused with either blood,

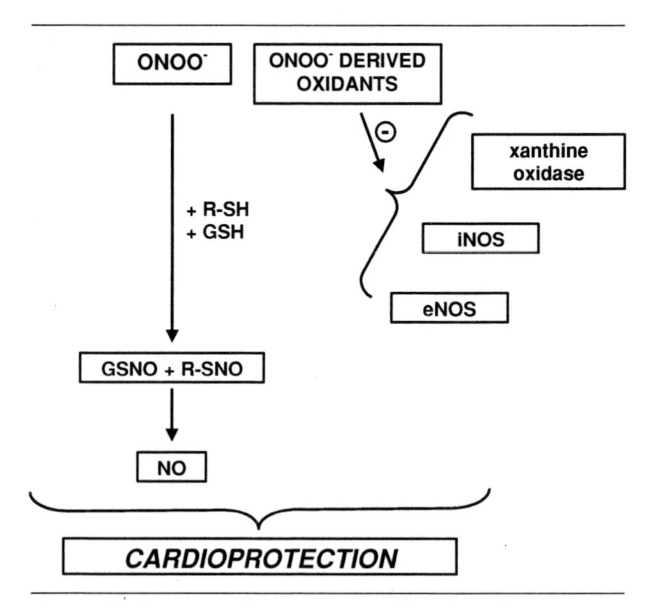

Fig. 3. Mechanisms of peroxynitrite induced cardioprotection: Peroxynitrite primarily exerts its protective effects through reactions with antioxidants, particularly thiol containing compounds (R-SH). A significant amount of peroxynitrite can be detoxified by gluatathione (GSH), which is consequestly transformed into the NO donor nitrosoglutathione (GSNO). The NO released from nitrosoglutathione has cytoprotective effects. Alternatively, peroxynitrite may diminish its own production by inhibiting NO and $O_2^{\bullet-}$ producing enzymes.

neutrophils, or a crystalloid solution containing antioxidants (for example, glutathione). Lefer et al. [97] demonstrated that infusion of 0.8 μM of peroxynitrite in rat hearts perfused with a neutrophil-enriched crystalloid buffer significantly improved cardiac function following reperfusion. This improved function was associated with decreased neutrophil infiltration and improved coronary flow, suggesting that the exogenous addition of peroxynitrite decreased neutrophil adhesion and subsequent vascular plugging. Peroxynitrite was beneficial in this investigation and others [38,98–100] due to antioxidants and/or thiol containing proteins (albumin, uric acid, glutathione) present in the blood/leukocytes/perfusion buffer used. These antioxidants not only effectively reduce the concentration of peroxynitrite, but also convert it to NO donor compounds, thereby converting a potentially toxic molecule into a cytoprotective one [101,102].

In order to investigate the dependence of peroxynitrite's protective effects on such antioxidants, Ma et al. [39] perfused isolated rabbit hearts with either blood or a crystalloid buffer solution. The hearts were subjected to 30 minutes of ischemia and then infused with either 1, 3, 30, or 100 μM of peroxynitrite at the onset of reperfusion. In crystalloid buffer perfused hearts, peroxynitrite produced a concentration dependent aggravation of cardiac mechanical function. However, in blood perfused hearts, peroxynitrite produced a bell shaped response, showing significant protection at lower concentrations, but none at all at 100 μM. This high concentration of peroxynitrite could have overwhelmed the antioxidants (such as glutathione) present in the blood. Furthermore, this high concentration of peroxynitrite was likely capable of penetrating into the cardiac myocytes. A similar study demonstrated that exogenous peroxynitrite was beneficial *in vivo* only during blood cardioplegia, and not crystalloid buffer cardioplegia [99]. These investigations clearly reveal that the dichotomous nature of peroxynitrite is fully dependent on the perfusion environment.

The cardioprotective effects of peroxynitrite can be mimicked if glutathione is supplemented into the crystalloid perfusion buffer. We showed that as little as 30 μM glutathione readily protected aerobically perfused rat hearts from functional damage induced either by 80 μM peroxynitrite or by ischemia-reperfusion [100]. Significantly, the improved functional recovery of glutathione treated hearts correlated with decreased dityrosine formation in the perfusate (that is, reduced peroxynitrite availability) as well as enhanced cGMP levels during reperfusion, suggesting that an NO donor intermediate was formed during reperfusion [100]. Furthermore, Ma et al. [54] demonstrated that the toxic effect of SIN-1 on isolated ischemia-reperfused rat hearts can be converted into a beneficial effect with glutathione. Finally, using an *in vivo* dog model, the addition of glutathione to crystalloid cardioplegia solution increased nitrosoglutathione formation in the cardioplegic solution and decreased subsequent neutrophil adhesion upon reperfusion [98].

Thus, studies which have used authentic peroxynitrite infused into hearts perfused with either blood, or crystalloid buffer supplemented with plasma or glutathione, have essentially shown the effects of the administration of an exogenous source of NO. It has long been recognized that the infusion of L-arginine [32,58] or NO donors [29,103,104] attenuates myocardial ischemia-reperfusion injury, regardless of the perfusion environment, by working through a number of cardioprotective mechanisms. First, although NO is a free radical it is also a potent antioxidant which can effectively neutralize a number of oxidant species [31]. Second, NO has potent anti-neutrophil and anti-platelet effects that can decrease vascular plugging, and neutrophil infiltration [105]. Thirdly, through the activation of cGMP dependent pathways, calcium overload of cells is diminished. Finally, NO is a potent vasodilator [106] that can counter the decreased coronary flow which is prevalent during reperfusion.

Other Mechanisms of Protection of Exogenously Administered Peroxynitrite

Aside from the formation of nitrosoglutathione, peroxynitrite administered exogenously may act through several other mechanisms to effect cardioprotection. For example, peroxynitrite can inhibit xanthine oxidase (a $O_2^{\cdot-}$ generator) [107] and both eNOS and iNOS [108] in a concentration dependent manner. If this were to occur *in vivo* during reperfusion, peroxynitrite could act in a negative feedback mechanism to inhibit its own production. There has also been a report that 0.4 μM peroxynitrite infused into crystalloid buffer perfused into rat hearts during reperfusion had no effect on cardiac function but significantly attenuated reperfusion-induced arrhythmias [109].

Despite all these potentially beneficial effects of exogenous peroxynitrite, one should not lose sight of the fact that 'endogenous' peroxynitrite formation is cytotoxic. Even when using *in vivo* models of ischemia-reperfusion (a model with a large pool of various antioxidants), extensive tissue nitrotyrosine formation has been noted [17], which is indicative of widespread protein modification. Thus, although nitrosoglutathione may be cytoprotective, peroxynitrite which is not detoxified during reperfusion is most likely harmful.

Conclusions

The contribution of peroxynitrite to myocardial reperfusion injury remains complicated and elusive in many ways. However, based on the experimental evidence presented in this review, it may be possible to classify peroxynitrite as a molecule with dichotomous effects in myocardial reperfusion. Peroxynitrite produced within the myocardium is likely cytotoxic, and contributes to the myocardial ischemia-reperfusion injury. Exogenously produced or infused peroxynitrite, however, may exert protective effects by inhibiting leukocyte adhesion and infiltration via its transformation into nitrosoglutathione. Thus, in order to design new rational therapeutics for reperfusion, researchers must try to inhibit the detrimental effects of "endogenous" peroxynitrite, while potentiating the beneficial aspects of "exogenous" peroxynitrite.

One approach may be to target the downstream effectors of peroxynitrite, such as MMPs and PARP. More research needs to be conducted to provide a better understanding of these targets. Alternatively, peroxynitrite decomposition catalysts, such as 5,10,15,20-tetrakis-[4-sulfonatophenyl]-porphyrinato-iron[III] (FeTPPs) may also be evaluated [110]. These drugs are able to catalyze the rapid decomposition of peroxynitrite into harmless byproducts. FeTPPS was shown to be effective against ischemia-reperfusion injury of the splanchnic artery [111] and in pro-inflammatory cytokine-induced cardiac dysfunction [62]. Regardless of the approach taken for future therapies, a better understanding of peroxynitrite must be developed in order to deal adequately with reperfusion injury.

Acknowledgments

Work from this lab cited here is supported by grants to RS from the Canadian Institutes of Health Research (MT-14741) and the Heart and Stroke Foundation of Alberta, Northwest Territories, and Nunavut. ML is a graduate trainee of the Alberta Heritage Foundation for Medical Research and the Canadian Institutes of Health Research. WW is a research trainee of the Heart and Stroke Foundation of Canada. RS is a Senior Scholar of the Alberta Heritage Foundation for Medical Research.

References

1. Opie LH. Oxygen lack: Ischemia and angina: *The Heart*. Philadelphia, PA: Lippincott-Raven, 1998:515–561.
2. Ambrosio G, Tritto I. Reperfusion injury: Experimental evidence and clinical implications. *Am Heart J* 1999;138:S69–S75.
3. Xu KY, Huso DL, Dawson TM, Bredt DS, Becker LC. Nitric oxide synthase in cardiac sarcoplasmic reticulum. *Proc Natl Acad Sci USA* 1999;96:657–662.
4. Schulz R, Smith JA, Lewis MJ, Moncada S. Nitric oxide synthase in cultured endocardial cells of the pig. *Br J Pharmacol* 1991;104:21–24.
5. Pollock JS, Forstermann U, Mitchell JA, Warner TD, Schmidt HH, Nakane M, Murad F. Purification and characterization of particulate endothelium-derived relaxing factor synthase from cultured and native bovine aortic endothelial cells. *Proc Natl Acad Sci USA* 1991;88:10480–10484.
6. Moncada S, Higgs A. The L-arginine-nitric oxide pathway. *N Engl J Med* 1993;329:2002–2012.
7. Quyyumi AA, Dakak N, Andrews NP, Gilligan DM, Panza JA, Cannon RO, 3rd. Contribution of nitric oxide to metabolic coronary vasodilation in the human heart. *Circulation* 1995;92:320–326.
8. Radomski MW, Palmer RM, Moncada S. Endogenous nitric oxide inhibits human platelet adhesion to vascular endothelium. *Lancet* 1987;2:1057–1058.
9. Kubes P, Suzuki M, Granger DN. Nitric oxide: An endogenous modulator of leukocyte adhesion. *Proc Natl Acad Sci USA* 1991;88:4651–4655.
10. Brown GC, Cooper CE. Nanomolar concentrations of nitric oxide reversibly inhibit synaptosomal respiration by competing with oxygen at cytochrome oxidase. *FEBS Lett* 1994;356:295–298.
11. Mohr S, Stamler JS, Brune B. Posttranslational modification of glyceraldehyde-3-phosphate dehydrogenase by S-nitrosylation and subsequent NADH attachment. *J Biol Chem* 1996;271:4209–4214.
12. Balligand JL, Ungureanu-Longrois D, Simmons WW, Kobzik L, Lowenstein CJ, Lamas S, Kelly RA, Smith TW, Michel T. Induction of NO synthase in rat cardiac microvascular endothelial cells by IL-1 beta and IFN-gamma. *Am J Physiol* 1995;268:H1293–H1303.
13. Schulz R, Nava E, Moncada S. Induction and potential biological relevance of Ca^{2+}-independent nitric oxide synthase in the myocardium. *Br J Pharmacol* 1992;105:575–580.
14. Balligand JL, Ungureanu-Longrois D, Simmons WW, Pimental D, Malinski TA, Kapturczak M, Taha Z, Lowenstein CJ, Davidoff AJ, Kelly RA. Cytokine-inducible nitric oxide synthase (iNOS) expression in cardiac myocytes. Characterization and regulation of iNOS expression and detection of iNOS activity in single cardiac myocytes *in vitro*. *J Biol Chem* 1994;269:27580–27588.
15. Behr-Roussel D, Rupin A, Sansilvestri-Morel P, Fabiani JN, Verbeuren TJ. Histochemical evidence for inducible nitric oxide synthase in advanced but non-ruptured human atherosclerotic carotid arteries. *Histochem J* 2000;32:41–51.
16. Sanchez de Miguel L, Arriero MM, Farre J, Jimenez P, Garcia-Mendez A, de Frutos T, Jimenez A, Garcia R, Cabestrero F, Gomez J, de Andres R, Monton M, Martin E, De la Calle-Lombana LM, Rico L, Romero J, Lopez-Farre A. Nitric oxide production by neutrophils obtained from patients during acute coronary syndromes: Expression of the nitric oxide synthase isoforms. *J Am Coll Cardiol* 2002;39:818–825.
17. Liu P, Hock CE, Nagele R, Wong PYK. Formation of nitric oxide, superoxide, and peroxynitrite in myocardial

ischemia-reperfusion injury in rats. *Am J Physiol* 1997;272:H2327–H2336.

18. Depre C, Havaux X, Renkin J, Vanoverschelde JL, Wijns W. Expression of inducible nitric oxide synthase in human coronary atherosclerotic plaque. *Cardiovasc Res* 1999;41:465–472.
19. Valen G, Paulsson G, Bennet AM, Hansson GK, Vaage J. Gene expression of inflammatory mediators in different chambers of the human heart. *Ann Thorac Surg* 2000;70:562–567.
20. Wang P, Zweier JL. Measurement of nitric oxide and peroxynitrite generation in the postischemic heart. Evidence for peroxynitrite-mediated reperfusion injury. *J Biol Chem* 1996;271:29223–29230.
21. Zweier JL, Flaherty JT, Weisfeldt ML. Direct measurement of free radical generation following reperfusion of ischemic myocardium. *Proc Natl Acad Sci USA* 1987;84:1404–1407.
22. Peterson DA, Asinger RW, Elsperger KJ, Homans DC, Eaton JW. Reactive oxygen species may cause myocardial reperfusion injury. *Biochem Biophys Res Commun* 1985;127:87–93.
23. Stewart JR, Crute SL, Loughlin V, Hess ML, Greenfield LJ. Prevention of free radical-induced myocardial reperfusion injury with allopurinol. *J Thorac Cardiovasc Surg* 1985;90:68–72.
24. Terada LS, Rubinstein JD, Lesnefsky EJ, Horwitz LD, Leff JA, Repine JE. Existence and participation of xanthine oxidase in reperfusion injury of ischemic rabbit myocardium. *Am J Physiol* 1991;260:H805–H810.
25. Becker LB, vanden Hoek TL, Shao ZH, Li CQ, Schumacker PT. Generation of superoxide in cardiomyocytes during ischemia before reperfusion. *Am J Physiol* 1999;277:H2240–H2246.
26. Vasquez-Vivar J, Kalyanaraman B, Martasek P, Hogg N, Masters BS, Karoui H, Tordo P, Pritchard KA, Jr. Superoxide generation by endothelial nitric oxide synthase: The influence of cofactors. *Proc Natl Acad Sci USA* 1998;95:9220–9225.
27. Xia Y, Dawson VL, Dawson TM, Snyder SH, Zweier JL. Nitric oxide synthase generates superoxide and nitric oxide in arginine- depleted cells leading to peroxynitrite-mediated cellular injury. *Proc Natl Acad Sci USA* 1996;93:6770–6774.
28. Xia Y, Tsai AL, Berka V, Zweier JL. Superoxide generation from endothelial nitric-oxide synthase. A Ca^{2+}/calmodulin-dependent and tetrahydrobiopterin regulatory process. *J Biol Chem* 1998;273:25804–25808.
29. Yasmin W, Strynadka KD, Schulz R. Generation of peroxynitrite contributes to ischemia-reperfusion injury in isolated rat hearts. *Cardiovasc Res* 1997;33:422–432.
30. Ma XL, Weyrich AS, Lefer DJ, Lefer AM. Diminished basal nitric oxide release after myocardial ischemia and reperfusion promotes neutrophil adherence to coronary endothelium. *Circ Res* 1993;72:403–412.
31. Beckman JS, Koppenol WH. Nitric oxide, superoxide, and peroxynitrite: The good the bad, and the ugly. *Am J Physiol* 1996;271:C1424–C1437.
32. Schulz R, Wambolt R. Inhibition of nitric oxide synthesis protects the isolated working rabbit heart from ischaemia-reperfusion injury. *Cardiovasc Res* 1995;30:432–439.
33. Schulz R, Panas DL, Catena R, Moncada S, Olley PM, Lopaschuk GD. The role of nitric oxide in cardiac depression induced by interleukin-1 beta and tumor necrosis factor-alpha. *Brit J Pharmacol* 1995;114:27–34.
34. Maulik N, Engelman DT, Watanabe M, Engelman RM, Maulik G, Cordis GA, Das DK. Nitric oxide signaling in ischemic heart. *Cardiovasc Res* 1995;30:593–601.
35. Uppu RM, Squadrito GL, Pryor WA. Acceleration of peroxynitrite oxidations by carbon dioxide. *Arch Biochem Biophys* 1996;327:335–343.
36. Denicola A, Freeman BA, Trujillo M, Radi R. Peroxynitrite reaction with carbon dioxide/bicarbonate: Kinetics and influence on peroxynitrite-mediated oxidations. *Arch Biochem Biophys* 1996;333:49–58.
37. Gow A, Duran D, Thom SR, Ischiropoulos H. Carbon dioxide enhancement of peroxynitrite-mediated protein tyrosine nitration. *Arch Biochem Biophys* 1996;333:42–48.
38. Nossuli TO, Hayward R, Jensen D, Scalia R, Lefer AM. Mechanisms of cardioprotection by peroxynitrite in myocardial ischemia and reperfusion injury. *Am J Physiol* 1998;275:H509–H519.
39. Ma XL, Gao F, Lopez BL, Christopher TA, Vinten-Johansen J. Peroxynitrite, a two-edged sword in post-ischemic myocardial injury- dichotomy of action in crystalloid- versus blood-perfused hearts. *J Pharmacol Exp Ther* 2000;292:912–920.
40. Schulz R, Dodge KL, Lopaschuk GD, Clanachan AS. Peroxynitrite impairs cardiac contractile function by decreasing cardiac efficiency. *Am J Physiol* 1997;272:H1212–H1219.
41. Ferdinandy P, Schulz R. Peroxynitrite: Toxic or protective in the heart? *Circulation Research* 2000;88:12e–13e.
42. Jordan JE, Zhao ZQ, Vinten-Johansen J. The role of neutrophils in myocardial ischemia-reperfusion injury. *Cardiovasc Res* 1999;43:860–878.
43. Grisham MB, Granger DN, Lefer DJ. Modulation of leukocyte-endothelial interactions by reactive metabolites of oxygen and nitrogen: Relevance to ischemic heart disease. *Free Radic Biol Med* 1998;25:404–433.
44. Ronson RS, Nakamura M, Vinten-Johansen J. The cardiovascular effects and implications of peroxynitrite. *Cardiovasc Res* 1999;44:47–59.
45. Moreno JJ, Pryor WA. Inactivation of alpha 1-proteinase inhibitor by peroxynitrite. *Chem Res Toxicol* 1992;5:425–431.
46. Radi R, Beckman JS, Bush KM, Freeman BA. Peroxynitrite-induced membrane lipid peroxidation: The cytotoxic potential of superoxide and nitric oxide. *Arch Biochem Biophys* 1991;288:481–487.
47. Moro MA, Darley-Usmar VM, Lizasoain I, Su Y, Knowles RG, Radomski MW, Moncada S. The formation of nitric oxide donors from peroxynitrite. *Br J Pharmacol* 1995;116:1999–2004.
48. Salgo MG, Bermudez E, Squadrito GL, Pryor WA. Peroxynitrite causes DNA damage and oxidation of thiols in rat thymocytes [corrected]. *Arch Biochem Biophys* 1995;322:500–505.
49. Rubbo H, Radi R, Trujillo M, Telleri R, Kalyanaraman B, Barnes S, Kirk M, Freeman BA. Nitric oxide regulation of superoxide and peroxynitrite-dependent lipid peroxidation. Formation of novel nitrogen-containing oxidized lipid derivatives. *J Biol Chem* 1994;269:26066–26075.
50. Radi R, Beckman JS, Bush KM, Freeman BA. Peroxynitrite oxidation of sulfhydryls. The cytotoxic potential of superoxide and nitric oxide. *J Biol Chem* 1991;266:4244–4250.

51. Ischiropoulos H, Zhu L, Chen J, Tsai M, Martin JC, Smith CD, Beckman JS. Peroxynitrite-mediated tyrosine nitration catalyzed by superoxide dismutase. *Arch Biochem Biophys* 1992;298:431–437.
52. Zhang Y, Bissing JW, Xu L, Ryan AJ, Martin SM, Miller FJ, Jr., Kregel KC, Buettner GR, Kerber RE. Nitric oxide synthase inhibitors decrease coronary sinus-free radical concentration and ameliorate myocardial stunning in an ischemia-reperfusion model. *J Am Coll Cardiol* 2001;38:546–554.
53. Hayashi Y, Sawa Y, Nishimura M, Tojo SJ, Fukuyama N, Nakazawa H, Matsuda H. P-selectin participates in cardiopulmonary bypass-induced inflammatory response in association with nitric oxide and peroxynitrite production. *J Thorac Cardiovasc Surg* 2000;120:558–565.
54. Ma XL, Lopez BL, Liu GL, Christopher TA, Ischiropoulos H. Peroxynitrite aggravates myocardial reperfusion injury in the isolated perfused rat heart. *Cardiovasc Res* 1997;36:195–204.
55. Sato H, Zhao ZQ, Jordan JE, Todd JC, Riley RD, Taft CS, Hammon JW, Jr., Li P, Ma X, Vinten-Johansen J. Basal nitric oxide expresses endogenous cardioprotection during reperfusion by inhibition of neutrophil-mediated damage after surgical revascularization. *J Thorac Cardiovasc Surg* 1997;113:399–409.
56. Williams MW, Taft CS, Ramnauth S, Zhao ZQ, Vinten-Johansen J. Endogenous nitric oxide (NO) protects against ischaemia-reperfusion injury in the rabbit. *Cardiovasc Research* 1995;30:79–86.
57. Beresewicz A, Karwatowska-Prokopczuk E, Lewartowski B, Cedro-Ceremuazynska K. A protective role of nitric oxide in isolated ischaemic/reperfused rat heart. *Cardiovasc Res* 1995;30:1001–1008.
58. Wang QD, Morcos E, Wiklund P, Pernow J. L-arginine enhances functional recovery and Ca(2+)-dependent nitric oxide synthase activity after ischemia and reperfusion in the rat heart. *J Cardiovasc Pharmacol* 1997;29:291–296.
59. Brunner F, Leonhard B, Kukovetz WR, Mayer B. Role of endothelin, nitric oxide and L-arginine release in ischaemia/reperfusion injury of rat heart. *Cardiovasc Res* 1997;36:60–66.
60. du Toit EF, McCarthy J, Miyashiro J, Opie LH, Brunner F. Effect of nitrovasodilators and inhibitors of nitric oxide synthase on ischaemic and reperfusion function of rat isolated hearts. *Brit J Pharmacol* 1998;123:1159–1167.
61. Depre C, Vanoverschelde JL, Goudemant JF, Mottet I, Hue L. Protection against ischemic injury by nonvasoactive concentrations of nitric oxide synthase inhibitors in the perfused rabbit heart. *Circulation* 1995;92:1911–1918.
62. Ferdinandy P, Daniel H, Ambrus I, Rothery R, Schulz R. Peroxynitrite is a major contributor to cytokine-induced myocardial contractile failure. *Circ Res* 2000;87:241–247.
63. Weinstein DM, Mihm MJ, Bauer JA. Cardiac peroxynitrite formation and left ventricular dysfunction following doxorubicin treatment in mice. *J Pharmacol Exp Ther* 2000;294:396–401.
64. Sakurai M, Fukuyama, N, Iguchi A, Akimoto H, Ohmi M, Yokoyama H, Nakazawa H, Tabayashi K. Quantitative analysis of cardiac 3-L-nitrotyrosine during acute allograft rejection in an experimental heart transplantation. *Transplantation* 1999;68:1818–1822.
65. Kooy NW, Lewis SJ, Royall JA, Ye YZ, Kelly DR, Beckman JS. Extensive tyrosine nitration in human myocardial inflammation: Evidence for the presence of peroxynitrite. *Crit Care Med* 1997;25:812–819.
66. Castro L, Rodriguez M, Radi R. Aconitase is readily inactivated by peroxynitrite, but not by its precursor, nitric oxide. *J Biol Chem* 1994;269:29409–29415.
67. Lopaschuk GD. Treating ischemic heart disease by pharmacologically improving cardiac energy metabolism. *Am J Cardiol* 1998;82:14K–17K.
68. Yamakura F, Taka H, Fujimura T, Murayama K. Inactivation of human manganese-superoxide dismutase by peroxynitrite is caused by exclusive nitration of tyrosine 34 to 3-nitrotyrosine. *J Biol Chem* 1998;273:14085–14089.
69. Klebl BM, Ayoub AT, Pette D. Protein oxidation, tyrosine nitration, and inactivation of sarcoplasmic reticulum Ca^{2+}-ATPase in low-frequency stimulated rabbit muscle. *FEBS Letters* 1998;422:381–384.
70. Zou M, Martin C, Ullrich V. Tyrosine nitration as a mechanism of selective inactivation of prostacyclin synthase by peroxynitrite. *Biol Chem* 1997;378:707–713.
71. Woessner JF. The matrix metalloproteinase family. In: *Matrix Metalloproteinases*. San Diego, CA: Academic Press, 1998:1–14.
72. Yu AE, Murphy AN, Stetler-Stevenson WG. 72-kDa gelatinase (gelatinase A): Structure, activation, regulation, and substrate specificity. In: *Matrix Metalloproteinases*. San Diego: Academic Press, 1998:85–114.
73. Cheung P-Y, Sawicki G, Wozniak M, Wang W, Radomski MW, Schulz R. Matrix metalloproteinase-2 contributes to ischemia-reperfusion injury in the heart. *Circulation* 2000;101:1833–1839.
74. Wang W, Sawicki G, Schulz R. Peroxynitrite-induced myocardial injury is mediated through matrix metalloproteinase-2. *Cardiovasc Res* 2002;53:165–174.
75. Murphy G, Willenbrock F, Crabbe T, O'Shea M, Ward R, Atkinson S, O'Connell J, Docherty A. Regulation of matrix metalloproteinase activity. *Ann NY Acad Sci* 1994;732:31–41.
76. Okamato T, Akaike T, Nagano T, Miyajima S, Suga M, Ando M, Ichimori K, Maeda H. Activation of human neutrophil procollagenase by nitrogen dioxide and peroxynitrite: A novel mechanism for procollagenase activation involving nitric oxide. *Arch Biochem Biophys* 1997;342:261–274.
77. Maeda H, Okamoto T, Akaike T. Human matrix metalloprotease activation by insults of bacterial infection involving proteases and free radicals. *Biol Chem* 1998;379:193–200.
78. Rajagopalan S, Meng XP, Ramasamy S, Harrison DG, Galis ZS. Reactive oxygen species produced by macrophage-derived foam cells regulate the activity of vascular matrix metalloproteinases *in vitro*. Implications for atherosclerotic plaque stability. *J Clin Invest* 1996;98:2572–2579.
79. Okamato T, Akaike T, Sawa T, Miyamoto Y, van der Vliet A, Maeda H. Activation of matrix metalloproteinases by peroxynitrite-induced protein S-glutathiolation via disulfide S-oxide formation. *J Biol Chem* 2001;276:29596–29602.
80. Frears ER, Zhang Z, Blake DR, O'Connell JP, Winyard PG. Inactivation of tissue inhibitor of metalloproteinase-1 by peroxynitrite. *FEBS Letters* 1996;381:21–24.
81. Lindsey M, Wedin K, Brown MD, Keller C, Evans AJ, Smolen J, Burns AR, Rossen RD, Michael L,

Entman M. Matrix-dependent mechanism of neutrophil-mediated release and activation of matrix metalloproteinase 9 in myocardial ischemia/reperfusion. *Circulation* 2001;103:2181–2187.

82. Danielsen CC, Wiggers H, Andersen HR. Increased amounts of collagenase and gelatinase in porcine myocardium following ischemia and reperfusion. *J Mol Cell Cardiol* 1998;30:1431–1442.
83. Mayers I, Hurst T, Puttagunta L, Radomski A, Mycyk T, Sawicki G, Johnson D, Radomski MW. Cardiac surgery increases the activity of matrix metalloproteinases and nitric oxide synthase in human hearts. *J Thorac Cardiovasc Surg* 2001;122:746–752.
84. McDonough JL, Labugger R, Pickett W, Tse MY, MacKenzie S, Pang SC, Atar D, Ropchan G, Van Eyk JE. Cardiac troponin I is modified in the myocardium of bypass patients. *Circulation* 2001;103:58–64.
85. Bolli R, Marban E. Molecular and cellular mechanisms of myocardial stunning. *Physiol Rev* 1999;79:609–634.
86. Wang W, Schulze CJ, Suarez-Pinzon WL, Dyck JR, Sawicki G, Schulz R. Intracellular action of matrix metalloproteinase-2 accounts for acute myocardial ischemia and reperfusion injury. *Circulation* 2002;106: 1543–1549.
87. Fernandez-Patron C, Radomski MW, Davidge SM. Vascular matrix metalloproteinase-2 cleaves big endothelin-1 yielding a novel vasoconstrictor. *Circ Res* 1999;85:906–911.
88. Fernandez-Patron C, Stewart KG, Zhang Y, Koivunen E, Radomski MW, Davidge ST. Vascular matrix metalloproteinase-2-dependent cleavage of calcitonin gene-related peptide promotes vasoconstriction. *Circ Res* 2000;87:670–676.
89. Ikai K, Ueda K. Immunohistochemical demonstration of poly(adenosine diphosphate-ribose) synthetase in bovine tissues. *J Histochem Cytochem* 1983;31:1261–1264.
90. Pieper AA, Verma A, Zhang J, Snyder SH. Poly (ADP-ribose) polymerase, nitric oxide and cell death. *Trends Pharmacol Sci* 1999;20:171–181.
91. Szabo C, Zingarelli B, O'Connor M, Salzman AL. DNA strand breakage, activation of poly (ADP-ribose) synthetase, and cellular energy depletion are involved in the cytotoxicity of macrophages and smooth muscle cells exposed to peroxynitrite. *Proc Natl Acad Sci USA* 1996;93:1753–1758.
92. Zingarelli B, Cuzzocrea S, Zsengeller Z, Salzman AL, Szabo C. Protection against myocardial ischemia and reperfusion injury by 3-aminobenzamide, an inhibitor of poly (ADP-ribose) synthetase. *Cardiovasc Res* 1997;36:205–215.
93. Thiemermann C, Bowes J, Myint FP, Vane JR. Inhibition of the activity of poly(ADP ribose) synthetase reduces ischemia-reperfusion injury in the heart and skeletal muscle. *Proc Natl Acad Sci USA* 1997;94:679–683.
94. Halmosi R, Berente Z, Osz E, Toth K, Literati-Nagy P, Sumegi B. Effect of poly(ADP-ribose) polymerase inhibitors on the ischemia-reperfusion-induced oxidative cell damage and mitochondrial metabolism in Langendorff heart perfusion system. *Mol Pharmacol* 2001;59:1497–1505.
95. Grupp IL, Jackson TM, Hake P, Grupp G, Szabo C. Protection against hypoxia-reoxygenation in the absence of poly (ADP- ribose) synthetase in isolated working hearts. *J Mol Cell Cardiol* 1999;31:297–303.
96. Zingarelli B, Salzman AL, Szabo C. Genetic disruption of poly (ADP-ribose) synthetase inhibits the expression of P-selectin and intercellular adhesion molecule-1 in myocardial ischemia/reperfusion injury. *Circ Res* 1998;83:85–94.
97. Lefer DJ, Scalia R, Campbell B, Nossuli T, Hayward R, Salamon M, Grayson J, Lefer AM. Peroxynitrite inhibits leukocyte-endothelial cell interactions and protects against ischemia-reperfusion injury in rats. *J Clin Invest* 1997;99:684–691.
98. Nakamura M, Thourani VH, Ronson RS, Velez DA, Ma XL, Katzmark S, Robinson J, Schmarkey LS, Zhao ZQ, Wang NP, Guyton RA, Vinten-Johansen J. Glutathione reverses endothelial damage from peroxynitrite, the byproduct of nitric oxide degradation, in crystalloid cardioplegia. *Circulation* 2000;102:III332–III338.
99. Ronson RS, Thourani VH, Ma XL, Katzmark SL, Han D, Zhao ZQ, Nakamura M, Guyton RA, Vinten-Johansen J. Peroxynitrite, the breakdown product of nitric oxide, is beneficial in blood cardioplegia but injurious in crystalloid cardioplegia. *Circulation* 1999;100:II384–III391.
100. Cheung PY, Wang W, Schulz R. Glutathione protects against myocardial ischemia-reperfusion injury by detoxifying peroxynitrite. *J Mol Cell Cardiol* 2000;32:1669–1678.
101. Balazy M, Kaminski PM, Mao K, Tan J, Wolin MS. S-Nitroglutathione, a product of the reaction between peroxynitrite and glutathione that generates nitric oxide. *J Biol Chem* 1998;273:32009–32015.
102. Wu M, Pritchard KA, Jr., Kaminski PM, Fayngersh RP, Hintze TH, Wolin MS. Involvement of nitric oxide and nitrosothiols in relaxation of pulmonary arteries to peroxynitrite. *Am J Physiol* 1994;266:H2108–H2113.
103. Pabla R, Buda AJ, Flynn DM, Blesse SA, Shin AM, Curtis MJ, Lefer DJ. Nitric oxide attenuates neutrophil-mediated myocardial contractile dysfunction after ischemia and reperfusion. *Circ Res* 1996;78:65–72.
104. Pabla R, Buda AJ, Flynn DM, Salzberg DB, Lefer DJ. Intracoronary nitric oxide improves postischemic coronary blood flow and myocardial contractile function. *Am J Physiol* 1995;269:H1113–H1121.
105. Nash GB. Adhesion between neutrophils and platelets: A modulator of thrombotic and inflammatory events? *Thromb Res* 1994;74:S3–S11.
106. Kelly RA, Balligand JL, Smith TW. Nitric oxide and cardiac function. *Circ Res* 1996;79:363–380.
107. Lee CI, Liu X, Zweier JL. Regulation of xanthine oxidase by nitric oxide and peroxynitrite. *J Biol Chem* 2000;275:9369–9376.
108. Pasquet JP, Zou MH, Ullrich V. Peroxynitrite inhibition of nitric oxide synthases. *Biochimie* 1996;78:785–791.
109. Altug S, Demiryurek AT, Cakici I, Kanzik I. The beneficial effects of peroxynitrite on ischaemia-reperfusion arrhythmias in rat isolated hearts. *Eur J Pharm* 1999;384:157–162.
110. Misko TP, Highkin MK, Veenhuizen AW, Manning PT, Stern MK, Currie MG, Salvemini D. Characterization of the cytoprotective action of peroxynitrite decomposition catalysts. *J Biol Chem* 1998;273:15646–15653.
111. Cuzzocrea S, Misko TP, Costantino G, Mazzon E, Micali A, Caputi AP, Macarthur H, Salvemini D. Beneficial effects of peroxynitrite decomposition catalyst in a rat model of splanchnic artery occlusion and reperfusion. *FASEB J* 2000;14:1061–1072.

The Role of the NO Axis and its Therapeutic Implications in Pulmonary Arterial Hypertension

Evangelos D. Michelakis
University of Alberta Hospitals, Edmonton, Canada

Abstract. **Pulmonary Arterial Hypertension (PAH) is a disease of the pulmonary vasculature leading to vasoconstriction and remodeling of the pulmonary arteries. The resulting increase in the right ventricular afterload leads to right ventricular failure and death. The treatment options are limited, expensive and associated with significant side effects. The nitric oxide (NO) pathway in the pulmonary circulation provides several targets for the development of new therapies for this disease. However, the NO pathway is modulated at multiple levels including transcription and expression of the NO synthase gene, regulation of the NO synthase activity, regulation of the production of cyclic guanomonophosphate (cGMP) by phosphodiesterases, post-synthetic oxidation of NO, etc. This makes the study of the role of the NO pathway very difficult, unless one uses multiple complementary techniques. Furthermore, there are significant differences between the pulmonary and the systemic circulation which make extrapolation of data from one circulation to the other very difficult. In addition, the role of NO in the development of pulmonary hypertension varies among different models of the disease. This paper reviews the role of the NO pathway in both the healthy and diseased pulmonary circulation and in several animal models and human forms of the disease. It focuses on the role of recent therapies that target the NO pathway, including L-Arginine, inhaled NO, the phosphodiesterase inhibitor sildenafil and gene therapy.**

Key Words. **Sildenafil, Viagra, inhaled NO, primary pulmonary hypertension, potassium channels**

Introduction

There are several challenges in the study of the role of the NO pathway in pulmonary hypertension. First, the role of NO among different segments of the same circulation or among different circulations varies and data from one cannot be extrapolated to the other. For example NO synthase (NOS) is mostly expressed in the proximal pulmonary arteries in the healthy pulmonary circulation but in pulmonary hypertension NOS is significantly expressed in the resistance pulmonary arteries, which actually control most of pulmonary vascular resistance. The redox environment of the pulmonary compared to the systemic circulation is different and this might be one of the reasons that in health, the role of NO in the control of vascular tone is more important in the systemic versus the pulmonary circulation.

Second, the NO pathway is controlled at multiple levels, from gene transcription of NOS to post-synthetic redox-based modulation of NO and NO oxidation products. The use of only one or two techniques is usually not adequate in assessing the role of the NO pathway in one vascular bed or disease model and has led to conflicting results in the literature. Multiple and complementary techniques are needed for a comprehensive study of the NO pathway.

Third, the role of the NO pathway is regulated differently in different models of pulmonary hypertension. For example, in chronic hypoxic pulmonary hypertension, hypoxia itself regulates NOS gene transcription or NOS enzyme activity. Hypoxia is absent in monocrotaline-induced pulmonary hypertension and this can make the comparison of the two models of pulmonary hypertension difficult in terms of the role of the NO pathway.

Following the discussion of these challenges, we will focus on the role of several new therapies for pulmonary hypertension that modulate the NO pathway and discuss the data on human studies in detail, where available.

The Pulmonary Circulation

There are significant differences between the pulmonary and the systemic circulation that make extrapolation of findings from one circulation to the other very difficult. At baseline (normoxia) the pulmonary circulation is low pressure, i.e. it is relatively vasodilated, compared to the systemic circulation. On the other hand, the thin-walled pulmonary arteries (PA) constrict in response to hypoxia (hypoxic pulmonary vasoconstriction, HPV), while the more muscular systemic arteries dilate. HPV is mediated by a redox mechanism

Address for correspondence: Evangelos D. Michelakis, MD, FACC, University of Alberta Hospitals, 2C2 Walter C McKenzie Health Sciences Centre, Edmonton, T6G2B7, Canada. Tel.: 780-407-1576; Fax: 780-407-6032; E-mail: emichela@cha.ab.ca

[1,2] and recently significant redox differences have been described between the pulmonary and systemic vascular beds (like the renal circulation) [3]. The pulmonary circulation is in a more reduced redox state compared to the renal circulation as reflected by the higher levels of activated oxygen species (AOS, superoxide and hydrogen peroxide) and reduced glutathione (GSH) [3]. These differences might be in part due to the fact that the PA smooth muscle cells (SMC) have different mitochondria (more depolarized, higher levels of mitochondria manganese superoxide dismutase) compared to renal artery SMC and they might in part explain the opposing effects of the pulmonary and renal circulations to hypoxia [3]. These redox differences might also be important in the interpretation of the role of several redox-sensitive second messengers and pathways in the vascular biology of the two circulations (for review see [4]). One very important redox-sensitive pathway is that of the vasodilatation due to NO, which is itself an AOS [4].

NO Chemistry

NO is an AOS and although not particularly unstable in low concentrations and in the absence of O_2, it is very unstable in biological models, primarily due to the presence of O_2, AOS and oxidizing agents such as oxyhemoproteins [5]. NO tends to exist as a gas and is poorly soluble in water. This property allows the small molecule to pass freely across cell boundaries. The fate of NO in the presence of O_2, depends on the phase in which the reaction occurs. In the gas phase, NO reacts with O_2 to give the brown gas nitrogen dioxide (NO_2^-) and N_2O_3:$2NO + O_2 \rightarrow 2NO_2^-$, $NO + NO_2^- \rightarrow N_2O_3$. In the liquid phase -in the absence of hemoproteins- NO is oxidized to NO_2^-:$4NO + O_2 \rightarrow 4HNO_2 \rightarrow 4NO_2^- + 4H^+$. In biological systems however, where oxidizing agents such as oxyhemoproteins and superoxide are present, NO is further oxidized to NO_3^- and peroxynitrite, respectively. NO_3^- is the dominant species of NO oxidation products (NO_x) *in vivo*. Furthermore, in the presence of strong oxidizing agents, NO_2^- is also oxidized to NO_3^-.

Because of its very short half-life and its susceptibility to postsynthetic oxidation, measuring NO is a challenge [5]. It is very important to use several methods, when one wants to study complex conditions that can affect the NO pathway at many levels, like hypoxia. For example, chronic hypoxia increases the NO and NOx levels (measured by *chemiluminescence*) but at the same time the activity of NO synthase (NOS) is decreased (measured by the *L-citrulline assay*). This apparent paradox appears to be explained at least in part by the finding that the total amount of the expressed NOS protein is increased in chronic hypoxia (shown by *immunohistochemistry*). Use of only one technique can lead to misleading or confusing results (for further discussion see [6]). The importance of this concept is also discussed later when the NO axis is compared among different models of pulmonary hypertension (PHT).

The NO Axis in the Vasculature

NO is formed by oxidation of a terminal guanidino nitrogen of L-arginine in the presence of a heme-containing enzyme, NOS [7]. The process is oxygen-dependent and important cofactors include reduced nicotinamide adenine nucleotide phosphate, flavin adenine dinucleotide, flavin mononucleotide and tetrahydrobiopterin [7]. There are 3 known NOS isozymes: Isozyme I is mostly expressed in neurons (neuronal NOS, nNOS) but also in epithelial and vascular cells including PASMC [8]. Isozyme II is induced (inducible NOS, iNOS) by several mediators of inflammation and is regulated at the level of gene expression. Once expressed, iNO produces NO at very high rates. In contrast to isozymes I and III, the activity of iNOS is independent of the levels of intracellular Ca^{++}. Isozyme III is constitutively expressed mostly in, but not restricted to, endothelial cells (endothelial NOS, eNOS). Although eNOS is the main isozyme involved in the regulation of vascular tone, both nNOS and iNOS have been reported to be involved in the production of pulmonary vascular NO in both disease states and normal development [9,10].

Whereas at very high levels NO reacts to O_2 and superoxide giving rise to NOx and highly toxic substances like peroxynitrite, at lower levels, as occurs within the normal vasculature, NO activates soluble guanylate cyclase, resulting in increased levels of cGMP within the target cells [7]. cGMP activates a cGMP-dependent protein kinase, which is responsible for most of the vasodilatory effects of NO [7]. A major pathway by which NO relaxes PAs is via cGMP-kinase-dependent phosphorylation of PASMC sarcolemmal potassium (K^+) channels (Fig. 1) [11,12].

K^+ channels are transmembrane proteins with a pore-forming unit that allows the selective efflux of K^+ ions from the cytoplasm [13]. Based on pharmacologic and molecular criteria, K^+ channels in the vasculature are separated into 3 families: the Kv, the Ca^{++}-activated (K_{Ca}) and the inward rectifier (Kir) [14]. When K^+ channels open there is an efflux of K^+ ions from the cells down a concentration gradient (intracellular / extracellular K^+ concentration = 140/5 mEq) and the interior of the

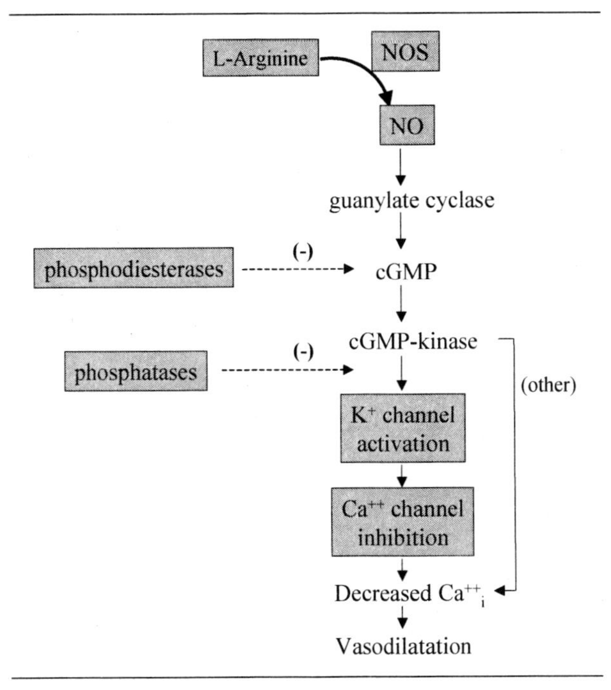

Fig. 1. *The NO axis and potential therapeutic targets in the treatment of pulmonary hypertension.*

cell becomes more negatively charged (hyperpolarization). In contrast, when K^+ channels close, the cell becomes depolarized. Depolarization beyond a certain level, due to failure of K^+ channels to open (whether because they are pharmacologically inhibited, mutated or downregulated) causes opening of the voltage-gated L-type Ca^{++} channels, influx of Ca^{++}, activation of the actin-myosin complex and contraction [14]. Thus, in blood vessels, K^+ channel openers are vasodilators and K^+ channel inhibitors are vasoconstrictors. For example, NO causes vasodilatation in part by opening the K_{Ca} channels (Fig. 2). This leads to PASMC hyperpolarization, inhibition of the voltage-gated L-type Ca^{++} channels, a decrease in the Ca_i^{++} levels and vasorelaxation.

In addition to the effects on K^+ channels, cGMP-kinase causes a decrease in Ca_i^{++} via effects on several types of Ca^{++} channels and transporters and the sarcoplasmic reticulum [15]. Furthermore, cGMP-kinase causes SMC relaxation by direct effects on the actin-myosin apparatus [16].

K^+ channels play a major role in the regulation of pulmonary vascular tone both in health and disease (for review see [14,17]). Kv channels control PASMC membrane potential and their inhibition by hypoxia leads to the very important HPV [14,18]. Their direct inhibition by anorectic agents like dexfenfluramine has been implicated in the pathogenesis of several epidemics of anorectic-induced pulmonary hypertension in the recent years [19]. Furthermore, a selective downregulation of Kv channels has been implicated in the pathogenesis of chronic hypoxic pulmonary hypertension [20,21] and in primary pulmonary hypertension in humans [22].

The pathway leading to K^+ channel activation in the pulmonary circulation was elegantly elucidated and pharmacologically dissected by Archer et al. and shown in Fig. 1 [11]. The levels of cGMP are regulated by the balance between production (by NO and other endothelial derived products) and degradation by phosphodiesterases. Also the activity of the K^+ channels in the PASMC is regulated in part by the balance between phosphorylation by the cGMP-kinase (which promotes activation) and the de-phosphorylation by phosphatases (which promote inhibition). Several levels of this pathway (shown in Fig. 1 in boxes) are the target of current standard or investigational pharmacologic interventions in the treatment of pulmonary hypertension and are discussed in detail later.

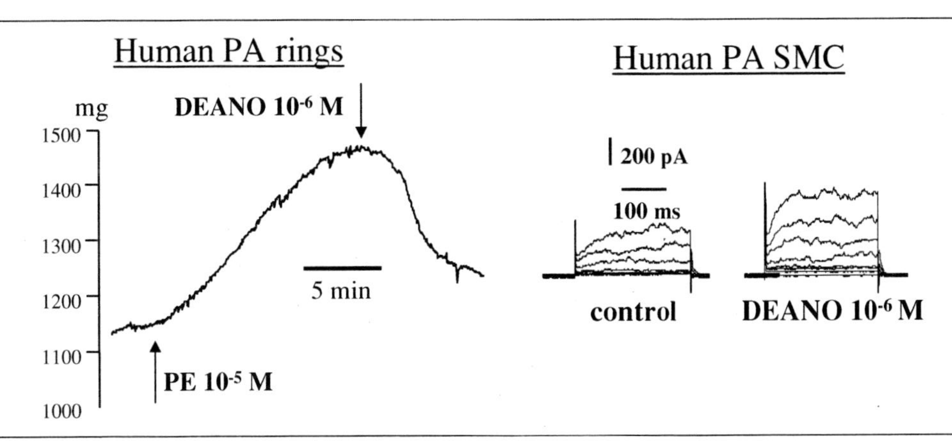

Fig. 2. *Left: The NO donor DEANO relaxes phenylephrine (PE)-constricted human PAs, obtained from transplant surgery. Right: DEANO activates K^+ currents in freshly isolated human PASMC.*

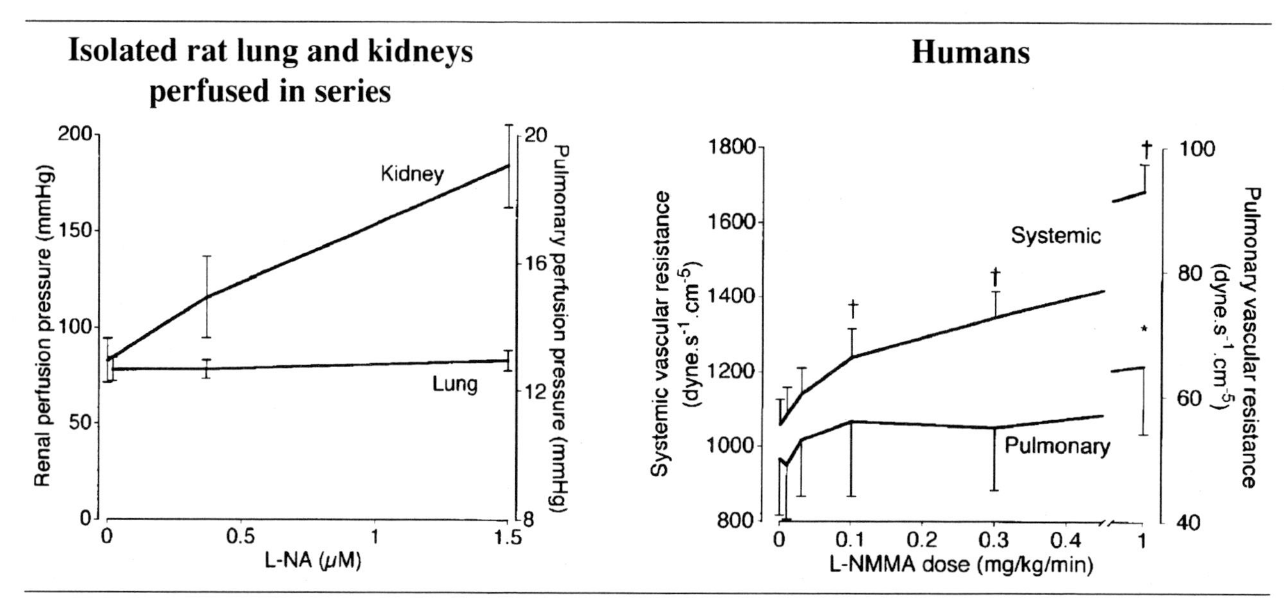

Fig. 3. NOS inhibitors increase the systemic but not the pulmonary circulation in isolated rat organs (left) and in humans (right). Obtained with permission from Hampl and Herget (2000).

NO and the Normal Pulmonary Circulation

NO does not appear to play an important role in the normal pulmonary circulation. This conclusion is mostly based on experiments using NOS inhibitors, showing that these drugs do not alter tone in a variety of models, including isolated PAs, perfused lungs and intact animals, nicely summarized and reviewed recently by Hampl and Herget [23]. This is in contrast to the systemic circulation, where NOS inhibitors result in the expected increase in tone in a variety of models [23]. Figure 3 shows that NOS inhibitors increase systemic vascular resistance but not PVR both in the rat and in humans. Note that in humans (on the right), PVR does not increase by doses of NOS inhibitors enough to inhibit systemic endothelial-derived NO and increase systemic vascular resistance. There is some increase in the PVR only at very high doses and this is probably due to the nonspecific effects of NOS inhibitors. These data suggest that under normal conditions, a basal tonic release of NO from the endothelium regulates vascular tone in the systemic but not the pulmonary circulation. There are several possibilities that could explain this difference.

First, expression of eNOS is low in normal resistance PAs, the vessels that essentially control PVR, and most of the eNOS is seen in the endothelium of the large-conduit PAs in both animals [24–27] and humans [28]. This is in contrast to pulmonary hypertension, where strong expression of eNOS is seen in the resistance and neomuscularized PAs (Fig. 4) [25–27], as discussed later.

Second, the biological effects of NO are modulated by the redox environment of the target cell, as NO itself is an AOS. The significant differences in the redox environments might differentially regulate the fate of NO*x* between the 2 circulations, but clearly this requires further research.

The fact that eNOS knockout mice develop mild pulmonary hypertension, compared to wild mice, at first might suggest that NO plays a role in controlling baseline pulmonary vascular tone in this model [29]. However, if the eNOS function is inhibited by NOS inhibitors in the wild control mice, they do not develop pulmonary hypertension. It has been postulated that the fact that the eNOS−/− mice have mild pulmonary hypertension reflects an abnormal transition from the fetal to the neonatal pulmonary circulation [23]. In contrast to the low-pressure adult pulmonary circulation, NO appears to be important in the control of tone in the high-pressure fetal pulmonary circulation [30,31]. In other words, NO contributes to the normal transition from the fetal to the low pressure neonatal pulmonary circulation and the lack of eNOS at this critical transition, results in persistence of the fetal remodeling and increased resistance in the pulmonary circulation. Therefore, the adult eNOS−/− mice do not lack NO in the pulmonary circulation and this is further supported by the fact that they do not decrease their PVR in response to exogenous NO [23].

Overall, it appears that NO is not important in the low-pressure normal pulmonary circulation. However, as discussed below, the NO axis becomes important in the high-pressure pulmonary

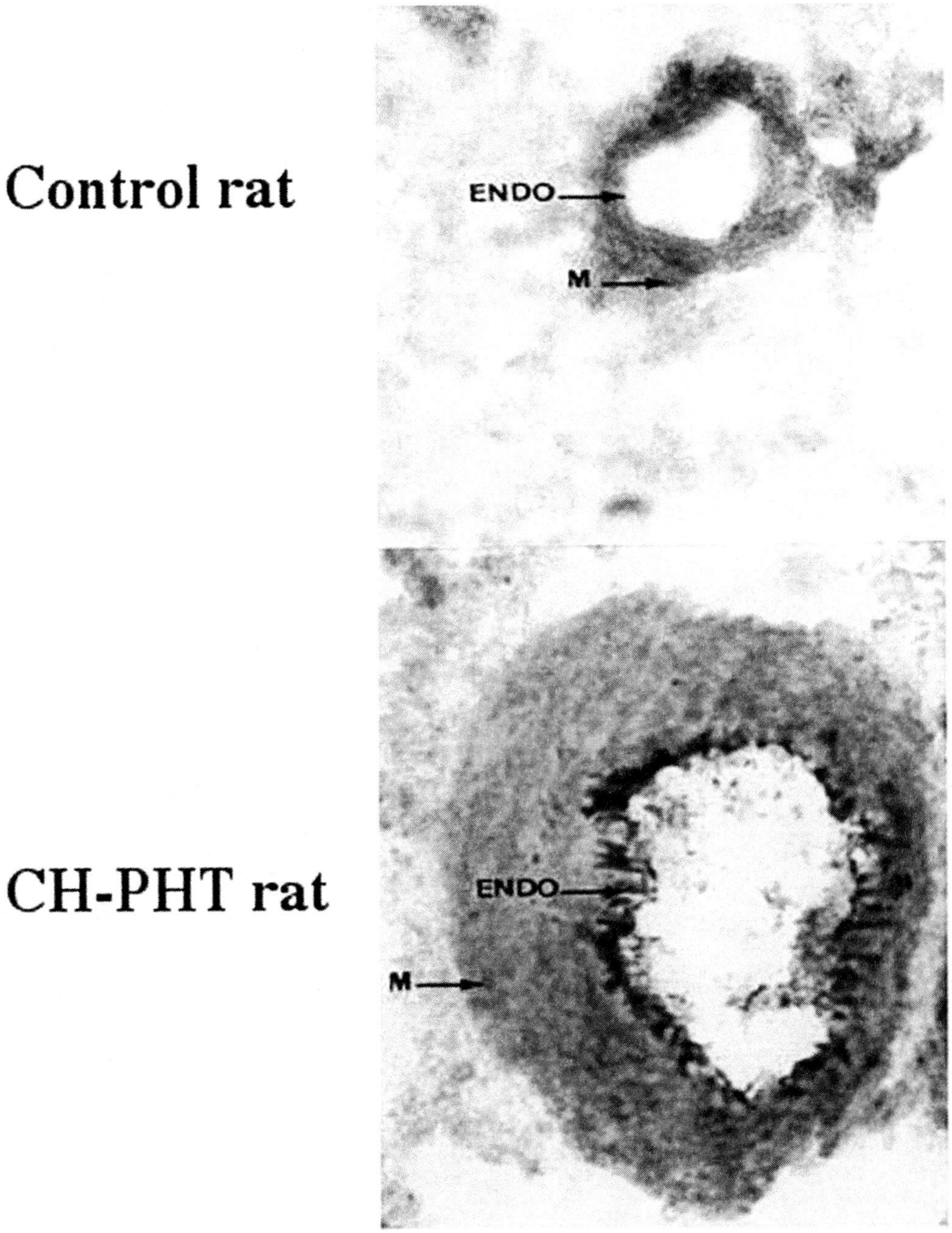

Control rat

ENDO

M

CH-PHT rat

ENDO

M

Fig. 4. *Immunohistochemistry (double staining technique) of a control and a CH-PHT rat intrapulmonary small PA. eNOS is shown in brown and the Kv channel Kv2.1 is shown in pink. There is very low expression of eNOS in the control PA endothelium in contrast to the very high eNOS expression in the CH-PHT. Note the medial thickening in the CH-PHT artery. In this black and white version brown shows in black and pink shows in gray.*

circulation seen in several models of pulmonary hypertension, as it is in the fetal pulmonary circulation.

The NO Axis in Pulmonary Hypertension

The importance of NO in the pathogenesis or maintenance of PHT varies between species, different models of PHT and different stages of the disease. The NO axis is rarely assessed comprehensively within a single study (i.e. NOS mRNA, NOS protein expression, NOS activity, NO and NO*x* levels) and this further complicates the assessment of the role of NO on this disease. The importance of using multiple complementary techniques and avoiding extrapolation of findings between the healthy and diseased pulmonary circulation or between the systemic and pulmonary circulation cannot be overemphasized.

Pulmonary hypertension, whether primary or secondary, is characterized by constricted and remodeled PAs, resulting in increased PVR and right ventricular afterload. The thin-walled right ventricle initially hypertrophies in an attempt to maintain cardiac output but soon dilates and fails. Patients with primary pulmonary hypertension often die within 5 years from the time of diagnosis and the available treatment options are very few, expensive and associated with significant side effects [32]. Exploration of the NO axis in pulmonary hypertension is very important for the understanding of current and future therapies of pulmonary hypertension. In general, the NO axis appears to be activated and play a more important role in the regulation of tone and remodeling of the hypertensive compared to the healthy pulmonary circulation. This activation is generally viewed as a compensatory mechanism, limiting the extent of PA vasoconstriction, which is often inadequate. It is perhaps the patients with inadequate NO-mediated compensation that might benefit the most from therapies that aim to enhance the endogenous NO axis or directly provide exogenous NO. On the other hand, it has also been speculated that the increased NO in the diseased pulmonary circulation has itself toxic effects and contributes to the maintenance of the disease [23].

Pulmonary Hypertension: Animal Models and the Human Disease

There are 3 commonly used animal models of pulmonary hypertension.

1. *Chronic hypoxia-induced pulmonary hypertension (CH-PHT).* This model is relevant to the PHT in humans with chronic obstructive pulmonary disease [33]. Rats develop pulmonary hypertension within ~3 weeks of placement in a hypoxic chamber.
2. *Monocrotaline-induced pulmonary hypertension* (MC-PHT). In this model rats develop severe pulmonary hypertension after a single intraperitoneal dose of monocrotaline, an alkaloid found in the weed crotalaria spectabilis. Monocrotaline is thought to initiate pulmonary hypertension via its toxic effects on the endothelium, but the mechanism remains unknown [34–36].
3. *The Fawn-hooded rat (FH).* These rats develop spontaneous pulmonary hypertension when exposed to Denver altitude and although serotoninergic mechanisms have been implicated, the pathogenesis of pulmonary hypertension in this model remains unknown as well [37–39].

Pulmonary Arterial Hypertension (PAH) in humans includes a number of diseases characterized by abnormalities intrinsic to the pulmonary vasculature, i.e. abnormalities in the endothelial cells, smooth muscle cells, fibroblasts and extracellular matrix, leading to PA vasoconstriction, remodeling and in situ thrombosis [32]. PAH includes essentially all forms of pulmonary hypertension, except those that are secondary to thromboembolic pulmonary vascular disease and secondary to abnormalities of the left ventricle. PAH includes primary pulmonary hypertension (PPH—no association with any identifiable cause) as well as pulmonary hypertension associated with the use of anorectic agents, rheumatologic diseases (like scleroderma or lupus erythematosus), congenital heart disease, HIV infection and cirrhosis (portopulmonary hypertension) [32].

While not significantly expressed in the normal resistance PAs, eNOS expression is increased in the endothelium of the resistance PAs in CH-PHT [25–27,40], MC-PHT [26,40] and the FH rats [26]. Interestingly, in addition to the increase in eNOS expression, in rats with CH-PHT there is an increase in PA iNOS expression as well [25,27,41]. However, the significance of this remains unclear since relatively specific inhibitors of the iNOS do not alter tone in CH-PHT PAs [27,41].

Whether eNOS expression is increased in human PAH remains unclear. Gaid and Saleh reported decreased eNOS expression [42], whereas Xue and Johns [43] reported increased and Tuder et al. [44] unaltered eNOS immunostaining. The discrepancies among the human studies and between the human versus animal data are likely due to methodological differences as well as differences in the stages and severity of the disease in the models studied. For example, animals tend to be studied early in the development of the disease, whereas human lungs tend to be studied at the end stages of the disease; biopsies are now rarely performed in the workup of pulmonary hypertension and most of the tissue is obtained during transplant surgery or postmortem.

Increases in the expression of NOS do not necessarily imply increase in the NO production, since the increased protein might have decreased enzyme activity. For example, Rengasamy et al. [45,46] using the citrulline assay, showed that the activity of eNOS is decreased under hypoxic conditions, whereas as discussed above, the protein expression of this enzyme is often increased in chronic hypoxia. These authors suggest that O_2 substrate limitation might regulate NOS activity under hypoxic conditions [45,46]. NOS activity is rarely studied in both animal and human studies. Isaacson et al., used the chemiluminescence technique and showed that NOx accumulation in the perfusate is negligible in the perfused lungs of

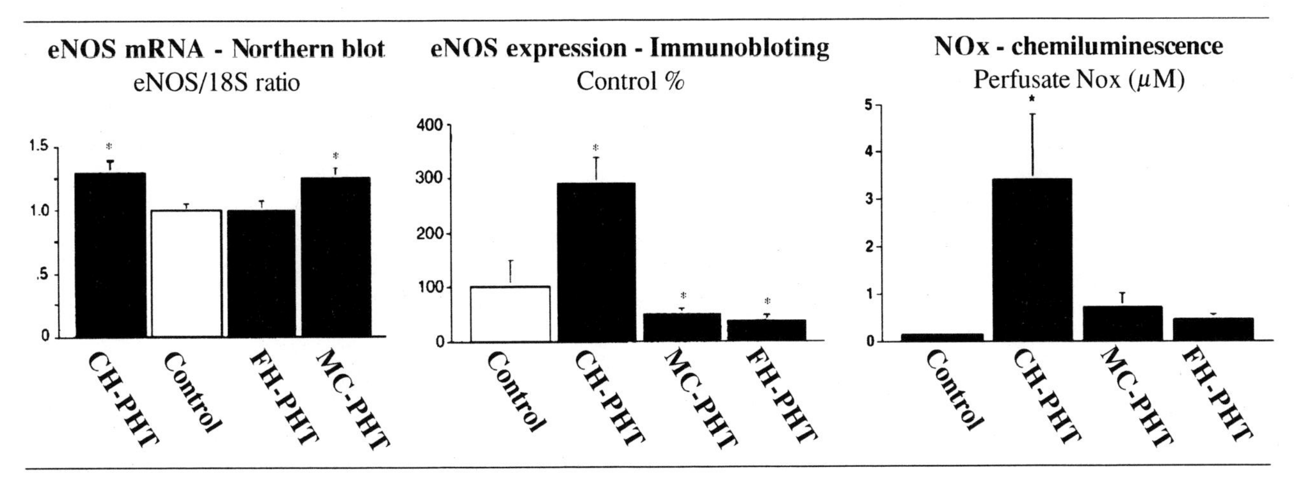

Fig. 5. *Comparison of eNOS mRNA levels, protein expression and activity in lungs from rats from 3 different pulmonary hypertension models. The findings are discussed in the text and show the importance of using multiple complementary techniques in the study of the NO axis in pulmonary hypertension.* Obtained with permission from Tyler et al. (1999).

normal and significantly elevated in the lungs from rats with CH-PHT [47] and this was confirmed by Tyler et al. [26]. In this later study, the investigators compared eNOS mRNA, protein expression and activity (NO, NO*x* levels) in all 3 models of rat pulmonary hypertension (Fig. 5) [26]. This study showed nicely the importance of using multiple complementary techniques in the study of the NO axis. While mRNA for eNOS was increased in both the CH-PHT and MC-PHT and was unaltered in the FH rats, protein expression was increased only in the CH-PHT and decreased in FH and MC-PHT rats [26]. Lung perfusate NO*x* were increased in the CH-PHT rats and not changed in the FH and MC-PHT rats, although there was a significant trend towards an increase in NO*x* in the MC-PHT rats [26].

The mechanism for the eNOS upregulation remains unclear. At least for CH-PHT it is possible that Hypoxia Inducible Factor (HIF) induces NOS, since HIF-1 expression is increased by hypoxia in PASMC and endothelial cells [48]. However, there is yet no evidence for a HIF-1 binding site in the human eNOS promoter, in contrast to iNOS [49] and nNOS [50].

There is now evidence suggesting that NO and ET-1 are involved in the regulation of each other through an autocrine feedback loop [51]. For example, in endothelial cells ETB receptor activation stimulates eNOS activity [52,53]. Therefore, it is possible that ET-1, which is known to be elevated in several models of pulmonary hypertension and in humans [54–56] is in part responsible for the eNOS upregulation. On the other hand, NO-cGMP inhibits ET-1 secretion and gene expression [51]. NO donors, such as molsidomine have been shown to inhibit the formation of ET-1 in the pulmonary circulation of rats with CH-PHT [57].

The NO axis might also be important in the remodeling of the PAs in pulmonary hypertension since NO inhibits SMC proliferation [58,59] and is known to induce apoptosis in vascular smooth muscle [60,61]. The role of apoptosis in the development of the PA remodeling in pulmonary hypertension is not clear, i.e. does remodeling require increased apoptosis or does apoptosis promote regression from remodeling and medial hypertrophy? Also, it is possible that apoptosis in the endothelial cells is different than the apoptosis in the smooth muscle cells in the PA media. Nevertheless, Yuan et al. recently showed that NO might induce PASMC apoptosis by activating Kv and KCa channels as well as depolarizing mitochondrial membrane potential ($\Delta\Psi m$) [20]. The opening of sarcolemmal K^+ channels would cause efflux of K^+ and therefore osmotic cell shrinkage, which is known to initiate apoptosis; $\Delta\Psi m$ depolarization has also been implicated in the initiation of apoptosis [62]. This elegant study raises the intriguing possibility that the loss of Kv channels associated with the development of pulmonary hypertension in animals [17,21] and humans [22], might contribute to the development of PA remodeling by suppressing PASMC apoptosis and thus promoting PASMC proliferation and medial hypertrophy and neomuscularization.

The role of the NO axis in the development and maintenance of pulmonary hypertension remains unclear and eNOS knockout models have not offered a definitive answer. In contrast to Steudel et al. [29] and Fagan et al. [63], Quinlan et al. [64] found that eNOS deficient mice show decreased muscularization and media thickness in resistance PAs in response to chronic hypoxia, compared to the control mice. They speculated that differences in their genetic background of the eNOS deficient and control mice might have accounted for the opposing results between their study and those from Streudel et al. and Fagaan et al.

In summary, there appears to be an overall enhancement of the NO axis in PAH and there appear to be 3 potential roles of the NO axis in the hypertensive pulmonary circulation.

NO in PHT—A Cause of Vascular Injury? Hampl and Herget in a recent review suggest that the high levels of NO in several models of pulmonary hypertension might in fact have detrimental effects, causing oxidative damage in the PA wall [23]. NO is 1000 times smaller than Cu, Zn SOD (copper, zinc superoxide dismutase), it diffuses fast and therefore reacts with superoxide at least 10 times faster than SOD can scavenge superoxide [65]. Because of this competitive advantage, a significant fraction of superoxide will produce peroxynitrite when NO production is increased and especially when exogenous NO is administered [65]. This is particularly important in PASMC since, as it was recently showed, PAs make much more superoxide and AOS at baseline, compared to systemic arteries like the renal artery [3]. Although peroxynitrite is a very potent oxidant, it tends to react rather selectively with a variety of vascular signaling pathways [65]. Peroxynitrite nitrates and inactivates prostacyclin synthase [66] and thus decrease the production of prostacyclin (PGI_2). PGI_2 is a very important vasodilator and antithrombotic factor in the pulmonary circulation and currently, exogenous PGI_2 is the only drug that has been shown to increase survival in patients with PAH [67]. Peroxynitrite can activate several redox sensitive inflammatory mediators, apoptosis related kinases, fibroblast growth factors, all of which can induce injury and remodeling in the pulmonary circulation [23,65]. Peroxynitrite formation has also been implicated in the rebound hypertensive phenomenon that often complicates even short use of inhaled NO, discussed later on [68].

NO Deficiency—A Risk Factor for the Development of PAH? Archer and Rich recently suggested a "multiple hit" hypothesis for the development of PAH [32]. They proposed that the development of PAH is a result of a toxic insult (the use of an anorectic agent, infection with HIV, etc) that takes place in the presence of a predisposing factor, like endothelial dysfunction, on a genetically favorable background (bone morphogenetic protein gene mutations, Kv channel abnormalities etc) [32]. The role of endothelial dysfunction as a predisposing factor in PAH is nicely shown in the story of the recent epidemics of anorectic-induced PAH [69]. Although millions of patients were exposed to anorectics (like Phen-Fen, dexfenfluramine, etc) only few developed PAH, suggesting the presence of one or more predisposing factors. Dexfenfluramine (and most of the recently widely used anorectic agents) have been shown to cause vasoconstriction by inhibiting Kv channels in both pulmonary and systemic vascular SMCs [19]. However, when dexfenfluramine is given in a perfused rat lung [70] and in an intact rat [71], it does not cause significant vasoconstriction unless the lungs or the rats are pretreated with a NOS inhibitor (Fig. 6 A, B). Furthermore, Archer et al. measured exhaled NO and NO*x* in patients with anorectic induced PAH and showed a relative deficiency of NO in these patients, long after the anorectic agents were discontinued, perhaps explaining their original susceptibility [72]. Patients with anorectic induced pulmonary hypertension had lower levels of exhaled NO, compared to patients with PPA, which in turn had higher levels compared to normal control patients. The exhaled NO levels in these patients correlated inversely with the hemodynamic severity of PHT (the lower the NO levels, the higher the PVR) [72] (Fig. 6 C,D).

NO in PAH—An Inadequate Compensatory Mechanism? This is supported by the beneficial effects of the enhancement of the NO axis, discussed below. However, only ~20% of patients respond acutely to NO, with significant decreases in the PVR [32]. It is possible that it is these patients that suffer from inadequate or incomplete compensatory enhancement of the NO axis in the pulmonary circulation, although it is also possible that these patients are simply diagnosed at an earlier stage of the disease.

Exogenous Enhancement of the NO Axis

Inhaled NO

There are 2 important challenges in the treatment of pulmonary hypertension using vasodilators: First, the lack of pulmonary vasculature selective vasodilators. Ca^{++} channel blockers are currently used in the treatment of PAH [73] and in theory they are very attractive drugs since, as can be seen in Figure 1, they are acting very distally in the pathway of several vasodilators, including NO. However, in order to be effective, very high doses of dihydropyridines have to be used and their tolerability is limited by systemic effects like systemic hypotension of edema. Selective pulmonary vasodilatation can be achieved with continuous infusion of short acting vasodilators, like adenosine or prostacyclin [74]. Second, even the selective pulmonary vasodilators act by dilating

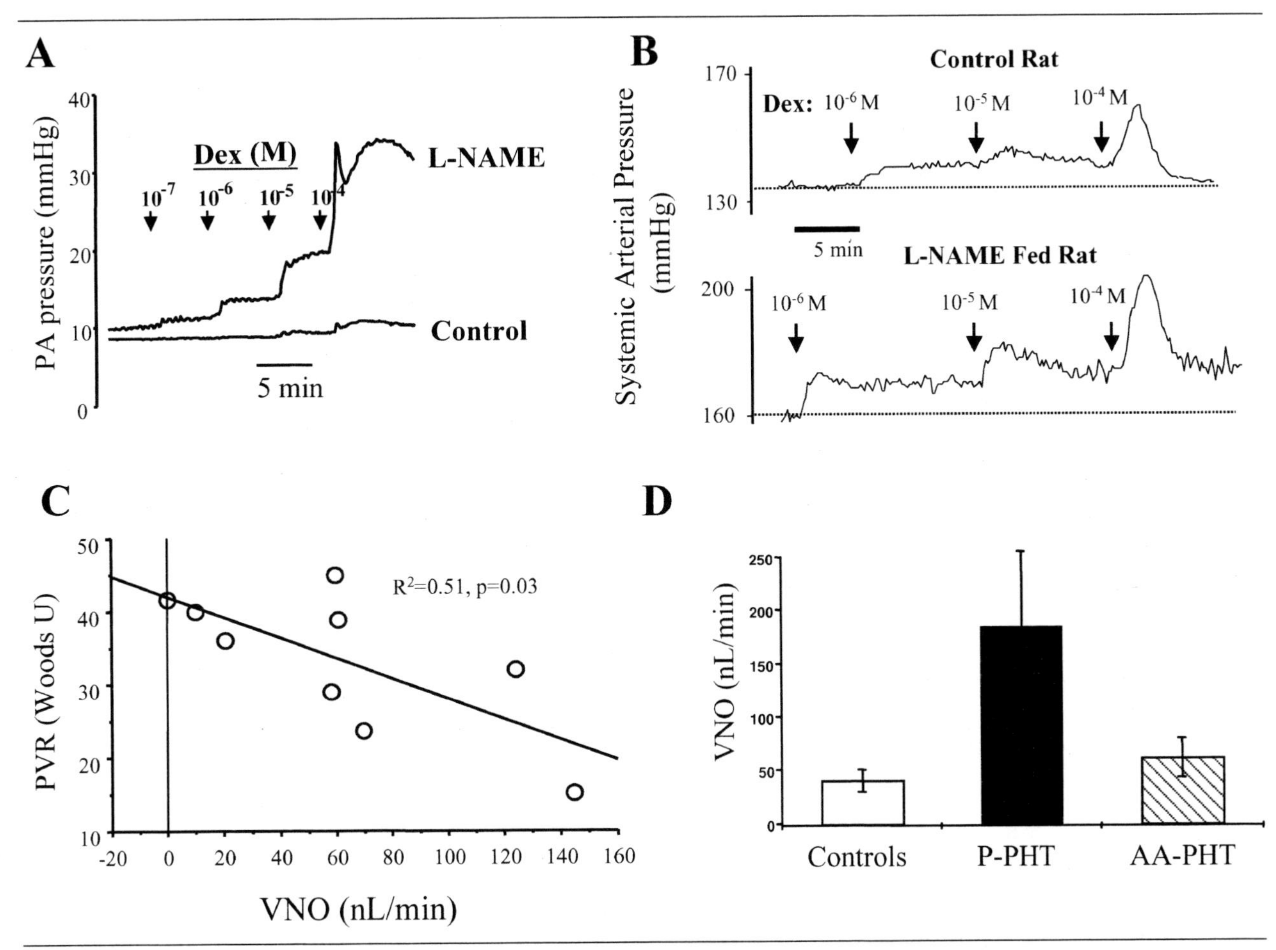

Fig. 6. *(A) Dexfenfluramine (Dex) significantly increases PA tone in an isolated perfused rat lung, but only after the lung is pretreated with a NOS inhibitor. Note that NOS inhibition does not increase the baseline tone in this lung from a healthy rat.* Obtained with permission Weir et al. (1996). *(B) Dex significantly increases arterial pressure when given intravenously in an intact ventilated rat and this is potentiated by pretreatment with a NOS inhibitor. Note that, in contrast to A, NOS inhibition in vivo resulted in an increase in the baseline pressure (a rise from ~135 to ~160 mmHg).* Obtained with permission from Michelakis et al. (1999). *(C and D) Exhaled NO (controlled for minute ventilation, VNO) correlates inversely with the PVR in patients with PAH (the higher the PVR, the lower the VNO). Patients with anorectic-associated pulmonary hypertension (AA-PHT) have lower VNO levels compared to the elevated levels seen in patients with primary pulmonary hypertension (P-PHT). This suggests that perhaps low NO levels in the pulmonary circulation of AA-PHT (due to endothelial dysfunction) predisposed them to the development of PHT after the ingestion of anorectic agents.* Obtained with permission from Archer et al. (1998).

the whole pulmonary circulation, often disrupting HPV, impairing V/Q matching and causing or worsening hypoxemia.

Both of these challenges can be overcome with the use of inhaled NO (iNO) [75,76]. Exogenous iNO can reach the PASMC through diffusion form the alveoli. After further diffusion into the lumen, NO reacts with hemoglobin and is inactivated avoiding any systemic effects. Furthermore, iNO will only be delivered in ventilated lobes and thus dilate only the vascular beds in well ventilated areas. The lack of vasodilatation in nonventilated areas will prevent the intrapulmonary shunting seen with the systemically administered pulmonary vasodilators and preserve V/Q matching.

Initiation of chronic therapy for PAH usually follows an acute hemodynamic trial to determine prognosis, assess safety of a proposed treatment and guide future medical therapy [73,77–79]. The acute hemodynamic study employs a selective pulmonary vasodilator, usually iNO, to evaluate the responsiveness of the pulmonary vasculature while avoiding systemic hypotension [77–80]. iNO is currently considered the gold standard for the evaluation of patients with PAH [77–80]. A positive response to iNO (>20% decrease in pulmonary artery pressure or pulmonary vascular resistance) predicts the response to conventional vasodilators, such as calcium channel blockers [77,79] and identifies patients with a better long-term prognosis than the non-responders [73]. In addition to patients with PAH, iNO is also used in patients with severe pulmonary hypertension due to left ventricular dysfunction as part of the preoperative assessment for cardiac transplantation. Because unresponsive and severe pulmonary hypertension is a contraindication to cardiac transplantation, the response to iNO is used to identify patients that require combined heart-lung transplantation [81,82]. iNO is also extensively used in the treatment of neonatal pulmonary hypertension [83].

Although NO is a potent and selective pulmonary vasodilator, chronic use is limited by its short half-life and, more recently, significant increases in the price of this gas. Even its use as an acute vasodilator is cumbersome, requiring an expensive medical form of NO gas, a complicated delivery system and monitoring equipment. Nevertheless there is some preliminary evidence that chronic outpatient therapy might be possible. In an uncontrolled pilot study of chronic iNO in 5 patients with PAH, using nasal cannulae and a gas pulsing device, improvement in PA pressure or cardiac output was shown after 12 weeks of treatment in 3 out of the 5 patients [84]. Chronic continuous exposure to iNO significantly prevents the monocrotaline-induced remodeling in the pulmonary circulation in rats [85].

There are 2 important potential complications of even the short-term use of iNO.

First, iNO causes increase in the pulmonary artery wedge pressure, especially in patients with left ventricular dysfunction [86,87], perhaps explaining occasional cases of *pulmonary edema* with this therapy [88]. It has been suggested that this is a result of the increased return of blood from the lungs to a noncompliant left ventricle. Recently, however pulmonary edema was reported in patients with PAH due to the CREST syndrome that normally have intact left ventricular function [89]. Pulmonary artery wedge pressure was increased by iNO in a cohort of 11 patients with PAH and 2 patients with left ventricular dysfunction [90]. However, wedge pressure was not increased in response to sildenafil (Fig. 7), a phosphodiesterase inhibitor (see discussion below) despite a more effective decrease in PVR (Fig. 8) and increase in cardiac index (Fig. 9) compared to iNO and similar increases in the cGMP in the pulmonary circulation [90]. This suggests that the effects of iNO on the wedge pressure (and perhaps the mechanism of iNO-induced pulmonary edema) are not mediated by cGMP but perhaps by a different mechanism proximal to cGMP or perhaps direct negative inotropic effects in the myocardium.

Second, sudden termination of iNO occasionally causes a potentially life threatening *hypertensive rebound*, even after treatment for a few hours and even in patients that showed no initial vasodilator

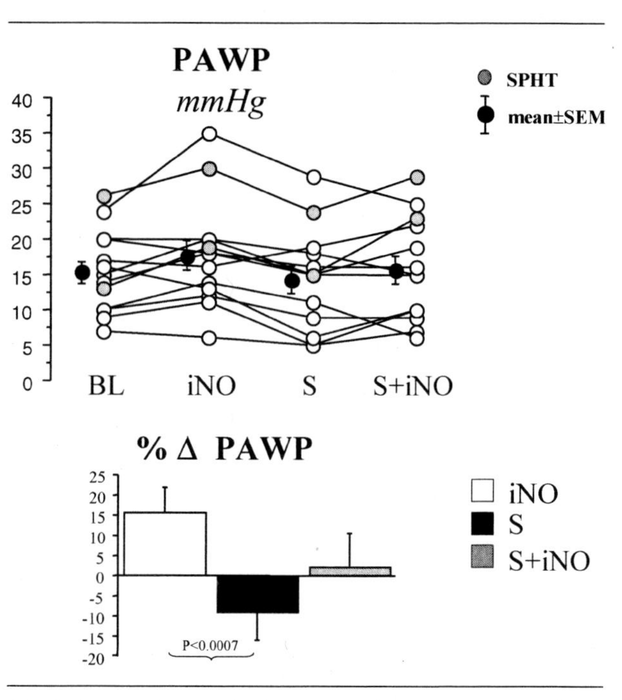

Fig. 7. Sildenafil acutely decreases pulmonary artery wedge pressure (PAWP) compared to iNO in a cohort of patients with PAH see text.

response [91]. Exogenous NO causes a decrease in eNOS activity [92,93] and an increase in endothelin levels [94]. In addition to the potent vasoconstrictor effect, endothelin induces superoxide production, which in the presence of NO causes the formation of peroxynitrite, with its known deleterious vascular effects [68]. This suggests that endothelin receptor blockers might be beneficial in the management iNO-rebound effect [68].

Another way of delivering NO selectively in the pulmonary circulation is using inhalation of aerosolized adenoviruses carrying the genes for eNOS or iNOS [95,96]or using cell-based gene transfer of eNOS [97]. These approaches are very promising forms of gene therapy but several challenges need to be overcome before their human application, like the immune reactions against the adenovirus or the transient nature of the expression of the transferred gene.

Enhancement of the Endogenous NO Axis

L-Arginine

It has been suggested that the production of NO can be limited by insufficient supply of NOS substrate, i.e. L-Arginine. Therefore L-Arginine has been given in a variety of cardiovascular diseases, in an attempt to optimize NO production. For example, oral L-Arginine improves endothelial function and exercise capacity in patients with

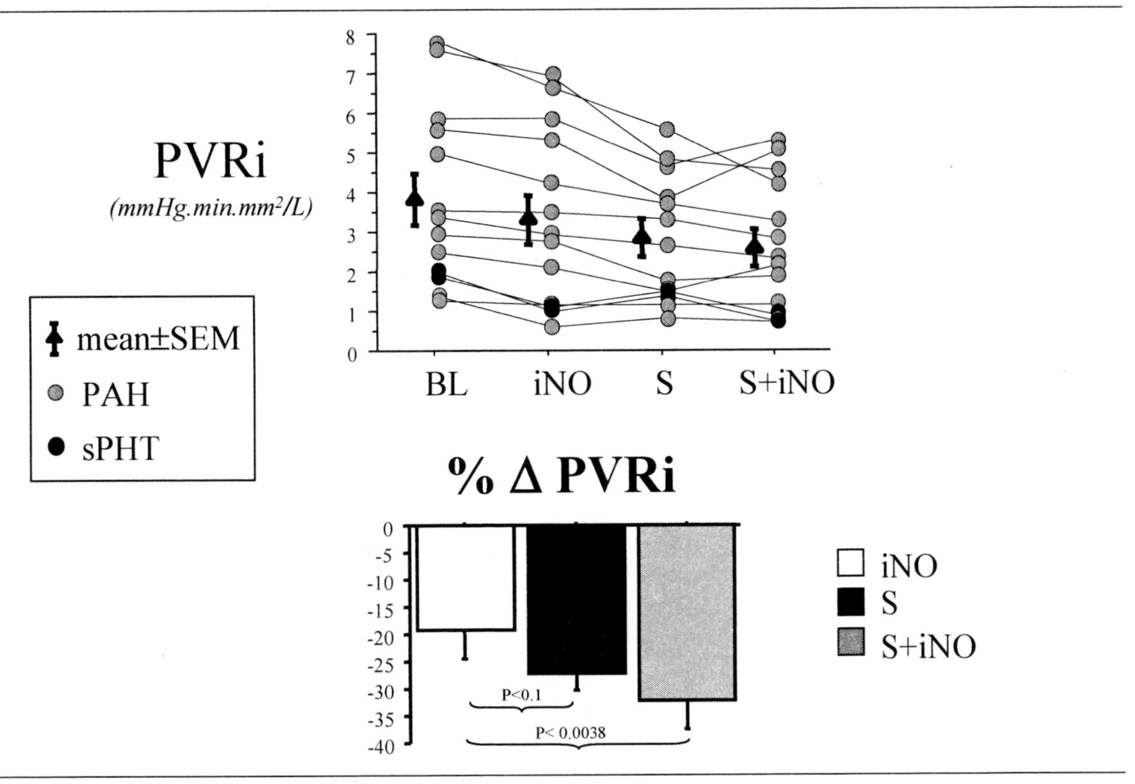

Fig. 8. *Sildenafil acutely decreases PVR more than maximal dose of iNO (80 ppm) in a cohort of patients with PAH see text.*

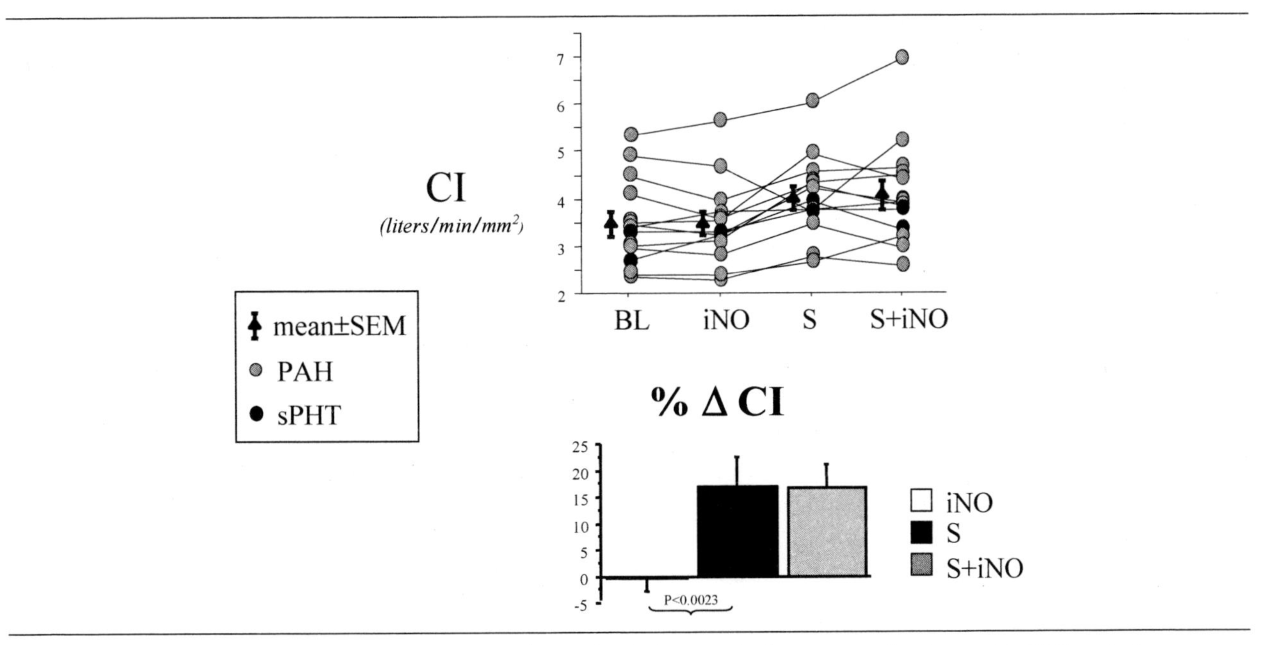

Fig. 9. *Sildenafil, but not iNO, significantly increase the cardiac index acutely, in a cohort of patients with PAH see text.*

congestive heart failure [98]. Intraperitoneal injections of L-Arginine have also been shown to reduce mean PA pressure, PA remodeling (% muscularization) and right ventricular hypertrophy in both rats with CH-PHT and MC-PHT [99]. In humans with both primary and secondary pulmonary hypertension, intravenous administration of L-Arginine has been shown to decrease pulmonary vascular resistance acutely [100]. Although systemic vascular resistance was slightly decreased in both the patients and healthy controls, PVR was not decreased in patients with congestive heart failure and no pulmonary hypertension, suggesting a relatively selective effect of L-Arginine in the hypertensive pulmonary vasculature. The same group later showed that systemic intravenous L-Arginine increases expired NO [101]. Oral administration of L-Arginine (0.5 g capsule/10 Kg body weight) in 19 patients with pulmonary arterial hypertension acutely decreased PVR by 16% and, after 1 week or treatment, significantly improved exercise capacity compared to placebo [102]. More studies are needed to establish the role of this drug in the treatment of patients with pulmonary hypertension, especially in combination with other drugs that enhance the NO axis, like sildenafil.

Phosphodiesterase Inhibitors

The main effector of NO's vasoactive effects is cGMP, which, like NO, is also short-lived due to the rapid degradation by phosphodiesterases [103] (Fig. 1). There are numerous phosphodiesterases but the isoform that is active in degrading cGMP in the lung is cyclic nucleotide phosphodiesterase-5 [104]. Phosphodiesterase-5 inhibitors cause pulmonary vasodilatation by promoting an enhanced and sustained level of cGMP, which in turn promotes K^+ channel activation, PASMC hyperpolarization and vasodilatation (Fig. 10) [11]. There have been recent anecdotal reports and preliminary studies indicating that sildenafil, a specific phosphodiesterase-5 inhibitor widely used in the treatment of erectile dysfunction [105], decreases PVR in humans with primary pulmonary hypertension (PPH) [106,107], in normal volunteers with hypoxic pulmonary vasoconstriction [108] and in animals with experimental PAH [109,110]. We hypothesized that sildenafil would be as effective in decreasing pulmonary vascular resistance as iNO in the acute assessment of patients with severe pulmonary hypertension. We directly compared the effects of iNO with a single dose of oral sildenafil as well as their combination, on pulmonary and systemic hemodynamics in patients with severe pulmonary hypertension [90].

We studied 13 consecutive patients with a mean (±SEM) age of 44 ± 2 years referred to the University of Alberta Hospital cardiac catheterization laboratory over a period of one year for evaluation of suitability for transplantation or medical therapy. Eleven patients had PAH and two patients had pulmonary hypertension, which although it was associated with left ventricular dysfunction, was disproportionate to their pulmonary wedge pressure. We showed that a single dose of oral sildenafil is a potent and selective pulmonary vasodilator [90]. Compared to the gold standard, iNO, sildenafil is superior in decreasing the mean pulmonary artery pressure and equally effective and selective in reducing pulmonary vascular resistance (Fig. 8), the primary end point of this study [90]. In contrast to iNO, sildenafil causes a significant increase in the cardiac index (Fig. 9). Both iNO and oral sildenafil are selective pulmonary vasodilators, as neither lowers the mean arterial pressure [90].

Our finding that sildenafil tends to decrease the wedge pressure (Fig. 7) suggests that sildenafil might be superior to iNO in the evaluation of the patients with severe pulmonary hypertension. This might have important safety implications both for the acute study and for eventual long-term use of this drug in patients with left ventricular dysfunction.

The preferential effect of sildenafil on the pulmonary circulation probably reflects the high expression of this isoform in the lung. However, phosphodiesterase 5 is also found in the myocardium,

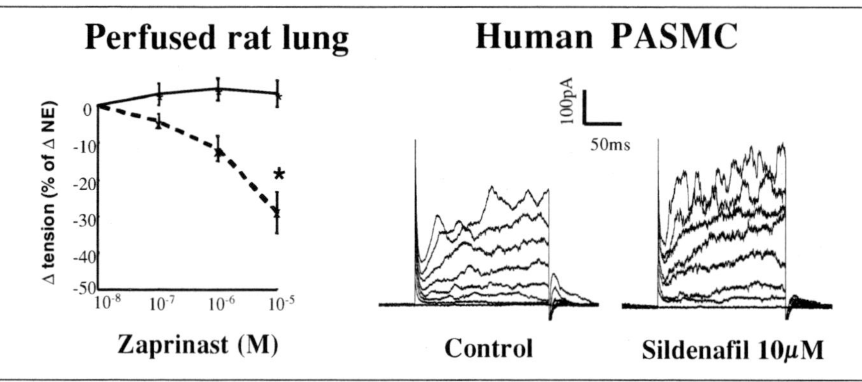

Fig. 10. *Left: Zaprinast, a phosphodiesterase inhibitor, dilates the norepinephrine preconstricted pulmonary vasculature in isolated perfused rat lungs (solid line = vehicle).* Obtained with permission from Archer et al. (1994). *Right: The phosphodiesterase 5 inhibitor sildenafil activates K^+ currents in freshly isolated human pulmonary artery smooth muscle cells (PASMC), studied with the whole-cell patch clamping technique. As discussed in the text, this K^+ channel activation explains, at least in part, the pulmonary vasodilatory properties of this drug.*

where it maybe downregulated in heart failure [111]. The finding that sildenafil decreases the wedge pressure and increases the cardiac index suggests that it does not have negative inotropic effects, at least in the patients studied. Phosphodiesterase 5 has been implicated in modulation of sympathetic tone [111] and sildenafil has recently been shown to cause sympathetic nervous system activation in normal volunteers [112]. However, the fact that the heart rate did not change after sildenafil in our study suggests that sympathetic activation is not the basis for the observed increase in the cardiac index [90]. The data suggest that sildenafil increases cardiac index because of its selective pulmonary vasodilatory effects and the resulting reduction in right ventricular afterload.

Another important finding of this study is that iNO and sildenafil have additive vasodilatory effects in the pulmonary but not the systemic circulation. The mechanism for this might be related to the synergistic effects of iNO and sildenafil on serum cGMP levels [90]. The data suggest that NO 80 ppm, which currently is usually the maximum dose used in acute vasodilator testing, is not as effective in lowering pulmonary vascular resistance alone, as it is in combination with sildenafil (Fig. 8).

The dose of sildenafil used in this study was smaller than the 100 mg that has recently been used in acute hemodynamic studies involving sildenafil [108,113], although it is in the range used for erectile dysfunction (50–100 mg) [105]. Wilkens et al. very recently showed that the maximal hemodynamic effects of sildenafil on the human pulmonary circulation were achieved with a 25 mg dose [107]. They also showed that maximal hemodynamic effects were achieved within 30 minutes after intake [107]. Furthermore, newer phosphodiesterase-5 inhibitors that are perhaps more potent and specific than sildenafil are currently under development [114,115].

The simplicity and safety of the acute administration of sildenafil versus iNO and its possible superiority over iNO in terms of its effects on cardiac index and wedge pressure, suggest a role for sildenafil in the evaluation and treatment of patients with pulmonary hypertension and support the need for further studies of its chronic use.

Conclusion

The treatment of pulmonary hypertension is difficult and currently limited. Because PAH is an inhomogeneous syndrome and because several abnormalities can be present in the pulmonary arteries of a single patient, the ultimate treatment of the disease will likely consist of combination treatments and cocktails of drugs. The NO axis will be a nice example of this concept. The ease and the safety of sildenafil administration make this drug an attractive addition to such cocktail treatments. One can envision a patient treated with both L-Arginine and sildenafil or an NO donor and sildenafil or sildenafil and a phosphatase inhibitor. Several such trials will surface in the field of pulmonary hypertension in the next few years.

Acknowledgments

Dr Michelakis is funded by the Canadian Institutes for Health Research, the Alberta Heritage Foundation for Medical Research, the Canadian Heart and Stroke Foundation and the Canadian Foundation for Innovation.

References

1. Michelakis ED, Archer SL, Weir EK. Acute hypoxic pulmonary vasoconstriction: A model of oxygen sensing. *Physiol Res* 1995;44:361–367.
2. Weir EK, Archer SL. The mechanism of acute hypoxic pulmonary vasoconstriction: The tale of two channels. *FASEB J* 1995;9:183–189.
3. Michelakis ED, Hampl V, Nsair A, Wu X, Harry G, Haromy A, Gurtu R, Archer SL. Diversity in mitochondrial function explains differences in vascular oxygen sensing. *Circ Res* 2002;90:1307–1315.
4. Wolin MS. Interactions of oxidants with vascular signaling systems. *Arterioscler Thromb Vasc Biol* 2000;20:1430–1442.
5. Archer SL. Measurement of nitric oxide in biological models. *FASEB J* 1993;7:349–360.
6. Michelakis E TD-X, Djaballah K, Souil E, Archer S. Measurement of NO and NO synthase activity. In: Mathie RT, Griffith TM, eds. *Hemodynamic Effects of NO*. Singapore: World Scientific, 1999:161–183.
7. Moncada S, Higgs A. The L-arginine-nitric oxide pathway. *N Engl J Med* 1993;329:2002–2012.
8. Sherman TS, Chen Z, Yuhanna IS, Lau KS, Margraf LR, Shaul PW. Nitric oxide synthase isoform expression in the developing lung epithelium. *Am J Physiol* 1999;276:L383–L390.
9. Rairigh RL, Le Cras TD, Ivy DD, Kinsella JP, Richter G, Horan MP, Fan ID, Abman SH. Role of inducible nitric oxide synthase in regulation of pulmonary vascular tone in the late gestation ovine fetus. *J Clin Invest* 1998;101:15–21.
10. Boulanger CM, Heymes C, Benessiano J, Geske RS, Levy BI, Vanhoutte PM. Neuronal nitric oxide synthase is expressed in rat vascular smooth muscle cells : Activation by angiotensin II in hypertension [In Process Citation]. *Circ Res* 1998;83:1271–1278.
11. Archer SL, Huang JM, Hampl V, Nelson DP, Shultz PJ, Weir EK. Nitric oxide and cGMP cause vasorelaxation by activation of a charybdotoxin-sensitive K channel by cGMP-dependent protein kinase. *Proc Natl Acad Sci USA* 1994;91:7583–7587.
12. Robertson BE, Schubert R, Hescheler J, Nelson M. cGMP-dependent protein kinase activates Ca-activated K channels in cerebral artery smooth muscle cells. *Am J Physiol* 1993;265:C299–C303.

13. Archer SL, Rusch NJ (eds). *Potassium Channels in Cardiovascular Biology*, 1st edn. New York: Kluwer Academic/Plenum Publishers, 2001.
14. Archer SL, Weir EK, Reeve HL, Michelakis E. Molecular identification of O_2 sensors and O_2-sensitive potassium channels in the pulmonary circulation. *Adv Exp Med Biol* 2000;475:219–240.
15. Murad F. The 1996 Albert Lasker Medical Research Awards. Signal transduction using nitric oxide and cyclic guanosine monophosphate. *JAMA* 1996;276:1189–1192.
16. Wu X, Haystead TA, Nakamoto RK, Somlyo AV, Somlyo AP. Acceleration of myosin light chain dephosphorylation and relaxation of smooth muscle by telokin. Synergism with cyclic nucleotide-activated kinase. *J Biol Chem* 1998;273:11362–11369.
17. Michelakis ED, Weir EK. The pathobiology of pulmonary hypertension. Smooth muscle cells and ion channels. *Clin Chest Med* 2001;22:419–432.
18. Weir EK, Archer SL. Hypoxic pulmonary vasoconstriction: A tale of two channels. *FASEB*, in press.
19. Michelakis ED, Weir EK. Anorectic drugs and pulmonary hypertension from the bedside to the bench. *Am J Med Sci* 2001;321:292–299.
20. Yuan X-J, Goldman W, Tod ML, Rubin LJ, Blaustein MP. Hypoxia reduces potassium currents in cultured rat pulmonary but not mesenteric arterial myocytes. *Am J Physiol* 1993;264:L116–L123.
21. Michelakis ED, McMurtry MS, Wu XC, Dyck JR, Moudgil R, Hopkins TA, Lopaschuk GD, Puttagunta L, Waite R, Archer SL. Dichloroacetate, a metabolic modulator, prevents and reverses chronic hypoxic pulmonary hypertension in rats: Role of increased expression and activity of voltage-gated potassium channels. *Circulation* 2002;105:244–250.
22. Yuan XJ, Wang J, Juhaszova M, Gaine SP, Rubin LJ. Attenuated K^+ channel gene transcription in primary pulmonary hypertension [letter]. *Lancet* 1998;351:726–727.
23. Hampl V, Herget J. Role of nitric oxide in the pathogenesis of chronic pulmonary hypertension. *Physiol Rev* 2000;80:1337–1372.
24. Kawai N, Bloch DB, Filippov G, Rabkina D, Suen HC, Losty PD, Janssens SP, Zapol WM, de la Monte S, Bloch KD. Constitutive endothelial nitric oxide synthase gene expression is regulated during lung development. *Am J Physiol* 1995;268:L589–L595.
25. Le Cras TD, Xue C, Rengasamy A, Johns RA. Chronic hypoxia upregulates endothelial and inducible NO synthase gene and protein expression in rat lung. *Am J Physiol* 1996;270:L164–L170.
26. Tyler RC, Muramatsu M, Abman SH, Stelzner TJ, Rodman DM, Bloch KD, McMurtry IF. Variable expression of endothelial NO synthase in three forms of rat pulmonary hypertension. *Am J Physiol* 1999;276:L297–L303.
27. Xue C, Rengasamy A, Le Cras TD, Koberna PA, Dailey GC, Johns RA. Distribution of NOS in normoxic vs. hypoxic rat lung: Upregulation of NOS by chronic hypoxia. *Am J Physiol* 1994;267:L667–L678.
28. Kobzik L, Bredt DS, Lowenstein CJ, Drazen J, Gaston B, Sugarbaker D, Stamler JS. Nitric oxide synthase in human and rat lung: Immunocytochemical and histochemical localization. *Am J Respir Cell Mol Biol* 1993;9:371–377.
29. Steudel W, Ichinose F, Huang PL, Hurford WE, Jones RC, Bevan JA, Fishman MC, Zapol WM. Pulmonary vasoconstriction and hypertension in mice with targeted disruption of the endothelial nitric oxide synthase (NOS3) gene. *Circ Res* 1997;81:34–41.
30. Steinhorn RH, Millard SL, Morin FC, III. Persistent pulmonary hypertension of the newborn. Role of nitric oxide and endothelin in pathophysiology and treatment. *Clin Perinatol* 1995;22:405–428.
31. Steinhorn RH, Morin FC, III, Fineman JR. Models of persistent pulmonary hypertension of the newborn (PPHN) and the role of cyclic guanosine monophosphate (GMP) in pulmonary vasorelaxation. *Semin Perinatol* 1997;21:393–408.
32. Archer S, Rich S. Primary pulmonary hypertension : A vascular biology and translational research "work in progress." *Circulation* 2000;102:2781–2791.
33. Higenbottam T, Cremona G. Acute and chronic hypoxic pulmonary hypertension. *Eur Respir J* 1993;6:1207–1212.
34. Schultze AE, Roth RA. Chronic pulmonary hypertension–the monocrotaline model and involvement of the hemostatic system. *J Toxicol Environ Health B Crit Rev* 1998;1:271–346.
35. Molteni A, Ward WF, Ts'ao CH, Port CD, Solliday NH. Monocrotaline-induced pulmonary endothelial dysfunction in rats. *Proc Soc Exp Biol Med* 1984;176:88–94.
36. van Suylen RJ, Smits JF, Daemen MJ. Pulmonary artery remodeling differs in hypoxia- and monocrotaline- induced pulmonary hypertension. *Am J Respir Crit Care Med* 1998;157:1423–1428.
37. Le Cras TD, Kim DH, Gebb S, Markham NE, Shannon JM, Tuder RM, Abman SH. Abnormal lung growth and the development of pulmonary hypertension in the Fawn-Hooded rat. *Am J Physiol* 1999;277:L709–L718.
38. Le Cras TD, Kim DH, Markham NE, Abman AS. Early abnormalities of pulmonary vascular development in the Fawn-Hooded rat raised at Denver's altitude. *Am J Physiol Lung Cell Mol Physiol* 2000;279:L283–L291.
39. Stelzner T, Hofmann TA, Brown D, Deng A, Jacob HJ. Genetic determinants of pulmonary hypertension in fawn-hooded rats. *Chest* 1997;111:96S.
40. Resta TC, Gonzales RJ, Dail WG, Sanders TC, Walker BR. Selective upregulation of arterial endothelial nitric oxide synthase in pulmonary hypertension. *Am J Physiol* 1997;272:H806–H813.
41. Resta TC, O'Donaughy TL, Earley S, Chicoine LG, Walker BR. Unaltered vasoconstrictor responsiveness after iNOS inhibition in lungs from chronically hypoxic rats. *Am J Physiol* 1999;276:L122–L130.
42. Giaid A, Saleh D. Reduced expression of endothelial nitric oxide synthase in the lungs of patients with pulmonary hypertension. *N Engl J Med* 1995;333:214–221.
43. Xue C, Johns RA. Endothelial nitric oxide synthase in the lungs of patients with pulmonary hypertension. *N Engl J Med* 1995;333:1642–1644.
44. Tuder RM, Cool CD, Geraci MW, Wang J, Abman SH, Wright L, Badesch D, Voelkel NF. Prostacyclin synthase expression is decreased in lungs from patients with severe pulmonary hypertension. *Am J Respir Crit Care Med* 1999;159:1925–1932.
45. Rengasamy A, Johns RA. Characterization of endothelium-derived relaxing factor/nitric oxide synthase from bovine cerebellum and mechanism of modulation

by high and low oxygen tensions. *J Pharmacol Exp Ther* 1991;259:310–316.
46. Rengasamy A, Johns RA. Determination of Km for oxygen of nitric oxide synthase isoforms. *J Pharmacol Exp Ther* 1996;276:30–33.
47. Isaacson TC, Hampl V, Weir EK, Nelson DP, Archer SL. Increased endothelium-derived NO in hypertensive pulmonary circulation of chronically hypoxic rats. *J Appl Physiol* 1994;76:933–940.
48. Yu AY, Frid MG, Shimoda LA, Wiener CM, Stenmark K, Semenza GL. Temporal, spatial, and oxygen-regulated expression of hypoxia-inducible factor-1 in the lung. *Am J Physiol* 1998;275:L818–L826.
49. Melillo G, Musso T, Sica A, Taylor LS, Cox GW, Varesio L. A hypoxia-responsive element mediates a novel pathway of activation of the inducible nitric oxide synthase promoter. *J Exp Med* 1995;182:1683–1693.
50. Forstermann U, Boissel JP, Kleinert H. Expressional control of the 'constitutive' isoforms of nitric oxide synthase (NOS I and NOS III). *FASEB J* 1998;12:773–790.
51. Kourembanas S, McQuillan LP, Leung GK, Faller DV. Nitric oxide regulates the expression of vasoconstrictors and growth factors by vascular endothelium under both normoxia and hypoxia. *J Clin Invest* 1993;92:99–104.
52. Hirata Y, Emori T, Eguchi S, Kanno K, Imai T, Ohta K, Marumo F. Endothelin receptor subtype B mediates synthesis of nitric oxide by cultured bovine endothelial cells. *J Clin Invest* 1993;91:1367–1373.
53. Wong J, Vanderford PA, Winters J, Soifer SJ, Fineman JR. Endothelin B receptor agonists produce pulmonary vasodilation in intact newborn lambs with pulmonary hypertension. *J Cardiovasc Pharmacol* 1995;25:207–215.
54. Frasch HF, Marshall C, Marshall BE. Endothelin-1 is elevated in monocrotaline pulmonary hypertension. *Am J Physiol* 1999;276:L304–L310.
55. Giaid A, Yanagisawa M, Langleben D, Michel RP, Levy R, Shennib H, Kimura S, Masaki T, Duguid WP, Stewart DJ. Expression of endothelin-1 in the lungs of patients with pulmonary hypertension. *N Engl J Med* 1993;328:1732–1739.
56. Stewart DJ, Levy RD, Cernacek P, Langleben D. Increased plasma endothelin-1 in pulmonary hypertension: Marker or mediator of disease? *Ann Intern Med* 1991;114:464–469.
57. Blumberg FC, Wolf K, Sandner P, Lorenz C, Riegger GA, Pfeifer M. The NO donor molsidomine reduces endothelin-1 gene expression in chronic hypoxic rat lungs. *Am J Physiol Lung Cell Mol Physiol* 2001;280:L258–L263.
58. Sharma RV, Tan E, Fang S, Gurjar MV, Bhalla RC. NOS gene transfer inhibits expression of cell cycle regulatory molecules in vascular smooth muscle cells. *Am J Physiol* 1999;276:H1450–H1459.
59. Garg UC, Hassid A. Nitric oxide-generating vasodilators and 8-bromo-cyclic guanosine monophosphate inhibit mitogenesis and proliferation of cultured rat vascular smooth muscle cells. *J Clin Invest* 1989;83:1774–1777.
60. Fukuo K, Hata S, Suhara T, Nakahashi T, Shinto Y, Tsujimoto Y, Morimoto S, Ogihara T. Nitric oxide induces upregulation of Fas and apoptosis in vascular smooth muscle. *Hypertension* 1996;27:823–826.
61. Pollman MJ, Yamada T, Horiuchi M, Gibbons GH. Vasoactive substances regulate vascular smooth muscle cell apoptosis. Countervailing influences of nitric oxide and angiotensin II. *Circ Res* 1996;79:748–756.
62. Duchen MR. Mitochondria and Ca(2+) in cell physiology and pathophysiology. *Cell Calcium* 2000;28:339–348.
63. Fagan KA, Fouty BW, Tyler RC, Morris KG, Jr., Hepler LK, Sato K, LeCras TD, Abman SH, Weinberger HD, Huang PL, McMurtry IF, Rodman DM. The pulmonary circulation of homozygous or heterozygous eNOS-null mice is hyperresponsive to mild hypoxia. *J Clin Invest* 1999;103:291–299.
64. Quinlan TR, Li D, Laubach VE, Shesely EG, Zhou N, Johns RA. eNOS-deficient mice show reduced pulmonary vascular proliferation and remodeling to chronic hypoxia. *Am J Physiol Lung Cell Mol Physiol* 2000;279:L641–L650.
65. Beckman JS.—OONO: Rebounding from nitric oxide. *Circ Res* 2001;89:295–297.
66. Zou M-H, Ullrich V. Peroxynitrite formed by simultaneous generation of NO and suproxide selectively inhibits bovine aortic prostacyclin synthase. *FEBS Lett* 1996;382:101–104.
67. Barst RJ, Rubin LJ, Long WA, McGoon MD, Rich S, Badesch DB, Groves BM, Tapson VF, Bourge RC, Brundage BH et al. A comparison of continuous intravenous epoprostenol (prostacyclin) with conventional therapy for primary pulmonary hypertension. The Primary Pulmonary Hypertension Study Group. *N Engl J Med* 1996;334:296–302.
68. Wedgwood S, McMullan DM, Bekker JM, Fineman JR, Black SM. Role for endothelin-1-induced superoxide and peroxynitrite production in rebound pulmonary hypertension associated with inhaled nitric oxide therapy. *Circ Res* 2001;89:357–364.
69. Abenhaim L, Moride Y, Brenot F, Rich S, Benichou J, Kurz X, Higenbottam T, Oakley C, Wouters E, Aubier M, Simonneau G, Begaud B. Appetite-suppressant drugs and the risk of primary pulmonary hypertension. International Primary Pulmonary Hypertension Study Group. *N Engl J Med* 1996;335:609–616.
70. Weir EK, Reeve HL, Huang J, Michelakis E, Nelson DP, Hampl V, Archer SL. Anorexic agents Aminorex, Fenfluramine and Dexfenfluramine inhibit potassium current in rat pulmonary vascular smooth muscle and cause pulmonary vasoconstriction. *Circulation* 1996;94:2216–2220.
71. Michelakis ED, Weir EK, Nelson DP, Reeve HL, Tolarova S, Archer SL. Dexfenfluramine elevates systemic blood pressure by inhibiting potassium currents in vascular smooth muscle cells. *J Pharmacol Exp Ther* 1999;291:1143–1149.
72. Archer S, Djaballah K, Humbert M, Weir E, Fartoukh M, Dall'Ava-Santucci J, Mercier J-C, Simonneau G, Dinh-Xuan A. Nitric oxide deficiency in fenfluramine- and dexfenfluramine-induced pulmonary hypertension. *Am J Respir Crit Care Med* 1998;158:4:1061–1067.
73. Rich S, Kaufmann E, Levy PS. The effect of high doses of calcium-channel blockers on survival in primary pulmonary hypertension. *N Engl J Med* 1992;327:76–81.
74. Barst R. Pharmacologically induced pulmonary vasodilatation in children and young adults with primary pulmonary hypertension. *Chest* 1986;89:497–503.
75. Lunn RJ. Inhaled nitric oxide therapy. *Mayo Clin Proc* 1995;70:247–255.

76. Steudel W, Hurford WE, Zapol WM. Inhaled nitric oxide: Basic biology and clinical applications. *Anesthesiology* 1999;91:1090–1121.
77. Ricciardi MJ, Knight BP, Martinez FJ, Rubenfire M. Inhaled nitric oxide in primary pulmonary hypertension: A safe and effective agent for predicting response to nifedipine. *J Am Coll Cardiol* 1998;32:1068–1073.
78. Rich S. Primary Pulmonary Hypertension: Executive Summary from the World Symposium—Primary Pulmonary Hypertension 1998. In: World Health Organization, 1998.
79. Sitbon O, Humbert M, Jagot JL, Taravella O, Fartoukh M, Parent F, Herve P, Simonneau G. Inhaled nitric oxide as a screening agent for safely identifying responders to oral calcium-channel blockers in primary pulmonary hypertension. *Eur Respir J* 1998;12:265–270.
80. Pepke-Zaba J, Higenbottam TW, Dinh-Xuan AT, Stone D, Wallwork J. Inhaled nitric oxide as a cause of selective pulmonary vasodilatation in pulmonary hypertension. *Lancet* 1991;338:1173–1174.
81. Pagano D, Townend JN, Horton R, Smith C, Clutton-Brock T, Bonser RS. A comparison of inhaled nitric oxide with intravenous vasodilators in the assessment of pulmonary haemodynamics prior to cardiac transplantation. *Eur J Cardiothorac Surg* 1996;10:1120–1126.
82. Adatia I, Perry S, Landzberg M, Moore P, Thompson JE, Wessel DL. Inhaled nitric oxide and hemodynamic evaluation of patients with pulmonary hypertension before transplantation. *J Am Coll Cardiol* 1995;25:1656–1664.
83. Abman SH. Pathogenesis and treatment of neonatal and postnatal pulmonary hypertension. *Curr Opin Pediatr* 1994;6:239–247.
84. Channick RN, Rubin LJ. New and experimental therapies for pulmonary hypertension. *Clin Chest Med* 2001;22:539–545.
85. Roberts JD, Jr., Chiche JD, Weimann J, Steudel W, Zapol WM, Bloch KD. Nitric oxide inhalation decreases pulmonary artery remodeling in the injured lungs of rat pups. *Circ Res* 2000;87:140–145.
86. Loh E, Stamler JS, Hare JM, Loscalzo J, Colucci WS. Cardiovascular effects of inhaled nitric oxide in patients with left ventricular dysfunction. *Circulation* 1994;90:2780–2785.
87. Semigran MJ, Cockrill BA, Kacmarek R, Thompson BT, Zapol WM, Dec GW, Fifer MA. Hemodynamic effects of inhaled nitric oxide in heart failure. *J Am Coll Cardiol* 1994;24:982–988.
88. Bocchi EA, Bacal F, Auler Junior JO, Carmone MJ, Bellotti G, Pileggi F. Inhaled nitric oxide leading to pulmonary edema in stable severe heart failure. *Am J Cardiol* 1994;74:70–72.
89. Preston IR, Klinger JR, Houtchens J, Nelson D, Mehta S, Hill NS. Pulmonary edema caused by inhaled nitric oxide therapy in two patients with pulmonary hypertension associated with the CREST syndrome. *Chest* 2002;121:656–659.
90. Michelakis E, Tymchak W, Lien D, Webster L, Hashimoto K, Archer S. Oral sildenafil is an effective and specific pulmonary vasodilator in patients with pulmonary arterial hypertension: Comparison with inhaled nitric oxide. *Circulation* 2002;105:2398–2403.
91. Miller OI, Tang SF, Keech A, Celermajer DS. Rebound pulmonary hypertension on withdrawal from inhaled nitric oxide [letter]. *Lancet* 1995;346:51–52.
92. Sheehy AM, Burson MA, Black SM. Nitric oxide exposure inhibits endothelial NOS activity but not gene expression: A role for superoxide. *Am J Physiol* 1998;274:L833–L841.
93. Black SM, Heidersbach RS, McMullan DM, Bekker JM, Johengen MJ, Fineman JR. Inhaled nitric oxide inhibits NOS activity in lambs: A potential mechanism for rebound pulmonary hypertension. *Am J Physiol* 1999;277:H1849–H1856.
94. McMullan DM, Bekker JM, Johengen MJ, Hendricks-Munoz K, Gerrets R, Black SM, Fineman JR. Inhaled nitric oxide-induced rebound pulmonary hypertension: Role for endothelin-1. *Am J Physiol Heart Circ Physiol* 2001;280:H777–H785.
95. Budts W, Pokreisz P, Nong Z, Van Pelt N, Gillijns H, Gerard R, Lyons R, Collen D, Bloch KD, Janssens S. Aerosol gene transfer with inducible nitric oxide synthase reduces hypoxic pulmonary hypertension and pulmonary vascular remodeling in rats. *Circulation* 2000;102:2880–2885.
96. Janssens SP, Bloch KD, Nong Z, Gerard RD, Zoldhelyi P, Collen D. Adenoviral-mediated transfer of the human endothelial nitric oxide synthase gene reduces acute hypoxic pulmonary vasoconstriction in rats. *J Clin Invest* 1996;98:317–324.
97. Campbell AI, Kuliszewski MA, Stewart DJ. Cell-based gene transfer to the pulmonary vasculature: Endothelial nitric oxide synthase overexpression inhibits monocrotaline-induced pulmonary hypertension [see comments]. *Am J Respir Cell Mol Biol* 1999;21:567–575.
98. Rector TS, Bank AJ, Mullen KA, Tschumperlin LK, Sih R, Pillai K, Kubo SH. Randomized, double-blind, placebo-controlled study of supplemental oral L-arginine in patients with heart failure. *Circulation* 1996;93:2135–2141.
99. Mitani Y, Maruyama K, Sakurai M. Prolonged administration of L-arginine ameliorates chronic pulmonary hypertension and pulmonary vascular remodeling in rats [see comments]. *Circulation* 1997;96:689–697.
100. Mehta S, Stewart DJ, Langleben D, Levy RD. Short-term pulmonary vasodilation with L-arginine in pulmonary hypertension. *Circulation* 1995;92:1539–1545.
101. Mehta S, Stewart DJ, Levy RD. The hypotensive effect of L-arginine is associated with increased expired nitric oxide in humans. *Chest* 1996;109:1550–1555.
102. Nagaya N, Uematsu M, Oya H, Sato N, Sakamaki F, Kyotani S, Ueno K, Nakanishi N, Yamagishi M, Miyatake K. Short-term oral administration of L-arginine improves hemodynamics and exercise capacity in patients with precapillary pulmonary hypertension. *Am J Respir Crit Care Med* 2001;163:887–891.
103. Gibson A. Phosphodiesterase 5 inhibitors and nitrergic transmission-from zaprinast to sildenafil. *Eur J Pharmacol* 2001;411:1–10.
104. Sanchez LS, de la Monte SM, Filippov G, Jones RC, Zapol WM, Bloch KD. Cyclic-GMP-binding, cyclic-GMP-specific phosphodiesterase (PDE5) gene expression is regulated during rat pulmonary development. *Pediatr Res* 1998;43:163–168.
105. Cheitlin MD, Hutter AM, Jr., Brindis RG, Ganz P, Kaul S, Russell RO, Jr., Zusman RM. ACC/AHA expert consensus document. Use of sildenafil (Viagra) in patients with cardiovascular disease. American College of Cardiology/American Heart Association. *J Am Coll Cardiol* 1999;33:273–282.

106. Prasad S, Wilkinson J, Gatzoulis MA. Sildenafil in primary pulmonary hypertension. *N Engl J Med* 2000;343:1342.
107. Wilkens H, Guth A, Konig J, Forestier N, Cremers B, Hennen B, Bohm M, Sybrecht GW. Effect of inhaled iloprost plus oral sildenafil in patients with primary pulmonary hypertension. *Circulation* 2001;104:1218–1222.
108. Zhao L, Mason NA, Morrell NW, Kojonazarov B, Sadykov A, Maripov A, Mirrakhimov MM, Aldashev A, Wilkins MR. Sildenafil inhibits hypoxia-induced pulmonary hypertension. *Circulation* 2001;104:424–428.
109. Holzmann A, Manktelow C, Weimann J, Bloch KD, Zapol WM. Inhibition of lung phosphodiesterase improves responsiveness to inhaled nitric oxide in isolated-perfused lungs from rats challenged with endotoxin. *Intensive Care Med* 2001;27:251–257.
110. Ichinose F, Erana-Garcia J, Hromi J, Raveh Y, Jones R, Krim L, Clark MW, Winkler JD, Bloch KD, Zapol WM. Nebulized sildenafil is a selective pulmonary vasodilator in lambs with acute pulmonary hypertension. *Crit Care Med* 2001;29:1000–1005.
111. Senzaki H, Smith CJ, Juang GJ, Isoda T, Mayer SP, Ohler A, Paolocci N, Tomaselli GF, Hare JM, Kass DA. Cardiac phosphodiesterase 5 (cGMP-specific) modulates beta-adrenergic signaling in vivo and is down-regulated in heart failure. *FASEB J* 2001;15:1718–1726.
112. Phillips BG, Kato M, Pesek CA, Winnicki M, Narkiewicz K, Davison D, Somers VK. Sympathetic activation by sildenafil. *Circulation* 2000;102:3068–3073.
113. Herrmann HC, Chang G, Klugherz BD, Mahoney PD. Hemodynamic effects of sildenafil in men with severe coronary artery disease. *N Engl J Med* 2000;342:1622–1626.
114. Padma-Nathan H, McMurray JG, Pullman WE, Whitaker JS, Saoud JB, Ferguson KM, Rosen RC. On-demand IC351 (Cialis) enhances erectile function in patients with erectile dysfunction. *Int J Impot Res* 2001;13:2–9.
115. Porst H, Rosen R, Padma-Nathan H, Goldstein I, Giuliano F, Ulbrich E, Bandel, The Vardenafil Study Group T. The efficacy and tolerability of vardenafil, a new, oral, selective phosphodiesterase type 5 inhibitor, in patients with erectile dysfunction: The first at-home clinical trial. *Int J Impot Res* 2001;13:192–199.

Nitric Oxide and Cardioprotection During Ischemia-Reperfusion

Bodh I. Jugdutt
Cardiology Division, Department of Medicine, University of Alberta, Edmonton, Alberta, Canada

Abstract. **Coronary artery reperfusion is widely used to restore blood flow in acute myocardial infarction and limit its progression. However, reperfusion of ischemic myocardium results in reperfusion injury and persistent ventricular dysfunction even when achieved after brief periods of ischemia. Normally, small amounts of nitric oxide (NO) generated by endothelial NO synthase (eNOS) regulates vascular tone. Ischemia-reperfusion triggers the release of oxygen free radicals (OFRs) and a cascade involving endothelial dysfunction, decreased eNOS and NO, neutrophil activation, increased cytokines and more OFRs, increased inducible NO synthase (iNOS) and marked increase in NO, excess peroxynitrite formation, and myocardial injury. Low doses of NO appear to be beneficial and high doses harmful in ischemia-reperfusion. eNOS knock-out mice confirm that eNOS-derived NO is cardioprotective in ischemia-reperfusion. iNOS overexpression increases peroxynitrite but did not cause severe dysfunction. Increased angiotensin II (AngII) after ischemia-reperfusion inactivates NO, forms peroxynitrite and produces cardiotoxic effects. Beneficial effects of angiotensin-converting-enzyme inhibition and AngII type 1 (AT_1) receptor blockade after ischemia-reperfusion are partly mediated through AngII type 2 (AT_2) receptor stimulation, increased bradykinin and NO. Interventions that enhance NO availability by increasing eNOS might be beneficial after ischemia-reperfusion.**

Key Words. **nitric oxide, ischemia-reperfusion, peroxynitrite, angiotensin, nitroglycerin**

Introduction

Myocardial infarction is the major cause of heart failure. Over the last two decades, coronary reperfusion therapy with thrombolytic and fibrinolytic agents, coronary angioplasty or coronary artery bypass surgery has become established for the management of acute myocardial infarction [1]. However, restoration of blood flow to previously ischemic myocardium results in injury (Table 1) to viable myocardium [2–4] and a "mismatch" between flow and recovery of mechanical function even in the absence of irreversible damage [3,4]. Post-infarction survivors with persistent left ventricular dysfunction are prone to develop heart failure and progressive ventricular remodeling. Since the timing and adequacy of reperfusion are major determinants of functional recovery, every effort is made to achieve as early and as complete reperfusion as possible [5]. Oxygen free radicals (OFRs) and oxidative stress are major contibutors to reperfusion injury [6–10]. Although mammalian cells including cardiomyocytes express superoxide dismutase and catalase which act as scavengers of oxygen free radicals [11], these systems may be overwhelmed after ischemia-reperfusion. Reperfusion after 2 hours of ischemia is associated with significant necrosis [12] but still limits early [13] and late [14,15] remodeling. Even early reperfusion results in persistent left ventricular dysfunction and stunning [9,10] and does not guarantee that function will improve [14–17]. A priority in research is therefore to find adjunctive therapies that limit ischemia-reperfusion injury and speed functional recovery [18,19]. Potential candidates include nitric oxide (NO) or NO donors such as nitroglycerin [20–23]. This review focuses on the role of NO in myocardial reperfusion injury and, paradoxically, cardioprotection after ischemia-reperfusion.

Vascular Injury After Ischemia-Reperfusion and NO

A common problem in patients undergoing reperfusion after myocardial infarction is myocardial perfusion-function mismatch (Fig. 1). Thus, a significant number of patients have persistent ventricular dysfunction despite restoration of flow to the optimal so called "TIMI-3" level in epicardial vessels [24,25]. This dysfunction, due to myocardial stunning or hibernation, was shown to be associated with poor intramural microcirculatory flow which jeopardizes myocardial viability [24,25]. The underlying mechanism appears to be vascular dysfunction at the microcirculatory level.

Address for correspondence: Dr. Bodh I. Jugdutt, 2C2.43 Walter Mackenzie Health Sciences Centre, Division of Cardiology, Department of Medicine, University of Alberta, Edmonton, Alberta, Canada T6G 2R7. Tel.: (780) 407-7729; Fax: (780) 437-3546; E-mail: bjugdutt@ualberta.ca

Table 1. Mechanisms for myocardial stunning and reperfusion injury

- Increased oxygen derived free radicals
- Decreased antioxidant reserve
- Increased peroxynitrite formation
- Vascular injury
- Uncoupling of excitation-contraction
- Calcium overload
- Impaired metabolism
  - Decreased mitochondrial energy production
  - Impaired energy utilization by myofibrils
- Impaired sympathetic neural responsiveness
- Decreased sensitivity of myofibrils to calcium
- Impaired perfusion
- Damage to extracellular collagen matrix
- Myocardial hibernation

After ischemia-reperfusion, flow in the reperfused area is characterized by hyperemia, low-reflow or no-reflow [26–32]. These phenomena can be readily visualized at the bedside using myocardial contrast echocardiography [24,25]. The cause of the low-reflow or no-reflow phenomena in intramural perfusion is thought to be due to capillary obstruction by leucocytes and/or tissue edema associated with myocardial necrosis [33,34]. However, even when perfusion appears to be preserved in areas containing dead cardiomyocytes, abnormalities in microvascular reserve can be demonstrated [31,32].

Endothelial Dysfunction After Ischemia-Reperfusion

The time course of vascular injury following ischemia-reperfusion has been reviewed by Lefer's group which contributed significantly to this field [35,36]. Experimentally, vascular injury, after reperfusion, begins very early with an endothelial triggering phase followed by a neutrophil amplification phase [35]. Within 2.5 to 5 minutes of reperfusion, the endothelium becomes dysfunctional and NO formation decreases (endothelial triggering phase). By 20 minutes, leucocyte adhesiveness increases, leucocytes adhere to endothelium, and neutrophils migrate across the endothelium into reperfused tissue. The activated neutrophils release cytotoxic and chemotactic substances, such as cytokines, proteases, leukotrienes, and OFRs. These substances cause injury to surrounding cells and attract more neutrophils (neutrophil amplification phase). In contrast to experimental ischemia-reperfusion, ischemia without

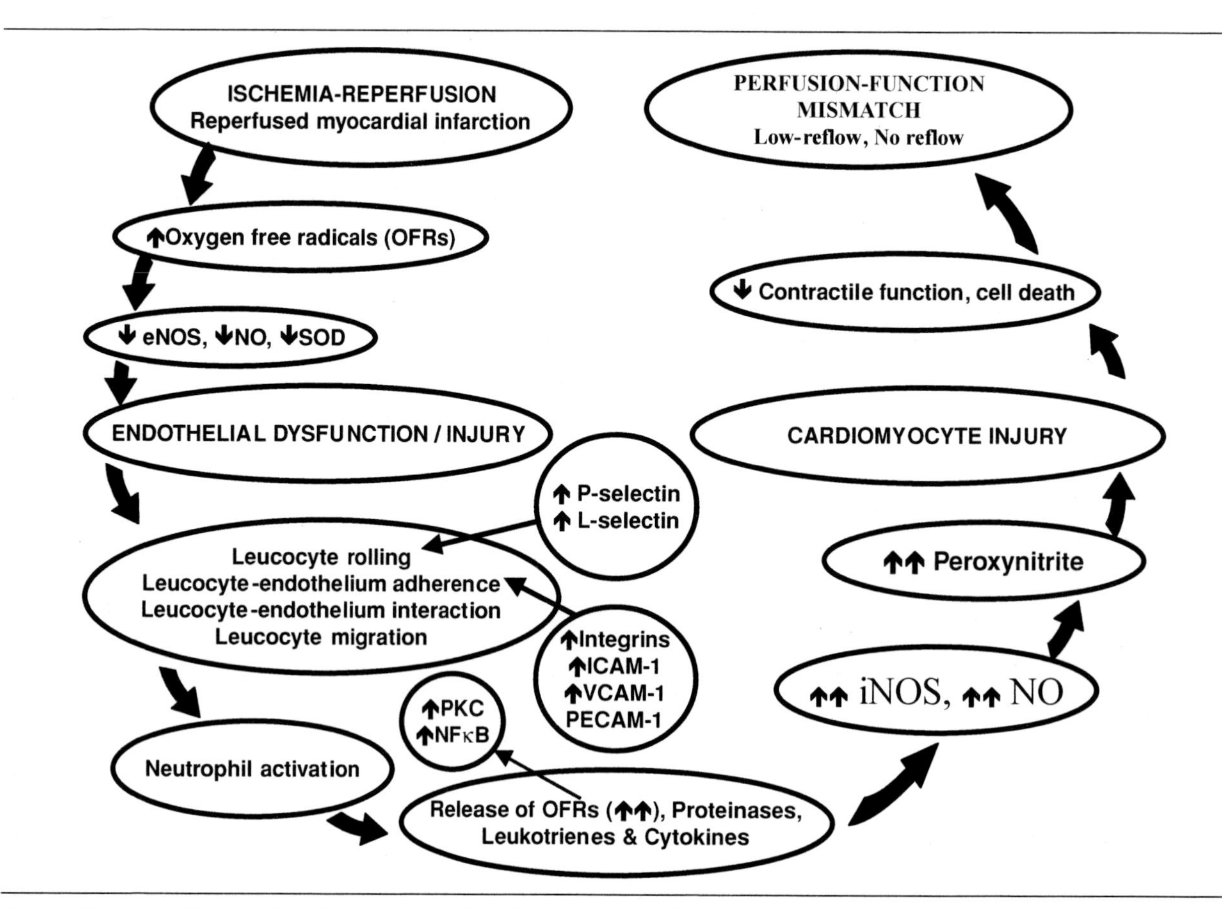

Fig. 1. Diagram of the postulated mechanism of flow-function mismatch after reperfusion of ischemic myocardium.

reperfusion is associated with little change in vascular responses to endothelium-dependent vasodilators for up to 2 to 3 hours [37,38] and histologic evidence of endothelial injury cannot be detected until after 4 to 6 hours [38,39].

The early attenuation of endothelium-dependent vasodilation has been considered to be a sensitive marker of endothelial dysfunction after ischemia-reperfusion in the heart and other vascular beds [35,37,40]. However, it should be noted that patients with acute coronary syndromes in the clinical setting differ from the experimental animal setting in that they already have dysfunctional endothelium and reduced NO levels before undergoing reperfusion procedures.

Experimental evidence suggests that after ischemia-reperfusion, early and severe endothelial dysfunction is due largely to the production of OFRs arising from endothelial cell xanthine oxidase activity [41,42]. Thus, during ischemia, ATP is degraded to hypoxanthine, and xanthine dehydrogenase is converted to xanthine oxidase, and upon reperfusion, xanthine oxidase converts hypoxanthine to uric acid and produces OFRs or superoxide [43,44]. Excessive OFRs, derived from xanthine oxidase or other sources, were shown to be present in reperfused tissues [41,45,46] and to contribute to endothelial damage [47] and dysfunction [48]. It was shown, after ischemia-reperfusion in rat hearts, that endothelial dysfunction from OFRs developed in the absence of neutrophils and was attenuated by recombinant human superoxide dismutase (rhSOD) [49]. The findings suggested that endothelial superoxide inhibits vasorelaxation either by quenching NO and forming peroxynitrite or by inhibiting the endothelial generation of NO [50].

Endothelial Injury and Neutrophils

Several studies have linked neutrophils to reperfusion injury since depletion of neutrophils is associated with less cellular damage after ischemia-reperfusion [51]. Dysfunctional endothelium is associated with a sequence of increased adhesiveness to neutrophils, increased neutrophil-endothelial cell interactions, release of inflammatory mediators (e.g. OFRs and platelet-activating factor), activation of circulating neutrophils, release of various harmful substances (cytokines, proteases, leukotrienes, more OFRs) and tissue injury [52–54]. The OFRs can cause cellular damage in reperfused tissue by inducing lipid peroxidation, altering membrane permeability, inactivating endothelial NO, impairing function of the sarcolemma or sarcoplasmic reticulum and altering calcium handling [55]. Oxygen free radical scavengers such as rhSOD were shown to limit myocardial infarct size and endothelial dysfunction [19,56] whereas the hydroxyl radical scavenger MPG [N-2-mercaptoproprionyl-glycine] did not induce cardioprotection [40,49].

Endothelial NO After Ischemia-Reperfusion

The beneficial effects of the small concentrations of NO produced by endothelial cells [57,58] have been reviewed [59]. Evidence suggests that small quantities of NO regulate vascular tone, inhibit platelet aggregation and adhesion, prevent leucocyte adhesion to the endothelium, scavenge OFR, maintain normal vascular permeability, inhibit smooth muscle proliferation, and stimulate endothelial cell regeneration. Inhibition of endogenous NO using NO synthase inhibitors such as L-NAME (N^G-nitro-L-arginine methyl ester) or L-NMMA (N^G-monomethyl-L-arginine) result in increased blood pressure and vasoconstriction [60], and increased leucocyte rolling and adherence [61].

After ischemia-reperfusion, vascular relaxation in response to endothelium-dependent vasodilators is impaired, suggesting endothelial dysfunction [35,40]. Coronary arteries show blunting of the relaxation response to acetylcholine and failure to constrict with L-NAME, suggesting diminished basal NO release [62]. Administration of L-arginine, the substrate for NO synthesis by NO synthase, induces cardioprotection [63], improves endothelial dysfunction in cardiac veins [64], reduces leucocyte adherence to post capillary venular endothelium [65], and reduces leucocyte adherence and rolling [61]. Impaired vasodilation is restored by tetrahydrobiopterin [66], a cofactor for NO synthase. Decreased NO promotes vasoconstriction [53,62] and vasospasm [67], and increases neutrophil adherence [62], thereby increasing tissue injury. The decreased NO can also promote the release of platelet derived pro-inflammatory mediators (e.g. thromboxane A_2, platelet activating factor), thereby inducing platelet aggregation.

Leucocyte-Endothelium Interaction and NO

Endogenous endothelium-derived relaxing factor or NO [57,58] and exogenous NO can suppress inflammation by regulating cell adhesion molecules (CAMs) after ischemia-reperfusion. The migration of polymorphonuclear neutrophils from the vessel lumen to sites of tissue inflammation involves 3 steps [68]: (i) recognition and capture of the leucocytes by endothelium, which involves leucocyte rolling; (ii) strengthening of leucocyte-endothelium adherence; and (iii) migration of neutrophils between endothelial cells into the reperfused area. Leucocyte rolling is regulated by the family of P-, E- and L-selectins. Endothelial P-selectin expression peaks within 10 to 20 minutes of reperfusion and persists for up to 270 minutes [69]. Neutrophils, which are stimulated during ischemic-reperfusion, show increased expression followed by a shedding of L-selectin [70,71]. After ischemia-reperfusion, both P- and L-selectins contribute to leucocyte rolling and can be attenuated by specific monoclonal antibodies [72]. However, E-selectin which is not upregulated on endothelial cells until much later (4 to 6 hours after reperfusion) seems to play a lesser role in leucocyte rolling [69].

After ischemia-reperfusion, superoxide mediates leucocyte-endothelial interactions [73]. Leucocyte adherence to endothelium is regulated by integrins (e.g. CD11/CD18 complex coupled to intercellular adhesion molecule-1 [ICAM-1] and members of the immunoglobulin [Ig]-like family of adhesion proteins). CD18-specific monoclonal antibody attenuates leucocyte adherence and reduces accumulation of neutrophils in the reperfused area [74]. ICAM-1 and platelet-endothelial CAMs (PECAM-1) also mediate injury [75–78] so that anti-ICAM-1 antibodies [75,76] and anti-PECAM-1 antibodies [77,78] limit leucocyte-mediated damage and induce cardioprotection.

After ischemia-reperfusion, evidence suggests that NO regulates cell adhesion in both the short- and long-term. Thus, NO can downregulate expression of CAMs within minutes and inhibit transcription of proteins involved in inflammation over hours [36]. Ischemia-reperfusion decreases endothelial NO and promotes early leucocyte rolling [61] and adherence [62]. P-selectin expression increases after 90 minutes of myocardial ischemia and peaks within 10–20 minutes after reperfusion in the cat [69] and neutrophil adherence increases 3-fold by 20 minutes after reperfusion [61]. An NO donor given before reperfusion inhibits neutrophil adherence and decreases the number of myocardial vessels expressing P-selectin [79].

Several studies [36] support the concept of an inverse relationship between NO and expression of CAMs. Reduced NO in ischemia-reperfusion upregulates CAMs (such as ICAM-1) whereas increased NO (as after the NO donor S-nitroso-N-acetyl penicillamine [SNAP]) downregulates CAMs [80]. The regulation of CAMs by NO early after reperfusion is attributed to interactions with OFRs. Thus, increased OFRs following ischemia-reperfusion [39,41,45,48] stimulate CAMs on neutrophils [81] and endothelial cells [82]. Normally, NO inhibits OFRs and attenuates the upregulation of CAMs and limits leucocyte-endothelial interactions. However, during ischemia-reperfusion (associated with decreased NO and increased OFRs), CAMs are upregulated and leucocyte-endothelial interactions are enhanced. Decreasing free radical production therefore reduces ICAM-1 expression [80].

NO also plays a regulatory role on the cytosolic enzyme protein kinase C (PKC) which is involved in the translocation of P-selectin to the cell surface. PKC activation leads to rapid upregulation of P-selectin on platelets and endothelial cells, which leads to increased cell-to-cell interactions [83]. NO inhibits PKC directly [84] or via stimulation of cGMP [85], which would be expected to result in inhibition of P-selectin expression and neutrophil adherence during ischemia-reperfusion. Indeed, inhibition of PKC with TMS (N,N,N-trimethyl sphingosine) results in (i) inhibition of leucocyte rolling and adherence induced by L-NAME in mesenteric microvessels and attenuation of P-selectin expression on platelets and endothelial cells, and (ii) attenuation of neutrophil accumulation following myocardial ischemia-reperfusion [86].

NO and Inflammation

Several studies suggest that NO can suppress inflammation and regulate the synthesis of pro-inflammatory proteins and CAMs. Following stimulation, endothelial cells in culture show decreased expression of VCAM-1 protein (and mRNA) with NO donors and increased expression of VCAM-1 with NO inhibition using L-NMMA [87]. Following stimulation of endothelial cells with TNFα, NO donors attenuate surface expression of VCAM-1, ICAM-1 and E-selectin [88]. Incubation of cultured human iliac vein endothelial cells with L-NAME results in increased expression of P-selectin (protein and mRNA) which are reversed by L-arginine or an NO donor [89]. Studies of ischemia-reperfusion in the small intestine showed that expression of P-selectin increased after 20 minutes of ischemia and 10 minutes of reperfusion, peaked after 5 hours of reperfusion and normalized by 24 hours [90].

Other studies showed that NO regulation of pro-inflammatory proteins after ischemia-reperfusion involves transcription factors such as the nuclear factor-κB (NF-κB) and re-oxygenation is a key stimulus [80,91]. Thus, NF-κB induces the rapid expression of several pro-inflammatory genes involved in ischemia-reperfusion, including TNFα, interleukins, chemokines and CAMs [91]. Stimulation of cultured endothelial cells with pro-inflammatory cytokines results in increased NF-κB activity, degradation of the inhibitory protein IκBα, and surface expression of VCAM-1, ICAM-1 and E-selectin while NO donors lead to decreased NF-κB activity, degradation of IκBα, and expression of VCAM-1, ICAM-1 and E-selectin [88].

Several studies have shown that OFRs stimulate PKC and activate NF-κB [92] and this response is attenuated by NO donors [80]. It appears that NO acts as an antioxidant and inactivates OFRs, increases levels of inhibitory proteins, and prevents the formation of substances that activate NF-κB and enhance the activity of proteins that inhibit NF-κB [93]. Decreased NO after ischemia-reperfusion has been suggested to have the opposite effect, resulting in increased formation of proinflammatory enzymes, cytokines and adhesion molecules.

NO and NO Donors in Ischemia-Reperfusion

Since ischemia-reperfusion is characterized by an NO deficit, a logical therapeutic approach is to replace or restore physiological levels of NO. However, how much NO is good and safe remains to be defined.

Infusion of NO Gas. Infusion of a solution of NO gas was first shown to limit reperfusion injury of the splanchnic bed and decrease myocardial depressant factor in cats [94]. In the feline model of ischemia-reperfusion (90 minutes of ischemia, 4.5 hours of reperfusion), infusion of NO solution beginning 10 minutes before reperfusion and continued throughout reperfusion limited necrosis and neutrophil infiltration without changes in arterial pressure or heart rate [95]. No significant negative inotropic effect was associated with infusion of NO or NO donors *in vivo* [96,97]. Inhaled NO at low concentrations (20–80 ppm) is beneficial in hypoxic pulmonary hypertension [98] and inhibits leucocyte adherence in the non-ischemic bed after ischemia-reperfusion [99].

NO Donors. NO donors, which release NO in solution, include organic nitrates (e.g. nitroglycerin; sodium nitroprusside), sydnonimines (e.g. 3-morpholinosydnonimine or SIN-1), cysteine-containing NO donors (e.g. SPM-5185), and NONOates (e.g. DEA/NO; SPER/NO). Intravenous nitroglycerin infusion given in low dose before, during and after coronary reperfusion (intracoronary streptokinase and/or angioplasty) to patients after anterior myocardial infarction was shown to recruit left ventricular function, and accelerate recovery of left ventricular function, suggesting decreased myocardial 'stunning' [20,21,100]. These studies suggested that early administration of NO donors before reperfusion might widen the therapeutic window for reperfusion therapy and "buy more time" until therapy can be applied [20,21,100].

Experimentally in the cat model, acidified $NaNO_2$ limited necrosis and polymorphonuclear neutrophil infiltration *in vivo* and platelet aggregation *in vitro* after ischemia-reperfusion [101]. The NO donors SIN-1 and C87-3754 in low doses also limited necrosis in the cat [102]. The beneficial effects of the NO donors was associated with an attenuation of coronary vasoconstriction [103]. The S-nitrosated tissue plasminogen activator (NO-tPA) also limited injury in the cat model [104]. NO donors were shown to suppress arrhythmias after myocardial infarction in the pig model [105]. Inhibition of NO synthesis with N^G-nitro-L-arginine (L-NNA) enhances myocardial stunning in conscious dogs [106]. The overall findings suggest that low doses of NO donors are beneficial after ischemia-reperfusion.

L-Arginine. L-arginine, which provides the substrate for NO synthesis, has also been shown to exert beneficial effects in ischemia-reperfusion. Thus, infusion of L-arginine limits myocardial necrosis and neutrophil infiltration after ischemia-reperfusion in cats [61] and dogs [107]. The L-arginine-induced improvement in myocardial contractility in isolated rat hearts is lost in the absence of neutrophils, suggesting that the protection is due to prevention of neutrophil-induced contractile dysfunction [108]. Since L-arginine reduces ventricular arrhythmias after myocardial infarction [109], it may also reduce reperfusion arrhythmias. Long-term arginine supplementation in humans also improves endothelial function in small coronary vessels [110].

Excessive NO and Peroxynitrite. Formation of the highly reactive peroxynitrite ($ONOO^-$) oxidant from NO and superoxide [111] can be a two-edged sword. This quenching of NO by superoxide might in part explain the decrease in myocardial NO after ischemia-reperfusion [61]. However, evidence indicates that high concentrations of peroxynitrite are cytotoxic [112]. Indeed, peroxynitrite has been suggested to mediate ischemia-reperfusion injury in the splanchnic bed [113] and myocardium [114,115]. In isolated rat hearts *in vitro*, peroxynitrite concentrations ranging from 30 to 100 μM have been shown to cause reperfusion injury [115]. In both rat and dog hearts with ischemia-reperfusion, the concentration of nitrotyrosine, regarded as a "footprint" of peroxynitrite formation, also increase [115–117].

The idea that peroxynitrite mediates reperfusion injury has generated considerable controversy. There are 5 main opposing arguments [36]: (i) the levels of NO (1–10 nM) [118,119] and superoxide (0.4–5 μM) [120], even when inducible NO synthase (iNOS) is maximally activated, are low and the half-life of superoxide is short (<1 second), making *in vivo* concentrations of 1–2 μM unlikely; even with ischemia-reperfusion, the peroxynitrite concentration is about 100 nM [114,121]; (ii) NO and superoxide exist in nearly equimolar concentrations and an imbalance leads to feedback inhibition of peroxynitrite generation [122]; (iii) the lack of a direct *in vivo* method to measure peroxynitrite has led to the use of the immunohistochemistry method for the detection of nitrotyrosine. However, this method also detects nitrotyrosine from other sources including chloride ions, hypochlorous acid, myeloperoxidase, and other nitrogenous radicals [123,124]; (iv) physiologic concentrations of peroxynitrite (400 nM to 2 μM) are associated with cardioprotection and limit leucocyte endothelium interactions after ischemic injury in rats [125] and cats [126]; (v) peroxynitrite releases NO *in vivo*, as NO donors [127], and NO is protected by nitrosation onto carrier plasma proteins [128,129].

The arguments in favor of the detrimental effects of peroxynitrite on ischemia-reperfusion injury include: (i) studies in the rat using electron paramagnetic resonance spin demonstrating peroxynitrite formation [114]; (ii) early detection of nitrotyrosine following ischemia-reperfusion in the isolated rat heart [115], *in vivo* rat and dog models of ischemia-reperfusion [116,117] and the human heart during cardiopulmonary bypass [130]; (iii) blocking peroxynitrite is cardioprotective [115,116]; (iv) NOS inhibitors at low concentrations improve mechanical function after ischemia-reperfusion [115,131]; (v) infusions of peroxynitrite (40 and 80 μM, respectively) in isolated working rat hearts result in mechanical dysfunction [132,133].

Whether peroxynitrite is protective or harmful appears to depend on the dose. Cumulative evidence indicates that there are other sources of NO than from the endothelium. Thus, there are 2 classes of NO synthase (NOS) enzymes [134]: two constitutive NOS isoforms found mainly in endothelial cells (eNOS) and neurons (nNOS) release small amounts of NO for short periods of time, while an inducible isoform (iNOS), found in macrophages releases large amounts of NO for prolonged periods of time. Both eNOS and iNOS are also found in cardiomyocytes [135,136]. An excess of NO is cytotoxic and is expected to increase reperfusion injury. Recently, iNOS mRNA was shown to be present in the myocardium of patients with coronary artery disease [137]. During reperfusion, excess NO and OFRs, peroxynitrite formation not only increases but may be sustained, especially as superoxide dismutase activity has been suggested to react with glutathione or other endogenous thiols to produce NO donors that can induce cardioprotection [135,138].

Genetic Models of NO Synthases and Ischemia-Reperfusion. Studies using eNOS gene knockout mice and gene transfer, support the role of endothelial NO in preventing leucocyte-endothelial interaction and ischemia-reperfusion injury. Mice with eNOS deletion are hypertensive and fail to respond to endothelium-dependent vasodilators [139–141], develop adverse vascular remodeling (intimal proliferation, decreased lumen size and flow) [142], and increased necrosis after ischemia-reperfusion and increased expression of P-selectin [143]. Several recent studies have shown that eNOS-derived NO is cardioprotective during ischemia-reperfusion [144] and decreases infarct size [145,146]. Enhanced eNOS by transfection of the eNOS gene using Sendai-virus vector is also beneficial and inhibits neointimal lesions after balloon injury [147].

Interestingly, iNOS knock-out mice show altered immune responses but no abnormalities in baseline phenotype [148–150]. In addition, iNOS knock-out mice show decreased vascular remodeling [151,152] but no significant role of iNOS in early ischemia-reperfusion [153]. Overexpression of iNOS in transgenic mice was associated with increased peroxynitrite generation as well as increased brady-arrhythmias and sudden deaths [154]. However, another study suggested that iNOS overexpression did not result in severe cardiac dysfunction [155]. These paradoxes continue to be investigated.

NO Synthesis Inhibition and Ischemia-Reperfusion. Studies using inhibitors of NO synthesis by the 3 NOS isoforms (eNOS, nNOS and iNOS) support the concept of the protective role of NO on vascular function and cardioprotection during ischemia-reperfusion. In isolated rabbit hearts with ischemia-reperfusion (35 minutes of ischemia; 30 minutes of reperfusion; leucocyte-free perfusate), L-NAME induces cardioprotection [131]. In isolated rat hearts perfused with polymorphonuclear neutrophils containing perfusate, L-NAME increases cardiac dysfunction during ischemia-reperfusion and an NO donor improved function [108]. The NOS inhibitor L-NNA worsened hypoperfusion in the dog model with coronary stenosis [156]. Also in the dog, NOS inhibition with L-NNA worsened stunning after ischemia-reperfusion [106].

Angiotensin II, Bradykinin and NO During Ischemia-Reperfusion

Angiotensin II (AngII) is the major effector molecule of the renin-angiotensin system (RAS) and is upregulated during ischemia-reperfusion [157], myocardial infarction [158] and heart failure [159]. Cumulative evidence indicates that the effects of excess AngII are deleterious and contribute substantially to the pathophysiology of these conditions so that blocking the effects of AngII might be beneficial [160]. The endothelium contains the complete RAS and mediates vascular AngII formation [161]. AngII stimulates the membrane-bound NAD/NADPH oxidase system to generate superoxide which inactivates NO and leads to the formation of peroxynitrite, thereby causing impaired endothelial-dependent relaxation, thrombosis, inflammation and other deleterious effects [162–164]. The effects of AngII are mediated mainly through the AngII type 1 (AT_1) and type 2 (AT_2) receptors [165]. The AT_2 receptor is normally abundant in the fetus but is re-expressed in adult hearts after ischemia-reperfusion [166–172], myocardial infarction [173] and heart failure [174]. The AT_2 receptor antagonizes the effects of the AT_1 receptor, thereby exerting vasodilatory, pro-apoptotic, antigrowth, and antihypertrophic effects [165].

Extensive evidence suggests that the beneficial effects of angiotensin-converting enzyme (ACE) inhibition are not only due to inhibition of AngII formation but in large part to increased bradykinin levels and bradykinin-induced release of NO and prostacyclin [175–182]. AT_1 receptor antagonists block the effects of AngII mediated via this specific receptor and additionally results in unopposed AT_2 receptor stimulation, which leads to increased bradykinin, NO and cGMP [168,183–185] and PKCε activation [168,185]. This increase in NO is probably derived from eNOS. Moreover, in some studies, AT_1 receptor blockade did not increase flow despite an overall cardioprotective effect after ischemia-reperfusion [186] or failed to improve function [187,188] or decrease infarct size [189]. However, AT_2 receptor blockade during *in vitro* ischemia-reperfusion also upregulated AT_2 receptor expression [166,188], triggered an increase in PKCε expression and cGMP levels [166] and induced cardioprotection [166,188]. Cardioprotection during ischemic preconditioning was also suggested to involve NO and PKC signaling [190]. NO was recently shown to downregulate the AT_1 receptor [191].

The hypothesis that AT_2 receptor activation and signaling through bradykinin, PKCε, NO and cGMP might contribute to cardioprotection during AT_1 receptor blockade was first tested in an *in vivo* dog model of ischemia-reperfusion [171]. AT_1 receptor blockade was produced by candesartan (CV-11974, 1 mg/kg over 30 min i.v.) given before ischemia-reperfusion and was confirmed by inhibition of AngII pressor responses (0.25 μg/kg I.V.) at baseline and after drug infusions. The animals were randomized to 7 groups: sham controls; ischemia-reperfusion controls; ischemia-reperfusion and candesartan; AT_2 receptor blockade (PD 123319, 3 mg/kg/min i.c. for 30 min) followed by candesartan and ischemia-reperfusion; PKC inhibitor (chelerythrine, 17 μg/kg/min i.c. for 30 min) followed by candesartan and ischemia-reperfusion; NOS inhibition (L-NMMA, 75 mg/kg/min i.c. for 30 min) followed by candesartan and ischemia-reperfusion; bradykinin inhibition (HOE-140, 10 ng/kg/min i.c. for 30 min) followed by candesartan and ischemia-reperfusion. The intracoronary (i.c.) injections were made via a catheter into the occluded bed distal to the site of the occlusion site of the left anterior descending coronary artery. Ischemia for 90 minutes was then produced by coronary occlusion and followed by reperfusion for 120 minutes. The results showed that candesartan alone improves global systolic and diastolic function, limits acute LV remodeling, decreases infarct size and regionally increases AT_2 receptor and PKCε proteins as well as cGMP in the ischemic zone. Importantly, the AT_2 receptor antagonist and the inhibitors of BK, PKCε and NOS attenuated these beneficial effects of candesartan as well as the regional increase in AT_2 receptor and PKCε proteins. The overall findings support the hypothesis that AT_2 receptor activation and downstream signaling through bradykinin, PKCε, NO and cGMP play a significant role in the cardioprotective effect of AT_1 receptor blockade during ischemia-reperfusion (Fig. 2).

AngII Blockade and NO in Heart Failure

The major ongoing clinical trials of AT_1 receptor blockers have been reviewed elsewhere [192]. A major drawback of the clinical trials is that AT_1 receptor blockers are being tested on top of a background therapy, including ACE inhibitors, reperfusion and other agents. The ELITE [Evaluation of Losartan in the Elderly II] [193] and the RESOLVD [Randomized Evaluation of Strategies for Left Ventricular Dysfunction] study using candesartan [194] in heart failure patients did not show a dramatic advantage of AT_1 receptor blockade over ACE inhibition. The Val-HeFT [Valsartan in Heart Failure Trial] study [195] showed that 'valsartan significantly reduces the combined end point of mortality and morbidity and improves clinical signs and symptoms in patients with heart failure, when added to prescribed therapy'.

Heart Failure and NO

Heart failure is associated with a host of abnormalities which involve NO, such as increased oxidative stress, oxygen free radicals, inflammation and cytokines. Cytokines and increased iNOS activity in heart failure [196,197] have been linked to the attenuated response to β-adrenergic stimulation [198]. The negative inotropic effect of NO during β-adrenergic stimulation [199] appears to be due to cGMP-mediated inhibition of Ca^{++} influx

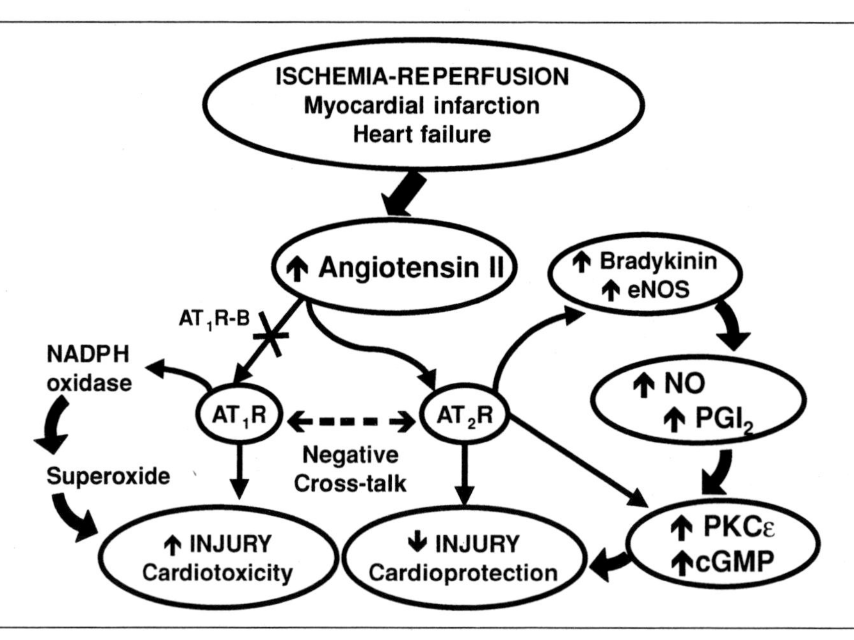

Fig. 2. Diagram of the postulated role of AT_2 receptor (AT_2R) signaling and NO in cardioprotection with AT_1 receptor (AT_1R) blockade (AT_1R-B) during ischemia-reperfusion. PGI_2 = prostacyclin.

by the L-type voltage-dependent Ca^{++} channels [200]. In a recent study, left ventricular eNOS and iNOS gene expression correlated with the severity of left ventricular dysfunction [201]. In end-stage heart failure, NO production in the microcirculation was shown to be reduced [202], suggesting impaired NO formation by eNOS. However, high and sustained levels of NO, produced by iNOS stimulated by cytokines, appear to interact with OFRs to form peroxynitrite. This interaction has been suggested to lead to an imbalance in cellular Bak and Bcl-xL and promote apoptosis [203]. In contrast, pulses of low amounts of NO were shown to suppress apoptosis [204]. The role of apoptosis in cardiovascular disease has been reviewed [205] and cardiomyocyte apoptosis has been linked to ventricular dysfunction after ischemia-reperfusion [206], although its role in reperfused myocardial infarction needs further study. It is important to note that β-blockers and ACE-inhibitors, which are commonly used for the management of myocardial infarction and heart failure, increase eNOS expression [197]. The effects of such therapies on iNOS and apoptosis also deserve further study.

Conclusion

Although it is not possible to acknowledge all the publications on the effects of NO in ischemia-reperfusion, it is clear that whether NO is cardioprotective or detrimental depends on the dose as well as spatial and temporal factors. Ischemia-reperfusion is associated with the activation of a multitude of proteins, peptides, cytokines, growth factors, active free radical and signaling molecules, several of which interact with NO. The interactions between NO and angiotensin II, oxygen free radicals and inflammatory cytokines are complex. In general, small amounts of NO for short periods from eNOS are beneficial whereas large amounts of NO for sustained periods from iNOS are harmful. The NO molecule is characterized by considerable versatility and diversity and these properties lead to biphasic or bifunctional responses depending on the conditions. Some of the controversial effects of NO can be explained by differences in the severity and extent of the ischemia-reperfusion injury. Thus, large areas of severe injury, associated with very low flow and more prolonged ischemia, would be expected to induce more severe microvascular injury and inflammatory responses, greater release of OFRs and cytokines, more peptides such as AngII, greater activation of iNOS, formation of more peroxynitrite and matrix metalloproteinases, and more cellular and structural disruption. This hypothesis is supported by the correlation found between expression of iNOS and eNOS genes with the degree of left ventricular dysfunction in dilated cardiomyopathy patients [201] and the bifunctional regulation of apoptosis by NO [204]. However, this hypothesis needs to be verified in ischemia-reperfusion. Since small pulses of NO are beneficial, especially during reperfusion after brief periods of ischemia as in ischemic preconditioning [190], therapies that enhance NO availability by increasing endothelial eNOS might be effective in recruiting left ventricular function in reperfused infarction and heart failure. Whether decreasing iNOS-derived NO might be beneficial also deserves to be studied.

Acknowledgments

We are grateful for the assistance Catherine Jugdutt for manuscript preparation.

This study was supported in part by a grant from the Canadian Institutes for Health Research, Ottawa, Ontario, Canada.

References

1. Ryan TJ, Antman EM, Brooks NH, et al. Update: ACC/AHA guidelines for the management of patients with acute myocardial infarction. A report of the American College of Cardiology/American Heart Association Task Force on practice guidelines (Committee on management of acute myocardial infarction). *J Am Coll Cardiol* 1999;34:890–911.
2. Hearse DJ. Reperfusion of the ischemic myocardium. *J Mol Cell Cardiol* 1997;9:605–616.
3. Braunwald E, Kloner RA. The stunned myocardium: Prolonged, post-ischemic ventricular dysfunction. *Circulation* 1982;66:1146–1149.
4. Braunwald E, Kloner RA. Myocardial reperfusion: A double-edged sword? *J Clin Invest* 1985;76:1713–1719.
5. Topol EJ. Early myocardial reperfusion: An assessment of current strategies in acute myocardial infarction. *Eur Heart J* 1996;17(Suppl. E):42–48.
6. Kloner RA, Jennings RB. Consequences of brief ischemia: Stunning, preconditioning, and their clinical implications. Part 1. *Circulation* 2001;104:2981–2989.
7. Kloner RA, Jennings RB. Consequences of brief ischemia: Stunning, preconditioning, and their clinical implications. Part 2. *Circulation* 2001;104:3158–3167.
8. Przyklenk K, Kloner RA. 1986. Superoxide dismutase plus catalase improve contractile function in the canine model of the "stunned myocardium." *Circ Res* 1986;58:148–156.
9. Bolli R, Jeroudi MO, Patel BS, et al. Marked reduction of free radical generation and contractile dysfuncrtion by antioxidant therapy begun at the time of reperfusion: Evidence that myocardial "stunning" is a manifestation of reperfusion injury. *Circ Res* 1989;65:607–622.
10. Bolli R. Mechanisms of myocardial "stunning." *Circulation* 1990;82:723–738.
11. Beckman JS, Koppenol WH. Nitric oxide, superoxide, and peroxynitrite: The good the bad, and the ugly. *Am J Physiol* 1996;271:C1424–C1437.
12. Becker LC, Jeremy RW, Schaper J, Schaper W. Ultrastructural assessment of myocardial necrosis occurring during ischemia and 3-h reperfusion in the dog. *Am J Physiol* 1999;277:H243–H252.
13. Hochman JS, Choo H. Limitation of myocardial infarct expansion by reperfusion independent of myocardial salvage. *Circulation* 1987;75:299–306.
14. Boyle MP, Weisman HF. Limitation of infarct expansion and ventricular remodeling by late reperfusion. Study of time course and mechanism in a rat model. *Circulation* 1993;88:2872–2883.
15. Jugdutt BI. Effect of reperfusion on ventricular mass, topography and function during healing of anterior infarction. *Am J Physiol (Heart and Circulatory Physiology)* 1997;272: H1205–H1211.
16. Kim CB, Braunwald E. Potential benefits of late reperfusion of infarcted myocardium. The open artery hypothesis. *Circulation* 1993;88:2426–2436.
17. Jugdutt BI, Khan MI, Jugdutt SJ, Blinston GE. Impact of left ventricular unloading after late reperfusion of canine anterior myocardial infarction on remodeling and function using isosorbide-5-mononitrate. *Circulation* 1995;92:926–934.
18. Ellis SG, Henschke CL, Sandor T, Wynne J, Braunwald E, Kloner RA. Predictors of success for coronary angioplasty performed for AMI. *J Am Coll Cardiol* 1988;12:1407–1415.
19. Ambrosio G, Becker LC, Hutchins GM, Weisman HR, Weisfeldt ML. Reduction in experimental infarct size by recombinant human superoxide dismutase: Insights into the pathophysiology of reperfusion injury. *Circulation* 1986;74:1424–1433.
20. Jugdutt BI. Nitroglycerin. In: Bates E, ed. Thrombolysis and Adjunctive Therapy for Myocardial Infarction. New York: Dekker, 1992:119–144.
21. Jugdutt BI, Schwarz-Michorowski BL, Tymchak WJ, Burton JR. Prompt improvement of left ventricular function and topography with combined reperfusion and intravenous nitroglycerin in acute myocardial infarction. *Cardiology* 1997;88:170–179.
22. Jugdutt BI, Humen DP. Limitation of left ventricular hypertrophy by ACE inhibition after anterior Q-wave myocardial infarction. *Cardiology* 1998;89:283–290.
23. Jugdutt BI, Tymchak WJ, Humen DP, Gulamhusein S, Tang SB. Effect of thrombolysis and prolonged captopril and nitroglycerin on infarct size and remodeling in transmural myocardial infarction. *J Am Coll Cardiol* 1992;19(Suppl. A):205A (Abstract).
24. Ragosta M, Camarano G, Kaul S, Powers ER, Sarembock IJ, Gimple LW. Microvascular integrity indicates myocellular viability in patients with recent myocardial infarction. New insights using myocardial contrast echocardiography. *Circulation* 1994;89:2562–2569.
25. Ito H, Okamura A, Iwakura K, Mauyama T, Hori M, Takiuchi S, Negoro S, Nakatsuchi Y, Taniyama Y, Higashino Y, Fuji K, Minamino T. Myocardial perfusion patterns related to thrombolysis in myocardial infarction grades after coronary angioplasty in patients with acute anterior wall myocardial infarction. *Circulation* 1996;93:1933–1999.
26. Kloner RA, Ganote CE, Jennings RB. The "no-reflow" phenomenon after temporary coronary occlusion in the dog. *J Clin Invest* 1974;54:1496–1508.
27. White FC, Sanders M, Bloor CM. Regional redistribution of myocardial blood flow after coronary occlusion and reperfusion in the conscious dog. *Am J Cardiol* 1978;42:234–243.
28. Ito H, Tomooka T, Sakai N, Yu H, Higashino Y, Fujii K, Masuyama T, Kitabatake A, Minamino T. Lack of myocardial perfusion immediately after successful thrombolysis. A predictor of poor recovery of left ventricular function in anterior myocardial infarction. *Circulation* 1992;85:1699–1705.

29. Cobb FR, Bache RJ, Rivas F, Greenfield JC Jr. Local effects of acute cellular injury on regional myocardial blood flow. *J Clin Invest* 1976;57:1359–1368.
30. Ambrosio G, Weisman HF, Mannisi JA, Becker LC. Progressive impairment of regional myocardial perfusion after initial restoration of postischemic blood flow. *Circulation* 1989;80:1846–1861.
31. Heyndrickx GR, Amano J, Patrick TA, Manders WT, Rogers GG, Rosendorff C, Vatner SF. Effects of coronary artery reperfusion on regional myocardial blood flow and function in conscious baboons. *Circulation* 1985;71:1029–1037.
32. Willerson JT, Watson JT, Hutton I, Templeton GH, Fixler DE. Reduced myocardial reflow and increased coronary vascular resistance following prolonged myocardial ischemia in the dog. *Circ Res* 1975;36:771–781.
33. Johnson WB, Malone SA, Pantely GA, Anselone CG, Bristow JD. No reflow and extent of infarction during maximal vasodilation in the porcine heart. *Circulation* 1988;78:462–472.
34. Kloner RA, Rude RE, Carlson N, Maroko PR, DeBoer LW, Braunwald E. Ultrastructural evidence of microvascular damage and myocardial cell injury after coronary artery occlusion: Which comes first? *Circulation* 1980;62:945–952.
35. Tsao PS, Aoki N, Lefer DJ, Johnson G III, Lefer AM. Time course of endothelial dysfunction and myocardial injury during myocardial ischemia and reperfusion in the cat. *Circulation* 1990;82:1402–1412.
36. Lefer AM, Hayward R. The role of nitric oxide in ischemia-reperfusion. In: Loscalzo J, Vita JA, eds. *Contemporary Cardiology, vol 4: Nitric Oxide and the Cardiovascular System*. Humana Press Inc., 2000:357–380.
37. Hayward R, Lefer AM. Time course of endothelial-neutrophil interaction in splanchnic artery ischemia-reperfusion. *Am J Physiol* 1998;275:H2080–H2086.
38. Viehman GE, Ma XL, Lefer DJ, Lefer AM. Time course of endothelial dysfunction and myocardial injury during coronary arterial occlusion. *Am J Physiol* 1991;261:H874–H881.
39. Arminger LC, Gavin JB. Changes in the microvasculature of ischemia and infarcted myocardium. *Lab Invest* 1975;33:51–56.
40. Mehta JL, Nichols WW, Donnelly WH, Lawson DL, Saldeen TG. Impaired canine coronary vasodilator response to acetylcholine and bradykinin after occlusion-reperfusion. *Circ Res* 1989;64:43–54.
41. Lefer AM, Tsao P, Aoki N, Palladino MA Jr. Mediation of cardioprotection by transforming growth factor-beta. *Science* 1990;249:61–64.
42. Zweier JL, Kuppusamy P, Lutty GA. Measurement of endothelial cell free radical generation: Evidence for a central mechanism of free radical injury in postischemic tissues. *Proc Natl Acad Sci USA* 1988;85:4046–4050.
43. Chambers DE, Parks DA, Patterson G, Roy R, McCord JM, Yoshida S, Parmley LF, Downey JM. Xanthine oxidase as a source of free radical damage in myocardial ischemia. *J Mol Cell Cardiol* 1985;17:145–152.
44. McCord JM. Oxygen-derived free radicals in postischemic tissue injury. *N Engl J Med* 1985;312:159–163.
45. Bolli R, Patel BS, Jeroudi MO, Lai EK, McCay PB. Demonstration of free radical generation in "stunned" myocardium of intact dogs with the use of the spin trap α-phenyl N-tert-butyl nitrone. *J Clin Invest* 1988;82:476–485.
46. Zweier JL. Measurement of superoxide-derived free radicals in the reperfused heart. Evidence for a free radical mechanism of reperfusion injury. J Biol Chem 1988;263:1353–1357.
47. Lamb FS, King CM, Harrell K, Burkel W, Webb RC. Free radical-mediated endothelial damage in blood vessels after electrical stimulation. *Am J Physiol* 1987;252:H1041–H1046.
48. Stewart DJ, Pohl U, Bassenge E. Free radicals inhibit endothelium-dependent dilation in the coronary resistance bed. *Am J Physiol* 1988;255:H765–H769.
49. Tsao PS, Lefer AM. Time course and mechanism of endothelial dysfunction in isolated ischemic- and hypoxic-perfused rat hearts. *Am J Physiol* 1990;259:H1660–H1666.
50. Gryglewski RJ, Palmer RM, Moncada S. Superoxide anion is involved in the breakdown of endothelium-derived vascular relaxing factor. *Nature* 1986;320:454–456.
51. Litt MR, Jeremy RW, Weisman HF, Winkelstein JA, Becker LC. Neutrophil depletion limited to reperfusion reduces myocardial infarct size after 90 minutes of ischemia. Evidence for neutrophil-mediated reperfusion injury. *Circulation* 1989;80:1816–1827.
52. Granger DN, Benoit JN, Suzuki M, Grisham MB. Leukocyte adherence to venular endothelium during ischemia-reperfusion. *Am J Physiol* 1989;257:G683–G688.
53. Tsao PS, Ma XL, Lefer AM. Activated neutrophils aggravate endothelial dysfunction after reperfusion of the ischemic feline myocardium. *Am Heart J* 1992;123:1464–1471.
54. Ma XL, Tsao PS, Viehman GE, Lefer AM. Neutrophil-mediated vasoconstriction and endothelial dysfunction in low-flow perfusion-reperfused cat coronary artery. *Circ Res* 1991;69:95–106.
55. Werns SW, Lucchesi BR. Free radicals and ischemic tissue injury. *Trends Pharmacol Sci* 1990;11:161–166.
56. Tsao PS, Lefer AM. Recovery of endothelial function following myocardial ischemia and reperfusion in rats. *J Vasc Med Biol* 1991;3:5–10.
57. Ignarro LJ, Buga GM, Wood KS, Byrns RE, Chaudhuri G. Endothelium-derived relaxing factor produced and released from artery and vein is nitric oxide. *Proc Natl Acad Sci USA* 1987;84:9265–9269.
58. Palmer RM, Ferrige AG, Moncada S. Nitric oxide release accounts for the biological activity of endothelium-derived relaxing factor. *Nature* 1987;327:524–526.
59. Moncada S, Palmer RM, Higgs EA. Nitric oxide: Physiology, pathophysiology, and pharmacology. *Pharmacol Rev* 1991;43:109–142.
60. Rees DD, Palmer RM, Schulz R, Hodson HF, Moncada S. Characterization of three inhibitors of endothelial nitric oxide synthase *in vitro* and *in vivo*. *Br J Pharmacol* 1990;101:746–752.
61. Davenpeck KL, Gauthier TW, Lefer AM. Inhibition of endothelial-derived nitric oxide promotes P-selectin expression and actions in the rat microcirculation. *Gastroenterology* 1994;107:1050–1058.
62. Ma XL, Weyrich AS, Lefer DJ, Lefer AM. Diminished basal nitric oxide release after myocardial ischemia and reperfusion promotes neutrophil adherence to coronary endothelium. *Circ Res* 1993;72:403–412.

63. Weyrich AS, Ma XL, Lefer AM. The role of L-arginine in ameliorating reperfusion injury after myocardial ischemia in the cat. *Circulation* 1992;86:279–288.
64. Lefer DJ, Nakanishi K, Vinten-Johansen J, Ma XL, Lefer AM. Cardiac venous endothelial dysfunction after myocardial ischemia and reperfusion in dogs. *Am J Physiol* 1992;263:H850–H856.
65. Kubes P, Suzuki M, Granger DN. Nitric oxide: An endogenous modulator of leukocyte adhesion. *Proc Natl Acad Sci USA* 1991;88:4651–4655.
66. Tiefenbacher CP, Chilian WM, Mitchell M, DeFily DV. Restoration of endothelium-dependent vasodilation after reperfusion injury by tetrahydrobiopterin. *Circulation* 1996;94:1423–1429.
67. Kugiyama K, Yasue H, Okumura K, Ogawa H, Fujimoto K, Nakao K, Yoshimura M, Motoyama T, Inobe Y, Kawano H. Nitric oxide activity is deficient in spasm arteries of patients with coronary spastic angina. *Circulation* 1996;94:266–271.
68. Butcher EC. Leukocyte-endothelial cell recognition: Three (or more) steps to specificity and diversity. *Cell* 1991;67:1033–1036.
69. Weyrich AS, Buerke M, Albertine KH, Lefer AM. Time course of coronary vascular endothelial adhesion molecule expression during reperfusion of the ischemic feline myocardium. *J Leukoc Biol* 1995;57:45–55.
70. Spertini O, Kansas GS, Munro JM, Griffin JD, Tedder TF. Regulation of leukocyte migration by activation of the leukocyte adhesion module-1 (LAM-1) selectin. *Nature* 1991;349:691–694.
71. Kishimoto TK, Jutila M, Berg EL, Butcher EC. Neutrophil MAC-1 and MEL-14 adhesion proteins inversely regulated by chemotactic factors. *Science* 1989;245:1238–1241.
72. Kubes P, Jutila M, Payne D. Therapeutic potential of inhibiting leukocyte rolling in ischemia-reperfusion. *J Clin Invest* 1995;95:2510–2519.
73. Suzuki M, Inauen W, Kvietys PR, Grisham MB, Meininger C, Schelling ME, et al. Superoxide mediates reperfusion-induced leukocyte-endothelial cell interactions. *Am J Physiol* 1989;257:H1740–H1745.
74. Ma XL, Tsao PS, Lefer AM. Antibody to CD-18 exerts endothelial and cardiac protective effects in myocardial ischemia and reperfusion. *J Clin Invest* 1991;88:1237–1243.
75. Ma Xl, Lefer DJ, Lefer AM, Rothlein R. Coronary endothelial and cardiac protective effects of a monoclonal antibody to intercellular adhesion molecule-1 in myocardial ischemia and reperfusion. *Circulation* 1992;86:937–946.
76. Zhao ZQ, Lefer DJ, Sato H, Hart KK, Jeffords PR, Vinten-Johansen J. Monoclonal antibody to ICAM-1 preserves postischemic blood flow and reduces infarct size after ischemia-reperfusion in the rabbit. *J Leukoc Biol* 1997;62:292–300.
77. Gumina RJ, Schultz JE, Yao Z, Kenny D, Warltier DC, Newman PJ, et al. Antibody to platelet/endothelial cell adhesion molecule-1 reduces myocardial infarct size in a rat model of ischemia-reperfusion injury. *Circulation* 1996;94:3327–3333.
78. Murohara T, Delyani JA, Albeda SM, Lefer AM. Blockade of platelet endothelial cell adhesion molecule-1 protects against myocardial ischemia and reperfusion injury in cats. *J Immunol* 1996;156:3550–3557.
79. Murohara T, Guo JP, Lefer AM. Cardioprotection by a novel recombinant serine protease inhibitor in myocardial ischemia and reperfusion injury. *J Pharm Exp The* 1995;274:1246–1253.
80. Kupatt C, Weber C, Wolf DA, Becker BF, Smith TW, Kelly RA. Nitric oxide attenuates reoxygeneration-induced ICAM-1 expression in coronary microvascular endothelium: Role of NF-κB. *J Mol Cell Cardiol* 1997;29:2599–2609.
81. Gaboury JP, Anderson DC, Kubes P. Molecular mechanisms involved in superoxide-induced leukocyte-endothelial cell interactions *in vivo*. *Am J Physiol* 1994;266:H637–H642.
82. Mataki H, Inagaki T, Yokoyama M, Maeda S. ICAM-1 expression and cellular injury in cultured endothelial cells under hypoxia/reoxygenation. *Kobe J Med Sci* 1994;40:49–63.
83. Geng JG, Bevilacqua MP, Moore KL, McIntyre TM, Prescott SM, Kim JM, Bliss GA, Zimmerman GA, McEver RP. Rapid neutrophil adhesion to activated endothelium mediated by GMP-140. *Nature* 1990;343:757–760.
84. Gopalakrishna R, Chen ZH, Gundimeda U. Nitric oxide and nitric oxide-generating agents induce a reversible inactivation of protein kinase C activity and phorbol ester binding. *J Biol Chem* 1993;268:27180–27185.
85. Takai Y, Kaibuchi K, Matsubara T, Nishizuka Y. Inhibitory action of guanosine 3′,5′-monophosphate on thrombin-induced phosphatidylinositol turnover and protein phosphorylation in human platelets. *Biochem Biophys Res Commun* 1981;101:61–67.
86. Scalia R, Murohara T, Delyani JA, Nossuli TO, Lefer AM. Myocardial protection by N,N,N-trimethylsphingosine in ischemia reperfusion injury is mediated by inhibition of P-selectin. *J Leukoc Biol* 1996;59:317–324.
87. De Caterina R, Libby P, Peng HB, Thannickal VJ, Rajavashisth TB, Gimbrone MA Jr, Shin WS, Liao JK. Nitric oxide decreases cytokine-induced endothelial activation. Nitric oxide selectively reduces endothelial expression of adhesion molecules and proinflammatory cytokines. *J Clin Invest* 1995;96:60–68.
88. Spiecker M, Darius H, Kaboth K, Hubner F, Liao JK. Differential regulation of endothelial cell adhesion molecule expression by nitric oxide donors and antioxidants. *J Leukoc Biol* 1998;63:732–739.
89. Armstead VE, Minchenko AG, Schuhl RA, Hayward R, Nossuli TO, Lefer AM. Regulation of P-selectin expression in human endothelial cells by nitric oxide. *Am J Physiol* 1997;273:H740–H746.
90. Eppihimer MJ, Russell J, Anderson DC, Epstein CJ, Laroux S, Granger DN. Modulation of P-selectin expression in the postischemic intestinal microvasculature. *Am J Physiol* 1997;273:G1326–G1332.
91. Barnes PJ. Nuclear factor-κB. *Int J Biochem Cell Biol* 1997;29:867–870.
92. Fialkow L, Chan CK, Grinstein S, Downey GP. Regulation of tyrosine phosphorylation in neutrophils by the NADPH oxidase. Role of reactive oxygen intermediates. *J Biol Chem* 1993;268:17131–17137.
93. Peng HB, Libby P, Liao JK. Induction and stabilization of IκBα by nitric oxide mediates inhibition of NF-κB. *J Biol Chem* 1995;270:14214–14219.
94. Aoki N, Johnson G III, Lefer AM. Beneficial effects of two forms of NO administration in feline splanchnic artery occlusion shock. *Am J Physiol* 1990;258:G275–G281.

95. Johnson G III, Tsao PS, Lefer AM. Cardioprotective effects of authentic nitric oxide in myocardial ischemia with reperfusion. *Crit Care Med* 1991;19:244–252.
96. Pennington DG, Vezeridis MP, Geffin G, O'Keefe DD, Lappas DG, Daggett WM. Quantitative effects of sodium nitroprusside on coronary hemodynamics and left ventricular function in dogs. *Circ Res* 1979;45:351–359.
97. Crystal GJ, Gurevicius J. Nitric oxide does not modulate myocardial contractility acutely in *in situ* canine hearts. *Am J Physiol* 1996;270:H1568–H1576.
98. Frostell C, Fratacci MD, Wain JC, Jones R, Zapol WM. Inhaled nitric oxide. A selective pulmonary vasodilator reversing hypoxic pulmonary vasoconstriction. *Circulation* 1991;83:2038–2047.
99. Fox-Robichaud A, Payne D, Hasan SU, Ostrovsky L, Fairhead T, Reinhardt P, Kubes P. Inhaled NO as a viable antiadhesive therapy for ischemia/reperfusion injury of distal microvascular beds. *J Clin Invest* 1998;101:2497–2505.
100. Tymchak WJ, Michorowski BL, Burton JR, Jugdutt BI. Preservation of left ventricular function and topography with combined reperfusion and intravenous nitroglycerin in acute myocardial infarction (Abstr). *J Am Coll Cardiol* 1988;11:90A.
101. Johnson G III, Tsao PS, Mulloy D, Lefer AM. Cardioprotective effects of acidified sodium nitrite in myocardial ischemia with reperfusion. *J Pharmacol Exp Ther* 1990;252:35–41.
102. Siegfried MR, Erhardt J, Rider T, Ma XL, Lefer AM. Cardioprotection and attenuation of endothelial dysfunction by organic nitric oxide donors in myocardial ischemia-reperfusion. *J Pharmacol Exp Ther* 1992;260:668–675.
103. Siegfried MR, Carey C, Ma XL, Lefer AM. Beneficial effects of SPM-5185, a cysteine-containing NO donor in myocardial ischemia-reperfusion. *Am J Physiol* 1992;263:H771–H777.
104. Delyani JA, Nossuli TO, Scalia R, Thomas G, Garvey DS, Lefer AM. S-nitrosylated tissue-type plasminogen activator protects against myocardial ischemia/reperfusion injury in cats: Role of the endothelium. *J Pharmacol Exp Ther* 1996;279:1174–1180.
105. Wainwright CL, Martorana PA. Pirsidomine, a novel nitric oxide donor, suppresses ischemic arrhythmias in anesthetized pigs. *J Cardiovasc Pharmacol* 1993;22:S44–S50.
106. Hasebe N, Shen YT, Vatner SF. Inhibition of endothelium-derived relaxing factor enhances myocardial stunning in conscious dogs. *Circulation* 1993;88:2862–2871.
107. Nakanishi K, Vinten-Johansen J, Lefer DJ, Zhao Z, Fowler WC III, McGee DS, Johnston WE. Intracoronary L-arginine during reperfusion improves endothelial function and reduces infarct size. *Am J Physiol* 1992;263:H1650–H1658.
108. Pabla R, Buda AJ, Flynn DM, Blesse SA, Shin AM, Curtis MJ, Lefer DJ. Nitric oxide attenuates neutrophil-mediated myocardial contractile dysfunction after ischemia and reperfusion. *Circ Res* 1996;78:65–72.
109. Fei L, Baron AD, Henry DP, Zipes DP. Intrapericardial delivery of L-arginine reduces the increased severity of ventricular arrhythmias during sympathetic stimulation in dogs with acute coronary occlusion: Nitric oxide modulates sympathetic effects on ventricular electrophysiological properties. *Circulation* 1997;96:4044–4049.
110. Lerman A, Burnett JC Jr, Higano ST, McKinley LJ, Holmes DR Jr. Long-term L-arginine supplementation improves small-vessel coronary endothelial function in humans. *Circulation* 1998;97:2123–2128.
111. Beckman JS, Beckman TW, Chen J, Marshall PA, Freeman BA. Apparent hydroxyl radical production by peroxynitrite: Implications for endothelial injury from nitric oxide and superoxide. *Proc Natl Acad Sci USA* 1990;87:1620–1624.
112. Radi R, Beckman JS, Bush KM, Freeman BA. Peroxynitrite-induced membrane lipid peroxidation: The cytotoxic potential of superoxide and nitric oxide. *Arch Biochem Biophys* 1991;288:481–487.
113. Szabo C, Salzman AL, Ischiropoulos H. Peroxynitrite-mediated oxidation of dihydrorhodamine 123 occurs in early stages of endotoxic and hemorrhagic shock and ischemia-reperfusion injury. *FEBS Lett* 1995;372:229–232.
114. Wang P, Zweier JL. Measurement of nitric oxide and peroxynitrite generation in the postischemic heart. Evidence for peroxynitrite-mediated reperfusion injury. *J Biol Chem* 1996;271:29223–29230.
115. Yasmin W, Strynadka KD, Schulz R. Generation of peroxynitrite contributes to ischemia-reperfusion injury in isolated rat hearts. *Cardiovasc Res* 1997;33:422–432.
116. Liu P, Hock CE, Nagele R, Wong PYK. Formation of nitric oxide, superoxide, and peroxynitrite in myocardial ischemia-reperfusion injury in rats. *Am J Physiol* 1997;272:H2327–H2336.
117. Zhang Y, Bissing JW, Xu L, Ryan AJ, Martin SM, Miller FJ Jr, Kregel KC, Buettner GR, Kerber RE. Nitric oxide synthase inhibitors decrease coronary sinus-free radical concentration and ameliorate myocardial stunning in an ischemia-reperfusion model. *J Am Coll Cardiol* 2001;38:546–554.
118. Kelm M, Schrader J. Control of coronary vascular tone by nitric oxide. *Circ Res* 1990;66:1561–1575.
119. Guo JP, Murohara T, Buerke M, Scalia R, Lefer AM. Direct measurement of nitric oxide release from vascular endothelial cells. *J Appl Physiol* 1996;81:774–779.
120. Grisham MB, Granger DN, Lefer DJ. Modulation of leukocyte-endothelial interactions by reactive metabolites of oxygen and nitrogen: Relevance to ischemic heart disease. *Free Radic Biol Med* 1998;25:404–433.
121. Wang P, Samouilov A, Kuppasamy P, Zweier JL. Quantitation of superoxide, nitric oxide, and peroxynitrite generation in the postischemic heart *Circulation* 1996;94:I-467 (Abstract).
122. Miles AM, Bohle DS, Glassbrenner PA, Hansert B, Wink DA, Grisham MB. Modulation of superoxide-dependent oxidation and hydroxylation reactions by nitric oxide. *J Biol Chem* 1996;271:40–47.
123. Eiserich JP, Cross CE, Jones AD, Halliwell B, van der Vliet A. Formation of nitrating and chlorinating species by reaction of nitrite with hypochlorous acid. A novel mechanism for nitric oxide-mediated protein modification. *J Biol Chem* 1996;271:19199–19208.
124. Eiserich JP, Hristova M, Cross CE, Jones AD, Freeman BA, Halliwell B, van der Vliet A. Formation of nitric oxide-derived inflammatory oxidants by myeloperoxidase in neutrophils. *Nature* 1998;391:393–397.
125. Lefer DJ, Scalia R, Campbell B, Nossuli T, Hayward R, Salamon M, Grayson J, Lefer AM. Peroxynitrite inhibits leukocyte-endothelial cell interactions and protects

against ischemia-reperfusion injury in rats. *J Clin Invest* 1997;99:684–691.

126. Nossuli TO, Hayward R, Scalia R, Lefer AM. Peroxynitrite reduces myocardial infarct size and preserves coronary endothelium after ischemia and reperfusion in cats. *Circulation* 1997;96:2317–2324.
127. Moro MA, Darley-Usmar VM, Lizasoain I, Su Y, Knowles RG, Radomski MW, Moncada S. The formation of nitric oxide donors from peroxynitrite. *Br J Pharmacol* 1995;116:1999–2004.
128. Nossuli TO, Hayward R, Jensen D, Scalia R, Lefer AM. Mechanisms of cardioprotection by peroxynitrite in myocardial ischemia and reperfusion injury. *Am J Physiol* 1998;275:H509–H519.
129. Stamler JS, Jaraki O, Osborne J, Simon DI, Keaney J, Vita J, Singel D, Valeri CR, Loscalzo J. Nitric oxide circulates in mammalian plasma primarily as an S-nitroso adduct of serum albumin. *Proc Natl Acad Sci USA* 1992;89:7674–7677.
130. Hayashi Y, Sawa Y, Nishimura M, Tojo SJ, Fukuyama N, Nakazawa H, Matsuda H. P-selectin participates in cardiopulmonary bypass-induced inflammatory response in association with nitric oxide and peroxynitrite production. *J Thorac Cardiovasc Surg* 2000;120:558–565.
131. Schulz R, Wambolt R. Inhibition of nitric oxide synthesis protects the isolated working rabbit heart from ischaemia-reperfusion injury. *Cardiovasc Res* 1995;30:432–439.
132. Schulz R, Dodge KL, Lopaschuk GD, Clanachan AS. Peroxynitrite impairs cardiac contractile function by decreasing cardiac efficiency. *Am J Physiol* 1997;272:H1212–H1219.
133. Wang W, Sawicki G, Schulz R. Peroxynitrite-induced myocardial injury is mediated through matrix metalloproteinase-2. *Cardiovasc Res* 2002;53:165–174.
134. Nathan C, Xie QW. Nitric oxide synthases: Roles, tolls, and controls. *Cell* 1994;78:915–918.
135. Kelly RA, Balligand JL, Smith TW. Nitric oxide and cardiac function. *Circ Res* 1996;79:363–380.
136. Balligand JL, Ungureanu-Longrois D, Simmons WW, Pimental D, Malinski TA, Kapturczak M, Taha Z, Lowenstein CJ, Davidoff AJ, Kelly RA. Cytokine-inducible nitric oxide synthase (iNOS) expression in cardiac myocytes. Characterization and regulation of iNOS expression and detection of iNOS activity in single cardiac myocytes *in vitro*. *J Biol Chem* 1994;269:27580–27588.
137. Depre C, Havaux X, Renkin J, Vanoverschelde JL, Wijns W. Expression of inducible nitric oxide synthase in human coronary atherosclerotic plaque. *Cardiovasc Res* 1999;41:465–472.
138. Maulik N, Engelman DT, Watanabe M, Engelman RM, Maulik G, Cordis GA, Das DK. Nitric oxide signaling in ischemic heart. *Cardiovasc Res* 1995;30:593–601.
139. Huang PL, Huang Z, Mashimo H, Bloch KD, Moskowitz MA, Bevan JA, Fishman MC. Hypertension in mice lacking the gene for endothelial nitric oxide synthase. *Nature* 1995;377:239–242.
140. Shesely EG, Maeda N, Kim HS, Desai KM, Krege JH, Laubach VE, Sherman PA, Sessa WC, Smithies O. Elevated blood pressures in mice lacking endothelial nitric oxide synthase. *Proc Natl Acad Sci USA* 1996;93:13176–13181.
141. Godecke A, Decking UK, Ding Z, Hirchenhain J, Bidmon HJ, Godecke S, Schrader J. Coronary hemodynamics in endothelial NO synthase knockout mice. *Circ Res* 1998;82:186–194.
142. Moroi M, Zhang L, Yasuda T, Virmani R, Gold HK, Fishman MC, Huang PL. Interaction of genetic deficiency of endothelial nitric oxide, gender, and pregnancy in vascular response to injury in mice. *J Clin Invest* 1998;101:1225–1232.
143. Lefer DJ, Girod WG, Jones SP, Palazzo AJ, Fishman MC, Huang PL, et al. Myocardial ischemia-reperfusion injury s exacerbated in ecNOS deficient mice. *FASEB J* 1998;12:A325.
144. Sumeray MS, Rees DD, Yellon DM. Infarct size and nitric oxide synthase in murine myocardium. *J Mol Cell Cardiol* 2000;32:35–42.
145. Yang XP, Liu YH, Shesely EG, Bulagannawar M, Liu F, Carretero OA. Endothelial nitric oxide gene knockout mice: Cardiac phenotypes and the effect of angiotensin-converting enzyme inhibitor on myocardial ischemia/reperfusion injury. *Hypertension* 1999;34: 24–30.
146. Bell RM, Yellon DM. The contribution of endothelial nitric oxide synthase to early ischaemic preconditioning: The lowering of the preconditioning threshold. An investigation in eNOS knockout mice. *Cardiovasc Res* 2001;52:274–780.
147. von der Leyen HE, Gibbons GH, Morishita R, Lewis NP, Zhang L, Nakajima M, Kaneda Y, Cooke JP, Dzau VJ. Gene therapy inhibiting neointimal vascular lesion: *In vivo* transfer of endothelial cell nitric oxide synthase gene. *Proc Natl Acad Sci USA* 1995;92:1137–1141.
148. MacMicking JD, Nathan C, Hom G, Chartrain N, Fletcher DS, Trumbauer M, Stevens K, Xie QW, Sokol K, Hutchinson N, et al. Altered responses to bacterial infection and endotoxic shock in mice lacking inducible nitric oxide synthase. *Cell* 1995;81:641–650.
149. Wei XQ, Charles IG, Smith A, Ure J, Feng GJ, Huang FP, Xu D, Muller W, Moncada S, Liew FY. Altered immune responses in mice lacking inducible nitric oxide synthase. *Nature* 1995;375:408–411.
150. Laubach VE, Shesely EG, Smithies O, Sherman PA. Mice lacking inducible nitric oxide synthase are not resistant to lipopolysaccharide-induced death. *Proc Natl Acad Sci USA* 1995;92:10688–10692.
151. Chyu KY, Dimayuga P, Zhu J, Nilsson J, Kaul S, Shah PK, Cercek B. Decreased neointimal thickening after arterial wall injury in inducible nitric oxide synthase knockout mice. *Circ Res* 1999;85:1192–1198.
152. Tolbert T, Thompson JA, Bouchard P, Oparil S. Estrogen-induced vasoprotection is independent of inducible nitric oxide synthase expression: Evidence from the mouse carotid artery ligation model. *Circulation* 2001;104:2740–2745.
153. Xi L, Jarrett NC, Hess ML, Kukreja RC. Myocardial ischemia/reperfusion injury in the inducible nitric oxide synthase knockout mice. *Life Sci* 1999;65:935–945.
154. Mungrue IN, Gros R, You X, Pirani A, Azad A, Csont T, Schulz R, Butany J, Stewart DJ, Husain M. Cardiomyocyte overexpression of iNOS in mice results in peroxynitrite generation, heart block, and sudden death. *J Clin Invest* 2002;109:735–743.
155. Heger J, Godecke A, Flogel U, Merx MW, Molojavyi A, Kuhn-Velten WN, Schrader J. Cardiac-specific overexpression of inducible nitric oxide synthase does not result in severe cardiac dysfunction. *Circ Res* 2002;90:93–99.

156. Duncker DJ, Bache RJ. Inhibition of nitric oxide production aggravates myocardial hypoperfusion during exercise in the presence of a coronary artery stenosis. *Circ Res* 1994;74:629–640.
157. Youhua Z, Shouchun X. Increased vulnerability of hypertrophied myocardium to ischemia and reperfusion injury. Relation to cardiac renin-angiotensin system. *Chin Med J* 1995;108:28–32.
158. Sun Y, Weber KT. Angiotensin II receptor binding following myocardial infarction in the rat. *Cardiovasc Res* 1994;28:1623–1628.
159. Francis GS, McDonald KM, Cohn JN. Neurohumoral activation in preclinical heart failure. Remodeling and the potential for intervention. *Circulation* 1993;87(5 Suppl):IV90–96.
160. Brunner HR. Experimental and clinical evidence that angiotensin II is an independent risk factor for cardiovascular disease. *Am J Cardiol* 2001;87:3C–9C.
161. Hilgers KF, Veelken R, Muller DN, Kohler H, Hartner A, Botkin SR, Stumpf C, Schmieder RE, Gomez RA. Renin uptake by the endothelium mediates vascular angiotensin formation. *Hypertension* 2001;38:243–248.
162. Griendling KK, Sorescu D, Ushio-Fukai M. NAD(P)H Oxidase: Role in cardiovascular biology and disease. *Circ Res* 2000;86:494–501.
163. Cai H, Harrison DG. Endothelial dysfunction in cardiovascular diseases: The role of oxidant stress. *Circ Res* 2000;87:840–844.
164. Dzau VJ. Tissue angiotensin and pathobiology of vascular disease. A unifying hypothesis. *Hypertension* 2001;37:1047–1052.
165. Matsubara H. Pathophysiological role of angiotensin II type 2 receptor in cardiovascular and renal diseases. *Circ Res* 1998;83:1182–1191.
166. Xu Y, Clanachan AS, Jugdutt BI. Enhanced expression of AT_2R, IP_3R and PKC_ε during cardioprotection induced by AT_2R blockade. *Hypertension* 2000;36:506–510.
167. Xu Y, Menon V, Jugdutt BI. Cardioprotection after angiotensin II type 1 blockade involves angiotensin II type 2 receptor expression and activation of protein kinase C-ε in acutely reperfused myocardial infarction. Effect of UP269-6 and losartan on AT_1 and AT_2 receptor expression, and IP_3 receptor and PKC_ε proteins. *J Renin-Angiotensin Aldosterone System* 2000;1:184–195.
168. Jugdutt BI, Xu Y, Balghith M, Moudgil R, Menon V. Cardioprotection induced by AT_1R blockade after reperfused myocardial infarction: Association with regional increase in AT_2R, IP_3R and PKC_ε proteins and cGMP. *J Cardiovasc Pharmacol Ther* 2000;5:301–311.
169. Moudgil R, Xu Y, Menon V, Jugdutt BI. Effect of chronic pretreatment with AT_1 receptor antagonism on postischemic functional recovery and AT_1/AT_2 receptor proteins in isolated working rat hearts. *J Cardiovasc Pharmacol Ther* 2001;6:183–188.
170. Jugdutt BI, Xu Y, Balghith M, Menon V. Cardioprotective effects of angiotensin II type 1 receptor blockade with candesartan after reperfused myocardial infarction: Role of angiotensin II type 2 receptor. *J Renin-Angiotensin Aldosterone System* 2001;2:S162–S166.
171. Jugdutt BI, Balghith M. Enhanced regional AT_2 receptor and PKCε expression during cardioprotection induced by AT_1 receptor blockade after reperfused myocardial infarction. *J Renin-Angiotensin Aldosterone System* 2001;2:134–140.
172. Moudgil R, Menon V, Xu Y, Musat-Marcu S, Jugdutt BI. Postischemic apoptosis and functional recovery after angiotensin II type 1 receptor blockade in isolated working rat hearts. *J Hypertension* 2001;19:1121–1129.
173. Nio Y, Matsubara H, Murasawa S, Kanasaki M, Inada M. Regulation and gene transcription of angiotensin II receptor subtypes in myocardial infarction. *J Clin Invest* 1995;95:46–54.
174. Haywood GA, Gullestad L, Katsuya T, Hutchinson HG, Pratt RE, Horiuchi M, Fowler MB. AT_1 and AT_2 angiotensin receptor gene expression in human heart failure. *Circulation* 1997;95:1201–1206.
175. Garg UC, Hassid A. Nitric-oxide vasodilators and 8-bromo-cyclic guanosine monophosphate inhibit mitogenesis and proliferation of cultured rat vascular smooth muscle cells. *J Clin Invest* 1989;83:1774–1777.
176. Unger T, Gohlke P. Tissue renin-angiotensin systems in the heart and vasculature: Possible involvement in the cardiovascular actions of converting enzyme inhibitors. *Am J Cardiol* 1990;65:31–101.
177. Linz W, Schölkens BA. A specific β_2-bradykinin receptor antagonist HOE 140 abolishes the antihypertrophic effect of ramipril. *Br J Pharmacol* 1992;105:771–772.
178. Wiemer G, Schölkens BA, Becker RHA, Busse R. Ramprilat enhances endothelial autacoid formation by inhibiting breakdown of endothelium-derived bradykinin. *Hypertension* 1992;18:558–563.
179. Wiemer G, Schölkens BA, Wagner A, Heitsch H, Linz W. The possible role of angiotensin II subtype AT_2 receptors in endothelial cells and isolated ischemic rat hearts. *J Hypertens* 1993;11:S234–S235.
180. Zanzinger J, Zheng X, Bassenge E. Endothelium dependent vasomotor responses to endogenous agonists are potentiated following ACE inhibition by a bradykinin dependent mechanism. *Cardiovasc Res* 1994;28:209–214.
181. Hall AS, Tan L-B, Ball SG. Inhibition of ACE/kininase-II, acute myocardial infarction, and survival. *Cardiovasc Res* 1994;28:190–198.
182. McDonald KM, Mock J, D'Aloia A, Parrish T, Hauer K, Francis G, Stillman A, Cohn JN. Bradykinin antagonism inhibits the antigrowth effect of converting enzyme inhibition in the dog myocardium after discrete transmural myocardial necrosis. *Circulation* 1995;91:2043–2048.
183. Liu YH, Yang XP, Sharov VG, Nass O, Sabbah HN, Peterson E, Carretero OA. Effects of angiotensin-converting enzyme inhibitors and angiotensin II type 1 receptor antagonists in rats with heart failure: Role of kinins and angiotensin type 2 receptors. *J Clin Invest* 1997;99:1926–1935.
184. Jalowy A, Schulz R, Dorge H, Behrends M, Heush G. Infarct size reduction by AT_1 receptor blockade through a signal cascade of AT_2receptor activation, bradykinin and prostaglandins in pigs. *J Am Coll Cardiol* 1998;32:1787–1796.
185. Bartunek J, Weinberg EO, Tajima M, Rohrbach S, Lorell BH. Angiotensin II type 2 receptor blockade amplifies the early signals of cardiac growth response to angiotensin II in hypertrophied hearts. *Circulation* 1999;99:22–25.
186. Dörge H, Behrends M, Schulz R, Jalowy A, Heusch G. Attenuation of myocardial stunning by the AT_1 receptor antagonist candesartan. *Basic Res Cardiol* 1999;94:208–214.
187. Ford WR, Clanachan AS, Jugdutt BI. Opposite effects of angiotensin receptor antagonists on recovery of

mechanical function after ischemia-reperfusion in isolated working rat hearts. *Circulation* 1996;94:3087–3089.

188. Xu Y, Dyck J, Ford WR, Clanachan AS, Lopaschuk GD, Jugdutt BI. Angiotensin II type 1 and type 2 receptor protein after acute ischemia-reperfusion in isolated working rat hearts. *Am J Physiol Heart Circ Physiol* 2002;282:H1206–H1215.
189. Liu Y-H, Yang X-P, Sharov VG, Sigmon DH, Sabbah HN, Carretero OA. Paracrine systems in the cardioprotective effect of angiotensin-converting enzyme inhibitors on myocardial ischemia/reperfusion injury in rats. *Hypertension* 1996;27:7–13.
190. Ping P, Takano H, Zhang J, Tang X-L, Qiu Y, Li RCX, Banerjee S, Dawn B, Balafonova Z, Bolli R. Isoform-selective activation of protein kinase C by nitric oxide in the heart of conscious rabbits. A signaling mechanism for both nitric oxide-induced and ischemia-induced preconditioning. *Circ Res* 1999;84:587–604.
191. Ichiki T, Usui M, Kato M, Funakoshi Y, Ito K, Egashira K, Takeshita A. Downregulation of angiotensin II type 1 receptor gene transcription by nitric oxide. *Hypertension* 1998;31:342–348.
192. Jugdutt BI. New advances in the use of AT_1 receptor blockers (ARBs). In: *Proceedings, 2nd International Congress on heart Disease: New Trends in Research, Diagnosis and Treatment*. Medimond A. Kimchi ed. New Jersey: Medical Publishers, 2001:531–538.
193. Pitt B, Poole-Wilson PA, Segal R, Martinez FA, Dickstein K, Camm AJ, Konstam MA, Riegger G, Klinger GH, Neaton J, Sharma D, Thiyagarajan B, on behalf of the ELITE II investigators. Effect of losartan compared with captopril on mortality in patients with symptomatic heart failure: Randomized trial—The Losartan Heart Failure Survival Study ELITE II. *Lance* 2000;355:1582–1587.
194. McKelvie RS, Yusuf S, Pericak D, Avezum A, Burns RJ, Probstfield J, Tsuyuki RT, White M, Rouleau J, Latini R, Maggioni A, Young J, Pogue J. Comparison of candesartan, enalapril, and their combination in congestive heart failure: Randomized evaluation of strategies for left ventricular dysfunction (RESOLVD) pilot study. The RESOLVD Pilot Study Investigators. *Circulation* 1999;100:1056–1064.
195. Cohn JN, Tognoni G, Valsartan Heart Failure Trial Investigators. A randomized trial of the angiotensin-receptor blocker valsartan in chronic heart failure. *N Engl J Med* 2001;345:1667–1675.
196. de Belder AJ, Radomski MW, Why HJ, Richardson PJ, Bucknall CA, Salas E, Martin JF, Moncada S. Nitric oxide synthase activities in human myocardium. *Lancet* 1993;341:84–85.
197. Fukuchi M, Hussain SN, Giaid A. Heterogeneous expression and activity of endothelial and inducible nitric oxide synthases in end-stage human heart failure: Their relation to lesion site and beta-adrenergic receptor therapy. *Circulation* 1998;98:132–139.
198. Hare JM, Givertz MM, Creager MA, Colucci WS. Increased sensitivity to nitric oxide synthase inhibition in patients with heart failure: Potentiation of beta-adrenergic inotropic responsiveness. *Circulation* 1998;97:161–166.
199. Drexler H, Kastner S, Strobel A, Studer R, Brodde OE, Hasenfuss G. Expression, activity and functional significance of inducible nitric oxide synthase in the failing human heart. *J Am Coll Cardiol* 1998;32:955–963.
200. Méry PF, Lohmann SM, Walter U, Fischmeister R. Ca^{2+} current is regulated by cyclic GMP-dependent protein kinase in mammalian cardiac myocytes. *Proc Natl Acad Sci USA* 1991;88:1197–1201.
201. Heymes C, Vanderheyden M, Bronzwaer JG, Shah AM, Paulus WJ. Endomyocardial nitric oxide synthase and left ventricular preload reserve in dilated cardiomyopathy. *Circulation* 1999;99:3009–3016.
202. Kichuk MR, Seyedi N, Zhang X, Marboe CC, Michler RE, Addonizio LJ, Kaley G, Nasjletti A, Hintze TH. Regulation of nitric oxide production in human coronary microvessels and the contribution of local kinin formation. *Circulation* 1996;94:44–51.
203. Ing DJ, Zang J, Dzau VJ, Webster KA, Bishopric NH. Modulation of cytokine-induced cardiac myocyte apoptosis by nitric oxide, Bak, and Bcl-x. *Circ Res* 1999;84:21–33.
204. Kim YM, Bombeck CA, Billiar TR. Nitric oxide as a bifunctional regulator of apoptosis. *Circ Res* 1999;84:253–256.
205. Haunstetter A, Izumo S. Apoptosis: Basic mechanisms and implications for cardiovascular disease. *Circ Res* 1998;82:1111–1129.
206. Fliss H, Gattinger D. Apoptosis in ischemic and reperfused rat myocardium. *Circ Res* 1996;76:949–956.

Index